104/- NET

6507

TRAFFIC ENGINEERING

TRAFFIC ENGINEERING

THEODORE M. MATSON

Late Director, Bureau of Highway Traffic, Yale University

WILBUR S. SMITH

Associate Director, Bureau of Highway Traffic, Yale University

FREDERICK W. HURD

Director, Bureau of Highway Traffic, Yale University

McGRAW-HILL BOOK COMPANY, INC.

New York   Toronto   London

1955

TRAFFIC ENGINEERING

Library of Congress Catalog Card Number 55-5692

VII

40910

THE MAPLE PRESS COMPANY, YORK, PA.

PREFACE

Traffic engineering is one of the newer branches of engineering, having developed largely within the past two decades. This book is one of the first attempts to bring together into a simple matrix the general knowledge of this new field. A basic purpose in preparing the book was to provide a text for traffic engineering training and a reference for those engaged in highway traffic planning, operations, and administration.

The objective of the authors throughout this work has been to explain highway-traffic phenomena by simple and elementary analysis and to show how highway-traffic treatments are founded in restrictive methods, on the one hand, and constructive measures, on the other, both of which take place in the environment of administration and planning.

The plan and content of the book have grown out of the training program of the Bureau of Highway Traffic in Yale University. It is based on teaching experiences gained since 1938. The sections of the book conform with the courses of study in the Bureau, except that the materials in Section 1 are dispersed throughout the training. The principal responsibilities for sections corresponded with the teaching duties of the authors, i.e., Matson, Section 1, Characteristics, and Section 3, Control Devices and Aids; Smith, Section 2, Regulations, and Section 5, Administration and Planning; and Hurd, Section 4, Design.

We wish to acknowledge the abundant encouragement, assistance, and support of those who have helped make this work possible. While it is impracticable to set forth here the numerous works of others from whom we have drawn heavily, every attempt has been made to cite references at appropriate points throughout the manuscript. This volume would never have been undertaken without the many contributions in traffic engineering and related fields.

The authors wish to express their gratitude to the authorities of Yale University, and particularly to the members of the University's Transportation Committee. We are deeply appreciative of the inspiration and encouragement which has been given to us by Mr. Kent T. Healy, Chairman of the Committee on Transportation in Yale University.

We wish to express appreciation to Mr. Richard O'Neill, who patiently labored over the original manuscript to reduce the total materials to conform to the confines of this volume. His many helpful suggestions and detailed work have contributed greatly to the final manuscript.

In the preparation of this work, valuable assistance was given by other members of the staff of the Bureau of Highway Traffic. The authors wish to thank especially Mr. William McGrath, Mr. Martin Wallen, Mr. Sherman Hasbrouck, and Mr. Charles Pratt, who were so helpful in the compilation of many data and technical details. Mrs. Cele Kagan and Mrs. Katherine Cassidy of the Yale Transportation Library rendered patient and essential assistance. For typing, proofreading, and the many sundry details necessary in preparation of the manuscript, the authors are especially thankful to Miss Irene Bouzoucos, Miss Mary Lou Pedroni, and others of the Bureau's clerical staff.

The authors recognize that it has not been possible to develop fully many phases of traffic engineering. To contain the subject within one volume, it has been necessary to abbreviate. In discussion of some areas, which might be considered as allied only to the over-all subject, it is realized that many entire volumes are available and that it is necessary in this book only to suggest the proper integration of these to traffic engineering.

THEODORE M. MATSON
WILBUR S. SMITH
FREDERICK W. HURD

Shortly after completion of the manuscript for this book the field of traffic engineering suffered a deep loss through the death of Theodore M. Matson. He was a pioneer in the development and establishment of the profession of traffic engineering. He was one of the founders of the Institute of Traffic Engineers. He was a leader in the formation of national traffic policies and has been responsible for the initiation of numerous research activities in the field. His leadership in the training of traffic engineers has been a major influence in the development of better highway transportation facilities.

To Mr. Matson, who provided leadership in the development of this work and who was a great source of inspiration to the other authors in the preparation of their portion of the manuscript, this book is dedicated.

CONTENTS

Section 5. Administration and Planning

CHAPTER 1

INTRODUCTION

Motor vehicles have made sweeping changes in transportation that have placed at the command of the individual greatly increased facility, distance, and speed of travel. Since the turn of the century, there has been enormous increase in the number, speed, and length of trips of road vehicles. The social and economic benefits rendered to the public at large through this improved means of highway transportation are prodigious.

While the gains to society are immeasurable, losses which waste community resources have been incurred. Problems of traffic congestion, which thwart the very benefits of motor vehicle use, are found on many arteries of travel and are prevalent especially in urban areas.

Furthermore, the mechanized development in highway traffic has brought with it a shocking toll of human casualties. Congestion and accidents which arise as a consequence of motor transport result in loss to the public at large. The net gains establish the utility and place of highway transportation and its economic and social advantages to society.

So great have been the net advantages of motor transport since its invention and development that the increase in motor vehicle use has revolutionized transportation. The application of mechanical power to highway vehicles at the turn of the twentieth century, together with the development of hard-surfaced, all-weather roads, have given services to society, the demands for which at times seem insatiable.

The demand for individual, high-speed, powerful (in comparison with the horse and wagon) vehicles has created enormous amounts of new highway traffic. Since World War I traffic mileage has increased nearly tenfold, and the rate of increase has accelerated since World War II. The average rate of motor vehicle use is in the magnitude of 10,000 miles per year. The demand for vehicles is so great that there is now about one vehicle for each three persons. Hence, there are developed about 3,500 vehicle-miles per person per year (Fig. 1-1). Passenger automobiles alone develop nearly 5,000 passenger miles per person per year, since the average passenger vehicle ordinarily carries between one and two persons.

This phenomenal growth in highway traffic has brought serious amounts of loss through congestion and casualty. In the year 1951 the one-millionth fatality from motor vehicle accidents was incurred. Death rates from traffic accidents hover about 20 to 25 fatalities per 100,000 population per annum, or about 7 per 100 million vehicle-miles. Injuries from traffic accidents are incurred at the rate of about 35 times that of fatalities. Accident costs are estimated at nearly 4 billion dollars per annum. The losses due to congestion are not readily estimated, but it is clear that the direct wastes of congestion itself coupled with the indirect effects on city growth and development are enormous.

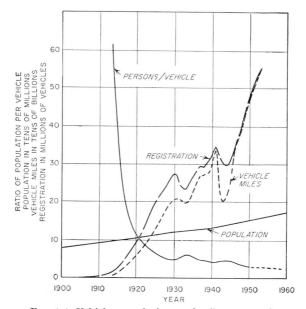

Fig. 1-1. Vehicle, population, and mileage growth.

1-1. The Nature of Highway-traffic Problems. The problems of highway traffic, then, are concerned with the achievement of a more efficient system of highway transportation, on the one hand, and the conservation of community resources, on the other. These two aspects of highway traffic are inseparable and derive from the three elements: (1) human beings as road users, (2) vehicles with their loads, and (3) fixed facilities for the accommodation of traffic movements and vehicle storage. All traffic problems and all traffic improvements are concerned with these three components and must recognize the qualifications and limitations imposed by them. Highway traffic is enormously complex, for it involves the interrelationships of human nature, on the one hand, and physical laws of time, space, and motion, on the other.

In striving for highway-traffic improvements, therefore, two fundamental avenues of approach suggest themselves. The first would seek to adjust human behavior to the environment of the traffic stream and the fixed facilities which carry it. The second avenue of approach would accept human behavior but would seek to contrive ways and means of better traffic accommodation. Experience has shown that both methods are required and that each complements the other.

There is frequently heard, among highway-traffic professionals, an expression of the three "E's," i.e., Education, Enforcement, and Engineering. Obviously, educational methods are concerned with attitudes and skills and would seek to improve road-user behavior through persuasion, explanation, drill, and practice. Enforcement methods are also directed at the human element. These methods would change road-user attitudes and behavior through supervision of road use, apprehension of those road users who disobey the rules, and punishment of violators. On the other hand, engineering is concerned basically with the qualities of the vehicles and fixed facilities and seeks to design and build these to fit road-user demands and limitations more ideally.

1-2. Branches of Traffic Engineering Work. Traffic engineering is that branch of engineering which is devoted to the study and improvement of the traffic performance of road networks and terminals. Its purpose is to achieve efficient, free, and rapid flow of traffic; yet, at the same time, to prevent traffic accidents and casualties. Its procedures are based on scientific and engineering disciplines. Its methods include regulation and control, on the one hand, and planning and geometric design, on the other.

Through the studied use of traffic characteristics, operation improvements are achieved by effective application of regulatory methods to existing facilities and by proper planning and design of the construction of new facilities.

Factual studies of traffic operations provide the foundations for intelligent development of ways and means to improve performance. In this phase of traffic engineering work, the central purpose is to explore the true nature and characteristics of traffic phenomena. This is the experimental and analytical aspect of the work and requires the making of measurements and the collection of facts. From analysis of these facts, the diverse qualities, the magnitudes and the interrelationships of traffic and its environment become known, and the performance characteristics of highway traffic are established.

The principal characteristics of traffic concern the abilities, requirements, and performance of (1) human beings as *road users* and (2) *highway vehicles*. In vehicular traffic these two components, the man and his machine, comprise the discrete unit of traffic. The primary characteristics of traffic movement, then, are concerned with *speed* (and delay), *volume* or time rates of flow, and *origin and destination* or the location, distance, and direction of movement. The secondary phenomena of *stream flow* and *intersection flow* establish capacity and other performance values of traffic facilities. As an integral aspect of highway traffic, *parking* and other terminal-operations characteristics are examined. Finally, the characteristic occurrence of *accidents* is explored, since these are the measure of failure.

The characteristics of traffic become, then, the foundation on which the improvement of operations is founded. On this foundation the art and skill of formulating policies, programs, and procedures toward better operations are exercised. It is in these aspects of traffic engineering work that the ultimate objectives, the achievement of efficient, free, and rapid flow with minimum allowable hazard, take place. The accomplishment of these improvements becomes the proving ground of practice.

Traffic Regulations. The *regulation* of highway traffic stems from the police power of the state and must be confined within the boundaries of public safety, welfare, and convenience. Of prime concern are the limitations which are imposed on road users and their vehicles; that is, who may be given the privilege of operating motor vehicles, and what kind of vehicles will be permitted to be operated in the public way. (Note, for example, no restrictions are ordinarily placed on who may use public ways as pedestrians.)

The next phase of traffic regulations sets up the basic general rules of road use and conduct, including lateral placement, speed, overtaking and passing, starting, turning and stopping, right of way, pedestrians' rights and duties, general parking rules, and similar regulations. Finally, to complete the requirements of traffic regulations, there are developed those rules which apply at only particular times and places. These include speed control, one-way regulations, curb-parking controls, and miscellaneous regulations such as stop rules and turning regulations.

Control Devices and Aids. The class of regulations especially applied to particular times and places, together with the warnings required for unusual road or traffic conditions as well as the furnishing of useful information about road use to the traveling public, require *traffic control devices* and aids for their implementation. Accordingly, an important branch of traffic engineering work involves (1) signs and markings, (2) signals, and (3) lighting. The attention-compelling values, the carrying of messages which are clear at a glance, the provision of adequate time to respond to these devices or traffic conditions, and the respect which such devices command are vital in the accomplishment of the desired goals of such equipment.

Traffic Design. The construction of fixed facilities is of special concern to the traffic engineer, since these facilities vitally affect traffic operations. Reconstruction or new construction should therefore be *designed* so as to fit traffic requirements. There are many traffic factors which influence design. For example, composition, volume, and speed, as well as starting, turning, and stopping abilities and performance values are significant. It is these dynamic traffic values that become critical to the suitability of a static structure in the accomplishment of its function. Hence, this aspect of construction is frequently referred to as *functional* or *geometric* traffic design.

Traffic design then is concerned with (1) road surfaces because of their functional and light-reflecting properties which affect not only starting, turning, stopping, and laning of traffic movements but also visibility conditions, (2) cross section, (3) sight distance, (4) horizontal alignment, (5) vertical alignment, (6) the geometry of elemental maneuver areas, (7) the integrated design, and (8) the practical form of intersections and junctions. Finally, traffic movements and vehicle requirements are of fundamental concern in (9) the design of parking and terminal structures not only for the acceptance and discharge of vehicles but also for internal movements and storage arrangements of these facilities.

Traffic Administration and Planning. As in all areas of public work, the success of traffic programs is largely dependent upon the administrative support given to them by key officials. Except perhaps for finance, there is no area of public work which involves as many agencies and interests of state and municipal government as traffic. This is of course particularly true at the city level. In one way or another practically all key departments of the government have a responsibility and interest in traffic affairs. Because traffic is so intimately related to economic and social values, it is inevitable that work in highway traffic touches upon and influences a wide diversity of interests and views. Traffic administration is complicated thereby.

The administration of traffic engineering work requires a broad knowledge of community affairs and a perspective which encompasses the desires of the different groups concerned with traffic, as well as the limitations and abilities of the public agencies to cope with the peculiar problems and situations which arise. The formulation of administrative machinery which will provide the proper attention and emphasis on traffic is sometimes difficult.

Traffic engineering administration requires careful planning and well-timed programming. While the organization might take on varied forms and different governmental arrangements, there are fundamental objectives and basic approaches which cut across all adequate programs of traffic engineering activities. Care must be exercised to see that proper balance is given to the constructive and to the regulatory approaches; both immediate and long-range problems must be considered. The movement of traffic must be recognized as essential as the safety of traffic operations. It is especially important that the functions of traffic engineering be properly integrated with all other activities having to do with traffic. It is apparent, therefore, that a first essential in any traffic engineering organization is the provision of an adequate technical staff; trained and experienced traffic engineers are essential to the sound development and direction of a traffic engineering program. While it is a common saying that "everyone is a traffic expert," care must be exercised that essential programs such as the program of traffic engineering are not left to the whims and responsibilities of untrained, inexperienced, and nontechnical persons.

It is not only essential in matters of traffic engineering to have a sound technical organization with properly authorized functions and responsibilities, but consideration must also be given to the many opportunities to collect and use traffic facts in administrative decisions having to do with highways, traffic control, and over-all governmental planning. The traffic administrator must have a thorough understanding of the problems of highway finance and the potential sources of funds for traffic improvement activities. Actually, the procurement of funds to support the traffic engineering program is as essential as the procurement of monies for all other public activities. Fortunately, it is a relatively simple matter to show direct returns from most traffic activities and improvements, so that studies of highway needs and over-all budgetary processes can be augmented in an objective manner through traffic engineering data.

The attention which is now being directed at off-street parking places is in a peculiar position with regard to traffic engineering administration. In some cases important administrative units are being formed to carry on off-street parking activities entirely apart from the conventional traffic activities. In other cases the parking and terminal affairs of cities are being made a part of the basic traffic engineering organization.

There are numerous ways in which traffic engineering must be associated with planning. The generation of traffic is dependent on population and land use. Conversely, land use and the distribution of population are affected by the availability and character of highway routes and the degree of accessibility and efficiency which such routes provide. It follows, therefore, that traffic values are of great importance in planning the future growth and development of the city, the region, the state, and perhaps even the nation.

City traffic engineers normally work very closely with city planners. Traffic is a primary factor in urban planning, and it is closely associated with all land uses. The matters of movement must be given equal consideration with the matters of parking and loading. There are many excellent opportunities to coordinate the interests of the traffic engineer and the local planner through current programs of public housing, urban redevelopment, and major street planning. The principal traffic values employed in planning work are (1) composition of the traffic streams, which determines the character of the routes required, (2) traffic volumes, which determine the capacity required, and (3) traffic origins and destinations, which indicate the proper locations for routes and terminal facilities.

It is not only in the urban areas that integration of traffic engineering and planning are desirable. The metropolitan problem is one which has been widely recognized in all governmental circles and which poses very challenging actions for proper traffic planning and highway development. The planning of land-access routes and the "farm-to-market" roads, as well as the development of the secondary and primary highway networks, are of vital concern to the economic development of the state and nation. The location, design, and operation of the rural highway system, with proper regard for present and future traffic movements, is an essential aspect of traffic in the total planning process. Even the minor street systems must be given careful study and must be properly related to the plans for major thoroughfares, expressways, and parkways in both urban and rural areas. The development and future growth of the city, the metropolitan community, the state, and the region are dependent upon the careful analysis of traffic requirements in the over-all plan.

1-3. A Perspective of Traffic Engineering. The revolution in highway transportation brought about by the motor vehicle and the hard-surface road has brought with it problems of accidents and congestion which require coordination of traffic regulation and construction in their solution. Beginning about 1925, when the "knee" in the curve of automobile ownership was reached, there was called into existence a new profession known as traffic engineering.

Since traffic is composed of "man and his machine," the traffic engineering field must recognize and be governed by both the social and physical sciences in the solution of its problems. While the strict disciplines of engineering are applicable to the physical aspects, the conventions and vagaries of the human element are ever present and are frequently found to be critical factors in the making of the final decisions.

Because both the physical and social sciences are involved, the research and theoretical problems as well as the practical applications require for their analysis and treatment mathematics, statistics, and physics, on the one hand, and city planning, psychology, government, and economics, on the other. While traffic engineering involves the strict disciplines of science, it is also subject to new departures and scholarly theorizing in many of its aspects which are not crystallized.

While the primary purposes and activities of traffic engineering are directly concerned with the operations of a road net and its terminal facilities, there are many aspects of traffic engineering knowledge which are a consideration in and influence on related matters. Thus, city planning and growth, some aspects of legislation, highway planning and finance, retailing, mass transit, trucking operations, casualty insurance, and even military operations directly or indirectly depend on or are affected by highway traffic in their purposes. Traffic and parking are recognized as fundamental in city growth and development. The decentralization of our mother cities finds traffic congestion and parking as a contributory factor. While motor vehicle codes and traffic ordinances require legislation directly, traffic values are of real influence in legislation creating parking, toll-road, and metropolitan-district authorities. It is traffic volumes and operations which form a major basis for highway programming and toll-road financing. Because accessibility and parking are of vital importance, the very location and operation of retail establishments must reckon with traffic values. Again, specialized highway-transportation services, such as truck and mass-carrier operations, require traffic data and analysis and evaluation. Finally, the underlying fundamental researches into the characteristics of traffic generation, movement, and parking will contribute to all military and civilian activities which rely partially, if not entirely, on highway transportation.

The study of traffic engineering is aimed toward the development of the skills required to carry out all aspects of the profession. It is essential to acquire factual knowledge of traffic characteristics and to develop ability to carry out studies and analyses which increase this fund of knowledge. The ability to plan for and effectively administer traffic improvement programs must be carefully cultivated. Continued study coupled with experience makes it possible to carry forward the objectives of traffic engineering and to enrich and adorn the profession.

SECTION 1

CHARACTERISTICS

CHAPTER 2

THE ROAD USER

The study of the road user is concerned with those things which activate him, the motivation and mechanism of his operation, and the patterns of his response. Human beings, either as pedestrians or drivers, taken individually and collectively, are an essential element to be understood and dealt with in highway traffic. Individual behavior in the traffic stream is frequently the critical determinant in many of the characteristics of highway traffic.

The following environmental conditions may affect road-user behavior: (1) land, its use and activities, (2) ambient atmosphere, weather, and visibility, (3) fixed facilities for traffic accommodation, including routes and terminals, and (4) the traffic stream and its characteristics which are manifest to the road user.

While these environmental conditions stimulate the road user from without, he is also affected by his own organic system. For example, alcohol, physical deficiencies, and even emotional disturbances work upon the human being and affect his conduct in the traffic stream.

Furthermore, the road user as a human being possesses natural desires and motivations. It is through the mechanism of his mind and body that he senses the total environment, integrates it with his purposes, and responds in a manner which is normally suited to his motives and the traffic stream.

The general qualities of speed and volume, as well as the origin and destination of traffic movements, result from road-user desires. Still further, the detailed time-space relationships found in traffic streams, in intersection maneuvers, or in the parking terminal, result directly or indirectly from human actions in response to the total environmental conditions, and in turn, these relationships largely determine the capacity of traffic facilities.

Reciprocally, traffic phenomena constitute an important and dynamic aspect of the environment. The very traffic conditions which grow out of human behavior feed back to affect the performance of road users.

2-1. Psychological Traits of Road Users. The drives and resistances, intelligence, learning, and emotion of road users are elements in their natures which are of profound significance in traffic operations.

11

Motivation. The primary motivating purposes of road users range throughout the social and economic purposes of the community in which traffic is found. People enter the traffic stream for business, social, or recreational purposes. They may be intent on going to a regular place of employment, to market, to rail, water, air, or other terminals. They may be hauling raw or finished materials of trade. They may be going to school, to visit, to the theater, or to dance. They may be going to games, to hunt and fish, or merely to enjoy the pleasure of a Sunday, holiday, or week-end drive. Whatever the reason for road use, all elements of society, under diverse motives of travel, enter the traffic stream as road users. The time, place, route, origin, and destination of travel are fundamentally the choice of road users in a free society.

But once the individual enters the traffic stream he is usually motivated by his desires for time-distance economy, on the one hand, and comfort or safety, on the other. These are the incentives of immediate concern to him, and they are ever present in road-user motivation. How much safety in how much time is a constant question demanding decision by individual road users in their normal traffic operations. While these two stimuli are nearly always present in the individual road user, social motivations are frequently manifest which arise from rivalry or cooperativeness in the use of the fixed facilities as a group. For example, there is a tendency for the heavy, fast stream of traffic movement to take the right of way requiring the individual to conform.

Intelligence. The capacity of the road user to be aware of all external factors pertinent to his behavior in traffic and to adapt and adjust himself in accord with his intents and motives requires a fair degree of social and mechanical intelligence. Drivers in the upper group of mechanical-comprehension tests have fewer traffic accidents, but persons with superior intelligence are not as mentally attentive to the task of road use and hence make poorer commercial drivers. However, conclusive correlations between human intelligence and traffic behavior have not been evaluated.

Learning. The *learning process*, while dependent on motivation, intelligence, and other modifying conditions, develops the skills, habits, and abilities of road users to respond properly to the total environment of traffic operations. The impression, retention, and recognition of traffic situations, as well as the development of reactions suitable to such situations, make for road-user habits which are not easily changed and which must be worked with. Through the impact of past experience, the road user is conditioned by his learning processes and continues to respond in patterns of behavior which he thus acquires. Learned responses are powerful factors which may be beneficial or harmful in traffic operations.

Emotion. Clearly, road-user motivation, intelligence, and learning are elements of the road user in his integration with the total traffic environment. But the road user's environment may stimulate *emotional response* which is essentially disintegrative. Usually emotion strongly motivates the road user to inefficient, random adjustment. Fear, anger, worry, and other similar emotional states tend to create disorganized reactions and behavior. The attitudes, moods, sentiments, and other personality qualities of road users are traits in which emotion is present. For example, the challenge of "me first," the spirit of "play," the desire to "show off," or the intent to "share the road" arise from emotional responses.

Individual Differences. Because of the modifying factors of motivation—intelligence, learning, and emotion—there is a great variation among users of the highway, since a full range of human behavior is to be found in every traffic stream. But just as there are limits to the range of variation in the height, weight, age, and other physical characteristics of human beings, there are limits in the range and there are central tendencies in the traffic performance qualities of the road users. Failure to recognize the wide range and variability of performance of road users leads to conflicting and erroneous conclusions. Sound traffic measures always recognize the inherent variability of the human factors.

2-2. Vision. More than any other single sensory perception, vision is significant for the traffic engineer in the design and operation of traffic facilities, because in traffic the road user is most dependent upon his sense of sight. It is the visual aspect of the road ahead that starts the chain of events causing the road user to take some positive action in any traffic situation. And the traffic engineer must realize that the road user's view of traffic is constantly changing and may at times be almost kaleidoscopic.

Vision is limited to the ability of the eye. Sight results from light striking the retina and creating an image, or sensation, which is transmitted through the optic nerve to the brain, causing perception of light, color, and form. In the formation of this image, the cornea, the lens, and certain optic fluids act as a refractive structure to bring light rays from external objects into focus at the retina.

When the eye is fixed in position, the area of most acute vision is subtended by a cone whose angle is 3°. For very distinct vision, the image of objects should fall on the retina within this cone. However, vision is quite sensitive within a visual cone of 5 or 6° and is still fairly satisfactory up to a range of 20°. In the vertical plane the angle of acute vision is only one-half to two-thirds what it is in the horizontal plane. For a given distance from the optical axis, acuity is generally better on the nasal portion of the retina.

As Fig. 2-1 indicates, for reading purposes visual acuity drops off rapidly outside of the 10 to 12° cone. To be accurately identified by the driver, words and symbols on traffic signs must fall within a visual cone of

10°. This is roughly the area covered by the width of the hand (4 in.) held at arm's length in line with the center of the road.

Eye Movement. Because the road user's field of vision is limited, it is necessary for him to shift his eyes about in traffic to scan areas significant for him. Speed of eye movement becomes increasingly important as speed of traffic increases. To obtain clear vision in highway traffic, the eye makes six different types of movement, all of which take time and consume travel distance.

SINGLE LETTER PLACED AT DISTANCE FROM OPTICAL AXIS SUCH THAT THE HORIZONTAL ANGLE, α, OF THE CONE OF VISON IS	PERCENTAGE OF CORRECT RESPONSES	
5.8°	98	
7.6°	95	
9.6°	90	
11.4°	84	
13.4°	74	
15.4°	66	

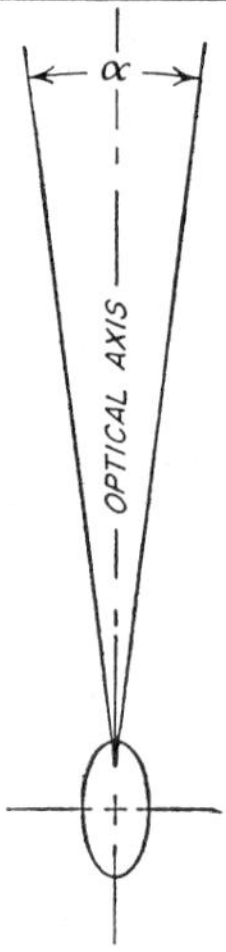

Fig. 2-1. Letter recognition by indirect vision. (From R. S. Woodworth, *Experimental Psychology*, Henry Holt and Company, Inc., New York, 1938, p. 719, quoting W. C. Ruediger, *Archives of Psychology*, 1907, No. 7.)

First, the eye must fixate on the object to be seen. A fixational pause averages 0.17 sec, ranging from 0.1 sec to 0.3 sec.[1]

Second, the eye "jumps" from one point of fixation to the next. The time for a jump ranges from 0.029 to 0.100 sec for movements of 5 to 40° respectively.[2] The reaction time required for such movements ranges from 0.125 sec to 0.235 sec, averaging nearly 0.2 sec. Thus the time required to shift the eye will vary from 0.15 to 0.33 sec when the shift is caused by a stimulus not in the field of clear vision.

Third, the eye must follow moving elements in the traffic stream.

Fourth, both eyes must move harmoniously for the pupils to converge or diverge in securing binocular vision on shifting objects in the road

[1] M. Luckeish and F. H. Moss, "Applied Science of Seeing," *Illuminating Engineering*, vol. 28, no. 10, 1933.

[2] R. S. Woodworth, *Experimental Psychology*, Henry Holt and Company, Inc., New York, 1938, p. 584, quoting R. Dodge and T. S. Clint, *Psychology Review*, no. 8, 1901, pp. 145–147.

ahead. The time required for the eyes to converge or diverge for binocular vision ranges from 0.3 sec to 0.5 sec.[1]

Fifth, the eye must move to compensate for movement of the head.

Sixth, the eye often moves involuntarily in response to noise and other stimuli.

For clear vision, these eye movements must occur constantly. Thus, in traffic, vision is not instantaneous. The road user requires significant amounts of time to see the continually changing aspect of road and traffic conditions. A driver whose eyes are fixated on the traffic scene at the right of an intersection may need as much as a full second to shift his eyes to the traffic scene at the left and back again to the right:

Shift to left.................... 0.15–0.33 sec
Fixate on left.................. 0.1 –0.30 sec
Shift to right.................. 0.15–0.33 sec
Fixate on right............... 0.1 –0.30 sec
Total time to see............ 0.5 –1.26 sec

These values show time for seeing only and do not include time for reaction.

When the driver is dependent upon artificial illumination, he averages a 20-ft loss in visibility distance for each 10-mph increase in speed. In night traffic, therefore, the road user requires 1.36 (20 ft/14.7 ft/sec) sec more, for each 10-mph increase in speed, to obtain his maximum visual perception of traffic conditions.[2]

Peripheral Vision. In addition to seeing a particular object or scene in traffic, the road user must be able to perceive conditions at the sides of the path ahead and at some distance outside his cone of clear vision. Normal sight can discern objects within the field of peripheral vision set forth in Fig. 2-2. But objects outside the 10° cone of acute vision will be seen without clear detail or color.

Any unusual movement or brightness in the peripheral field of vision attracts the road user's visual attention, setting up a reflex mechanism which tends to turn the eyes and head to bring the object noticed into the cone of clear vision. As previously pointed out, this action requires a reaction time of about 0.2 sec, plus time for the actual shifting of the head and time for the eye to jump and fixate. Assuming no head movement is made, an average of about 0.5 sec total time is required. However, experimental study shows that the best drivers do not rely on "seeing out of the corner of the eye" but turn the head and eyes to scan continuously the field ahead.[3]

[1] *Ibid.*, p. 582, quoting C. H. Judd, *Psychology Monthly*, vol. 8, no. 34, 1907.

[2] K. Reid, "Seeing on the Highway," *Proceedings of Highway Research Board*, Washington, 1937, vol. 17, p. 427.

[3] T. W. Forbes, *Accidents in Traffic and Industry as Related to the Psychology of Vision*, National Society for Prevention of Blindness, Inc., New York, 1936, p. 2.

Studies of drivers show that the total central angle of peripheral vision usually ranges from 120 to 160°. But because of visual concentration, the range of effective peripheral vision shrinks with increased speed from a central angle of about 100° at 20 mph to as little as 40° at 60 mph.[1]

There is no conclusive evidence that road users with impaired peripheral vision are accident-prone, as experience teaches them to compensate for such weakness. It is clear, however, that side vision is relied on in many

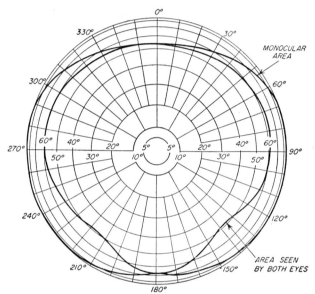

FIG. 2-2. Normal field of view of a pair of human eyes. (From H. L. Logan, "Normal Field of View of a Pair of Human Eyes," *Illuminating Engineering*, December, 1948, pp. 1–4.)

traffic situations. Nevertheless, good traffic regulation and design do not depend on the peripheral vision of road users, but on the cone of acute vision.

Visual Attention. A person with perfect eyesight may be effectively blind in areas extraneous to the point of visual concentration. In the location of traffic control devices and in the design of traffic facilities, lateral dimensions and other values should be chosen to fall within, and be effective at, the area of natural visual concentration of road users. This area on which the driver naturally focuses his eyes becomes farther from him as the speed of the vehicle increases[2] (Fig. 2-3).

The corollary of this basic principal of visual attention is clear: no

[1] J. R. Hamilton and L. L. Thurstone, *Safe Driving*, Doubleday & Company, Inc., New York, 1937, p. 23.

[2] *Ibid.*, p. 10.

traffic design or device should be so unusual or complicated as to compel undue visual concentration.

Visual Sensitivity to Light and Color. The approximate wavelengths of various colors of the spectrum and their relative visibilities at high levels of illumination are shown in Table 2-1.

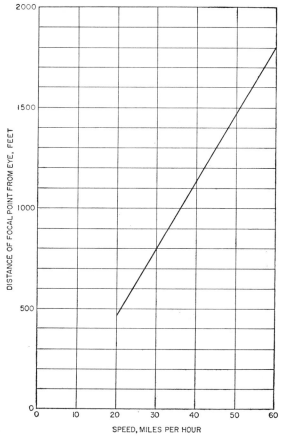

FIG. 2-3. Effect of speed on focal distance. (From J. R. Hamilton and L. L. Thurstone, *Safe Driving*, Doubleday & Company, Inc., New York, 1937, p. 10.)

There is very little sensitivity to light beyond the range of wavelengths varying from 420 to 700 millimicrons. The eye's maximum sensitivity at very low levels of illumination ranges from 510- to 520-millimicron wavelengths, and at higher levels of illumination from 520- to 550-millimicron wavelengths. The normal eye distinguishes color above intensities of 0.02 ft-c. At very low levels of illumination, all color sensation disappears.

Color blindness is a condition in which wavelengths and mixtures of wavelengths differ only in brilliance and not in hue. But total color blindness is extremely rare.   The most common type of color blindness is inability to distinguish between red and green; there is simply a sensation of difference in brilliance which grows less as the intensity is reduced or as the distance from the light is increased.   Inclusion of yellow in the red traffic signal and blue in the green signal is an aid to such color blindness. Moreover, the standardization of the relative positons of red and green in the signal is of real value.   However, color blindness has not been shown to be a significant factor in traffic accidents.

TABLE 2-1. RELATIVE VISIBILITY FOR AVERAGE NORMAL EYE

	Color	Wavelength, millimicrons	Relative visibility*
Shortest............	Violet	380	0.0
	Violet	400	0.0004
	Blue	450	0.038
	Green	520	0.710
	Yellow	570	0.952
	Red	650	0.107
Longest............	Red	750	0.0001

* Based on index of visibility of 556 millimicrons = 1.00.
SOURCE: National Bureau of Standards, Scientific Paper 475, Washington, 1923.

Glare Vision and Recovery.   Because of tunnels, street lighting, head-light glare, and so forth, heavy demands are placed on the road user's adaptability to light changes.   While the pupillary response to light changes can compensate for as much as a seventeenfold increase in external light, the light change from day to night varies in ratios of millions to one.   Residual adaptation to light change is a function of the retina.   Going from darkness into light, the eye adapts itself much faster than when going from light into darkness.   Traffic operation and lighting must take into account this problem of recovery to see again at much lower intensity of illumination after entering a tunnel or meeting a glaring headlight.

Figure 2-4 shows that the eye requires about 3 sec for most contractions of the pupil.   When oncoming headlights produce such contraction, visibility is reduced rapidly for the first few hundred candlepower of approaching glare but levels off as the intensity of the light is increased (Fig. 2-5).   Both increased age and the wearing of eyeglasses reduce the road user's ability to overcome glare difficulties.[1]

Perception of Space.   Space and time perception values based on vision allow the road user to form judgments of his own behavior as well as of the behavior of others in the traffic stream.   The angular sizes and shapes

[1] H. R. DeSilva and T. W. Forbes, *Driver Testing Results*, Works Progress Administration of Massachusetts, Boston, 1937, p. 26.   (Mimeographed.)

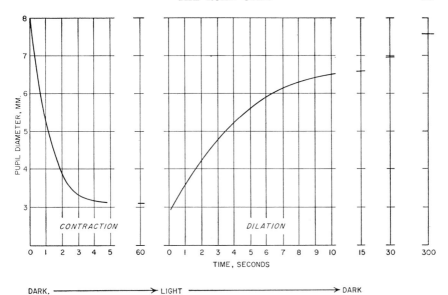

DARK. ──────────→ LIGHT ──────────────────→ DARK

Fig. 2-4. Adaptability to light changes. (From R. S. Woodworth, *Experimental Psychology*, Henry Holt and Company, Inc., New York, 1938, p. 548, quoting P. Reeves, *Psychology Review*, 1918, No. 25, pp. 330–340.)

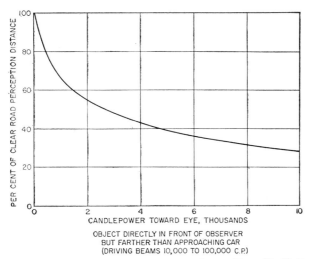

OBJECT DIRECTLY IN FRONT OF OBSERVER
BUT FARTHER THAN APPROACHING CAR
(DRIVING BEAMS 10,000 TO 100,000 C.P.)

Fig. 2-5. Visibility as affected by head-on illumination. (From K. M. Reid, "Seeing on the Highway," *Proceedings of Highway Research Board*, Washington, 1937, vol. 17, p. 429.)

of perceived details and their relative positions give the road user the ability to judge space. However, the judgment is subject to variation, owing to such factors as the convergence of the eyes for binocular accommodation, the sensation of straining to see through fog or smoke, and memory based on former experience.

Illustrative of the recognition of the driver's need for perception of space are the use of islands with two or more flashers spaced to form a sizable visual angle; road markings and delineators; parking guides; clearance lights; certain features on the vehicle itself; the layout of streets and junctions to obtain large visual angles; and adequate illumination.

While there are marked differences in the abilities of road users to perceive space, no significant correlation has been found between deficient depth perception, as an example of space judgment, and accident proneness.

2-3. Hearing. While the ear is also a distance and direction receptor, it is generally very inaccurate, and all evidence seems to show that deaf people compensate for deficiency in hearing. One study showed that of 600 deaf-mute licensed drivers, only one had an accident in a 2-year period, while nearly 4 per cent of all drivers are involved in accidents each year. The relative exposure for the two groups, however, is unknown.[1] Deafness has not been employed as a disqualification for a driver's license by any state.

2-4. Stability Sensations. Drivers usually react to their sense of what is comfortably stable: the rougher the road, the slower the speed. On horizontal curves the average driver limits his speed so that the induced centrifugal force on the vehicle gives a side-friction coefficient of 0.16 for speeds below 60 mph.[2]

If cross slopes, gradients of narrow ramps, roads on hills, and similar features of traffic design make the motorist feel unstable, he will react to gain a satisfactorily stable position if possible. Thus, on a road with a high crown and sharp cross slopes, motorists are inevitably led to the center of the road.

2-5. Time Factors in Response. Regardless of which sensory perceptions are involved, the road user's response to the stimulus of a particular traffic situation takes time. The time required for a response such as putting on the brake is taken up by the psychological processes of perception, intellection, emotion, and volition. In the analysis, interpretation, or treatment of any particular traffic phenomenon, the engineer should give thoughtful consideration to this time factor in response.

Perception. Sensations received through the eyes, ears, and body are transmitted to the brain and spinal column by the nervous system. When they are strong enough to be recognized, they become perceptions. The time required for this process may be called perception time.

Experience, habit, and other factors cause some sensations to result in

---

[1] "Deaf Mutes Are Safest Motorists on Pennsylvania's Highway System," American Association of Motor Vehicle Administrators, *News Bulletin*, vol. 5, no. 9, Washington, September, 1940, p. 15.

[2] J. Barnett, "Safe Side Friction Factors and Superelevation," *Proceedings of Highway Research Board*, Washington, 1936, vol. 16, p. 69.

reflex responses before they are perceived. However, the road user does not have a simple reflex response to the stimuli he encounters under most traffic conditions. In nearly every case he must perceive a complex and involved situation before he can react, and as the complexity of the situation increases, the time required to perceive conditions increases.

Little is definitely known about the exact time required for road users to perceive usual traffic situations. It is highly variable, dependent on the individual's particular psychological and physiological characteristics, as well as on the condition to be perceived.

Intellection. Perception may be simple recognition or it may result in intellection: the forming of new thoughts and ideas. The recognition of a red traffic light simply follows old memories and patterns for the experienced driver. The recognition of an unusual hole in the road will recall many similar old experiences, but will result in new ideas as well. The time required for comparing, regrouping, and registering new sensations (for further reference) is the intellection time. Intellection results in the sending of certain "messages" over the nervous system to the muscles and glands of the body. The message sent is based on previous experience coupled with the inherent drives and resistances of the particular individual and is only partially the result of the intellection process.

Emotion. Linked to the inseparable processes of perception and intellection are emotional sensations and disturbances dependent on the emotional traits of the individual. Emotion may vitally influence or affect the final message which is sent to the muscles to be carried out. Behavior growing out of fear or anger is not unusual in traffic accidents.

Volition. The *will* to take some act or produce some action is volition. While the intellection process may suggest many types of answers for a given problem, the final decision requires the resolution of all impulses received into a volitional outgoing message which results in some definite act or behavior. In this process the road user finally acts in accordance with his own memories, prejudices, beliefs, ideals, habits, weaknesses, desires, and attitudes.

Theory of PIEV. In his every act the road user is dependent on perception, intellection, emotion, and volition. This process of *PIEV* requires a time ranging from as low as perhaps 0.5 sec for simple problems to as much as 3 or 4 sec for a more complex problem, such as the decision to overtake and pass on a two-lane road. For any given condition of road use, the process of PIEV results in a normal pattern of behavior which can be timed and evaluated. In an abnormal individual the same process may result in some distorted behavior that becomes the problem of the police, the hospital, or the coroner. But normal PIEV qualities of road users should be taken into account by practicing traffic engineers.

Simple Reaction Time. Time values for the response to some of the simpler types of stimulus are known as *reaction time*. It has been found that response to visual stimulus is somewhat slower than to auditory or touch stimuli. Typical comparable values are shown in Table 2-2. As the task of response becomes more complicated, the time for response increases, as is shown in Table 2-3.

TABLE 2-2 REACTION TIME FOR VARIOUS STIMULI

Stimulus	Reaction time, sec
Light	0.18
Sound	0.14
Touch	0.14

SOURCE: Robert S. Woodworth, *Experimental Psychology*, Henry Holt and Company, Inc., New York, 1938, p. 324.

TABLE 2-3. REACTION TIME FOR VISUAL STIMULUS

Required action	Reaction time, sec
Release accelerator	0.205–0.247
Depress brake pedal	0.433–0.488
Depress brake pedal and steer simultaneously	0.604–0.707

SOURCE: H. R. DeSilva and T. W. Forbes, *Driver Testing Results*, Harvard University, Harvard Traffic Bureau, Cambridge, Mass., 1937. WPA Project 6246-12259.

Table 2-4 shows the results obtained when two vehicles were driven in tandem over the road and the rear driver was required to react with no stimulus other than his perception of the vehicle ahead. The increase in reaction time with increase of complexity of the situation is apparent.

Street-lighting conditions, signal and sign strength, simplicity of intersection design, and other traffic considerations are inextricably tied up with reaction time. Here again, however, no conclusive correlation has been found between accident proneness and slow reaction time. The individual seems to compensate for inability to react rapidly to traffic stimuli.

2-6. Modifying Factors. Modifying factors affecting reaction time in traffic are set forth below.

Fatigue involves drowsiness or sleep and usually stems from physical weariness (1 per cent of reported accidents are due to driver being asleep).

It reduces accuracy of judgment and lengthens PIEV time. Glare sources, monotony, and rhythmic effects can cause fatigue. Good traffic design, therefore, avoids excessive monotony or rhythmic and glare stimuli.

Disease, deformity, and *disabilities* are usually compensated for by the individual. However, they produce a tendency or proneness toward more intense emotional reaction, which has its implications in the traffic stream.[1]

[1] George M. Stratton, "Emotion and the Incidence of Disease," *Journal of Abnormal and Social Psychology*, vol. 31, 1926, pp. 19–23.

Alcohol, drugs, etc., have very noticeable effects upon mental and physical efficiency, usually producing poor attention, slower response, less dependable response, and less self-control. Secretion of the endocrine glands under emotional stress will create aberrated behavior of the individual.

TABLE 2-4. BRAKE REACTION TIMES UNDER VARYING CONDITIONS

Condition of:	Legend
Vehicle	
Standing	S
Moving (test condition)	T
Moving (normal road condition)	N
Stimulus	
Audible	AU
Bright light	BL
Stop light on vehicle	SL
No signal—stop light hidden	NO
Foot position	
On brake pedal	BR
On accelerator	AC

Conditions			Reaction time, sec
Vehicle	Stimulus	Foot	
S	AU	BR	0.24
S	BL	BR	0.26
S	SL	BR	0.36
S	AU	AC	0.42
S	BL	AC	0.44
N	AU	AC	0.46
S	SL	AC	0.52
T	SL	AC	0.68
N	SL	AC	0.83
T	NO	AC	1.34
N	NO	AC	1.65

SOURCE: *Report on Massachusetts Highway Accident Survey*, Massachusetts Institute of Technology and CWA-ERA Project, Cambridge, Mass., 1934, p. 57.

Climate, season, weather, time of day, altitude, ventilation, and *light* may produce complex responses in the individual. High tempers are more likely to flare at high temperatures; nervous irritability goes with the elements; mental and physical efficiency are tied up with amount of oxygen taken into the lungs, and the oxygen supply varies with altitude. Even the color of light has its psychological effects. The illumination of roads for traffic purposes should be governed by observed road-user performance under various lighting conditions.

Desires, habits, skills, and *attitudes* affect the *judgment* and *interpretive thinking* of individuals. Judgments, in turn, determine the road user's speed and his spatial relationship to traffic units in the traffic stream.

Fear and self-protection are fundamental in nearly every traffic maneuver. Anger is not uncommon. The competitive spirit of "me first" has resulted in the high-speed and high-acceleration qualities of the modern automobile and is manifest nearly every time the light changes to green, especially on multilane thoroughfares. All road users have experienced or observed the attitude of the "road hog" and those who refuse to be overtaken. The spirit of adventure and play is not without its effects on the performance of the individual in the traffic stream.

The basic process of thinking is "one thought at a time," and this rule is paramount in the placement of traffic signs. The design of traffic *paths* is best limited to a single choice. When the traffic situation grows complex, it may frustrate the road user, creating an emotional response.

Conditioned Response. Another fundamental trait of road users is their behavior upon being confronted with new traffic conditions after a period of stabilized operation. After driving a long distance at high speed, drivers tend to become *velocitated* and will not slow down until very strong stimulus for such slowing is manifest or until actual congestion impedes progress. Thus speeds entering a city tend to be higher than speeds leaving a city. This conditioning of motorists to speed, for example, may account for many accidents at the end of long tangent sections of roadway.

Drivers become accustomed to following a road marking or other delineation, and under some conditions, such as poor visibility, will follow a center line to their doom.

Motorists on traveling a *through* street notice at every intersection they pass that cross traffic is stopping for them. The conditioning of the motorist's mind when he comes to an intersection of two through streets, where he meets another motorist of the same mind, requires especially strong stimuli to get effective control.

This conditioned response of motorists is a characteristic to be recognized and treated in many traffic situations, as are the various modifying factors discussed above. Since traffic regulations and design are not determined by the behavior of abnormal road users, the traffic engineer seeks to determine the performance ability of the normal range of road users. Studies of the actual performance of road users will integrate the vagaries of human behavior into definite patterns of time-space relationships. These patterns should be used by the traffic engineer as actual performance specifications.

CHAPTER 3

THE VEHICLE

The dimensional and performance characteristics of vehicles are basic to the regulation of road use and to the design of routes and terminals. Although the limitations of traffic movement are generally due to a combination of vehicle and driver traits, at times they are due solely to the vehicle itself.

The purpose of vehicle use affects its design and operational qualities. Although speed and load capacity are basic, safety, convenience, comfort, and purpose of travel are important factors in passenger-vehicle design. The design of freight vehicles is influenced by all the passenger-vehicle design considerations, plus the nature, value, and protection requirements of the load to be carried. Unit size of load, distances and distributions of origins and destinations, as well as time factors, have resulted in extensive use of highway freight. However, the speed and maneuverability of larger freight vehicles are inferior to the private passenger car.

3-1. Vehicle Types and Dimensions. Common types of vehicles found in the heterogeneous traffic stream of today are shown in Table 3-1.

TABLE 3-1. PREDOMINANT TYPES OF HIGHWAY VEHICLES IN USE

Type	Number (000 omitted)	Year
Passenger		
Private automobile............	43,190	1952
Taxicabs.....................	96	1952
Buses........................	140	1952
Streetcars...................	11	1951
Subway and elevated cars......	10	1951
Trolley coaches..............	7	1951
Freight		
Trucks.....................	8,929	1952
Unclassified		
Bicycles (estimated)...........	14,000	Current
Motorcycles..................	411	1952
Trailers.....................	2,610	1952

SOURCES: *Automotive Industries*, vol. 108, no. 6, Mar. 15, 1953, pp. 107–110.
Transit Fact Book, American Transit Association, New York, 1952, p. 14.

25

Passenger vehicles predominate. There are over 180 makes and model types of private passenger automobiles, 500 types of trucks, and 74 types of buses for domestic use. Features of these vehicles significant in traffic regulation and design are set forth in Table 3-2. In addition to the single-unit motor vehicles listed, there are many combinations of tractor-truck, semitrailer, and truck-trailer units. These features affect lane and parking-stall dimensions, clearances, structural design, traffic performance, and geometric layouts.

TABLE 3-2. MOTOR VEHICLE DIMENSIONS, WEIGHT CAPACITIES, ETC.

	Passenger cars		Single trucks		Buses	
	Min.	Max.	Min.	Max.	Min.	Max.
Width, in...................	69	81	66	96	.....	96
Tread, in.*................	53	66	60	....	65	87
Over-all length, in..........	181	236	148	432	281	483
Height, in.................	58	69	69	150	96	114
Wheel base, in.............	100	147	80	255	146	270
Empty vehicle weight, lb.....	2,395	4,830	3,500	150,000	9,635	21,765
Brake horsepower†.........	68	210	72	350	101	220
Capacity..................	2 persons	8 persons	1,400 lb	103,500	21	52
Turning diameter, ft‡.......	36	58	40	97	30	87

* Treads measured on centers of tires.

† Horsepower is maximum brake horsepower of engine at specified rpm.

‡ Turning diameter is diameter of outside edge of front bumper for minimum turning circle.

SOURCES: *Automotive Industries*, vol. 108, no. 6, Mar. 15, 1953, pp. 93–163.

Turning Diameter; Overall Width Data, General Motors Corp., General Motors Proving Ground, Detroit, Mich., Jan. 15, 1940, pp. 1–6.

In urban traffic, streetcars and trolley coaches are frequently found to be part of the traffic stream. In single units these vehicles range in length from 35 to 60 ft (interurban); streetcars are sometimes also run in trains. The standard gauge (inside of rails) of streetcar treads is 4 ft 8½ in., and the width of car ranges from 8 ft to 8 ft 4 in. Many of the older models are very sluggish, and when making frequent stops, may constitute a hindrance to the traffic stream.

3-2. Resistance. In ordinary vehicle movement there are six resistances to be considered: (1) inertia, (2) rolling, (3) air, (4) gradient, (5) engine, and (6) braking.

Inertia. Since force is the product of mass and acceleration, the amount of motion a vehicle will attain in a given time is directly proportional to the amount of external force and inversely proportional to the weight of the vehicle. Ignoring all other resistances to movement, a force of 410 lb will be required to bring a 3,000-lb vehicle up to a speed of 30 mph in 10 sec.

A second but much smaller inertia effect arises from the rotary motion of the turning parts of the vehicle. In curvilinear motion the moment of force is a product of the moment of inertia and the angular acceleration of rotating parts. By analysis of the moments of inertia and angular acceleration given to wheels, axles, shafts, gears, etc., it can be shown that the force required to bring rotation of parts up to a speed sufficient to attain a linear translation amounts to 5 to 8 per cent of the total tractive resistance.

Rolling Resistance. Once the vehicle is in motion, a rolling resistance comes into play. The three components of rolling resistance are as follows:

1. *Impact resistance.* Irregularities of the road surface require the tires to flex and conform to such irregularities or, failing such conformation, cause *impact* of the vehicle's tires with the road surface. The magnitude of force from the impact of a given surface varies directly with the weight of the vehicle and speed of movement and is affected by the inflation and flexibility of the tires.

2. *Surface resistance.* Surface deformation of the road such as found on a very soft surface or in mud results in surface resistance.

3. *Internal resistance.* Friction inherent in the vehicle itself, due to the necessary slippage of moving parts, causes a resistance to forward motion. Lubrication, unit-bearing stress, temperature, tire construction and inflation, and speed of operation will affect this force.

Surface resistance and internal friction are relatively constant for all speeds. Impact resistance, however, will increase with speed. Total rolling resistance increases 25 to 30 per cent as the vehicle accelerates from speeds of less than 10 mph to speeds of 60 mph, and the increase results from impact and air losses due to wheel rotation only.[1,2] Rolling resistance increases as quality of surface becomes increasingly irregular or as it becomes more pliable. Typical values, at low speed, are set forth in Table 3-3. For most practical purposes it is satisfactory to assume a rolling resistance of 20 to 27 lb/ton, for all speeds, on smooth hard-surfaced roads.

Air Resistance. The air resistance of a vehicle depends on speed, air density, frontal area of the vehicle, and the stream flow of the air which is displaced. The formula for this resistance is $R = KAV^2$, where K is an empirical constant, A is the frontal area of the vehicle, and V is the velocity. Air resistance is reduced by the streamlining effects of the

[1] E. H. Lockwood, "Air Resistance of Automobiles," *Proceedings of Highway Research Board*, Washington, 1928, vol. 8, p. 146.

[2] R. G. Paustian, "Tractive Resistance Determinations with a Gas Electric Drive Automobile," *Proceedings of Highway Research Board*, Washington, 1932, vol. 12, p. 75.

vehicle's frontal area, and for vehicles without any unusual streamlining the constant has been found,[1] so that $R = 0.0017AV^2$. The air resistance to a vehicle with a 25-sq ft frontal area traveling at 50 mph becomes 106.25 lb. At higher speeds, air resistance becomes two or three times the magnitude of rolling resistance and is thus the larger component in power consumption. Wind-tunnel tests, however, show that streamlining may reduce the empirical constant from 0.0017 to as low as 0.0008.[2]

Gradient Resistance. On an inclined plane, the resisting (or assisting) force developed by gravity is proportional to the slope of the plane and

TABLE 3-3. ROLLING RESISTANCE ON VARIOUS SURFACES
(Determined with Studebaker test car)

Type surface	Rolling resistance (lb/ton)
Portland cement concrete................	19.0
Asphalt-filled brick......................	20.0
Minnesota oiled gravel...................	21.5
Bituminous macadam....................	23.0
Iowa oiled gravel........................	24.0
Untreated gravel (dry and firm)...........	27.0
Loose gravel...........................	50.0
Soft wet gravel........................	120.0
Iowa mud..............................	200.0

SOURCE: R. A. Moyer, "Motor Vehicle Power Requirements on Highway Grades," *Proceedings of Highway Research Board*, Washington, 1934, vol. 14, p. 168.

varies directly, as the sine of the angle for most purposes can be considered equal to the tangent, so that for each per cent of grade there is exerted a force equal to 1 per cent of the weight of the vehicle. This is frequently expressed as 20 lb/ton % grade.

Total Tractive Resistance. When constant speed and level grade prevail, the only resistances are air and rolling resistances. The sum of these is the total tractive resistance. The characteristic curves for rolling resistance and for the total tractive resistance are set forth in Fig. 3-1. At 50 mph, a 3,000-lb vehicle will meet a resistance of nearly 140 lb on level highway, allowing 27 lb/ton (see Table 3-3) for rolling resistance and assuming a 25-sq ft frontal area.

3-3. Power Requirements. Vehicle power is fundamental to speed, acceleration, and grade-climbing ability. The effects of power on vehicle movement must be considered by the engineer in traffic regulation, planning, and design.

The power required to overcome rolling resistance is a function of the weight of the vehicle in tons, the rolling resistance per ton, and the speed

[1] Lockwood, *op. cit.*, p. 148.
[2] C. Zeder, "Is it Practical to Streamline for Fuel Economy?" *Transactions of Society of Automotive Engineers*, vol. 49, no. 6, New York, 1941, p. 515.

of movement. In horsepower units (550 ft-lb/sec),

$$HP_{\text{roll}} = \frac{1.47R_r W_t V}{550},$$

where R_r is rolling resistance in pounds per ton, W_t is weight of vehicle in tons, and V is speed of vehicle in miles per hour. For a 3,000-lb vehicle

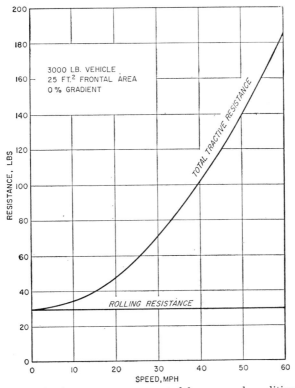

Fig. 3-1. Resistance curves computed for assumed conditions.

with a speed of 50 mph, 4 hp will be required to overcome a rolling resistance of 20 lb/ton.

The power required to overcome air resistance is a function of the frontal area of the vehicle and the speed of movement:

$$HP_{\text{air}} = \frac{0.0017A V^2 \times 1.47V}{550}$$

If for the above vehicle a 25-sq ft frontal area is assumed, 14.2 hp will be required to overcome air resistance at 50 mph. Characteristic curves for these values are set forth in Fig. 3-2 and are based on a vehicle which has the resistance curves shown in Fig. 3-1.

The horsepower required, or induced, by the gradient varies with the steepness of the slope, the weight of the vehicle, and the speed of movement:

$$HP_{\text{grade}} = \frac{W_t 20G \times 1.47V}{550}$$

where G is the per cent of grade. For the same 3000-lb vehicle, with a speed of 50 mph and an upgrade of 2 per cent, 8 hp is required to overcome gradient resistance. If the slope is downward, 8 hp less will be

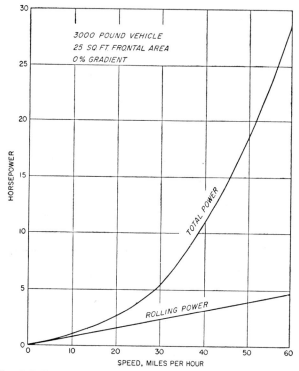

Fig. 3-2. Power requirements to overcome tractive resistance.

required. Combining the rolling-, air-, and gradient-resistance horsepower formulas produces the formula for total power required on an upgrade:

$$\frac{W_t V(R_r + 20G) + 0.0017A V^3}{375}$$

Characteristic curves of horsepower required for a test truck to overcome total resistance for various gradients are set forth in Fig. 3-3.

Figure 3-4 shows the difference between the power developed by the vehicle's engine and the power required to maintain speed on roads of two different grades. The difference between power developed and power required to maintain movement is the residual amount of power available for vehicle acceleration. The acceleration that may be obtained with a given residual power is $a = 550HP/MV$, where a is in ft/sec², HP is the residual horsepower, M is weight/32.2, and V is in feet per second. From

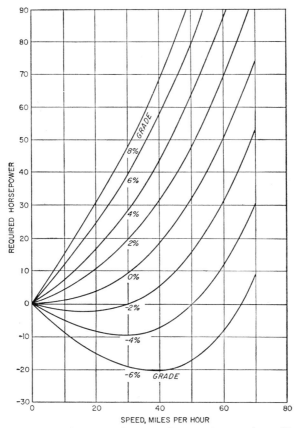

Fig. 3-3. Horsepower requirements of test truck on various grades. (From J. Beaky, "The Effect of Highway Design on Vehicle Speed and Fuel Consumption," Oregon State Highway Commission, Technical Bulletin 5, Salem, April, 1937, p. 28.)

the curves shown in Fig. 3-4, at 40 mph, the total power developed is 88 hp. Since the total resistance on the level requires 18 hp, a vehicle weighing 5,500 lb has available an instantaneous acceleration of 3.83 ft/sec².

Figure 3-5 shows average acceleration rates in different gears for several makes of passenger cars. These rates, shown by the solid curves, can be considered as the limitation of the vehicle in various gears. The dotted curve shows the average driver's normal acceleration of the vehicle up to 60 mph and clearly shows that throughout the acceleration process he has considerably more power available than he uses.

3-4. Stopping. Normal operation of the vehicle requires braking effort in addition to engine and other resistances to overcome inertia.

Engine Resistance. In addition to rolling, air, and gradient resistance, the engine, if properly engaged, will exert a force which tends to retard

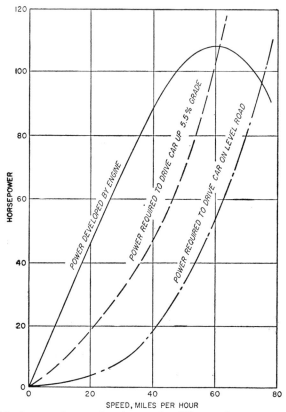

Fig. 3-4. Residual power for acceleration as the difference between power developed and that required to maintain motion. (From J. O. Eisinger and D. R. Barnard, "A Forgotten Property of Gasoline," *Transactions of Society of Automotive Engineers,* vol. 30, New York, August, 1935, p. 296.)

the vehicle's forward motion. Every driver uses this force in normal operation of a vehicle. The magnitude of engine resistance is shown in Table 3-4. The combined effect of rolling, air, and engine resistances in reducing motion on level, hard-surfaced roads has been explored through practical tests on passenger cars. Average values are set forth in Table 3-5.

Braking Resistance. Because the time required for engine, air, and rolling resistance to reduce motion from 50 mph to 10 mph amounts to as

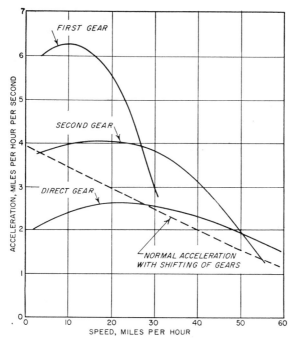

Fig. 3-5. Average accelerating ability in different gears found in several models of passenger cars. (Adapted from H. Tucker and M. Leager, *Highway Economics*, International Textbook Company, Scranton, Pa., 1942, p. 266.)

much as 35 to 45 sec, the driver will resort to his brakes to bring the vehicle to a stop. Where sharp rapid stops are required, brakes must be applied to halt the vehicle as rapidly as traffic conditions and the driver's judgment will allow.

For most practical conditions, especially at lower speeds, this required brake force is nearly equal and opposite to the linear inertia effects of the

TABLE 3-4. ENGINE RESISTANCE TO MOTION
(Passenger cars)

Gear	Speed	Lb/ton
High.............	10	17
High.............	40	100
Second..........	5	20
Second..........	25	150
Low.............	6	160

SOURCE: Ralph A. Moyer, "Power Requirements on Grades," *Proceedings of Highway Research Board*, Washington, 1934, vol. 14, p. 175.

vehicle. Neglecting the retarding forces caused by gradient, air, engine, and rolling resistances, the force required to stop a vehicle is dependent on the weight of the vehicle and the rate at which it is stopped or decelerated. That is, $F = Wd/32.2$, where F is braking force in pounds, W is weight in pounds, and d is the rate of stopping in ft/sec^2. The limiting value of

TABLE 3-5. TIME, RATE, AND DISTANCE OF DECELERATION

Speed change, mph	Time, sec	Distance, ft	Deceleration, mph/sec
70–65	2.0	198	2.50
65–60	2.2	202	2.30
60–55	2.4	203	2.10
55–50	2.6	203	1.90
50–45	2.9	202	1.70
45–40	3.2	199	1.55
40–35	3.6	196	1.40
35–30	4.0	191	1.25
30–25	4.5	182	1.10
25–20	5.2	172	0.95
20–15	6.2	159	0.80
15–10	7.8	143	0.65

SOURCE: D. W. Loutzenheiser, "Speed Change Rates of Passenger Vehicles," *Proceedings of Highway Research Board*, Washington, 1938, vol. 18, part 1, p. 93.

force F depends on the grip of the wheel with the road surface, since most modern vehicles have brakes with more power to hold the wheel than the road has to turn it. As a result, the driver can skid the tires over the road, so that the limiting force to retard motion is established by the *grip* of the tires with the road. The ratio of this grip to the weight of the vehicle is called *coefficient of friction*. That is, $F/W = f$, the coefficient of friction.

It follows then that

$$\text{Rate of deceleration which is possible } d = \frac{32.2F}{W} = 32.2f$$

$$\text{Time to slow or stop } t = \frac{1.47(V_1 - V_2)}{32.2f}$$

where V_1 and V_2 are the initial and final speed in miles per hour.

$$\text{Distance to slow or stop } s = \frac{V_1{}^2 - V_2{}^2}{30f}$$

$$\text{Work involved } W_o = \frac{W(V_1{}^2 - V_2{}^2)}{30}$$

$$\text{Horsepower required } HP = \frac{Wf(V_1 + V_2)}{750}$$

In the foregoing, wherever f is shown, the correction for gradient and tractive resistance can be made by algebraic addition of such forces expressed in pounds per pound of weight.

Numerous tests have been made to explore the practical values of coefficients of friction between tires and road surfaces. The nature of the road surface, its type of finish, the presence of moisture, mud, snow, and ice, and cinders or sand on ice are factors which greatly influence coefficients of friction. Again, the nature of the tire, its materials and design, the wheel load and unit-bearing stresses, the temperature of the atmosphere and tires, and the speed of the vehicle affect empirical values of the coefficients of friction. The values set forth in Table 3-6 are therefore representative values subject to considerable variation.

TABLE 3-6. FORWARD-SLIDING COEFFICIENTS OF FRICTION

Type of surface (1)	Dry		Wet	
	10 mph	40 mph	10 mph	40 mph
Portland cement concrete (smooth, sandy)....	0.78	0.42	0.66	0.35
Asphalt concrete (open texture).............	0.61	0.52	0.72	0.53
Bituminous macadam (rough)...............	0.60	0.41	0.62	0.46
Oil mat (open texture)....................	0.66	0.46	0.66	0.47

Condition of surface (2)	Dry
Packed snow or mud......................	0.20
Ice......................................	0.10
Water on ice............................	0.05

SOURCES: (1) J. Beaky, K. Klein, and W. J. Brown, *Skid Resistant Characteristics of Oregon Pavement Surfaces*, Oregon State Highway Department, Technical Report 39–5, Salem, 1939. (2) R. A. Moyer, *Skidding Characteristics of Automobile Tires on Roadway Surfaces and Their Relation to Highway Safety*, Iowa Engineering Experiment Station, Bulletin 120, Iowa State College, Ames, 1934.

Influence of Road User on Braking. Because the driver usually determines the amount of braking force which is utilized, it is significant to note coefficients of friction actually employed for different degrees of stopping. Results of studies made to determine these coefficients are set forth in Table 3-7.

3-5. Change of Direction. Normal operation in traffic requires steering and turning the vehicle to follow curved paths under a variety of conditions. As roads and vehicles have improved, riding qualities have improved and operating speeds have increased. These qualities, in turn, have magnified the importance of stability in vehicle maneuvers. Stability of operation is dependent on the driver as well as the vehicle. The skilled driver, experienced in the handling of a particular vehicle, will compensate for deficiencies in the vehicle itself.

TABLE 3-7. DRIVER INFLUENCE ON COEFFICIENTS OF FRICTION USED (FROM 70 MPH)

Description of condition	Acceleration, ft/sec;	f
(1) Engine alone at 70*...............................	3.2	0.10
(2) Comfortable for passenger. Driver's preference.......	8.55	0.26
(3) Undesirable for passengers. Driver would rather not use..	11.05	0.34
(4) Severe and uncomfortable for passengers. Objects slide off seats. Driver classes as emergency stop...........	13.90	0.43
(5) Maximum stop—skilled driver's car required to stop in 12-ft lane. Brakes in best condition.................	19.50	0.60

* Rates shown [except for (1)] are *average* deceleration rates employed in bringing a vehicle from 70 mph to a stop.

SOURCE: E. E. Wilson, "Deceleration Distances for High Speed Vehicles," *Proceedings of Highway Research Board*, Washington, 1940, vol. 20, pp. 393–397.

Minimum Turning Radius.   At slow speeds, the turning path of the vehicle is a function of wheel base and steering angle.   The path required by the vehicle when making its sharpest possible turn is especially important in maneuvering the vehicle in limited space, as in parking.   This path, in turn, affects the design of parking facilities, driveway, and intersection layouts, as well as curb-parking arrangements.

The minimum turning radius of a vehicle means the radius of the path described by the outer front wheel when the vehicle is making its sharpest possible turn at very slow speed.   The minimum turning radius for common types of vehicles will fall within the following ranges:

                Private passenger cars.......... 15.0–28.5 ft
                Buses......................... 20.5–45.0 ft
                Trucks....................... 18.5–47.5 ft

The paths described by "design" vehicles (see Chapter 25) when making sharpest turns are set forth in Fig. 3-6.   The higher values of turning radii of vehicles in common use have been employed for design vehicles, since these radii are the critical paths used in actual practice.

Off-tracking.   When a vehicle turns at slow speed, the rear wheels describe arcs which have smaller radii than those described by the corresponding front wheels.   This effect is termed *off-tracking* and is dependent on the turning radius and wheel-base dimensions of the vehicle. Under conditions shown in Fig. 3-6, off-tracking amounts to as much as 4.7 ft for the design truck and 2.7 ft for the design passenger car when making the sharpest turn, 90°, under very low speed conditions.

Centrifugal Forces in Turning. In general, turning a vehicle to change the direction of travel involves speed high enough to develop considerable centrifugal force. The centrifugal force must be overcome by a counterforce acting through the tires at the pavement surface, or the vehicle will continue in linear motion. If V is taken in miles per hour, W in pounds, and r (radius of the path) in feet, the centrifugal force in pounds becomes $F = 0.067WV^2/r$. In practice it has been found desirable partially to develop a counterforce by inclining the road surface toward the center of

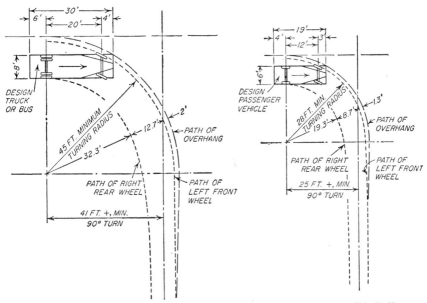

Fig. 3-6. Sharpest right turns of assumed-design vehicles. (From "A Policy on Intersections at Grade," *A Policy on the Geometric Design of Rural Highways,* American Association of State Highway Officials, Washington, 1940, p. 5.)

the curved path. (This inclination is called *superelevation.* See Chapter 28.)

Slip Angle. In Fig. 3-7, assume that the tire is mounted on a wheel with normal loading and is being supported by pavement surface. It is desired that the tire move in the path of automotive force OB, which is part of a very large arc to the right. However, because of centrifugal acceleration, there is another side force, DO, being imposed. If the tire were steered in the desired line of direction OB, the actual path of movement due to side force DO would be along OA. In order to maintain the desired path OB, therefore, it becomes necessary to steer the tire in direction OC. The angle produced by OB and OC is called the *slip angle.*

The frictional force between the tire and pavement needed to counteract the centrifugal force DO is called the *cornering force.* The amount of cornering force developed varies as the slip angle, load on the tire, and condition of the pavement (Fig. 3-8). Especially important is the peak noted at about 10° slip angle for dry pavement and 8° for slippery pavement. There is a definite limit to the total cornering force that may be

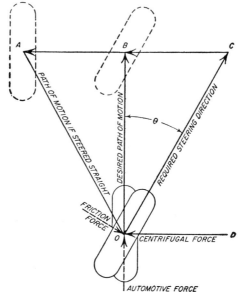

FIG. 3-7. Slip angle.

obtained by increasing the slip angle under any given condition of road and tire. If the cornering force required to balance the external forces becomes greater than this maximum, the tires will begin to skid and the vehicle will be out of control. Side thrusts developed by increasing the slip angle are quite noticeable to the driver's sense of stability and are liable to give the impression that skidding or overturning is imminent. Accordingly, under normal conditions the road user will tend to traverse curves in such a manner as to keep the required cornering force quite low. The normal values of slip angle actually employed rarely exceed 3°.

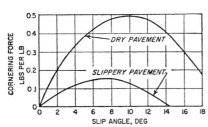

FIG. 3-8. Relationship of cornering force to slip angle. (From T. J. Carmichael, "Operating Characteristics of the Modern Automobile," *Proceedings of the Institute of Traffic Engineers*, New Haven, Conn., 1947, p. 76.)

Ordinarily, side thrust not more than 0.16 lb/lb is employed on curves, and at these lower slip angles a cornering force of 0.1 lb/lb is developed per degree of slip angle.

Tire design and construction may[1] vary the cornering force by as much

[1] H. W. Bull, "Tire Behavior in Steering," *Transactions of Society of Automotive Engineers*, vol. 34, New York, August, 1939, p. 347.

as 35 per cent. The cornering force reaches a maximum at the rated load of the tire and is reduced by either overload or underload. This is important in steering a vehicle around a turn, since there is a tendency (due to overturning moments) to overload the outside tires and underload the inside tires, thereby losing a part of the cornering force from all tires. This action is of increasing importance as the radius of turn is decreased or as the speed of operation is increased.

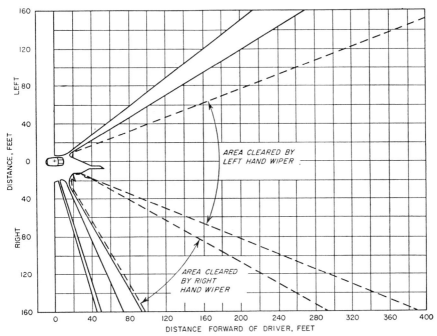

FIG. 3-9. Ground plan of driver's vision forward. (From *Visibility Test Procedure*, General Motors Corp., General Motors Proving Ground, No. PG-1, 1224A, Detroit, Mich., Mar. 26, 1945, plate 4.)

Again, the cornering force for a standard loading increases as tire inflation is increased within a range of 20 per cent of normal inflation.

In addition to the cornering force, other minor forces at work on the tire affect the stability of operation. The most important of these is a twisting moment, developed when the tire is on a slip angle, which tends to reduce slip angle and keep the tire steered in the direction of travel. This moment reaches a maximum at 4 to 6° of slip angle and can be as much as 60 ft-lb. It increases with tire load, decreases with increased inflation, and decreases as the power at the wheel changes from driving power to braking power.

3-6. Vision. In order that the driver may see all sources of potential traffic conflict in time to avoid accident, his field of vision should be as unobstructed as possible. Corner posts to support the top, radiator, and

hood; windshield wipers; and other essential features of the vehicle tend
to limit the field of vision.   These limitations are recognized by operators
and designers of vehicles and are important in the control of the traffic
stream and the design of facilities for traffic flow.   Overhead signals, for

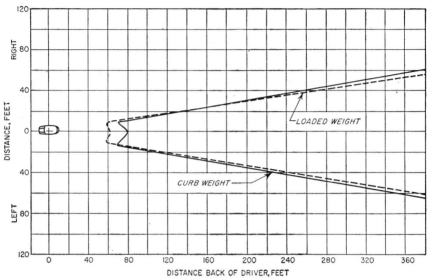

Fig. 3-10. Ground plan of driver's vision to rear. (From *Visibility Test Procedure*,
General Motors Corp., General Motors Proving Ground, No. PG-1, 1224A, Detroit,
Mich., Mar. 26, 1945, plate 5.)

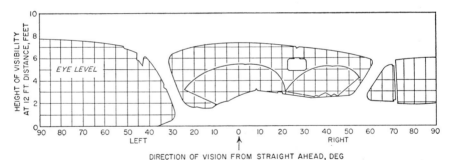

Fig. 3-11. Visibility from driver's position. (From *Visibility Test Procedure*, General
Motors Corp., General Motors Proving Ground, No. PG-1, 1224A, Detroit, Mich.,
Mar. 26, 1945, plate 3.)

example, are not readily seen by the driver of a vehicle waiting at the
intersection.   The glare from street lights is eliminated at certain vertical
angles.   Critical sight lines at intersections can be congruous with corner
posts at a given speed.

Figures 3-9 and 3-10 illustrate the ground-plan pattern of the driver's
forward and rear vision for present-day private passenger vehicles.   In

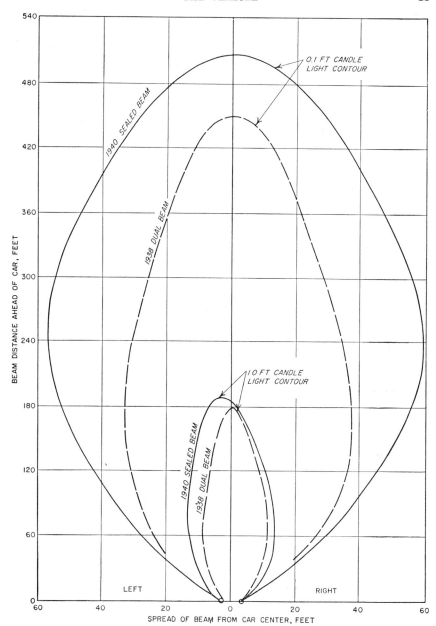

FIG. 3-12. Light contours as produced on the road ahead of a car by various head lamps. (From R. A. Moyer and D. S. Berry, "Making Highway Curves with Safe Speed Indications," *Proceedings of Highway Research Board*, Washington, 1940, vol. 20, p. 420.)

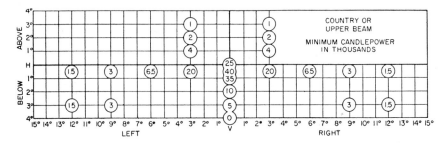

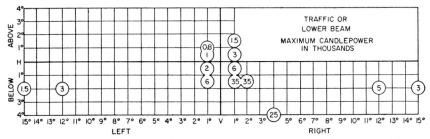

FIG. 3-13. Manufacturers' specifications for intensity of sealed-beam headlights. (From *SAE Handbook, 1947*, Society of Automotive Engineers, New York, 1947, p. 702.)

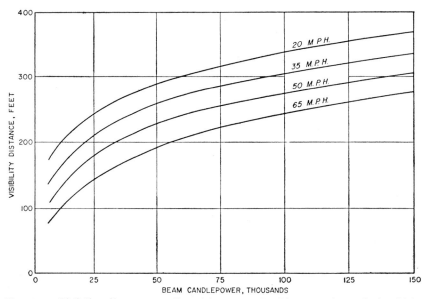

FIG. 3-14. Visibility distance as affected by strength of beam and speed of vehicle. (From V. J. Roper and E. A. Howard, "Seeing with Motor Car Headlamps," *Transactions of Illuminating Engineering Society*, vol. 33, New York, May, 1938, p. 422.)

Fig. 3-11, vertical as well as horizontal angles of vision to the front are shown.

Vision from private passenger vehicles has been greatly improved by slanting windshields to reflect light downward, away from the driver's eye.

3-7. Lighting. The lighting characteristics of vehicles are of extreme importance in night operation when there is no exterior source of illumination. In vision under headlight illumination, the light is reflected

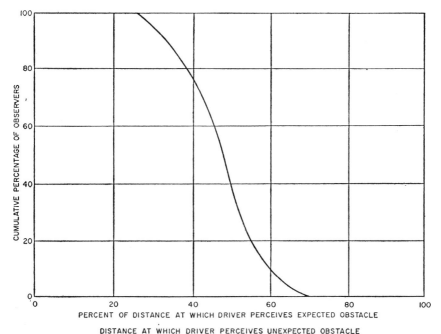

FIG. 3-15. Perception of unexpected obstacle. (From V. J. Roper and E. A. Howard, "Seeing with Motor Car Headlamps," *Transactions of Illuminating Engineering Society*, vol. 33, New York, May, 1938, p. 420.)

back from the road and the objects in the road. The pattern of light distribution on the road for modern head lamps is shown in Fig. 3-12.

At the road surface there is an incidence of 0.1 ft-c at a distance of as much as 500 ft ahead of the vehicle, and the maximum spread of beam, approximately 120 ft, at this or greater intensity occurs about 250 ft ahead of the vehicle. This pattern is produced by the relatively new sealed-beam head lamp which has a maximum beam intensity of 50,000 candlepower.

The vertical pattern of the sealed-beam head lamp as specified for manufacture and testing is set forth in Fig. 3-13. The upper pattern beam is intended for use on the open highway when not meeting other vehicles. In lower beam distribution, designed to reduce glare to oncoming drivers, the hot spot of intensity is shifted to the lower-right-hand quadrant. The basic relationship of visibility distance to beam candlepower and car speed is shown in Fig. 3-14. Visibility shows a 28-ft decrease in distance for each 10 mph increase in speed, owing to time required for seeing (see Chapter 2).

For perception of an unexpected object under head lamp illumination, the driver must be from 20 to 80 per cent nearer than for perception of an expected obstacle. See Fig. 3-15, a cumulative-frequency curve showing the proportion of drivers who were unaware of approaching a man-sized dummy in dark clothing. The distribution shown is based on observers who were operating under typical driving conditions with only head lamps for illumination. The average driver under such conditions sees only one-half the distance he sees under testing conditions, when he is warily looking for an object in the path ahead.

In order to augment the visibility of a given vehicle to identify it in darkness, tail lamps, stop lamps, clearance lamps, side-marker lamps, turn-signal lamps, auxiliary spot lamps, and special identification lamps for public vehicles have been employed[1] (see Chapter 13).

[1] *Uniform Act Regulating Traffic on Highways,* Uniform Vehicle Code, act V, art. XVI, secs. 124–149, 1948, pp. 32–41.

CHAPTER 4

SPEED

4-1. Distance and Time. Since speed determines the distance a road user may travel in a given time, it is vital in every traffic movement. A road user may wish to increase his speed to shorten time of travel or increase the distance of travel, or both. The time rate of change in speed is fundamental not only in starting and stopping but throughout the entire traffic stream.

Speed of movement is the ratio of distance traveled to time of travel. Where V is speed of travel, t is time of travel, and s is distance of travel, $V = s/t$. If t is fixed or held constant, the distance of travel varies directly with the speed. If V is held constant and t is made the variable, the distance of travel varies directly with the time of travel.

TABLE 4-1. REDUCTION IN TRAVEL TIME FOR 100-MILE TRIP THROUGH EQUAL INCREMENTS IN SPEED

Speed, mph	Travel time, hr	Reduction in travel time for preceding speed, hr	Total time savings from 10 mph, hr
10	10.00		
20	5.00	5.00	5.00
30	3.33	1.67	6.67
40	2.50	0.83	7.50
50	2.00	0.50	8.00
60	1.67	0.33	8.33
70	1.43	0.24	8.57
80	1.25	0.18	8.75

In many cases, such as travel from home to work or to shop, the origin and destination, and hence the distance of travel, are fixed. Time and speed then become the variables. Time of travel varies inversely with speed: as speed is increased, time of travel decreases. However, the relationship here is curvilinear, and equal increments of speed do not give equal decrements of time. The decrease in time of travel grows progressively smaller as speed is increased. For a trip of 100 miles, the reduction in time of travel through equal increments of speed is shown in Table 4-1.

The sharpest reductions in travel time for a given increment in speed occur at relatively low speeds, and they become progressively smaller as speeds are increased.   While a 50 per cent savings in time is effected by increasing speed from 10 to 20 mph, only a 20 per cent reduction in time is effected by increasing speed from 40 to 50 mph.   The significance of time savings to the road user is determined by length of trip: an increase from 40 to 50 mph will save 3 min for a 10-mile trip, 15 min for a 50-mile trip, and 2 hours for a day's trip of 400 miles.   The longer the trip, the higher the desired speed in order that sizable savings in time may be effected.

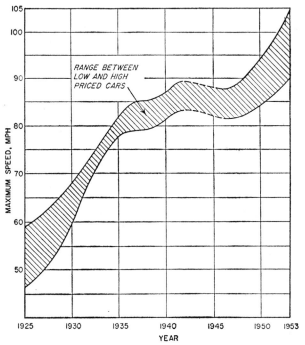

RANGE BETWEEN LOW AND HIGH PRICED CARS

Fig. 4-1. Growth in speed of standard passenger car.

4-2. Speed Trends.   Maximum motor vehicle speed attained over a single measured mile, employing a running start, increased from about 40 mph in 1898 to nearly 400 mph in 1948, averaging an increase of 7 mph per year.   Speedway records for races of 500 miles rose from about 75 mph in 1911 to nearly 129 mph in 1953,[1] averaging an annual increase of a little more than 1 mph.

These figures, however, do not reflect the development of the ordinary passenger vehicle, which is set forth in Fig. 4-1.   The average maximum speed of 31 mph in 1910 increased steadily at an average rate of almost 2 mph per year until 1937, when an average maximum rate of over 82 mph was attained.   From 1925 to 1935 the rate of growth was most rapid,

[1] H. Hansen (ed.), *The World Almanac*, 1954, p. 834.

averaging 2.5 mph per year. Since that period, the rate of growth of speed has been decreasing.

For a given roadway condition, maximum speed of a vehicle is a matter of vehicle design and the fuel employed, but actual speed attained is dependent largely on the road user's desires and environment. Statistics of free-driving speeds on flat, level, two-lane rural tangents during fair daylight hours show that average driving speeds have increased over the years. Average speeds of 21 mph were reported as early as 1910 in Iowa.[1] In 1925, average speeds of 25.6 mph, ranging from 14 to 61 mph, were found in a state-wide speed survey of Rhode Island.[2] Speeds of 37 mph were reported in Iowa for the same year.[3] By 1934, the average speed of motor vehicles in Connecticut had reached 39 mph.[4] Just prior to World War II, in 1941, the average free speed on rural roads was 47.1 mph. War conditions caused a drop in average speed to 36 mph by the end of 1942,[5] but since that time it has climbed back to values above prewar levels.

As Fig. 4-2 shows, for 15 years prior to World War II (1941), the average rate of growth in speed was about 1 mph per year, but the growth was leveling off under normal prewar conditions. It seems reasonable to suppose that any large increase in average free speed will be found only on wider roads of higher speed design, not on two-lane rural roads.

4-3. Classification of Speed Values. The actual speed of traffic over a given route may fluctuate widely. A given vehicle attains a certain speed at a given location under a set of conditions, but it may reduce or increase speed as required along the route. It may actually be temporarily at a standstill, yet will accomplish a known distance in a known time which yields a particular speed value. Accordingly, the term *speed* must be qualified.

Spot speed designates the instantaneous speed of a vehicle at a specified location. Spot speeds are used to measure speed of a traffic at an intersection, over a bridge, or at any other designated part of a given route. They are also used to compare diverse types of vehicles and drivers under specified conditions. Spot speeds are employed in describing the *crest* value of speed over a given course.

Running speed is the average speed maintained by a vehicle over a given course while the vehicle is in motion. Sustained running speed is

[1] H. F. Hammond, "Post-war Automobile Speeds," *Proceedings of Thirtieth Annual Highway Conference*, University of Michigan, Ann Arbor, 1944, p. 120.

[2] R. G. Paustian, *Speed Regulation and Control on Rural Highways*, Highway Research Board Special Investigation, Washington, 1940, p. 18.

[3] Hammond, *op. cit.*, p. 117.

[4] C. J. Tilden et al., *Motor Vehicle Speeds on Connecticut Highways*, Yale University, Committee on Transportation, New Haven, Conn., 1936, p. 12.

[5] *Traffic Speed Trends*, U.S. Public Roads Administration (now Bureau of Public Roads), Information Bulletin, Aug. 24, 1945, Table 1.

the *operating speed* along a route: the average speed at which a driver can travel under existing traffic and environmental conditions. The nominal running speed of an unimpaired vehicle is sometimes referred to as *cruising speed*. Running speed is obtained by dividing total distance covered by the total time the vehicle is in motion, exclusive of any time it is delayed while on a given test course.

Over-all speed is the effective speed with which a vehicle traverses a given route between two terminals. It is obtained by dividing the total

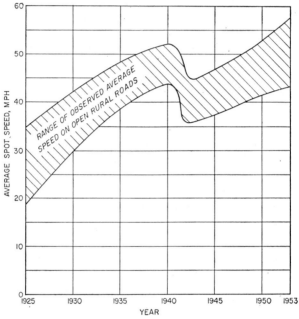

Fig. 4-2. Growth of average speeds on level tangent sections of open rural roads.

distance by the total time, including all delays incurred en route. The *system* speed of a transit fleet, for example, is expressed in over-all speed values.

4-4. Methods of Spot-speed Study. Most methods of spot-speed study depend on measuring the duration of time a vehicle requires to traverse a known, relatively short distance, usually less than 200 ft. Suitable points in the vehicle path, selected in accordance with the purpose of the investigation, are established at measured distances apart. As traffic flows past these points, the instants that a given vehicle enters and leaves the zone of measurement are procured, and the exact time units between entering and leaving are measured by chronometer or other means. The ratio of the distance to time required is the spot speed of the vehicle and may be recorded in feet per second, miles per hour, or other units of speed. Distance between points is often fixed at some multiple of 1.468, so that time values in seconds are quickly convertible to miles per hour. For example, if 176 ft (120 × 1.468) is the length of zone, the constant 120 may be divided directly by the number of seconds of elapsed time to give speed in miles per hour.

Mirror Box. A mirror box is often employed to eliminate errors due to parallax. This device bends the line of sight of the investigator to right angles to the path of movement of the vehicle as it enters or leaves the zone of measurement. The mirror box is placed opposite the point of entry (and/or exit), as shown in Fig. 4-3, and the investigator takes up a position so that he can see in the box. As a vehicle passes, it casts a shadow on the mirror, and the stop watch is started or stopped. If the investigator stations himself opposite one end of the zone, only one mirror

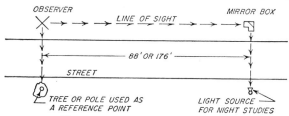

Fig. 4-3. Operation of mirror box.

box is needed. If two are used, the investigator may be anywhere between them.

The mirror box is not satisfactory in heavy, multilane traffic because of difficulty in associating an observed shadow flash with a particular vehicle. Also, it is sometimes difficult to conceal the mirror box and investigator from passing traffic.

Pressure-contact Strips. Pressure-contact strips, either pneumatic or electric, may be used to avoid errors due to parallax as well as those due to manually starting and stopping the chronometer. The passage of a vehicle actuates the air-impulse switch or, by direct pressure, closes the electric circuit by which the chronometer is started or stopped. Among electric spot-speed meters are solenoid-operated stop watches, resistor-condenser circuits discharged through ammeters calibrated in speed units, constant-speed motors with electromagnetic switches, and other devices sometimes equipped with graphic recorders.

Radar. A more recently developed method, and one of the most important because it automatically records speed, employs a radar transmitter-receiver unit. By transmitting a microwave radio frequency and taking advantage of the Doppler effect in the reflected wave, differences in potential may be amplified and indicated on a meter calibrated in miles per hour. A graphic recorder can be attached.[1]

[1] J. Barker, "Radar Measures Vehicle Speed," *Traffic Quarterly*, Eno Foundation for Highway Traffic Control, Saugatuck, Conn., July, 1948, p. 239.

Photography. Time series of photographs similar to those employed in motion pictures have been used in special cases concerned with detailed analysis of the time-space relationships in traffic. By projecting the film on especially prepared grid screens, the distance values or progress of a vehicle are read, and these linear values, divided by the constant time differential between exposures, give spot speed.

4-5. Statistical Interpretation of Spot-speed Studies. Though spot speeds for a given condition are frequently expressed by a single or average value, one figure cannot adequately describe the series of magnitudes found in a spot-speed study. The analysis of spot speeds, therefore, requires the application of some of the simpler statistical procedures.[1]

Frequency-distribution Tables. When spot-speed data for a given condition are collected and arranged in order of magnitudes, they form a significant frequency distribution. Table 4-2 shows the range and variation of spot speeds found in a given study.

TABLE 4-2. ILLUSTRATIVE FREQUENCY DISTRIBUTION, MID-BLOCK PASSENGER-CAR SPOT SPEEDS, URBAN CONDITIONS

Speed class, mph	Mean speed, mph	Number of vehicles (00 omitted)	Distributive per cent	Cumulative per cent
5– 9.9	7.5	0	0.0	0.0
10–14.9	12.5	1	0.2	0.2
15–19.9	17.5	34	6.2	6.4
20–24.9	22.5	146	27.1	33.5
25–29.9	27.5	178	32.9	66.4
30–34.9	32.5	130	24.1	90.5
35–39.9	37.5	31	5.7	96.2
40–44.9	42.5	16	3.0	99.2
45–49.9	47.5	3	0.5	99.7
50–54.9	52.5	2	0.3	100.0
55–59.9	57.5	0	0.0	
Total......	....	541	100.0	

Arithmetic Mean. The arithmetic mean or average speed of vehicles is computed from a frequency-distribution table by multiplying the number of vehicles in each speed class by the mean speed of that class, summing these amounts, and dividing their total by the total number of vehicles observed. In the examples furnished by Table 4-2, the average speed is 28 mph. However, this single figure gives no indication of range, variation, or even the number of vehicles.

[1] The student of traffic should understand the principles of statistical method and analysis. He will find a mastery of statistics an invaluable tool in the analysis and understanding of traffic characteristics.

Histogram, Frequency Curve, Mode, and Pace. It is frequently helpful to employ a histogram or frequency curve in analyzing the distribution of spot speeds and in determining the *modal average*, that speed at which the greatest number of vehicles travel. The modal average is quickly determined by the location of the peak on the frequency-distribution curve. In the example set forth in Fig. 4-4, a frequency-distribution curve for the data shown in Table 4-2, the modal average is around

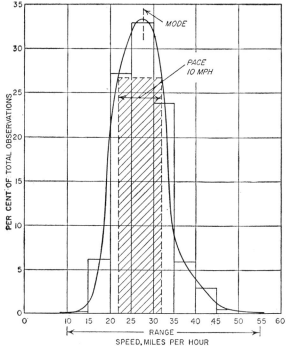

Fig. 4-4. A characteristic histogram and frequency-distribution curve of spot speeds. (Plotted from data shown in Table 4-2.)

28 mph. The frequency-distribution curve is also helpful in determining the *pace* of traffic: the range of speed which includes the greatest number of vehicles for some nominal increment in speed, usually 10 mph (Fig. 4-4). If the curve is symmetrical, the pace limits are equidistant from the modal value. In any event, it falls where the upper and lower limits intercept equal frequencies. In this example the pace is approximately 22 to 32 mph.

Cumulative-frequency Curve, Median and Percentile. The cumulative frequency curve is most useful in determining the median and the per cent of vehicles traveling at above or below a given speed. Such percentile values are of special significance in speed regulation and control as well as in the determination of design speed. The cumulative-frequency curve,

however, does not readily yield the distribution of spot speeds or the mode
or pace or any bimodal qualities.   Figure 4-5, based on the data in Table
4-2, is a typical cumulative-frequency curve.

4-6. Spot-speed Characteristics of Traffic.   Spot speeds of free-moving
vehicles are affected by the road user, the vehicle, the road itself, traffic
volume, weather, and other environmental influences.   When speed is

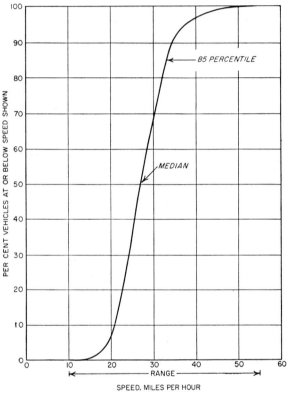

Fig. 4-5. A characteristic cumulative-frequency curve of spot speeds.   (Plotted from
data shown in Table 4-2.)

not limited by the characteristics of the highway or traffic interference
and the only limitation is that imposed by the individual driver or the
vehicle itself, spot speeds usually range from 20 to 80 mph.   On two-lane
rural highways, where speed is not restricted by extraneous conditions,
level-tangent maximum spot speeds are seldom higher than 70 mph.   On
main, high-speed roads, relatively free of curves and gradients, where the
only deterrents to speed are the driver and his vehicle, the maximum spot
speed is about 80 mph.

Figure 4-6 shows the distribution of spot speeds for free-flowing
vehicles, averaging 43 mph for an ordinary state highway, 48.5 mph on a

main two-lane high-speed highway, and 52.7 on a limited-access four-lane divided highway.

Characteristic cumulative-frequency curves for these free-moving vehicles, clearly showing percentile values, are set forth in Fig. 4.7. Of special interest to the traffic engineer are the 85 and 98 percentiles, the former being employed in speed regulation and the latter in design. For these free-moving vehicles the 85 percentile values range from 49 to nearly 58 mph, while 98 percentile falls between 55 and 70 mph. The

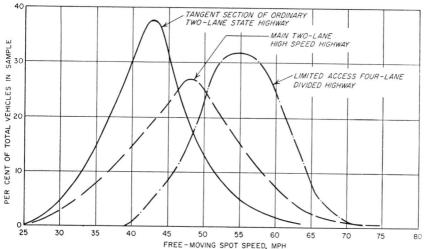

FIG. 4-6. Distribution curve of free-moving spot speeds. (From O. K. Norman, "The Influences of Alignment on Operating Characteristics," *Proceedings of Highway Research Board*, Washington, 1943, vol. 23, p. 331; and E. R. Ricker, *New Jersey Turnpike Traffic Statistics*, New Jersey Turnpike Authority, Trenton, 1953. Mimeographed.)

lower values are those prevailing on the majority of ordinary state highways, and the higher values are found on high-speed roads. For a particular location, considerable variation from these average values will be found.

Effect of Volume on Average Spot Speed. Volume has a pronounced effect on speed, particularly on two-lane rural highways where volumes as low as 200 vph (high for rural roads) reduce the average spot speed. On two-lane highways the average spot speed decreases linearly with increase in volume. The factor $0.009H$ (H is change in hourly volume of vehicles moving in both directions) gives the change in average speed.[1] The spot speed may be expressed as $V_s = 43 - 0.009h$, where the constant 43 is the average free-flowing speed for negligible volume. For high-speed highways the constant will be 48.5, (Fig. 4-8). The presence of increasing

[1] O. K. Norman, "Influence of Alignment on Operating Characteristics," *Proceedings of Highway Research Board*, Washington, 1943, vol. 23, p. 332.

numbers of trucks will reduce spot speeds at a slightly greater rate. When trucks represent 17 per cent of the traffic stream, the linear rate becomes $0.012H$.

Spot speed is intimately associated with the opportunity to overtake and pass. If a vehicle is to be considered free-flowing, it must be able to

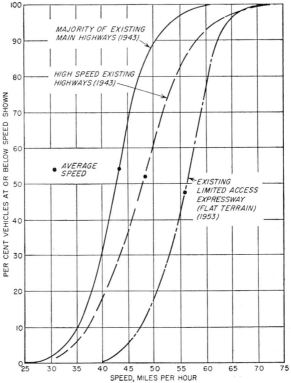

FIG. 4-7. Cumulative-frequency curves of free-moving spot speeds. (*A* and *B* from O. K. Norman, "The Influences of Alignment on Operating Characteristics," *Proceedings of Highway Research Board*, Washington, 1943, vol. 23, p. 331. *C* from E. R. Ricker, *New Jersey Turnpike Traffic Statistics*, New Jersey Turnpike Authority, Trenton, 1953, mimeographed.)

overtake and pass as frequently as desired without any reduction in speed. As the volume of traffic increases, it becomes difficult to carry out passing maneuvers, and greater numbers of vehicles are compelled to travel at the same speed. On two-lane highways the opportunity to overtake and pass will vary with the volume rate of the opposing stream of traffic flow. At about 2,000 vph total flow, the average speed is generally that of the slowest group of drivers, as the faster drivers find little or no opportunity to overtake and pass.

On four-lane highways, the opportunity to overtake and pass is not dependent on counterflow of vehicles but is related to volume movement in one direction. On high-speed four-lane highways, increased volume decreases the average spot speed in linear proportion to the increase. This rate of decrease in spot speed is similar to that of two-lane rural highways, though lesser in degree. For example, when average free-flowing spot speed of two lanes of a divided four-lane highway is approximately 49 mph, it will decrease at the rate of $0.0035H$, while an undivided four-lane highway with an average speed of 45 mph will decrease at the rate of $0.005H$ (H is the change in hourly volume from a given point on

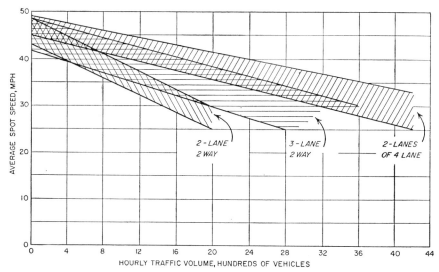

Fig. 4-8. Effect of volume on average spot speeds for two-, three-, and four-lane roads. (From O. K. Norman, "Results of Highway Capacity Studies," U.S. Bureau of Public Roads, *Public Roads*, vol. 23, no. 4, June, 1942, pp. 61, 63.)

the curve of spot-speed value for two lanes of the four-lane highway).[1] Any particular road may yield base values which vary from these typical constants.

On three-lane roads there is a decided difference in spot speed between the center and outer lanes. In the center lane it remains almost constant with increase in volume up to 2,000 vph total flow, while in the outer lane it decreases at a rate of about $0.004H$. The center lane carries a maximum of 15 per cent of the total volume and is used almost exclusively for passing. Even at heavy volumes, minimum speed in the center lane has been found to remain as high as 30 mph, owing to the necessity of completing the passing maneuvers and returning to the right lane as quickly as possible.

Effect of Volume on Spot-speed Distribution. As traffic volume increases, the *range* in spot speed as well as the average is often reduced

[1] O. K. Norman, "Results of Highway Capacity Studies," U.S. Bureau of Public Roads, *Public Roads*, vol. 23, no. 4, June, 1942, pp. 57–69.

because of the increasing difficulty of overtaking and passing and the tendency for faster-moving units to follow behind slower vehicles, even on multilane highways.[1] Figure 4-9 shows the relationship between volume, average speed, and range in speed for a two-lane highway, clearly indicating that as volume increases there is a lessening of range and a

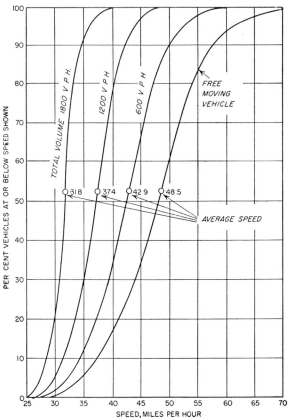

FIG. 4-9. Effect of volume on spot-speed distribution, two-lane road. (From O. K. Norman, "Influence of Alignment on Operating Characteristics," *Proceedings of Highway Research Board*, Washington, 1943, vol. 23, p. 324.)

sharp reduction in the percentage of vehicles traveling at high speeds. On four- and six-lane highways and on the outer lanes of three-lane highways, the effect is similar but less pronounced. The decrease in average and range of speeds is greatest in the lane of greatest density, except for the center lane of a three-lane highway, which reacts in the opposite manner. As the volume increases, the upper and lower limits of the range rise in the center lane, and there is a marked increase in the percentage of vehicles traveling over 55 mph, though the majority of vehicles

[1] Norman, "Influence of Alignment on Operating Characteristics," p. 332.

remain at the same speed. The increased speed is due to the need to complete the passing maneuver as quickly as possible.

Road-user Influence on Spot Speeds. Under ordinary conditions of free-moving traffic the individual road user has a wide choice of speed, influenced by such factors as trip length and presence of passengers in the vehicle. In general, the farther a road user travels on a given trip, the faster he goes. The increase in speed for long trips (over 100 miles) amounts to as much as 10 to 30 per cent over that employed for short trips, as Table 4-3 shows. The mean speeds of vehicles with out-of-state registration have been found to be from 3 to 8 mph faster than in-state

TABLE 4-3. INCREASE IN MEAN SPOT SPEEDS WITH TRIP LENGTH

Trip length, miles	Speed, mph		
	Connecticut	South Carolina	Average
0–19	40.1	36.8	38.4
20–49	41.1	39.6	40.3
50–99	42.6	42.6	42.6
100 and over	44.6	48.9	46.7

SOURCE: H. R. DeSilva, "Results from Speed Studies in Connecticut and South Carolina, *Proceedings of Highway Research Board*, Washington, 1940, vol. 20, p. 704.

vehicles. On the other hand, the maximum speeds of out-of-state vehicles are lower.

The lone driver tends to drive 2 or 3 mph faster than the driver carrying passengers. However, one study of Negro drivers in a Southern state found the driver with passengers traveling from 3 to 7 mph faster than the driver traveling alone.

Women drivers seem to travel at about the same or a slightly lower average speed than males[1] (Table 4-4).

Vehicle Influence on Spot Speeds. Table 4-4 illustrates the effect of vehicle type on spot speeds under free-moving traffic conditions. In general, buses are fastest, and heavy trucks are only slightly slower than buses. Vehicle speeds decrease about 1.5 mph for each additional year of age.

Effect of Road on Spot Speeds. Variation in type of hard-surface road does not greatly affect mean spot speed: the difference between mean spot speeds on concrete roads and macadam roads under similar conditions is only 3 to 4 mph.[2] Spot speeds on graveled surfaces, however, are about

[1] H. R. DeSilva, "Results from Speed Studies in Connecticut and South Carolina," *Proceedings of Highway Research Board*, Washington, 1940, vol. 20, pp. 702–706.

[2] Tilden *op. cit.*, p. 18.

10 mph below hard-surfaced roads.[1] Generally speaking, vehicle speeds are as high on 18-ft-wide pavement as on 20-ft.

The presence of sloping curbs does not affect speed, but vertical curbs on either inside or outside edge tend to reduce average speed 2 to 3 mph, unless these curbs are compensated for by wider lanes. Speeds on bridges are usually lower than on level tangents, though plate-girder bridges 24 ft or more wide for two-lane roads cause little or no reduction in speed.

TABLE 4-4. SPOT SPEEDS OF VEHICLES BY TYPE

Vehicle type	Mean spot speed, mph		Maximum spot speed, mph	
	Local	Out of state	Local	Out of state
Passenger:				
Male driver......................	47.2	49.9	76	69
Male driver with passenger.......	46.0	48.9	67	78
Passenger:				
Female driver alone..............	45.6	47.0	66	60
Female driver with passenger.....	42.8	46.6	63	62
Light truck......................	44.5	44.5	64	60
Heavy truck......................	42.1	44.0	67	62
Bus..............................	48.6	50.0	70	

SOURCE: *Highway Speed Study*, Connecticut State Highway Department, Traffic Division, Hartford, August, 1953.

Long, high truss bridges 24 ft wide, on the other hand, show reductions of about 7 mph below the speed on level tangents.[2]

Lengthy gradients up to 7 or 8 per cent spread the range of spot speeds uphill and narrow the range downhill from that found for level tangents. Average speeds downhill are 2 to 6 mph faster than uphill. The more intersections passed per unit of time or distance, the greater the decrease in speed: urban spot speeds average 25 to 30 mph, while rural spot speeds average 40 to 50 mph.

Curvature does not result in speed reduction until the cornering ratio required to offset centrifugal-force approaches 0.16 of the weight of the vehicle. Since the road user generally attempts to keep the required cornering ratio at or below 0.16, with a given superelevation, any increase in curvature (decrease of radius) which tends to raise this cornering ratio above 0.16 will instead result in lowered speed (see Chapters 27 and 28).

[1] Paustian, *op. cit.*, p. 23.

[2] *Speed and Placement Studies, 1941*, Iowa State Highway Commission and U.S. Bureau of Public Roads, Ames, Iowa, July, 1942, p. 27.

Effect of Time on Spot Speeds. The difference in mean spot speeds between daylight hours and darkness is small (Fig. 4-10). Free-moving spot speeds do not ordinarily change with the day of the week; many roads show lower speeds under crowded week-end conditions, but free-moving speeds themselves apparently are not affected by Sundays or

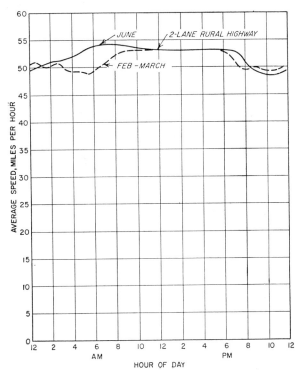

Fig. 4-10. Variation of average spot speed with time of day. (From R. C. Johnson and F. W. Hurd, *Rural Speed in Missouri*, Missouri State Highway Department, Safety Bureau, Jefferson City, Mo., 1941, p. 7; and R. L. Meyer, *Hourly and Daily Variation in Vehicle Speeds on a Rural Highway*, Yale University Bureau of Highway Traffic, New Haven, Conn., 1950.)

holidays. There is, however, a seasonal difference. Under good weather conditions, average spot speed in summer is 1.5 to 5 mph lower than in winter, as shown in Table 4-5. On four-lane concrete the reduction is about 5 mph.

Effect of Weather on Spot Speeds. Weather may influence spot speeds through reduction in visibility, through impairment of surface condition (as with snow), or through psychological effects on the road user. Table 4-6 shows reduction in speeds from 4 per cent to over 37 per cent due to weather conditions. Bad surface conditions create greater speed reduction than do lower visibility conditions.

Effect of Speed Regulations on Spot Speeds. Speed-control measures influence spot speed, as Fig. 4-11 illustrates. In the determination of speed regulations for sections of level-tangent roadways, care must be exercised to obtain the spot speed of the free-flowing traffic stream

TABLE 4-5. COMPARISON OF SEASONAL EFFECTS ON PASSENGER-CAR SPEEDS
(DAYLIGHT), 1934

Season	Type of road			
	2-lane concrete	2-lane macadam	4-lane concrete	All types
Winter............	42.8	41.5	42.2	42.4
Spring............	41.6	....	40.8	41.2
Summer...........	41.3	37.8	39.3	39.4
Max. difference.....	1.5	3.7	2.9	3.0

SOURCE: C. J. Tilden et al., *Motor Vehicle Speeds on Connecticut Highways,* Yale University, Committee on Transportation, New Haven, Conn., 1936, p. 18.

TABLE 4-6. EFFECT OF WEATHER ON SPOT SPEEDS

Type of weather	Speed, mph			
	Normal weather	Bad weather	Decrease	
			mph	Per cent
Visibility condition				
Dense fog, 100-yd visibility...............	40.8	31.4	9.4	23.0
Steady rain..........................	43.8	37.8	6.0	13.7
Light snow...........................	40.8	36.2	4.6	8.9
Light rain............................	42.6	39.4	3.2	7.5
Road surface and visibility				
Sleet storm, icy surface..................	46.3	35.4	10.9	23.5
Snow, rain, and slush...................	46.4	38.1	8.3	17.9
Snow flurries, snow on road..............	43.7	39.2	4.5	10.3
Road surface				
Clear, 3-in. hard-packed snow............	45.4	28.4	17.0	37.5
Clear, snow on road....................	43.7	38.8	5.1	11.7
Clear, 30% snow-covered................	36.8	35.2	1.6	4.4

SOURCE: C. J. Tilden et al., *Motor Vehicle Speeds on Connecticut Highways,* Yale University, Committee on Transportation, New Haven, Conn., 1936, p. 21.

uninfluenced by volume effects. For any given road there is an optimum speed limit which will have the greatest effect on spot speed. This value is usually between the 80 and 90 percentile of the free-flowing speed as plotted on a cumulative-frequency curve.

The effect of speed regulations on the maximum speed of single vehicles is influenced by the degree of enforcement, but the number of vehicles traveling at excessive speeds is sharply reduced when optimum speed limits are adopted. Furthermore, proper speed regulation causes vehicles in the lowest 10 to 25 per cent of the range to increase in speed, as road

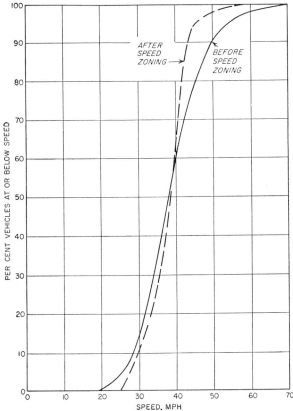

Fig. 4-11. Cumulative-frequency curve showing effect of speed zoning. (From Raymond Paustian, *Speed Regulation and Control on Rural Highways*, Highway Research Board Special Investigation, Washington, 1940, p. 45.)

users tend to accelerate, on the realization that they are traveling far below the stated safe allowable speed.

The over-all effect of proper speed regulation is that a greater number of vehicles travel closer to the average speed (see Chapter 15).

Summary of Spot-speed Characteristics of Traffic. Spot speeds, while simple in themselves, result from a combination of numerous factors, including desires of the road user, inherent speed and type of vehicle, road design, traffic volume, weather, season, and time of day. The average spot speed of normal highway traffic is usually the 55 per cent value in a frequency-distribution curve. The minimum speed is about 0.5 the average, and the maximum ranges from 1.5 times the average, for some limited-access roads, to 1.8 or 1.9 for rural two-lane highways. On rural highways almost one-half of the vehicles will travel in the intermediate third of the speed range. The highest third includes only a small number of vehicles, usually fewer than 5 per cent. In fact, the lower 50 per cent of the speed range includes about 85 per cent of the vehicles.

4-7. Methods of Speed and Delay Study. Though spot-speed studies are well suited to measuring the fluctuations in speed at a given point throughout a period of time, they are unsatisfactory in obtaining the fluctuation in speed or the delay incurred by the movement of traffic throughout the length of a given route. In order to obtain running speeds and over-all speeds between points spaced far apart and to measure fluctuations in speed with respect to consecutive locations along a course of travel, speed-and-delay studies are required. Investigations of this type show, with varying degrees of accuracy, the amount, location, duration, frequency, and causes of delay in the traffic stream, as well as the amount of time spent in actual motion from point to point along a given route.

Floating Car. If a vehicle is driven over a given course of travel with every effort to "float" with the traffic stream, it will approximate the average rate of speed which exists on the route. If the test vehicle overtakes as many vehicles as the test vehicle is passed by, the test vehicle should, with sufficient number of runs, approach the median speed of traffic movement on the route. In such a test vehicle, one passenger acts as observer while another records duration of delays and the actual elapsed time of passing control points along the route from start to finish of the run.

Control points may be intersections, bridges, or other fixed points, or, if an odometer is used, arbitrarily predetermined lengths of the route, such as $\frac{1}{10}$-, $\frac{1}{2}$-, or 1-mile points, may serve as control points. Between each pair of control points, if a stop watch is used, a faithful record may be made of causes and durations of delay for any particular run over the route. In addition, speedometer speed and mileage readings may be recorded every 10 sec or so. Causes of delay may be quickly entered on the log by predetermined code symbols.

Elevated Observer. In urban areas, it is sometimes possible to station observers in high buildings or other elevated points from which a considerable length of route may be observed. These investigators select vehicles at random and record time, location, and cause-of-delay information in similar fashion to that described under the floating-car method. Though considerable additional data may be gathered in a given time by this method, it is difficult to secure suitable points for observation throughout the length of the route to be studied.

License Plate. When the amount of turning off and on the route is not great and only over-all speed values are to be secured, the license-plate method of speed study may be satisfactorily employed. Investigators stationed at control points along the route enter, on a time-control basis, the license-plate numbers of passing vehicles. These are compared from point to point along the route, and the difference in time values, through use of synchronized watches, is computed. This method requires careful

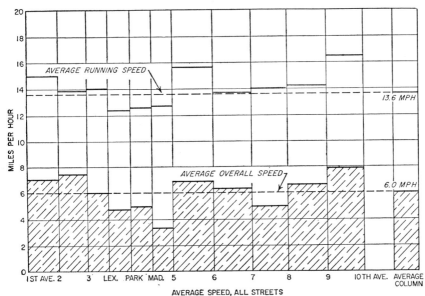

Fig. 4-12. Traffic speed and delay, eastbound movements in Manhattan, May, 1936. (From *Report of Traffic Speed and Delay in Manhattan,* New York Police Department, 1936.)

and time-consuming office work and does not show locations, cause, frequency, or duration of delay.

4-8. Analysis of Speed-and-delay Studies. Analysis of complete field data collected through thorough speed-and-delay studies will show (1) over-all and running speeds by location, (2) distribution of total, running, and delay time by location, (3) distribution of delays by cause, and (4) frequency and duration of delay. If speed-and-delay runs are made for the radial routes from a given point, a time-zone map may be prepared.

If several runs are made for a given route under specified conditions, the range and average of over-all and running speeds may be computed between each pair of control points and for the route as a whole. These values may be presented in tabular or graphical form (see Fig. 4-12) and will show the fluctuation in speeds throughout the length of the route. In addition, they furnish a basis for comparing one section of the route with another or with other routes.   As delays due to traffic flow disappear and speed is limited only by driver's choice, vehicle ability, and road conditions, the over-all speed approaches running speed.   However, such conditions are rarely achieved on urban streets.   Even if traffic interferences do not cause actual stoppage, they reduce the driver's desired speed.   Over-all speeds usually range from 3 to 15 mph in central business districts.

Relative amounts of time consumed in actual movement and in delay at various locations along the route are frequently compared by tabular or graphical means.   Graphical analysis is especially suited to studies

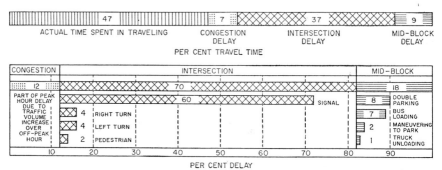

Fig. 4-13. Distribution of traffic delays.   (From *Traffic Survey of Spokane, Washington*, Washington State Department of Highways et al., Spokane, 1947, p. 18.)

based on control stations which are uniformly spaced along the route, in which case the slope of the line in a cumulative table or graph is a measure of speed.

A typical distribution of delay by causes is shown in Fig. 4-13.   The frequency of delay is a measure of its annoyance to the road user, but severity of delay depends on duration.   Table 4-7 is an illustration of tabular analysis showing delay type, frequency, and severity.

Time-zone Map.   The time-zone map is especially useful in comparing routes leading to a common destination or origin, for comparing facility of travel for different types of vehicles, or for comparing travel facility for different periods of the day (Fig. 4-14).   A time-zone map graphically depicts the time of accessibility of any section of the area surrounding a focal point in the traffic pattern.   It is particularly suited to analysis of marketing and commuting areas and for comparing routes of radial movement to and from a central business district.

4-9. Speed-and-delay Characteristics for Urban Areas.   There has been little investigation to date of over-all and running speeds and delays incurred by the average vehicle on diverse types of traffic routes.   Most speed-and-delay studies have been limited to a few streets in urban areas.

Average Over-all Speeds.   Available studies indicate that average over-all speeds in cities range from 3 mph in dense business areas to 35 or

40 mph in limited lengths of route in outlying residential territory. Over-all speed on city radial routes from center of the central business district to the outer periphery of urban development usually falls between 15 and 20 mph. A study of important San Francisco radials showed an over-all speed of 16.1 mph in the peak hours and 17.35 in the off-peak periods (midday).[1] An example of urban speed throughout a length of route is set forth in Fig. 4-12.

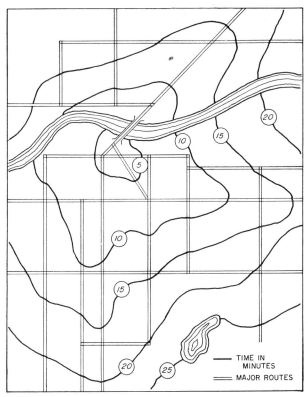

FIG. 4-14. Time-zone map.

Frequency of Delay. Delays in urban areas for city-wide travel occur on an average of 2 to 3 times per mile and as often as 10 to 12 times per mile in dense business districts. The average vehicle is delayed 40 to 60 times per hour of city-wide travel and as much as 70 to 80 times per hour of travel in central business districts.

Causes of Delay. Intersection delays, including cross traffic, signal, pedestrians, left and right turns, account for 70 to 75 per cent of the delays in ordinary central business districts, but in Manhattan they

[1] M. McClintock, *A Report on the San Francisco City-wide Traffic Survey*, San Francisco Department of Public Works and WPA Project 6108-5863, 1937, Appendix IV.

account for only slightly more than half the delays. Double parking, angle parking, maneuvering in parking, loading and unloading, slow-moving vehicles, and other mid-block interferences also cause delay.

Duration of Delay. Under usual urban conditions the duration of a single traffic delay is 0.25 to 0.40 min, ranging from 0.1 min to 0.6 min.

TABLE 4-7. TYPE, FREQUENCY, AND DURATION OF DELAYS FOUND IN AN
URBAN AREA
(Total of all streets and avenues)

Causes	Frequency, per 1,000 delays	Average duration, sec
Intersection delays		
Traffic crossing and signal....	422	40.3
Traffic clearing..............	55	11.2
Pedestrian...................	26	6.4
Left turn....................	60	7.5
Right turn..................	7	6.0
Total.....................	570	32.1
Mid-block delays		
Double parking.............	63	7.4
Angle parking..............	43	7.5
Maneuvering...............	90	9.1
Loading or unloading.......	51	9.2
Slow-moving vehicles........	68	11.4
Blockades.................	66	17.5
Miscellaneous..............	49	24.5
Total.....................	430	13.0
Grand total.............	1,000	23.9

SOURCE: *Report on Traffic Speed and Delay in Manhattan*, New York Police Department, 1936.

Shorter delays are caused by pedestrian and turning interference and longer delays by traffic signal or general blockades.

Distribution of Time in Travel. Total time lost through delay ranges from 15 to 16 per cent for city-wide travel and from 35 to 50 per cent for dense central business district conditions. Running time, therefore, is 50 to 65 per cent in dense business districts and about 85 per cent for city-wide travel on radial routes.

CHAPTER 5

VOLUME

A determination of traffic volume is basic to the evaluation of traffic movement. By counting the number of traffic units passing a given point or collecting in a given area during a known period, a *time rate* of traffic flow may be obtained. Pedestrians, passenger cars, buses, streetcars, light and heavy trucks, trucks with semitrailers, trucks with full trailers, bicycles, motorcycles, and "other vehicles" are commonly employed as *units* of traffic volume.

Because it furnishes a basic scale of comparison, a measure of traffic volume will show the relative importance of any route or facility. Volume data are needed in research, planning, designing, and regulation phases of traffic engineering and are also used in establishing priorities and schedules of traffic improvements. The traffic engineer must acquire general knowledge of traffic volume characteristics in order to measure and understand the magnitudes, composition, and time and route distributions of volume for each area under his jurisdiction.

Generally speaking, maximum hourly values and average daily values are those commonly employed in operation, design, and planning. Traffic demands in vehicles per hour per lane are of extreme significance in dealing with practical problems. In addition, routes are frequently specified, and comparisons may be made among routes on the basis of annual average daily traffic.

The factors shown in Table 5-1 are useful in distributing total traffic volumes into average volumes.

TABLE 5-1. CONVERSION FACTORS FOR CHANGING TOTAL TO AVERAGE VOLUME

Average total volume measured	Average volume expressed as a per cent of total volume				
	Av. year	Av. month	Av. week	Av. day	Av. hour
Year......	100	8.33	1.92	0.274	0.0114
Month....	1,200	100	23.1	3.29	0.137
Week.....	5,200	433	100	14.3	0.596
Day......	36,500	3,040	700	100	4.17
Hour......	876,000	73,000	16,800	2,400	100

If the total yearly volume of a traffic stream is 5,000,000, the average hourly will be 570, the average daily 13,700, the average weekly 96,000, and the average monthly 416,000.

5-1. Volume Counting. *Hand Tally.* To make volume counts by hand tally, the observer is equipped with suitable clip board, watch, pencils, and a supply of detail forms needed for his shift of duty. Detail forms should be designed to allow swift and accurate recording of all desired details of direction of movement, turning movements, and type of vehicle. A separate sheet is generally used for each counting period. Summary forms and diagrams are employed to recapitulate, on a single sheet or diagram, values obtained for each elementary unit of counting period and total values for the entire period of observation.

Manually Operated Counters. Manually operated counters obviate the need for detail counting sheets, so that summary entries for an entire count may be entered directly in the field from the totals on the counters at the end of each counting period. The amount of detail which may be collected is limited by the number of counters that can be operated.

Automatic Counters. One form of automatic volume counting employs actuation by interruption of a light beam falling on a photoelectric cell as traffic passes. Another depends on mechanical actuation by passage of vehicles over a treadle which closes an electric circuit directly. Another employs pneumatic action on an air switch attached to a flexible tube over which traffic passes. Magnetic detectors are also used; by clockwork and ratchet advance mechanisms, subtotal and total numerical values are registered on tapes or are recorded on dials. Use of these automatic mechanisms generally involves loss of detailed information such as classification of the diverse types or direction of traffic flow. The photoelectric, magnetic, and treadle types are especially suited to permanent installation at fixed counting stations. The pneumatic type lends itself more readily to portable requirements.

Counting Periods. If extremely short counting periods, such as 1 or 2 min of flow, are used, there is usually an unwarranted amount of numerical detail. However, if only daily totals are captured, there is frequently insufficient detail to lend the data value for analytical purposes.

The purpose of study and the rate of fluctuation in the traffic stream form the basis for determination of the duration of the elementary counting period. A volume study of pedestrian flow passing a given entry for transfer to buses may suggest periods of 5 or 10 min for rush-hour or change-of-shift conditions, but a study of the relative importance of two roads may be dealt with satisfactorily by daily or even annual totals. For most rural studies, 1-hr counting periods form a satisfactory basis for collecting volume data. For urban work, half-hour and sometimes quarter-hour periods are employed.

Counting Locations. For planning purposes, counts are made at selected strategic points throughout the network of routes under consideration. In urban areas it is desirable to obtain volume data at intersections which fall into the following classifications: (1) central business

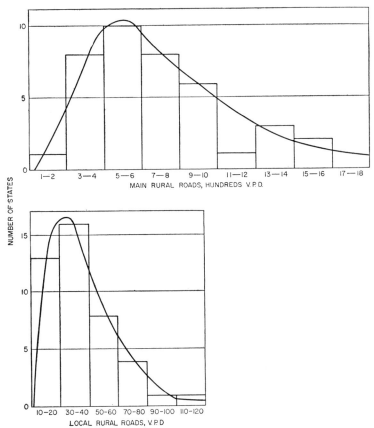

FIG. 5-1. Average daily volumes, rural. (Plotted from data presented by J. T. Lynch and T. B. Dimmick, "Amount and Characteristics of Trucking on Rural Roads," U.S. Bureau of Public Roads, *Public Roads*, vol. 23, no. 9, July, 1943, p. 215.)

district, (2) secondary business districts, (3) industrial areas, (4) high-accident points, (5) signalized intersections, (6) along major routes, (7) focal points in traffic stream such as bridges, ferries, natural passes, (8) irregular "problem" intersections. Satisfactory data in larger urban areas generally require one counting station to every 1 or 1½ miles of paved thoroughfare. Under rural conditions spaces of as much as 3 to 6 miles may be used.

5-2. Magnitudes of Traffic Volume. Traffic volumes range from a low of 1 or 2 vehicles per day in sparsely settled rural areas to more than

120,000 on a peak day on dense metropolitan expressways. Average volume on main rural primary state systems of roads throughout the nation is nearly 900 vehicles per day. Average volumes on individual state systems range from 200 to 1,800 vehicles per day, as Fig. 5-1 shows. Averages for local rural roads range from 5 to 125 vehicles per day, with a nationwide average of nearly 50 vehicles per day.

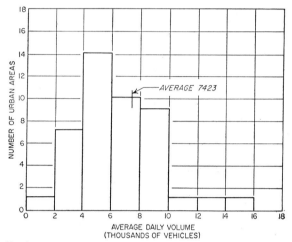

Fig. 5-2. Distribution of average daily volumes found in 44 urban areas. The volume for each urban area is the average volume of all streets carrying 200 or more vehicles per day. (Data from *Report of a Highway Traffic Survey in the County of Los Angeles,* County of Los Angeles Regional Planning Commission, 1937, p. 20.)

An examination of average traffic volumes in vehicles per day found on various types of rural roads shows the following values:[1]

Paved (concrete, brick, asphalt, etc.)........................ 1,232
Dustless (bituminous-treated)............................... 413
Nondustless (gravel, clay, and other nonbituminous)........... 77
Graded and drained, natural earth......................... 22
Unimproved... 13

As Fig. 5-2 shows, daily volume on major urban thoroughfares usually ranges from less than 1,000 to more than 15,000 vehicles per day. In rare instances, individual streams carry 30,000 and more, even as many as 80,000 vehicles per day. A distribution of over 1,400 individual locations in a single urban area is graphically shown in Fig. 5-3.

Pedestrian volumes in urban areas range from a few per day in sparsely developed areas to more than 100,000 per sidewalk per day. Business districts average 2,000 pedestrians per sidewalk per day, and 60,000 to 90,000 pedestrians per day pass the more important business fronts.

[1] J. T. Lynch and T. B. Dimmick, "State of Improvement of Rural Roads in Relation to Traffic and Dwellings Served," U.S. Bureau of Public Roads, *Public Roads,* vol. 21, no. 8, October, 1940, p. 142.

Though magnitudes of traffic volume furnish a relative scale of measurement for sections of more or less homogeneous characteristics, they state nothing concerning the quality of movement or the economic and social values involved.

5-3. Distribution by Routes of Traffic Volume. The volumetric measure of traffic movement is generally accepted as a true measure of the relative importance of any given traffic stream in a given area.

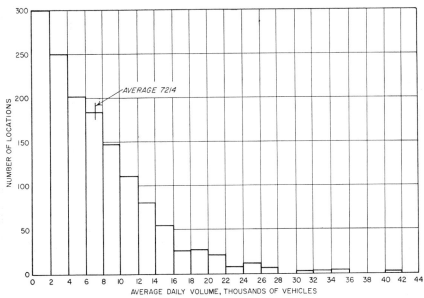

FIG. 5-3. Distribution of daily volumes recorded at 1,400 locations in one urban area. (Data from M. McClintock, *A Report on the San Francisco City-wide Traffic Survey,* San Francisco Department of Public Works and Works Progress Administration, 1937, Appendix 1.)

In analysis, it is customary to employ "flow maps" which show clearly the magnitude of traffic volume for any given location (Figs. 5-4 and 5-5). Such maps are prepared by developing the width of route in direct proportion to traffic volume handled for a fixed period of time, usually the average day, and may be rendered for trucks, passenger cars, transit systems, vehicles per lane, pedestrians, or passengers.

One of the basic characteristics of traffic, shown by the distribution of volumes by routes, is the concentration of traffic on relatively few routes and on a minor percentage of the total route mileage. This tendency is a vital factor in planning, design, regulation, and administration for traffic improvements. Figure 5-6 shows that more than 80 per cent of total travel occurs on 20 per cent of the route mileage. More than half of all vehicle-miles are traveled on city streets, though there are over 3,000,000 miles of rural roads and only about 300,000 miles of city streets.[1]

[1] *Highway Statistics, 1948,* U.S. Bureau of Public Roads, 1950, pp. 34, 116.

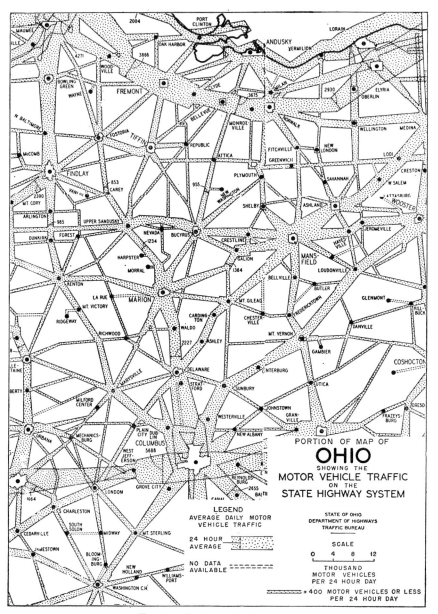

FIG. 5-4. State traffic flow map.

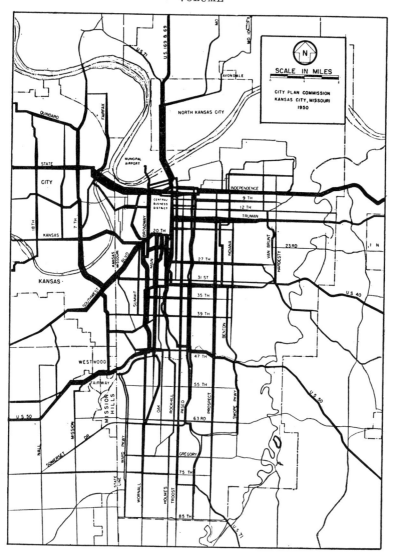

TRAFFIC FLOW 1948-1950

PORTION OF KANSAS CITY METROPOLITAN AREA

MISSOURI —— 24 HOUR COUNTS —— 1948
KANSAS —— 24 HOUR COUNTS —— 1950

20,000
40,000
60,000

FIG. 5-5. City traffic flow map.

5-4. Composition of Traffic Volumes. Table 5-2 shows the variability of the composition of the traffic stream. Because of the different individual characteristics involved, the average traffic stream does not have a high degree of homogeneity. This becomes especially significant as the density of traffic movement increases. For example, the frequent stops and sluggish performance of common-carrier passenger vehicles, in comparison with passenger autos, creates a condition enabling a small percentage of the total volume to hamper severely the capacity and speed of movement. Only under dense urban conditions are streetcars and high percentages of buses found.

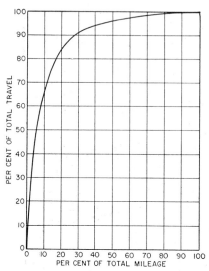

FIG. 5-6. Concentration of traffic flow. (Plotted from data presented in various state highway planning surveys, e.g., *West Virginia Highways, Preservation and Development,* 1940–1960, West Virginia State Road Commission, Charleston, 1940, p. 29.)

The most nearly uniform composition is found on parkways which limit vehicles to purely private passenger types, eliminating all commercial operations, and which do not allow pedestrians. In general, it is desirable to have only vehicles of

TABLE 5-2. COMPOSITION OF TRAFFIC BY VEHICLE TYPES BY
PERCENTAGE DISTRIBUTION

Vehicle type	Urban	Rural	Usual range
Passenger auto...........	80	77	74–86
Local................	..	59	60–90
Foreign..............	..	18	10–40
Truck...................	15	22	12–24
Single unit............	..	16	11–18
Combination..........	..	6	4–6
Transit................	5	1	0–7
Bus.................	3	1	1–7
Streetcar..............	2	0	0–5

SOURCE: Adapted from T. B. Dimmick, "Traffic Trends on Rural Roads in 1948," U.S. Bureau of Public Roads, *Public Roads,* vol. 25, no. 12, February, 1940, p. 290, and various urban traffic surveys.

the same type using a common route, but such conditions are scarcely ever achieved except on parkways in or near urban areas where traffic volumes justify special restrictions. The separation of pedestrian traffic from vehicle traffic is also desirable in the reduction of accidents and congestion.

5-5. Cyclical Variation in Traffic Volume. There are ordinarily three cyclical variations in the volume of traffic flow, the most pronounced being the fluctuation in hourly rates of flow throughout the day and night. In

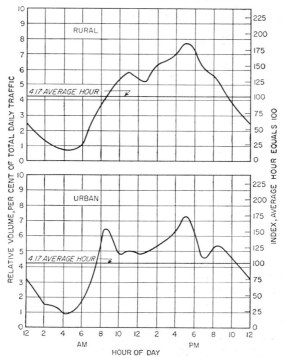

FIG. 5-7. Daily time patterns of traffic flow. (Plotted from various sources of hourly volume data, e.g., N. Cherniack, "Methods of Estimating Vehicular Traffic Volume with the Aid of Traffic Patterns," *Proceedings of Highway Research Board*, Washington, 1936, vol. 16, pp. 253–268.)

addition, daily volumes of most traffic streams vary considerably throughout the week, and there is a rise and fall of traffic flow from season to season.

Because of the volume variations inherent in the traffic stream, when a total volume is distributed into hourly, daily, or seasonal components, a *time pattern* of traffic flow results. Such patterns may be described in actual numbers of traffic units or in percentages that each component bears to the total or average volume.

Daily Time Patterns. Typical patterns of hourly traffic variation for rural and urban traffic are set forth in Fig. 5-7. Saturday, holiday, and Sunday patterns differ considerably from those for weekdays. The exact variation for any particular stream is a matter of individual determination. Radial flows differ from cross-town flows in urban areas; inbound varies from outbound movement. Routes serving industrial areas vary from those serving agricultural or retail land use. Transit, truck, or pedestrian elements will develop their individual time patterns of variation.

Studies of daily traffic patterns on a given stream over a number of years show a highly stable pattern, so that for specific days and seasons (autumn Wednesdays, for example) the relative volume of traffic during each hour of the day is highly consistent.

Most daily traffic patterns show approximately 70 per cent of total daily traffic movement occurring between 7 A.M. and 7 P.M. Typical variations from this value are set forth in Table 5-3.

TABLE 5-3. PERCENTAGE OF DAILY TRAFFIC VOLUMES MOVED IN
12 DAYLIGHT HOURS

Type of route	Per cent of 24-hr vol.	Max. deviation
Rural*		
Local....................	78.1	5.8
State highways, light.......	75.8	8.7
State highways, heavy......	71.0	7.7
Urban radials†		
Weekdays, heavy...........	71.0	8.5
Saturdays, heavy...........	67.0	5.2
Sundays, heavy............	64.0	10.6
General average..........	71.1	7.0

* 7 A.M. to 7 P.M. mean value for each type.
† 8 A.M. to 8 P.M. median value for each type.
SOURCES: L. E. Peabody and O. K. Norman, "Applications of Automatic Traffic Recorder Data in Highway Planning," U.S. Bureau of Public Roads, *Public Roads*, vol. 21, no. 11, January, 1941, pp. 220–221.
N. Cherniack, "Methods of Estimating Vehicular Traffic Volume with the Aid of Traffic Patterns," *Proceedings of Highway Research Board*, Washington, 1936, vol. 16, p. 267.

Peak hours of traffic volume are especially significant in the design and regulation of traffic routes. Daily patterns of traffic movement (Fig. 5-7) show that the peak hour carries 7 to 8 per cent of the 24-hr volume, or about 10 per cent of the 12-daylight-hours volume. The peak hour generally has almost twice the volume of flow found in the average hour. Because of weekly and seasonal variations, yearly patterns show the peak hour carrying 10 to 70 per cent of the average 24-hr volume. Steps taken to disperse movement into more uniform flow materially reduce congestion in peak hours and lessen the demand for increased capacity of route.

Minimum traffic flow usually occurs between 4 A.M. and 5 A.M., with the peak hour before 6 P.M. for weekday patterns. Saturday, Sunday, and holiday patterns usually develop peak hours earlier in the afternoon. In urban areas, secondary peaks usually occur just prior to 9 A.M. on weekdays, while in smaller cities a minor peak is frequently found during the luncheon hour.

Weekly Time Patterns. Characteristic curves of weekly traffic fluctuation are shown in Fig. 5-8. The pattern tends to sag on midweek days

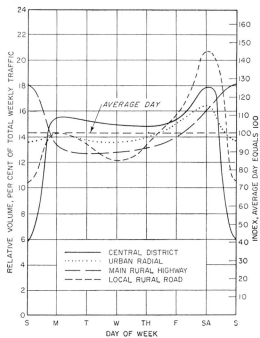

FIG. 5-8. Weekly time patterns of traffic flow. (Plotted from various sources of daily-volume data.)

and rise slightly above average on Monday, Friday, and Saturday. Monday and Friday variance is usually not more than 5 per cent from the average for urban areas but may be as great as 15 per cent on rural routes of minor importance. Saturday values range from 10 to 30 per cent above average on main thoroughfares, and on local rural roads may be as much as 50 per cent above average.

Sunday and holiday traffic may be either above or below average, depending on the type of area served. It is usually subnormal in central business districts, industrial zones, and other areas devoted primarily to production and selling, but ranges from 130 to 200 per cent above average on main rural highways. On routes leading to recreational areas it is especially high, with peaks amounting to two or three times the average daily volumes.

There is little or no relationship between Saturday, Sunday, or holiday traffic and ordinary weekday traffic. Independent measurements must be made for these days on any given route. On a holiday week-end, excessive peak values are found, and summer recreational activities also modify week-end traffic.

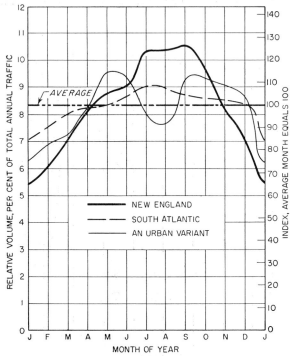

FIG. 5-9. Monthly time patterns of traffic flow. (From T. B. Dimmick and M. E. Kipp, "Traffic Trends on Rural Roads in 1946," U.S. Bureau of Public Roads, *Public Roads*, vol. 25, no. 3, March, 1948, p. 43.)

Seasonal Time Patterns. Characteristic variations of the seasonal pattern of traffic flow are shown in Fig. 5-9. This pattern is remarkably repetitive from year to year. It is closely related to the economic and social demands for highway transportation and reflects the effects of climate on these demands. In general, midwinter has the lowest volumes of traffic movement, and midsummer generates maximum volumes. A typical rural variation shows January or February traffic only 0.7 of normal and July or August traffic 1.3 times the average. May, June, October, and November volumes are usually nearly average.

The exact shape and amplitude of the seasonal pattern varies widely. In general, the amplitude decreases with the latitude of the location: as winter climate becomes milder, seasonal changes in volume become less extreme. In some Southern sections values are reversed, so that subnormal volumes occur in the summer months and above-normal values in winter. The recreational areas of Florida show this pattern.[1,2] Figure 5-9 illustrates another seasonal variation, a decline to subnormal values during July and August due to the exodus of city residents for the vacation months.

5-6. Distribution of Annual Traffic Volumes by Daily and Hourly Magnitudes. Because the capacity of a given route is relatively constant, in planning the route it is necessary to know the anticipated daily and hourly magnitudes of traffic volume throughout its expected life. It is also desirable to understand the extent and duration of peak demands so that regulation and control may be fitted to traffic demands as they arise. Since traffic volumes fluctuate by day, by week, and by season, relative magnitudes of volumes throughout the period of an entire year must be examined in determining the three basic cycles of fluctuation.

Relative Daily Magnitudes. In rural areas, the range of variation in daily volumes extends from a low of nearly 0 per cent to a high of 660 per cent of the annual average daily volume. However, each route has its own range of variation. In general, the minimum day's volume is 30 to 50 per cent and the peak day's volume 140 to 340 per cent of the annual average daily volume. The average relative level of peak volumes found on a variety of rural routes is about 230 per cent of the annual average daily volume. Peak days average 260 per cent in Northern latitudes and 180 per cent in Southern climates. As the stream of traffic grows heavier, ratios of peak-day volumes to average-day volumes are usually somewhat lower.[3]

Characteristic curves of volume developed throughout a year on a typical route are set forth in Fig. 5-10. When the total range is broken down into quarters, the per cent of the total traffic flowing within each quarter is as follows: first quarter, 17 per cent; second quarter, 59 per cent; third quarter, 19 per cent; fourth quarter, 5 per cent.

Although the total range of variation is 30 to 230 per cent of the annual average hourly volume, AAHV, over half the vehicles (59 per cent) travel on days that vary only 80 per cent to 130 per cent. When both the midquarters are considered together, it is seen that over three-fourths (78 per cent) of the vehicles move at daily volumes which constitute only one-half of the total variation in daily volumes.

[1] R. E. Craig, "An Automatic Recorder for Counting Highway Traffic," U.S. Bureau of Public Roads, *Public Roads*, vol. 19, no. 3, May, 1938, p. 49.

[2] L. E. Peabody and O. K. Norman, "Applications of Automatic Traffic Recorder Data in Highway Planning," U.S. Bureau of Public Roads, *Public Roads*, vol. 21, no. 11, January, 1941, p. 221.

[3] *Ibid.*, pp. 204–209.

If this highway were carrying 2,555,000 vehicles per year it would carry 434,000 vehicles at less than 5,600 per day; 1,507,000 vehicles at 5,600 to 9,100 per day; 486,000 vehicles at 9,100 to 12,600 per day; and 128,000

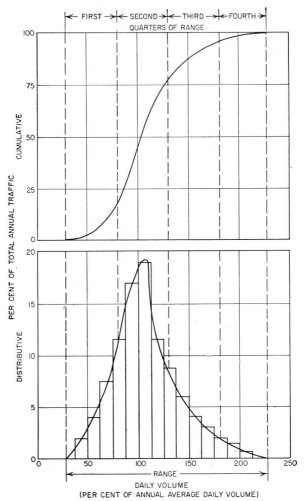

Fig. 5-10. Distribution of total annual traffic by daily volume. (Adapted from T. E. Peabody and O. K. Norman, "Applications of Automatic Traffic Recorder Data in Highway Planning," U.S. Bureau of Public Roads, *Public Roads*, vol. 21, no. 11, January, 1941, p. 209.)

vehicles at more than 12,600 per day. The over-all range of volume on this route would be from 2,100 to 16,100 vehicles per day.

Relative Hourly Magnitudes. Since the hourly volume is the practical working unit in actual design and operation of the road, the characteristic distribution of hourly volumes must be explored throughout the complete cycle of a year. For any given road, there will be one hour during the year which will carry more traffic than any other hour. This is the *maximum* hour, and its volume ranges from 10 to 70 per cent of the annual average daily volumes. The average relative level of the maximum hour on rural roads is about 25 per cent of the annual average daily volume. This figure is modified by climate, however, averaging 28 per

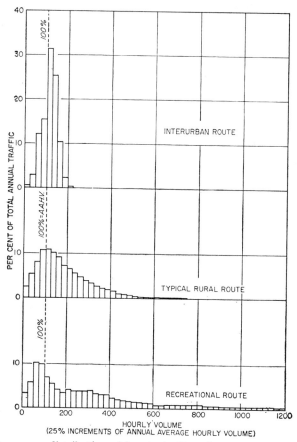

FIG. 5-11. Frequency distribution of hourly traffic volumes as a per cent of annual average hourly volume. (Plotted from data compiled by Connecticut State Highway Department for routes Conn. 156, U.S. 1, and Merritt Parkway, 1941.)

cent in the Northern and 20 per cent in the Southern portions of the country.[1]

The maximum hourly volume is usually about six times as great as the annual average hour. Ideally, a route should have sufficient capacity to accommodate six times as much traffic as will occur in the average hour.

The hourly magnitudes of traffic volumes found among diverse routes are subject to considerable variation. Characteristic histograms of the

[1] *Ibid.*, p. 210.

distribution of hourly volumes, expressed as a percentage of the annual
average hourly volume, are set forth in Fig. 5-11.   Figure 5-12 contains
characteristic curves of the data in Fig. 5-11, plotted to show the cumulative distribution of the hourly volumes.   The total range of variation
extends from 0 to over 12 times the average hourly volume.   A typical
variation is 0 to 750 per cent, and the minimum 0 to 225 per cent.   The
histogram for minimum variation is typical of dense interurban[1] traffic.

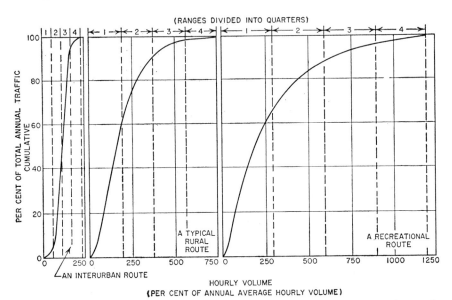

Fig. 5-12. Cumulative distribution of hourly traffic volume as a per cent of annual
average hourly volume.   (Plotted from data compiled by Connecticut State Highway
Department for routes Conn. 156, U.S. 1, and Merritt Parkway, 1941.)

This small range of variation is due to the naturally steady traffic demands
in urban areas and is probably kept down in peak hours by capacity
limitation of the route itself.   The histogram for maximum variation is
illustrative of a route employed primarily in the summer months in a
Northern latitude, as a route to a bathing beach.   The histogram showing
average range of variation is associated with a primary rural route devoted
to all types of traffic demand (including social as well as business) and
typifies the great majority of such routes as are not inhibited by physical
limitations on capacity.

It is clear from these characteristic histograms that the hourly distribution of traffic volumes over the period of a year may be skewed considerably owing to small percentages of the total traffic volume developing

---

[1] Interurban as used here refers to routes primarily handling short-distance trips
between urban areas.

abnormally low or abnormally high volumes which are of great concern to the traffic engineer.

From the viewpoint of traffic planning, design, and operation, abnormally high peaks are of great concern. From the viewpoint of gaining maximum economy from the road plant, abnormally low hourly volumes are especially significant.

When the various quarters of the distributions in Fig. 5-12 are examined, the following values are found:

PER CENT OF TOTAL TRAFFIC WITHIN VARIOUS QUARTERS OF RANGE

Type route	Total range, % AAHV	First quarter	Second quarter	Third quarter	Fourth quarter	Peak quarter	Range limits of peak quarters, % AAHV
Interurban...	0– 225	5	39	52	4	68	90–146
Rural.......	0– 750	60	33	6	1	64	63–250
Recreational.	0–1,200	64	25	7	4	66	25–325

Of great significance is the fact that over 90 per cent of traffic on the interurban route travels in the two mid-quarters of the range. On routes of greater variation less than 40 per cent of the annual travel is within this middle half of the range. If this latter figure is expressive of the natural characteristics of traffic demands, fully half the traffic using the interurban route is artificially congested into volume densities contrary to its basic nature. Much higher peak volumes might be developed if road design or traffic control permitted.

Of major interest is the relative amount of time that given hourly rates prevail. Characteristic curves of time consumed by different rates of volume are shown in Figs. 5-13 and 5-14, plotted for the same data as Figs. 5-11 and 5-12. When the various quarters of the cumulative-distribution curves in Fig. 5-14 are examined, the following volumes are found:

PER CENT OF TOTAL TIME THAT VARIOUS QUARTERS OF VOLUME RANGE OCCUR

Type route	Total range, % AAHV	First quarter	Second quarter	Third quarter	Fourth quarter	Peak quarter	Range limits of peak quarters, % AAHV
Interurban...	0– 225	12	43	42	3.0	56	80–136
Rural.......	0– 750	85	12.5	2.4	0.1	85	0–187
Recreational.	0–1,200	92	6.5	1.1	0.4	92	0–300

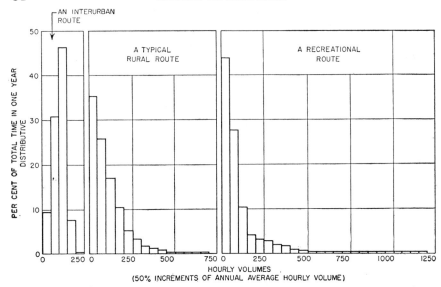

Fig. 5-13. Frequency distribution of per cent of total time that routes carry various hourly volumes. (Plotted from data compiled by Connecticut State Highway Department for routes Conn. 156, U.S. 1, and Merritt Parkway.)

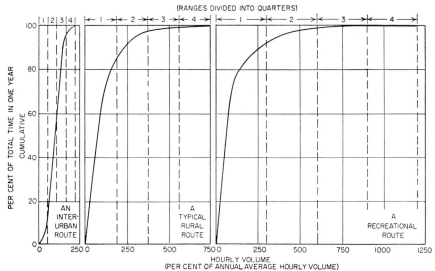

Fig. 5-14. Cumulative distribution of per cent of total time that routes carry various hourly volumes. (Plotted from data compiled by Connecticut State Highway Department for routes Conn. 156, U.S. 1, and Merritt Parkway.)

Compared with data shown previously for the percentage of total traffic, these data show that for the route of average variation only 60 per cent of the vehicles are carried during 85 per cent of the time, while 1 per cent demand passage in 0.1 per cent of the time. This leaves 39 per cent to be carried in about 15 per cent of the time. An average rural route carrying 5,000,000 vehicles per year or 570 per annual average hour varies from 0 vehicles during some hours to 570 × 7.5 or 4,275 during the maximum hour. Approximately 3,000,000 (60 per cent) of these vehicles consume 7,446 (85 per cent) of the 8,760 hr in a year, traveling at 1,070 or fewer vehicles per hour. Of the 2,000,000 remaining, 20,000 demand passage at the rate of 3,200 or more per hour, but this condition prevails for only 8.8 hr (0.1 per cent). The question for the traffic engineer is: Should the peak hour which can be expected during a year of route operation be provided for? Or should a lesser capacity be considered as more practical?

A significant factor in answering such questions is the amount of time that congestion would be experienced and tolerated. Characteristic curves of the hourly volumes found during the hours of highest volumes throughout the year are shown in Fig. 5-15. In almost all cases where the rate of decrease changes rapidly, the knee of the curve occurs around the 20th and 50th highest hour. Accordingly, the 30th and 50th highest hours are frequently employed for design purposes in determining the capacity to be provided.

Suppose, for example, that the main rural route was needed to carry an average hourly value of 600 vph. Its peak hour of demand would be about 3,100 vph (5.20 × 600), but if the design was made for the 30th highest hour, allowing 30 hr of congestion, the design would need to be for only 2,200 vph. On the other hand, if an attempt is made to further decrease the design volume by doubling the congestion time (using the 60th highest hour) the design value would only drop to about 2,000 vph. For each route an optimum design point must be found to give the best combination of low design volume and low congestion time.

5-7. Theory of Short Counts for Traffic Volume. Because of the cost of collecting volume data, there have been developed various methods of *short counts* for determining the entire pattern of any traffic stream. The validity of short counts is based on the consistency of traffic fluctuation patterns. If h per cent of a given traffic stream moves in a certain hour of the day and n is the actual number of vehicles observed in such an hour, then daily, weekly, monthly, seasonal, and even yearly volumes of traffic may be projected from the values of h taken over several days scattered throughout the year.

In making short counts for the purpose of establishing volume data for long periods, it is desirable to develop a schedule of actual counts which will sample the traffic stream in such a manner as to yield the maximum of accuracy for the cost of counting. For instance, 4 hr of count in a given day will sample the traffic stream in such a manner as to yield the maximum of accuracy for the cost of counting. Thus, 4 hr of count in a given day will usually yield greater accuracy than 2 hr of count. Again, if the total count is spread so that a few minutes are taken for each hour, greater accuracy will result than if all counting is taken in a single sample.[1]

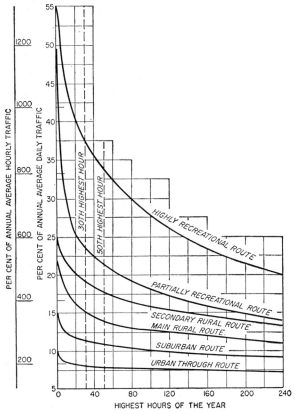

FIG. 5-15. Hourly volumes on various types of routes as found during hours of highest volume on these routes. (*Adapted from Hourly Traffic Volumes as a Basis for Design*, Connecticut State Highway Department, Hartford, 1947.)

Actual schedules found in practice will employ from 40 to more than 150 hr of count as a basis of estimating a yearly value or an annual average day. This total time of count may be expended in intervals of 1, 2, 4, or more hr consecutively and arranged to include sampling on each day of the week and each month of the year.

In making short counts it is sometimes possible to borrow known patterns of variation and apply them to locations where such patterns are

[1] Peabody and Norman, *op. cit.*, pp. 214–218.

not established. When this practice is used, however, the investigator should be reasonably sure that the borrowed pattern is based on traffic serving the same or nearly identical social and economic purposes, or great inaccuracies will result.

The accuracies obtained with short-count technique are of wide range. Two 5-min counts taken each hour at city intersections have yielded daily average totals accurate within 3 per cent; heavily traveled locations produced the best results, but one lightly traveled intersection showed an error of 26 per cent.[1] The use of widely spread 40-hr counts on rural routes to determine annual average day have yielded average weighted accuracies within $3\frac{1}{2}$ to $5\frac{1}{2}$ per cent, but individual stations showed as much as 18 per cent error.[2] In general, greater accuracy is obtained in equal total counting times by taking larger numbers of short counts distributed through the entire pattern of cyclical variation than by longer, but less frequent, counts.

In determining the best short-count schedule for a particular investigation the following important factors must be kept in mind: (1) desired degree of accuracy, (2) allowable cost of the investigation, (3) magnitude of traffic streams to be observed, (4) to what extent data must be broken down into details such as vehicle types and turning movements, (5) whether peak-hour recordings are necessary, (6) that portion of the counter's time which will be consumed in movement from check point to check point, (7) that portion of the total time allotted for the work which actually consists of counting time.

5-8. Induced Traffic. A new traffic facility, such as a bridge or nonstop route in urban areas, usually induces traffic volumes which cannot be accounted for on the basis of cyclical variations or trends. However, certain basic reasons are responsible for the new volume of traffic.

In the first place, some traffic will be *diverted* to the new facility, finding it more convenient or attractive than the former course of travel. Second, there may be an actual change in mode of transportation: some persons who formerly rode by common carrier may be attracted to the new facility in their own private automobiles. Third, there appears an entirely new traffic which never existed in the area before. This latter class of traffic volume is referred to as induced traffic and may be assumed to be the effect of making certain focal areas more accessible to larger numbers of people. Induced traffic appears on almost every new traffic facility, especially in urban areas and under conditions where the facility creates a new accessibility between areas. This factor calls for extreme care in estimating traffic volume.

---

[1] R. O. Swain, "Traffic Aids to Texas Municipalities," *The American City*, vol. 55, no. 7, July, 1940, p. 85.

[2] Peabody and Norman, *op. cit.*, pp. 216–217.

5-9. Directional Characteristics of Traffic. On two-way traffic routes, the volume of traffic in each direction is usually found to be nearly equal at the close of each cycle of variation (day, week, or year). However, there occur certain hours or days when unbalanced flows exist, and in certain special cases an almost rotary direction flow is found on portions of road network.

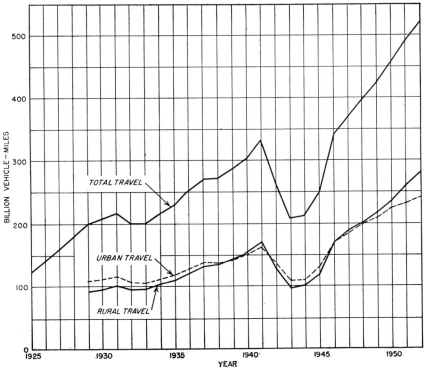

FIG. 5-16. Growth of motor vehicle travel. (From T. B. Dimmick, "Traffic Trends on Rural Roads in 1947," U.S. Bureau of Public Roads, *Public Roads*, vol. 25, no. 7, March, 1949, p. 159.)

Hourly Variation. In the morning there is a surge of traffic to places of employment, and in the evening this flow is reversed. The directional effect of traffic volume is a matter to be determined for each route, but in peak hours the traffic volume in one direction usually ranges from 1¼ to 1¾ times the counterflow for radial streets from central business districts.

Again, on those radial routes which take the urban dweller to and from recreational areas in the country for holidays and week-ends, the outbound movement does not reach as great a ratio to its counterflow as is found for inbound movement. This is due to the exodus from the city taking place over a large period of time, whereas the demand to return to the workaday world is scheduled for a fixed hour on the day following the holiday. When this hourly variation in inbound and outbound movement is distributed for the period of a year, the maximum flow in one direction during the annual peak hour will amount to as much as 80 per cent of the total hourly volume.

Other examples produce even greater directional effects. For example, a football stadium in a location which permits exit and entry from one direction only will produce extremely unbalanced flow before and after the game.

Rotary Effects. On some road nets an unbalanced volume between the two directions of flow may be observed over protracted periods of time, usually because of route layout. For example, a railroad fill running diagonally through a grid street system with few crossings will collect inbound traffic on one side of the fill and outbound traffic on the other side, producing a rotary or unbalanced directional effect in traffic movement.

5-10. Trends in Traffic Volumes. In addition to the cyclical variation in volume, there is a long-time trend in total traffic flow which has, except for the depression and war years, shown an increase from year to year. The estimated total of motor vehicle travel is shown in Fig. 5-16. Approximately 50 per cent occurs on city streets.

From 1925 to 1952 the total vehicle-miles of travel increased from 122 billion to 522 billion, an annual average growth of nearly 15 billion vehicle-miles per year.

Trends in vehicle registration and gasoline consumption are frequently applied in estimating future traffic flows. The standard of living, improvements in the vehicle, and the extent of the highway network and its capacity are all important in the trend of traffic volume. In estimating peak hours of demand over a period longer than a year, the trend of traffic growth or decline is usually taken into account by assuming an annual distribution similar to the one of the past. However, annual distributions of traffic are affected by route capacity and other controlling factors.

CHAPTER 6

ORIGIN AND DESTINATION

Origin and destination determine the direction of travel, selection of route, and length of trip. The terminal points of a trip also affect choice of method of travel and storage place of the vehicle when not in use. Origin-and-destination (O-D) studies and data are of prime importance in the planning, design, and operation of highway-traffic routes and terminals, since they determine the preferred location of traffic-handling facilities, the amount of the demand which will occur on any route or terminal, and the optimum route which may be used in expediting a given movement of traffic over an existing network.

6-1. Methods of Collecting O-D Data. To determine the origin and destination of traffic movement in a given area, the road users themselves must be questioned about their demands for movement, or each movement must be observed and identified. The road user may be interviewed (1) at some well-chosen point along his route of travel, (2) at the origin, or (3) at the destination of the trip. If inquiry is made along the route, it must be made during the time of travel and, for apparent reasons, must be hurried and brief. More time is usually available at origin or destination, and accordingly more questions may be more fully answered.

Forms of inquiry are widely diversified, because of the above-mentioned conditions of interview and the objectives of the study, which may include the gathering of ancillary data. Methods range from the simple recording of a license-plate number, keyed to a particular point and time of observation, to an intensive interview, sometimes lasting as much as an hour, covering a number of travel habits and requirements.

License Plate. In the license-plate method of study, observers are simultaneously stationed at selected points of entry and exit to form a complete cordon around a route, a complex intersection, or a larger area such as a business center or entire city.[1] By employing synchronized time pieces, license-plate numbers are recorded in a time series as vehicles enter or leave the cordoned area or route. Each sheet of recorded data is keyed to a location, direction, and time period of movement. The collected data are analyzed by tracing each license-plate number through

---

[1] For an example, see *Waco License Tag Origin and Destination Survey, Apr. 12, 1945*, Texas State Highway Department et al., Waco, Tex., 1945.

the complete record from point and time of entry to point and time of exit. These points are taken as the origin and destination of movement.

While this method has been applied over city-wide areas, it is primarily suited to smaller areas with few points of entry and exit, such as a complex interchange area or a small business center. It has the disadvantages of difficulty in tracing license-plate numbers and incomplete knowledge of origins and destinations, arising from the assumptions that O-D's are described by the points of entry and exit.

To simplify analysis, colored cards prestamped by time clocks at points of entry are sometimes affixed to bumpers or door handles for vehicle-identification purposes. The cards are removed at the point of exit and again time-stamped. In addition, check marks on the cards may be placed at intermediate points along the routes of travel.

Again, cards in the form of the business-machine punch card for analytical uses may be handed to the driver or rider for later collection at exit points. Dictating machines have also been used in the field to record the observer's voice calling off license-plate numbers. The observer dictates into each record his name, location, direction of traffic flow observed, and each 5 min of time lapse as well as the license number of each vehicle.

Post Card. Prepaid business-reply postal cards, with return address and questionnaire to be filled in by the road user, may be distributed at some selected point along the route of travel or at the destination. They usually carry a brief plea for cooperation, over the signature of a recognized public official, and request locations of origin and destination plus indications, in some cases, of purpose of trip and type of vehicle. Time, location, and direction of travel are precoded on such cards.

Post cards may be distributed to road users on the route, to workers, shoppers, or clientele at focal points of employment or business, or they may be mailed to homes or offices of registered vehicle owners.[2] Response depends largely on prepublicity and the importance of the particular route of travel to the individual, and usually ranges from 10 to 50 per cent. A check of postal response with actual flows in the field indicates accuracies of 80 to 98 per cent. Investigations of passenger cars usually yield higher accuracy than trucks. The post-card method does not guarantee a random sample, since there probably is underreporting in the response from persons who do not use writing in their daily lives.

As a variation of this method, cards may be collected in ballot boxes in stores, offices, shops, and other focal points. Prepared cards or forms distributed at such points may be filled in by pencil, or the origin and

---

[1] C. S. Cunningham, "O and D by Transcription," *Traffic Engineering*, vol. 12, no. 16, August, 1942, pp. 436, 437.

[2] For example see "Canton Counts Its Traffic by Postcard," *Engineering News-Record*, vol. 142, no. 1, Jan. 6, 1949, pp. 39, 55, 56.

mode of travel may be designated by tearing off certain perforated portions.

Route Interview. In the route-interview method,[1] mass-carrier patrons may be questioned on the vehicle while en route, but private passenger vehicles and trucks must be stopped long enough for the interview, which should take but a fraction of a minute. Carefully selected and trained personnel are stationed at points on the route where sight distance, gradient, and prevailing speeds assure safe operation. Stations selected are at natural passes in the topography, bridges, tunnels, ferries, and other funneling points of traffic flow. Thus in urban areas all major radials and significant cross-town routes are checked.

Traffic is filtered through the questionnaire lane by prewarning signs and police officers in uniform, so that each driver is interviewed. On heavy thoroughfares a multilane arrangement is necessary to avoid accumulation of traffic during peak periods: one lane is used for interviewing, and excess traffic is bypassed in the extra lane. Total-volume count is taken by hourly or half-hourly periods to determine the percentage of interviewed sample for each period of the study and to permit proper expansion factors to be applied to the collected sample.

This method provides a direct, quick, and efficient way of obtaining origin, destination, and a third point in the trip of each driver interviewed. A minimum of office work is required if precoded cards are used for the collection of data. It is well suited to points where major routes enter urban areas and to other focal points in the traffic pattern, but in dense urban areas with a multiplicity of routes it is less satisfactory because of the number of stations required and the difficulty of meeting qualifications for safety and traffic capacity.

When applied at the periphery of an urban area, this method is sometimes referred to as the *external* survey of O-D data. An example of a typical recording form is shown in Fig. 6-1.

Home Interview. A more recently developed means of procuring O-D data in urban areas, called the home-interview method,[2,3] samples the total highway-traffic demand for an entire urban area, including all modes of transportation and primary routes employed. By carefully controlled sampling procedures, thoroughly satisfactory O-D data for cities of 50,000 to 300,000 population can be collected by obtaining the

---

[1] For example see *The Denver-Area Study,* Colorado State Highway Department and the U.S. Bureau of Public Roads, Denver, Colo., 1940, pp. 10–16, or see *Metropolitan Atlanta Special Origin-Destination Study,* Georgia State Highway Board et al., Manual 9, Atlanta, Ga., 1940, pp. 1–5 and inserts.

[2] Method developed by U.S. Bureau of Public Roads in cooperation with U.S. Bureau of the Census. See *Manual of Procedures for the Metropolitan Area Traffic Studies,* U.S. Bureau of Public Roads, 1946 (mimeographed).

[3] For example see *1946 Portland Metropolitan Area Traffic Survey,* Oregon State Highway Commission, Technical Report 49-2, Salem, 1949, sec. II and Appendix B.

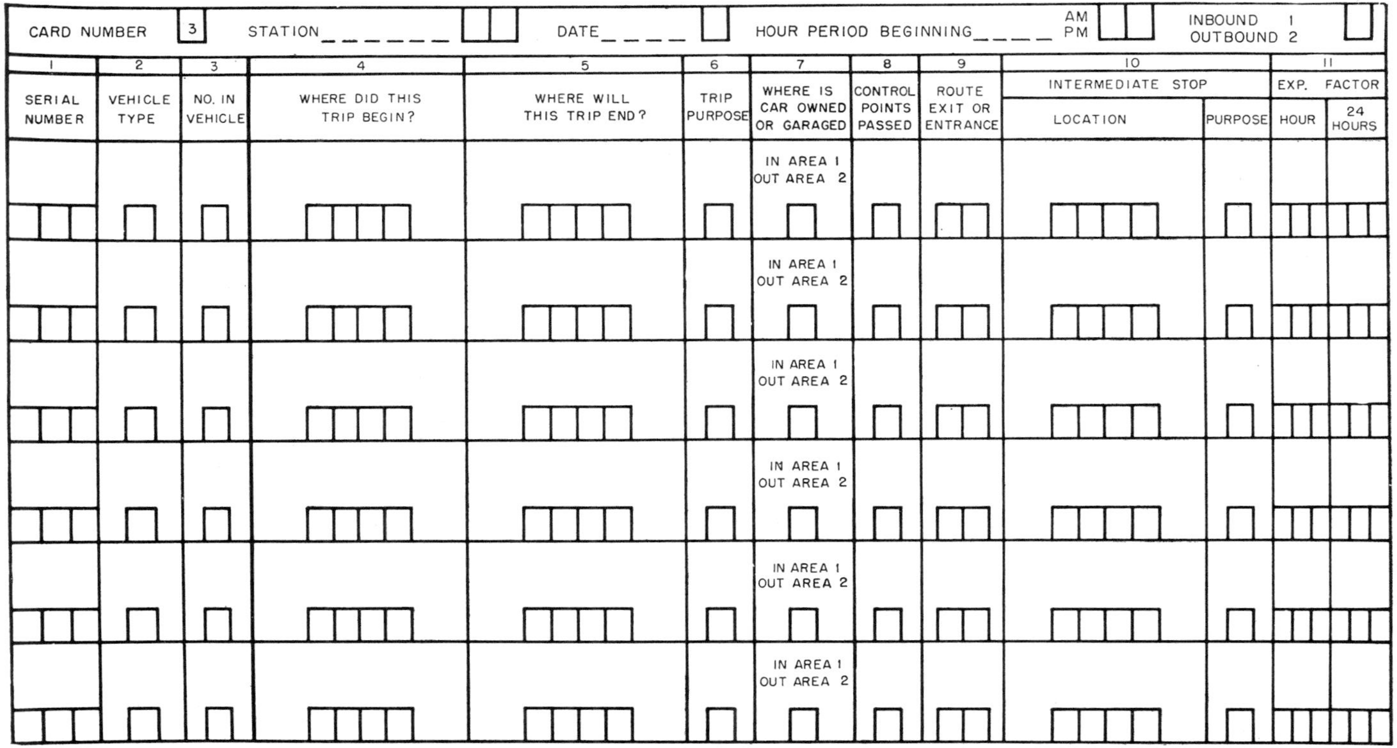

Fig. 6-1. External interview survey form. (From *Manual of Procedures for the Metropolitan Area Traffic Studies*, U.S. Bureau of Public Roads, Form P-4404, 1946.)

93

METROPOLITAN AREA TRAFFIC SURVEY

DWELLING UNIT SUMMARY

Preceding number _____

Interview address _____

Succeeding number _____

Card _____ [1]

Tract No. _____

Block No. _____

Sample No. _____

Subzone No. _____

Date of travel _____

A. How many passenger cars are owned by persons living at this address? _____

B. How many persons live here? _____

C. How many are 5 years of age or older? _____

D. Household information:

Person No.	Sex and Race	Person Identification	Code	Occupation and Industry	Trips Yes	Trips No
01						
02						
03						
04						
05						
06						
07						
08						
09						
10						

E. Total number of trips reported at this address _____

  1. Number of persons 5 years of age or older making trips _____

  2. Number of persons 5 years of age or older making no trips _____

  3. Number of persons 5 years of age or older with trips unknown _____

F. Comments and reason if complete information was not obtainable _____

G. Factor _____

Administrative Record

Interviewer _____

CALLS

Date _____ Time

(1) _____

(2) _____

(3) _____

(4) _____

REPORT SUBMITTED INCOMPLETE

Date _____

Reason _____

Supervisor's comment _____

Remarks _____

Report completed _____ (Date) _____ (Initial)

Interviews checked _____ (Initial)

Coded by _____ (Initial)

P-4290

94

INTERNAL TRIP REPORT

CARD ☐ TRACT ☐ BLOCK ☐ SAMPLE No. ☐ SUBZONE ☐ DAY OF TRAVEL ☐

Sheet _____ or _____

1	2	3	4	5	6	7	8		9	10	11	12
							Time of—		Purpose of Trip	No. in Car Including Driver		
Occupation And Industry	Person No.	Trip No.	Sex And Race	Where Did This Trip Begin?	Where Did This Trip Begin?	Mode of Travel	Starting	Arrival	From / To		Kind of Parking	Control Points
			1			1 Auto Driver	A.M.	A.M.	1 Work 1		1 Street free	1 Memorial Bridge
			2			2 Auto Pass.			2 Business........ 2		2 Street meter	2 Maine Avenue Viaduct
			3			3 Street-car-Bus	P.M.	P.M.	3 ...Med.-Den. 3		3 Lot free	3 Sixteenth Street Underpass
			4			4 Taxi Pass.			4 School 4		4 Lot paid	0 None
			5			5 Truck Pass.			5 ...Social. Rec. ... 5		5 Garage free	
			6						6 Ch. travel mode .. 6		6 Garage paid	
									7 ...Eat meal 7		7 Service or repairs	
									8 Shopping 8		8 Res. property	
									9 ...Serve pass 9		9 Cruised	
									0 Home		0 Not parked	
			1			1 Auto Driver	A.M.	A.M.	1 Work 1		1 Street free	1 Memorial Bridge
			2			2 Auto Pass.			2 Business........ 2		2 Street meter	2 Maine Avenue Viaduct
			3			3 Street-car-Bus	P.M.	P.M.	3 ...Med.-Den. 3		3 Lot free	3 Sixteenth Street Underpass
			4			4 Taxi Pass.			4 School 4		4 Lot paid	0 None
			5			5 Truck Pass.			5 ...Social Rec. 5		5 Garage free	
			6						6 Ch. travel mode .. 6		6 Garage paid	
									7 ...Eat meal 7		7 Service or repairs	
									8 Shopping 8		8 Res. property	
									9 ...Serve pass 9		9 Cruised	
									0 Home		0 Not parked	

Source – Manual of Procedures for the Metropolitan Area Traffic Studies. United States Public Roads Administration, Washington, D.C., 1946, Form P-4290

Fig. 6-2. Home interview survey forms. (From *Manual of Procedures for the Metropolitan Area Traffic Studies*, U.S. Bureau of Public Roads, Form P-4290, 1946.)

actual street or highway use on a typical day from 5 to 20 per cent of the population. To obtain a random 10 per cent sample, every tenth dwelling unit on some arbitrarily controlled listing or map is visited by a trained interviewer who collects travel data for the last previous weekday from each member of the household. A similar selection basis is employed for trucks and taxicabs.

Information may be obtained by this method on number and kinds of vehicles owned, number of persons, sex, race, occupation, industry, number of trips made by each person, purpose of trip, original starting point and ultimate destination for each trip, mode of transportation employed, and the time of day during which each trip was made. Data regarding parking of private passenger vehicles may also be collected (Fig. 6-2).

Though expensive and time-consuming, the home-interview method generally yields good dividends in obtaining comprehensive data on all types of highway travel, road users, and vehicles. It is not well suited to small or limited areas or for a limited number of routes and does not yield as large a sample of parkers as would be required for accurate analysis of parking demand in a central business district. Direct interviews at parking lots and garages seem better suited to this purpose. Because very small areas, and hence relatively small subtotals of demand, are used in such analysis, a large total number of interviews is required.

6-2. Planning for O-D Studies. Origin-and-destination studies are extremely valuable in gaining a fuller understanding of the traffic demand of a given area, and the engineer must understand that it is essential to plan well in organizing the study and analyzing and applying the data collected.

Choice of Method of Study. The selection of method is dependent on the objective, which usually involves the location of a route or a terminal. If the location of a simple river crossing is the objective, a simple route-interview method on existing bridges and ferries may be most satisfactory. If the location of a bus terminal is the objective, direct inquiry of bus passengers en route to or from, or even at present points of termination, is indicated. If the purpose is to gather facts for use in the design or operation of a complex interchange or intersectional area, the license-tag or car method is recommended. If data are needed to plan a network system of major routes for an urban area, the home-interview plan may prove to be the better method.

Background Information. It is important to have a full understanding of the road net, topography, natural passes, bridges, tunnels, ferries, and other basic factors vital in shaping traffic flow. Full consideration should be given also to political boundary lines (for reasons of administration and finance) as well as to the intensity and character of land use in the area to be examined. Land use largely determines the character and quantity of traffic which is generated.

Miscellaneous Factors in Planning. In addition, the investigator should understand the influence of time of year and dates of study on type and amount of traffic demand. Good judgment is needed to determine the extent of data to be collected as well as the design of the form on which data is entered. Extreme care is required in *selecting a sample* for home-interview purposes.

The record of trip origin and destination should be in tune with the purpose. For example, in a parking survey of a business district, great accuracy is required in recording the destination of a person leaving a car on foot, because pedestrian walking distances are small, but rather rough descriptions, such as nearest intersection, will serve for a motor trip. Origin and destination are the fundamental facts sought, and other data collected should not interfere with the basic purpose of the study. Finally, a neat balance of effort between field and office staffs should be worked out. The use of precoding, color combinations, pretiming, and other identification of field cards may simplify office work, but if carried too far will impose hardship and frustration on field forces.

6-3. Analysis of O-D Data. *Zoning.* Regardless of collection method, in analyzing data it is necessary to establish zones of O-D: that is, to group trips with more or less common origins or destinations. The size and location of each detail zone of origin or destination must be decided so that the complete area of investigation is resolved into a mosaic of lesser areas which lend themselves to desirable grouping. By combining a group of the smallest zones, combinations of land area may be made to form districts or segments, thus furnishing a comprehensive system for the control and summation of individual trips.

Figure 6-3 shows a hypothetical city resolved into 11 districts for O-D purposes. It will be noted from this figure that District 4 is made up of 13 zones. For purposes of clarity the zoning of other districts has been omitted, but in practice all districts would require zoning. Bodies of water, ridges and mountains, highways, railroads, and land uses are shown, for these features vitally affect origins, destinations, and routing of traffic, and are significant in the development of local *traffic sheds* or traffic-drainage areas.

In the establishment of zone and district boundaries, a compromise must be struck between the linear dimensions of a zone or district and its relative capacity as a generator of traffic. Zones comparable in traffic generation are desirable but are seldom achieved because of the modification of linear dimensions. The linear dimensions of a zone or district must be small enough to give accuracy in the direction and location of traffic movement. It is usually desirable to resolve zones or districts to land uses of similar nature, and, in general, zones and districts should straddle important collection streets and major arteries, for it is here that the rivulets of traffic flow collect in any trip.

The boundaries of districts are usually defined by main topographical features or major areas of population, industry, commercial land use, or traffic drainage into important streets. These are further segmented by railways or controls. Finally, local zones of origin and destination may be reduced to minor business districts and even to blocks or sections of blocks for parking or pedestrian movement.

Experience develops facility in the zoning control of O-D data. Attempts have been made to reduce zoning to an arbitrary system of polar

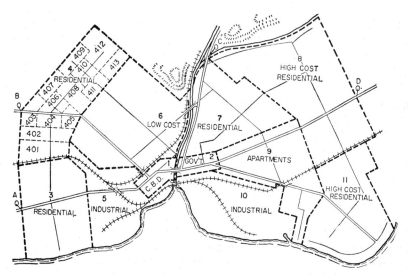

FIG. 6-3. Districts and zones of O-D for a hypothetical city.

or Cartesian coordinates, political wards, census tracts, or other arbitrary zones which disregard traffic generation and the natural topographic features as well as existing man-made developments on that topography. Such methods may be properly considered where uniform conditions prevail throughout the total area to be analyzed, but, in general, urban areas are not uniformly developed, nor do they lie on uniform topography. Accordingly, these irregular conditions usually affect the location of a route or terminal.

Purpose of study must be considered in the determination of zone size. For determining location of parking facilities, and in cases where walking is the mode of transportation, very small zones are needed to determine satisfactorily line and location. Larger dimensions may be satisfactorily employed where vehicular methods of movement are involved. Thus a city block or even quarter-block may be used in parking studies, while several city blocks or larger zones may be employed for motor vehicle trips.

As an example of the use of the data shown in Fig. 6-4, the total number of river crossings may be computed by summing the trips with origin on one side of the river and destination on the other side. Of the 6,647 trips with origin in District 4, there are 2,563 which cross the river. This number is the sum of all destinations on line 4 of origins and under columns 7, 8, 9, 10, 11, C, and D of destination. For the area as a whole, 96,773 or 43.7 per cent of all trips cross the river. Other similar calculations can be quickly made. For example, 48,006 vehicles or 21.7 per cent

TOTAL PRIVATE PASSENGER VEHICLE TRIPS
DISTRICTS OF DESTINATION

	1	2	3	4	5	6	7	8	9	10	11	SUB TOTAL	A	B	C	D	SUB TOTAL	GRAND TOTAL
1	9526	1143	849	862	4002	1028	804	3195	5310	3692	3157	33568	2066	1689	989	1066	5810	39378
2	1092	2786	579	645	888	231	455	825	1735	1139	892	11267	467	962	196	157	1791	13058
3	761	469	155	355	1110	134	147	84	358	834	698	5105	498	667	231	181	1577	6682
4	769	653	329	326	1161	343	103	284	255	931	694	5848	149	354	161	135	799	6647
5	4046	761	1188	961	3122	723	612	1920	3697	2107	1558	20695	479	1128	273	222	2102	22797
6	1004	281	139	337	908	500	503	485	286	951	822	6216	170	302	154	134	760	6976
7	826	391	123	110	657	394	401	473	913	1203	897	6388	158	219	197	156	730	7118
8	3026	928	111	240	2022	480	475	1255	3678	3281	2990	18486	135	301	321	310	1067	19553
9	5233	1718	345	279	3604	420	941	3080	1090	4733	3390	24833	303	538	236	329	1406	26239
10	4006	1129	850	988	2041	903	1149	3645	4507	2856	1984	24058	427	1030	215	403	2075	26133
11	3201	961	602	595	1675	815	803	3527	3444	2260	2070	19953	296	571	212	311	1390	21343
SUB TOTAL	33490	11220	5270	5698	21190	5971	6393	18773	25273	23987	19152	176417	5157	7761	3185	3404	19507	195924
A	2199	484	463	172	386	195	172	180	315	527	352	5445		86	592	632	1310	6755
B	1687	845	570	337	1125	276	144	346	555	977	364	7226	126		1135	650	1911	9137
C	1010	214	232	160	259	147	173	322	248	350	178	3293	420	937		472	1829	5122
D	1003	166	184	137	214	133	142	332	404	267	336	3318	300	630	454		1384	4702
SUB TOTAL	5899	1709	1449	806	1984	751	631	1180	1522	2121	1230	19282	846	1653	2181	1754	6434	25716
GRAND TOTAL	39389	12929	6719	6504	23174	6722	7024	19953	26795	26108	20382	195699	6003	9414	5366	5158	25941	221640

DISTRICTS OF ORIGIN

FIG. 6-4. Tabular array of O-D data by district and station interchange.

of the total trips with urban origin or destination travel to or from the central business district (designated as District 1), whereas only 11,709 or 5.3 per cent of the total having a rural origin or destination travel to or from the central business district.

In addition to these basic tabulations, other facts gathered during the survey may also be tabulated and analyzed. Trip lengths may be compared with trip purposes, trip modes with time of day, trip purposes with trip modes, and so forth.

Graphical Analysis. Since tables fail to develop a mental picture of concentrations and relative locations of origins, destinations, and their paths of connection, O-D information is frequently interpreted graphically. Concentrations at sites of traffic generation are usually shown by pie charts, spot maps, or isometric columnar figures. Concentration of lines of traffic movements are portrayed by flow bands, desire lines, and desire charts.

Sites of Traffic Generation. Figure 6-5 illustrates one method of portraying not only the relative magnitudes of generated traffic but also the geometrical relationships of the diverse zones involved. In Fig. 6-5 each zone has its own "pie" of a diameter scaled to the zone's total traffic generation and broken down into segments of rural and urban origins or destinations. Another variation is illustrated by Fig. 6-6, a chart

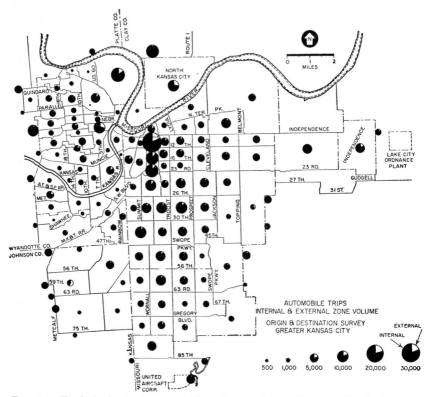

FIG. 6-5. Typical pie-chart presentation of O-D data. (From *A Traffic Survey of Greater Kansas City*, Missouri State Highway Department et al., Jefferson City, 1945, plate 2.)

showing the parking demand of 14 zones as determined from an O-D study of parkers in a central business district. For reasons of comparison, parking supply and actual usage are included. Zone 1, for example, shows very high demand and low supply, while Zone 12 shows low demand and a large supply. Of all the parkers in Zone 12 (as shown by the usage column), only about one-third (the demand column) were destined to Zone 12, and the other two-thirds had to walk to some other zone where demand exceeded supply. Spots, shading, coloring, and other variations are also used to depict density, relative magnitudes, and geometrical relations developed from O-D study.

Lines of Traffic Movement. Since lines of traffic movement which join the various zones may be concentrated into more or less common routes of travel in the desired line of movement, *desire-line* maps and similar devices are employed. An example of this type of analysis is shown in Fig. 6-7, which portrays only major directional desire lines. The lower

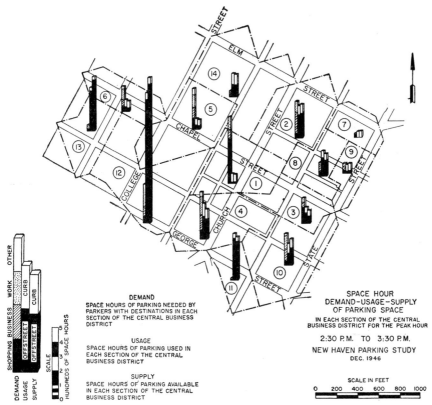

FIG. 6-6. Typical columnar presentation of O-D data. (From *New Haven Parking Study*, Connecticut State Highway Department, Hartford, 1947, Fig. 9.)

the number of trips per desire line portrayed, the more complex such a map becomes; if a desire line were shown for each single trip, the chart would become unintelligible. In practice, larger numbers of trips with more or less common origins and destinations are combined into common desire lines of movement, which disregard topography and street development and show only an ideal location of desirable routes.

If locations are selected where route developments exist or are practical, interzone trips can be assigned from the tributary zones to these practical routes, and comparisons between ideal desire line and practical routes may be made (Fig. 6-8).

A variation of desire-line analysis may be prepared by joining with desire lines each zone or point of origin and its corresponding zone or point of destination. The number of lines which overlap or cross within each small unit area of the map may be added together, regardless of their directions of movement, to produce magnitudes of traffic volume.

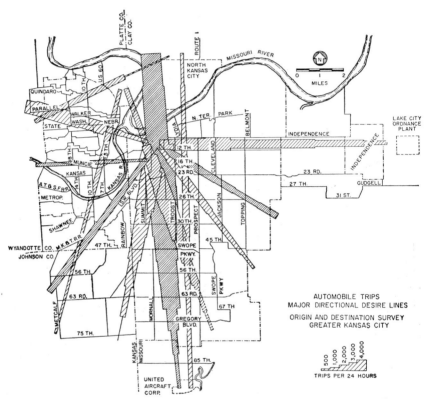

FIG. 6-7. Typical desire-line presentations of O-D data. (From *A Traffic Survey of Greater Kansas City*, Missouri State Highway Department et al., Jefferson City, 1945, plate 15.)

Such magnitudes may be considered *depth* values, and a resultant contour map may be developed. This *desire chart* develops a theoretical contour map of traffic density on which, however, one basic value of O-D study, *direction* of movement, is lost.[1]

Analysis by Method of Moments. The same principles used in the determination of neutral axes and centroids of physical bodies are also applicable to O-D data plotted on scale maps, showing amounts and zones of traffic generation. Called the method of moments, this type of

[1] For example see *Traffic Survey of the Sacramento Area*, State of California, Department of Public Works, and U.S. Bureau of Public Roads, Sacramento, 1948, pp. 20–21.

analysis may be used to show the optimum location of a common terminal or a straight-line route of a given bearing.

Terminal Location. If the ultimate origins and destinations of patrons of a given traffic area are known, the ideal location of a passenger terminal,

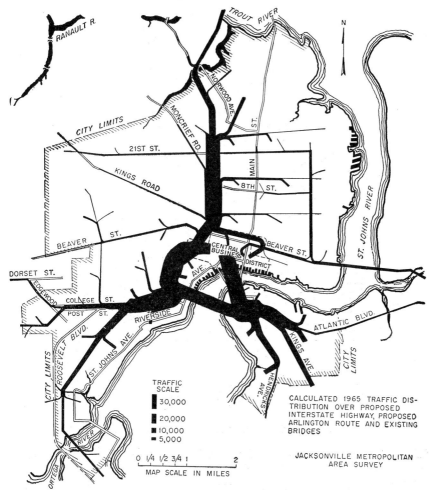

FIG. 6-8. Typical flow-band presentation of O-D data. (From *Interstate Highway Plan for Jacksonville, Florida,* Florida State Road Department, Tallahassee, 1945, plate 29.)

parking lot, or other traffic terminal may be found by the method of moments by locating the centroid of O-D data plotted on a scale map. Figure 6-9 shows the O-D data distributed into small zones for all bus passengers originating in or destined to the central business district (District 1) of the hypothetical city of Fig. 6-3. The total of 1,410 intercity bus passengers in this area, when distributed into the 11 zones as indicated, ranges from 20 to 300 passengers per zone.

Passengers for each zone were assumed to be centered at the spot shown on the map. The X and Y axes of each passenger group are shown in the

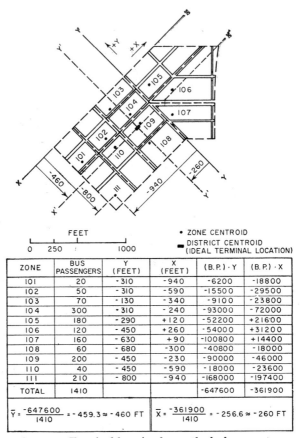

ZONE	BUS PASSENGERS	Y (FEET)	X (FEET)	(B.P.) · Y	(B.P.) · X
101	20	-310	-940	-6200	-18800
102	50	-310	-590	-15500	-29500
103	70	-130	-340	-9100	-23800
104	300	-310	-240	-93000	-72000
105	180	-290	+120	-52200	+21600
106	120	-450	+260	-54000	+31200
107	160	-630	+90	-100800	+14400
108	60	-680	-300	-40800	-18000
109	200	-450	-230	-90000	-46000
110	40	-450	-590	-18000	-23600
111	210	-800	-940	-168000	-197400
TOTAL	1410			-647600	-361900

$$\bar{Y} = \frac{-647600}{1410} = -459.3 \approx -460 \text{ FT} \qquad \bar{X} = \frac{-361900}{1410} = -256.6 \approx -260 \text{ FT}$$

FEET
0   250   1000

• ZONE CENTROID
▪ DISTRICT CENTROID (IDEAL TERMINAL LOCATION)

Fig. 6-9. Terminal location by method of moments.

table. The neutral axes for X moments falls at -460 ft from the arbitrary X axis, and the neutral axis for Y moments lies at -260 ft from the arbitrarily chosen Y axis. The intersection of these neutral axes is the centroid of the bus passenger O-D's and indicates the optimum location for an intercity bus terminal in this central business district.

Since spots chosen as typical points of passenger O-D's must be arbitrarily assumed in the light of physical development of each zone or located by the method of moments (ordinarily not justified for each elementary zone), this method has limitations in practice. When the centroid is finally located, it may fall on land which is already occupied or unattainable, or where street and traffic developments indicate problems which would better be avoided. Nevertheless, the inconvenience to terminal patrons increases as the distance of an alternate location from the computed site increases. Such inconvenience must be weighed against prevailing practical conditions. The method of moments gives a good starting point in locating traffic terminals, but usually requires modification because of street, traffic, land use, and other practical considerations.

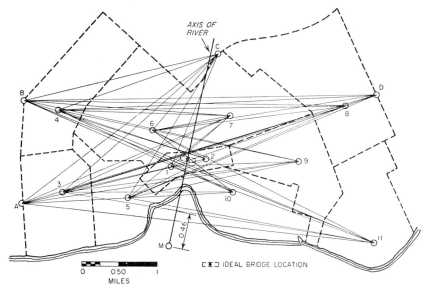

FIG. 6-10. Bridge location by method of moments.

Bridge Location. The method of moments may also be used in locating the optimum site of a bridge or ferry crossing, as Fig. 6-10 illustrates. Using the same hypothetical city and data from Figs. 6-3 and 6-4, it is proposed to create a single river crossing best suited to all traffic movements. Centroids of all districts have been located either by method of moments or by practical judgment, based on land use and street pattern and conditions, and are joined by transriver desire lines so that all districts and stations on one side of the river are joined to all districts and stations on the other side. Volume of traffic for each desire line is shown.

Since all traffic must cross the river at one point, the straight line which most closely follows the course of the river becomes an axis and may be used as a reference line. The point at which each desire line crosses this axis is the ideal river-crossing location for the particular volume of traffic represented.

Moments of each river-crossing volume are taken about some arbitrarily shown point M on the river axis. For example, the total interzone movement between Districts 3 and 11 is 1,300 vehicles. The desire line of this river-crossing movement lies 0.46 mile from the reference point M, and its moment about M is therefore 598 vehicle-miles.   The sum of all such moments is 106,547 vehicle-miles.   Since the tabular array showed the total of river-crossing vehicles to be 96,773, the ideal river crossing for this traffic is 106,547/96,773, or 1.10 miles up the river from reference point M.

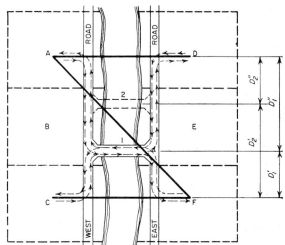

Fig. 6-11. Practical considerations applied to bridge locations.

Relative traffic volumes

Along desire line: $AD = 1,000$
$$CF = 2,000$$
$$AF = 3,000$$

Bridge locations

(1) West road and east road equally desirable for use
Disregard $\overline{AF}$ traffic

$$\Sigma M \text{ bridge} = 0 = \overline{AD} \times \text{distance} - \overline{CF} \times \text{distance}$$
$$= 1,000 D_1'' \qquad\qquad - 2,000 D_1'$$
$$D_1'' = 2D_1'$$

(2) West road undesirable for use
Consider $\overline{AF}$ traffic along line $\overline{AD}$

$$\Sigma M \text{ bridge} = 0 = \overline{CF} \times \text{distance} - (\overline{AD} + \overline{AF}) \text{ distance}$$
$$= 2,000 D_2' \qquad\qquad - (1,000 + 3,000) D_2''$$
$$D_2' = 2D_2''$$

In practice, the pattern of the street and road net, the traffic conditions on this net, and other considerations frequently modify the application of the method of moments in analyzing O-D data for a river-crossing problem.   Figure 6-11, a sketch of 6 zones of O-D abutting a river, shows desire lines and actual paths of travel between Zones A and D, A and F, and C and F.   The computed location for the bridge is shown in position 1.   In this case, roads and traffic conditions along the river banks are identical, and it is of little consequence where the bridge is located as far as travel between A and F is concerned. But for traffic flow A–D and C–F there is an obvious single ideal solution.

In such a case the traffic movements computed from flow A–F, and all other diagonal movements, should not be considered. Only those traffic streams which will form a U, J, or hook-shaped pattern should be employed. On the other hand, if roads along the river banks are unequal for travel movements, it is necessary to evaluate the degree of inequality and to consider the desires of diagonal traffic to the same degree.

In some cases time of travel may be employed in lieu of distance of travel, so that moments are expressed in vehicle-minutes rather than vehicle-miles. The bridge site is then established on the basis of minimum travel time rather than minimum travel distance.

Route Location. If general bearing for a route is determined by natural topography, road-net development, bridges, or other basic influences, a neutral axis of such bearing may be determined from the interdistrict O-D movements. In assigning the probable volume of traffic from each zone, however, certain practical factors must be considered, among them (1) travel distance or time on existing road net in comparison with similar values on a new route, (2) trip lengths and relative amount of each trip length which may be accommodated on a new facility, (3) comparison of travel conditions on the existing network with the new route. Such factors, singly and in combination, frequently complicate the use of the method of moments for route selection to such a degree that it is impracticable. Experienced judgment checked by use of moments will yield the better choice of route.

6-4. Adequacy of Sample. Since the conclusions of an O-D study are based on the observed behavior or expressed desire of relatively few of the vehicle drivers in the traffic stream, it is essential to determine how many drivers are required for a sample which will result in a satisfactory estimate or prediction of the total movement. If in a given instance a total of X trip movements are involved, is it necessary to capture every individual trip? Will 50 per cent be enough? Or would 25 per cent, 10 per cent or only 5 per cent be sufficient for the purpose in mind?

In the discussion which follows, some of the special problems involved in establishing sample adequacy have been developed in the simplest terms possible. It should be understood, however, that the treatment given here is only a beginning and is not intended to be a comprehensive study of the theory of sampling.[1]

It is desirable to keep the tests for sample adequacy as simple as possible. The simplest sampling problems can be reduced to only two kinds of trips, those trips that are going to or from a particular zone or

---

[1] The sampling theory will be found fully developed in the following work: J. G. Smith and A. J. Duncan, *Sampling Statistics and Applications,* McGraw-Hill Book Company, Inc., New York, 1945.

area and those that are not. The former may be said to have *attribute A* and the latter *attribute B*. Most O-D data can be reduced to such simple form for the purpose of testing.

In a survey of the home-interview type, the area will probably be subdivided into 20 or more numbered zones of traffic generation. Suppose that 10 per cent of all the homes in the area are visited and that the sample of trips generated by Zone I amounts to 2,000 movements. To test the reliability of that portion of the sample generated between Zones I and II, which we will suppose to be 160 trips, the traffic flowing between Zones I and II is designated attribute A and the remaining 1,840 sample trips generated by Zone I attribute B. Certain conditions must be met in the collection and preparation of sample data before statistical treatment can be undertaken.

Data Must Be Collected at Random. Every driver must stand an equal chance of being selected for interview. If there is a preponderance of one type of driver (truck drivers, for example), the sample is "biased" and will show a distribution of trip movements disproportionately weighted in favor of the practices of that particular driver type and not representative of the over-all traffic pattern.

Most of the current methods of driver interviewing usually attain random selection. At roadside-interview stations a very high percentage of passing traffic is sometimes interviewed because (1) the investigator is ignorant of sampling methods or (2) he wishes to remove even a slight possibility of bias. If the method of sample selection can safely be considered random, such complete samples are seldom warranted.

Sampling for the home-interview survey is usually conducted by *ordinal selection:* every 5th, 10th, or 20th dwelling in the survey area is designated a unit of the sample. If samples selected by this method are carefully inspected for bias, they may readily be tested for adequacy and can be kept to a reasonable size. Modified methods of ordinal selection may be applied to roadside interviews or to parking surveys in which fewer than 100 per cent of the parkers are to be interviewed.

Sample Data Must Be Complete. Though complete data can usually be obtained at roadside stations or parking spaces, home-interview information is sometimes incomplete. Adjustment of incomplete data is a delicate operation, and the method used depends on the individual case. The more complete the data, the more dependable will be the trip estimates made from it.

Sample Must Be Statistically Stable. Most samples of O-D data are composed of trips generated in many different areas or zones. Data from zones that generate many trips are usually of more interest to the investigator than data from zones which generate few trips. Since larger samples are more stable and easier to analyze, two or more zones of minor trip-generating power are often combined during analysis, provided the zones are adjacent and compatible in character. If sample data for each of the attributes A and B exceed certain minimum volumes, it is reasonable to assume that the data represent a *normal distribution;* that is, if attribute A amounts to 8 per cent of the sample, the chance that the actual number of attribute A trips in the total movement *exceeds* 8 per cent is just as great as the chance that it *falls short* of 8 per cent. The normal distribution should consists of no fewer than 30 trip movements. Samples of less than 30 movements may be used with satisfactory results, however, if techniques other than those assuming normal distribution are employed.[1]

Attributes which constitute a very small proportion of the sample (5 per cent or less) will also tend to deviate from normal distribution. However, if the smaller attribute includes less than 5 per cent but more than 30 trips, fairly good results will still be obtained for data amounting to less than 1 per cent of the sample.

In general, reasonably large samples are more satisfactory for estimating a traffic movement, but they are more costly to produce, and their greater accuracy may not be worth the additional expense.

If the 10 per cent sample of home-interview data described above meets the requirements of randomness, completeness, and stability, statistics can be applied to describe the underlying pattern of trips from which the sample was drawn.

Maximum-likelihood Estimate. A *single estimate* may be made of the percentage of all trips generated in Zone I which have either origin or destination in Zone II. This is the *maximum-likelihood estimate* and is equal to the sample percentage. In the example described above, the 160-trip movements between Zones I and II amount to 8 per cent of the 2,000 trips generated in the sample homes interviewed in Zone I. The sample represents 10 per cent of all trips generated in the area, or a total of 20,000 trips. The maximum-likelihood estimate of the total trip movement between Zones I and II is therefore 8 per cent of 20,000 trips, or 1,600 trips. This estimate, however, includes no qualification as to dependability.

Confidence Interval. It is impossible to determine the exact number of trips in any movement if the sample is less than 100 per cent. Instead, a range of values is determined within which the true value will fall with a stated degree of probability. The range is called a *confidence interval,* and the degree of probability is expressed as a decimal fraction termed the *coefficient of confidence.* In a normal distribution the maximum-likelihood estimate falls at the exact center of the confidence interval.

In the maximum-likelihood estimate of 1,600 trips between Zones I and II stated above, a range of values extending above and below 1,600 within which the true value might fall should be indicated. If the interval

[1] Full development of such techniques may be found in *ibid.*

of 0.9 confidence is Y trips, the chances are 9 to 1 that the true value for total trip movement between Zones I and II will fall within the range $1600 \pm Y/2$.

Degree of Risk. The chance that the true volume of trips will exceed the confidence interval is called the *degree of risk* and may also be expressed by a decimal fraction called the *coefficient of risk*. In the above example, the risk is 1 chance in 10 (coefficient of risk 0.1) that the true value will fall outside the range $1600 \pm Y/2$.

To establish the confidence interval for a particular sample, it is first necessary to decide on the degree of risk that can be tolerated. The greater the risk, the narrower the confidence interval will become. If a 0.2 coefficient of risk were allowed in the above problem, the confidence interval would include a narrower range of values. The allowable amount of risk is usually established on the basis of previous studies of a similar nature which have given satisfactory results or on an arbitrary value that appears to be safe.

Sample Size. Traffic data are usually related to some definite period of time, such as an afternoon peak hour, an average 24-hour weekday, or a total annual vehicle flow. The sample is usually taken from some finite volume of trip movements, known or estimated. Total volume of traffic expected to pass a roadside-interview station is determined by preliminary traffic counts. An estimate of total trips involved in a home-interview survey is based on expected trips per family, using census data to determine the number of family units. The total number of people parking each day in an area might be estimated by determining the total number of spaces available and assuming a likely rate of turnover from studies conducted elsewhere.

When the total volume of trips is very great, such as the annual total of vehicles moving along a heavily traveled highway, a reasonably large sample constitutes a very small percentage of the annual total, which, for practical purposes, can be considered infinite. In testing sample data for adequacy under these conditions, sample size is of prime importance. As size of sample increases, the maximum-likelihood estimate approaches the true value of the underlying traffic movement, and the confidence interval for every coefficient of confidence becomes narrower, with the result that the probability of any given finite deviation from the true value becomes progressively less.

When a sample is taken from a relatively small volume of trips, such as a 1-hr flow of traffic across a bridge, it constitutes a large part of the total movement and the confidence interval is markedly narrower. The number of trips may be a very small percentage of the annual flow, but a very high percentage of the hourly flow. A higher degree of confidence can be placed in such a sample *as representative of the 1-hr volume* than in the same sample representing the daily, weekly, or annual flow.

Testing the Sample. The statistical tests for sample adequacy are practical applications of the theory of probability. One easily applied method for calculating the confidence interval of normally distributed samples (which may be drawn from either infinitely large or moderately small volumes of traffic) is as follows:[1]

$$I_{p1} = \frac{N_1}{N} \pm M \sqrt{\frac{P_1 P_2}{N}} \times \left[1 - \frac{N}{S} \right]$$

where I_{p1} = confidence interval for p_1

N = number of trips in sample

N_1 = number of trips with attribute A

S = total number of trips

$P_1 = N_1/N$ = per cent of trips with attribute A

$P_2 = (N - N_1)/N$ = per cent of trips with attribute B (per cent not having attribute A)

M = confidence interval expressed in standard deviation units (see Table 6-1)

TABLE 6-1. VALUES OF M FOR SEVERAL COEFFICIENTS OF CONFIDENCE*

Coefficient of confidence	M	Coefficient of confidence	M	Coefficient of confidence	M
0.99	2.57	0.85	1.44	0.65	0.94
0.98	2.32	0.80	1.28	0.60	0.84
0.95	1.96	0.75	1.15	0.55	0.76
0.90	1.65	0.70	1.04	0.50	0.68

* These values require a normal distribution of data.

The upper limit of the confidence interval is found when the sign preceding M is positive; the lower limit when the sign is negative.

The quantity $\sqrt{P_1 P_2/N} \times [1 - (N/S)]$ is an estimate of the standard deviation of the total trips having attribute A.

The quantity $1 - (N/S)$ under the radical approaches a value of 1 as S becomes increasingly larger than N. This quantity is inserted in the formula as a modifying factor which narrows the confidence interval when N is large in relation to S. When N is relatively small (in general, if N/S is less than 0.01), this quantity need not be included in the formula.

The above formula can be used to show whether or not any portion of sample data is sufficiently dependable as a basis for analysis, after an acceptable confidence interval has been established.

Checking Adequacy of Existing Sample. Suppose it is desired to know the dependability of the 1 per cent sample of trips generated between Zones I and II in the example already discussed. The sample consisted of 2,000 trips generated in Zone I, of which 160 were movements between Zones I and II. The maximum-likelihood estimate of traffic between these zones is 1,600 trips. The investigator may decide he requires an

---

[1] The development and proof of this formula will be found in *ibid.*, chaps. 8, 9.

interval of 0.9 confidence and with true value within a range of 10 per cent of the maximum-likelihood estimate; that is, the interval of 0.9 confidence must not exceed the range $1,600 \pm 160$.

This problem of adequacy would be determined as follows:

Let $S = 20,000$ trips
$\quad N = 2,000$ trips
$\quad N_1 = 160$ trips

then

$$P_1 = \frac{N_1}{N} = \frac{160}{2,000} = 8 \text{ per cent of trips}$$

$$P_2 = \frac{N - N_1}{N} = \frac{1,840}{2,000} = 92 \text{ per cent of trips}$$

$$M = 1.65, \text{ (0.9 coefficient of confidence)}$$

Substituting in the formula

$$I_{p1} = 0.08 \pm 1.65 \sqrt{\frac{0.08 \times 0.92}{2,000} \left[1 - \frac{2,000}{20,000}\right]}$$

$$= 0.08 \pm 0.0095$$

Therefore the upper limit of the confidence interval equals

$$20,000 \times (0.08 + 0.0095) = 1790$$

and the lower limit is

$$20,000 \times (0.08 - 0.0095) = 1410$$

The interval $1,600 \pm 190(1,410 - 1,790)$ is greater than the imposed limits $1,600 \pm 160(1,440 - 1,760)$. The formula shows in this case that the sample did not include enough trips of attribute A to meet the needs of the investigation.

Extending Survey for Additional Data. To make the sample usable by obtaining more data, upper and lower limits of confidence can be established at 1,760 and 1,440 respectively, and the equation can be solved for N. (Assume that the proportions of $p_1 = p_2$ will not change appreciably.) Thus

$$I_{p1} = \frac{1,760}{20,000} = 0.088$$

and by solving for N in the original equation, $N = 2,707$.

A minimum of 2,707 trips, or $13.5 +$ per cent of the total movement, is required to achieve the dependability of sample required. The sample would probably be raised to 14 or 15 per cent for a small margin of safety.[1]

[1] A method of cut and try might also be used, instead of solving for N. Arbitrary values could be assigned N and the formula applied repeatedly for upper and lower limits of p_1 until a value was found to satisfy requirements. This method is less precise, though sometimes faster, than solving for N.

Preliminary Estimates of Sample Needed. The test for sample adequacy may be used advantageously in planning an O-D survey if certain features of the underlying body of data are known or can be approximated. If the total number of movements involved can be estimated accurately, the number of sample movements can be determined in advance for attributes which represent various proportions of the whole. Suppose an investigator has estimated that some 50,000 vehicles park within the central business district of a city during an average day and he wishes to know the purposes for which they park. He suspects, from data at hand for similar cities, that the smallest group of trip purposes for which he will test will constitute no less than 10 per cent of the total trip movement. He desires to know how large a sample he must take to obtain a 0.9 coefficient of confidence for an interval not to exceed ±5 per cent of the actual value (5,000 ± 250). He may substitute his assumed values in the equation given above and solve for N. The required sample amounts to about 8,200 vehicles, or 16.4 per cent. If the investigator is almost positive that his smallest group of data will exceed 10 per cent, he may sample only 8,200 vehicles. If he is less sure, he may increase the indicated size of sample, but in any case he has developed a reference point to guide him.

In planning a survey, a series of estimates may be prepared for sample sizes needed to achieve any of several degrees of confidence and for attributes of various size. The combination of conditions best suited to the needs and budget of the survey may then be selected.

6-5. O-D Characteristics. For obvious reasons, most O-D investigations are made in and near regions of dense population. Such regions may be readily subdivided into three principal areas of traffic origin and destination: the dense central business districts, surrounding urban development, and less populous adjacent rural lands. Since traffic movements take place within and between these areas, any trip may be classified by its origin and destination into one of the six classes shown in Table 6-2.

When all auto, truck, and taxi trips (excluding buses and streetcars) are sorted into these six O-D classes, the relative percentage that each class bears to the total is as shown. Graphical presentation of this data is set forth in Fig. 6-12. The concentration of traffic movement within the urban area is apparent. In general, trips which have both origin and destination within the city constitute more than three-fourths of all trips. All trips which have at least one terminus outside the urban area constitute the remainder. Traffic with at least one rural terminus is ordinarily designated as *external* traffic, while traffic with both termini in the city is called *internal* traffic.

Influence of Population Size. The population size of the city studied has a strong influence on the percentage of total traffic found in each class of origin and destination. This factor is largely responsible for the

TABLE 6-2. HIGHWAY TRAFFIC CLASSIFICATIONS BY AREAS OF ORIGIN
AND DESTINATION

Class of trip	Trip origin (or destination)	Trip destination (or origin)	Percentage of total vehicle traffic	
			Average	Range
RR	Rural	Rural	1.8	0.6– 5.3
RU	Rural	Urban	14.8	4.3–31.6
RC	Rural	Central	5.2	1.1–11.1
UU	Urban	Urban	53.7	37.3–66.2
UC	Urban	Central	22.0	13.5–33.1
CC	Central	Central	2.5	0.1– 9.2

SOURCE: Based on O-D data obtained in studies of 28 cities of 23,000 to 540,000 population. All studies were made between 1944 and 1949.

range in values shown in Table 6-2. The influence of population size on the distribution of traffic trips is set forth graphically in Fig. 6-13.

The increase in urban-urban vehicle trips, the constancy of urban-central and central-central, and the decrease in all other O-D classes of trips with increase in population size is significant. External trips become less important as population increases; they loom large in the movements of the small town but have little significance in the larger population pools. The central business district tends to generate a con-

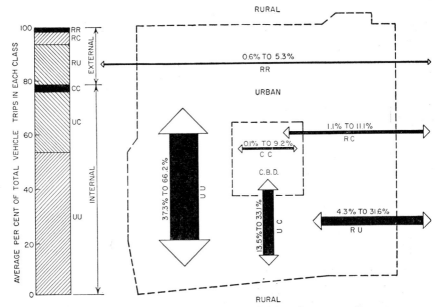

FIG. 6-12. Six basic classes of traffic in the vicinity of a metropolitan area.

stant per cent, averaging 30 per cent, of total city traffic regardless of population. Urban-central and central-central trips remain relatively constant, and rural-central decreases only at the rate of all rural trips.

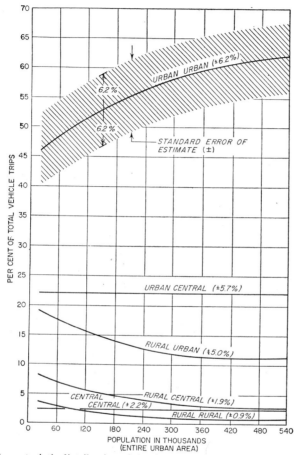

Fig. 6-13. Characteristic distribution of the six basic classes of traffic in cities of various sizes. (Based on analysis of data presented in O-D surveys made in 28 cities of 23,000 to 540,000 population from 1944 to 1949.)

Considerable variation from the actual numerical values shown in Fig. 6-13 may be found in a particular city. The data shown are rough approximations of the relative amount of traffic in each class of O-D and show the changing importance of each class with population size.

O-D Characteristics in Relation to Route Usage. O-D characteristics of traffic movement are significant in posting, locating, and designing routes to best serve the demand of traffic flow. Existing routes are frequently poorly located, inadequate, or in other ways unsuited to the diverse and incompatible traffic demands developed by the six different

O-D classes operating in the same stream. Trips in the central business district are short, slow-moving, and demand loading-unloading operations or parking. Trips of the rural-rural class are longer, faster, and require no terminal provision at points in the central districts. The separation of central-district movements from rural-rural traffic is desirable, not only because of the route inadequacies in the central business district, but also because of the differing traffic needs of each class of O-D. Similarly, urban-urban movements may be improved by avoiding routes used by central-district traffic movements.

The separation of different O-D classes of traffic movement by route location or limited-access design may result in advantages which justify the costs. In the route-location method, a bypass route is provided around areas which lie between the origin and destination of the bypass traffic. Regardless of method of segregation used or the area avoided, the separation of one or more classes of O-D to avoid interference with other classes must be justified by O-D analysis.

As Fig. 6-12 shows, in the vicinity of a city there are basically two areas which may be considered as bypassable, the central business district or the whole city. Contact with either or both of these areas may be eliminated for those vehicles which do not wish to enter them by the use of one or more of three basic route arrangements.

Case I. Bypass of City. A bypass which circumvents the entire urban area will serve primarily only the rural-rural class of movement, which averages only about 1.8 per cent of the total vehicle trips. A small portion of the rural-urban and rural-central classes might also find the bypass useful along some portion of its length. The entire rural-rural class, however, would not use it because of differences of direction between the route and the desired movement. Figure 6-13 shows that in smaller cities the percentage of rural-rural trips is highest, thus indicating a greater justification for this type of bypass for such cities. On the other hand, because of increased totals in larger cities, the actual number of bypassable trips may be larger than in smaller cities.

Bypassable traffic may be shown as a per cent of the total *external* traffic approaching the city from the rural area. The relationship that the rural-rural class bears to total external trips for different population sizes is shown in Table 6-3.

Case II. Bypass of Central Business District. A bypass created by design and location to circumvent the central business district accommodates portions of the rural-urban and urban-urban classes of movement which originate on one side of the central business district and terminate on the other side. In the absence of a Case I type bypass, it also accommodates the rural-rural class. These three classes of movement constitute an average of over 70 per cent of the city's trips, but not all of them can be accommodated by the central business district bypass.

Studies of parked vehicles in the central business district together with cordon counts at the boundaries of the central business district show that.

regardless of population size, 50 to 60 per cent of all vehicles which enter the central business district do not desire to stop there. Since, in the average case, trips between the central business district and other areas constitute about 27 per cent of the total, about an equal amount (27 per cent) pass through the central business district without desire to stop and therefore would benefit from a bypass.

TABLE 6-3. PROPORTION OF APPROACHING EXTERNAL TRAFFIC WHICH MAY BE BYPASSED AROUND A CITY

Cities by population size	Per cent bypassable traffic
Less than 2,500	50.7
2,500– 10,000	43.3
10,000– 25,000	21.9
25,000– 50,000	21.0
50,000– 100,000	16.2
100,000– 300,000	18.2
300,000– 500,000	7.2
500,000–1,000,000	4.2

SOURCE: *Interregional Highways*, U.S. Bureau of Public Roads, Interregional Highway Committee, 1944, p. 40. (Mimeographed.)

The geometry and arrangement of the city and its central business district, the exact location of the bypass, the degree of its accessibility to all traffic which it might serve, and the relative quality of transmission of traffic on such a bypass will affect the total load which is actually accommodated, however.

Case III. The Limited-access Route. By limiting access to certain routes to a few well-chosen points, it is possible to separate different O-D classes of traffic and thus effectively accomplish the basic purpose of bypass routes. Limited-access routes may traverse in radial or circumferential fashion the urban area surrounding the central business district, thereby separating trips of longer distance from those which are purely local. Radial routes which terminate at the central business district, on one hand, and the rural area, on the other, will serve to transmit all rural-central and nearly all of the urban-central classes of traffic. As Table 6-2 shows, in the average city this will amount to about 5 per cent of the total at the periphery of the city and about 27 per cent at the central business district. In addition to central business district–bound traffic, such radial routes may transmit all rural-rural and varying amounts of rural-urban, as well as those portions of the urban-urban class which have radial desire lines and lengths greater than the distance between interchanges. Though the total percentage of trips which will find such routes advantageous is unknown, the utility of radial streets is clear. Their daily volumes exceed all other routes in nearly every area of the traffic network.[1]

[1] For example see M. McClintock, *A Report on the San Francisco City-wide Traffic Survey*, Department of Public Works and WPA Project 6108-5863, San Francisco, 1937, pp. 52–54.

Trip Lengths.   The average length of trip made by the highway vehicle is relatively short.   Figure 6-14 shows that 85 per cent of all rural trips are 20 miles or less in length and that 40 per cent are 5 miles or less in

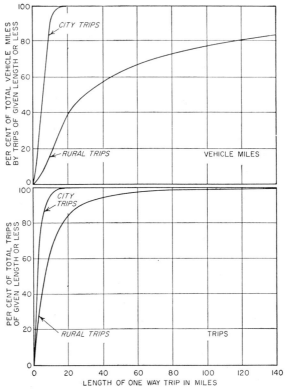

FIG. 6-14. Cumulative distribution by trip lengths of total trips and total mileage traveled.   (R. H. Paddock and R. P. Rodgers, "Preliminary Results of Road-use Studies," U.S. Bureau of Public Roads, *Public Roads*, vol. 20, no. 3, May, 1939, pp. 49, 52; *Traffic Survey of the Sacramento Area*, State Department of Public Works and U.S. Bureau of Public Roads, Sacramento, 1948, p. 24, Table 9.)

length.   This data is for all rural trips regardless of origin or destination, yet 66 per cent of such trips were found to be made by urban residents.[1]

Trip length of urban traffic averages about 3 miles.   A characteristic curve for city trip lengths is shown in Fig. 6-14.   Trip length varies with density of population: in sparsely populated rural regions where towns are relatively far apart, trip lengths will tend to become larger.   Rural trip lengths of vehicles registered in cities increases as the size of the city increases.   For example, the median trip length is about 6 miles for vehicles on rural routes from unincorporated areas and about 16 miles for

[1] R. H. Paddock and R. P. Rodgers, "Preliminary Results of Road-use Studies," U.S. Bureau of Public Roads, *Public Roads*, vol. 20, no. 3, May, 1939, pp. 46–52.

vehicles from population centers of 100,000 or more.[1] Median trip length within metropolitan areas is little more than 2 miles.

As Fig. 6-14 shows, 50 per cent of all rural trips are 7.5 miles or less in length, but such trips generate only 12.5 per cent of the vehicle-miles. For metropolitan traffic, 50 per cent of all trips are about 2 miles in length, but such trips generate only 20 per cent of the total vehicle-miles.

Trip Generation. In urban areas, the average resident makes one or two trips per day. Accordingly, each family dwelling unit in the city will create from two to eight trips per day by all modes of highway travel (Table 6-4). Data shown are for residents of the city only and do not include truck or taxi trips made by such vehicles in the course of a day's operation.

Table 6-4 shows that the average dwelling unit houses 3.35 persons, yet only 1.53 persons per dwelling make trips. Since the average dwelling unit generates 4.64 trips per day, those persons making trips will average 3.29 trips each. In addition to these trips generated by the individuals themselves, each dwelling unit generates a vehicle-trip demand not immediately reflected by its occupants' travel—the delivery of merchandise, callers from out of the city, and other vehicle trips. The number of vehicle trips is affected by the percentage of personal travel by common-carrier transit. Accordingly, the vehicle trips generated in an urban area will vary with population size, since a larger percentage of persons travel by transit in larger cities.

Illustrative values of vehicle trips generated by cities of various sizes are set forth in the column headed "vehicle trips per capita" in Table 6-4. While all other columns of this table are based on trips by residents of the area only, this column is based on vehicle trips regardless of origin, destination, or who made them. Thus it is a measure of the generating power of the population group rather than the travel of the individuals of that population group.

Trip Mode. Clear differentiation must be made in O-D studies between freight and passenger movements. Nearly all freight moved over the urban street network is carried by trucks, but for persons there are several different modes of transportation available. The truck component of the traffic stream ranges ordinarily between 12 and 24 per cent of total vehicles.[2] Many of these trucks, however, such as those engaged in home deliveries, generate a very high proportion of very short trips. Since each trip is enumerated in the home-interview methods of inquiry, the per cent of truck trips to total vehicle trips does not reflect relative road usage because of the undue number of short truck trips. Home interview O-D studies show that truck trips range from 16 to 33 per cent of all vehicle trips.[3]

Since passenger trips are of more comparable lengths, the mode of

[1] *Ibid.*, p. 54, fig. 4.
[2] See Chapter 5, Table 5-2.
[3] Based on reports from 28 city O-D surveys.

TABLE 6-4. RESIDENT TRIP GENERATION IN VARIOUS CITIES
(Based on home-interview data)

City	Approximate population	Date of survey	Automobile per dwelling unit	Persons per dwelling unit	Persons making trips per dwelling unit	Trips per dwelling unit	Person trips per capita	Vehicle trips per capita*
St. Louis, Mo.	1,480,000	1945	0.49	3.31	1.38	5.81	1.76	0.39
Baltimore, Md.	912,900	1945	0.45	3.32	1.77	4.50	1.35	0.65
Seattle, Wash.	518,600	1946	0.64	2.77	1.58	4.60	1.68	1.09
Portland, Oreg.	422,700	1947	0.68	2.97	1.77	6.10	2.07	1.48
Indianapolis, Ind.	411,600	1945	0.60	3.22	1.48	3.56	1.11	1.02
Denver, Colo.	360,000	1944	0.72	3.10	1.44	5.10	1.64	1.00
Memphis, Tenn.	278,500	1944	0.46	3.58	1.49	3.24	0.91	0.66
Nashville, Tenn.	195,500	1945	0.42	3.46	1.52	3.20	0.92	0.74
Tacoma, Wash.	138,700	1948	0.73	2.89	1.61	5.40	1.85	1.68
Spokane, Wash.	138,400	1946	0.61	2.85	1.32	3.40	1.18	1.34
Little Rock, Ark.	130,300	1944	0.56	3.34	1.37	2.62	0.78	0.99
Tucson, Ariz.	126,900	1949	0.85	3.28	1.88	7.50	2.28	1.74
Chattanooga, Tenn.	126,100	1946	0.63	5.42	2.14	4.48	0.83	0.81
Reading, Pa.	119,800	1946	0.45	3.16	1.27	3.49	1.10	1.26
South Bend, Ind.	119,400	1945	0.69	3.27	1.36	3.40	1.04	1.56
Ft. Wayne, Ind.	114,400	1944	0.76	3.34	1.55	3.84	1.15	1.16
Charlotte, N. C.	100,600	1945	0.51	3.51	1.57	3.80	1.10	1.39
Harrisburg, Pa.	95,000	1946	0.52	3.27	1.59	8.30	2.54	1.88
Columbus, Ga.	79,200	1947	0.43	3.90	1.52	3.70	0.95	1.97
St. Joseph, Mo.	75,700	1945	0.48	3.02	1.10	2.60	0.85	1.08
Port Huron, Mich.	33,800	1945	0.87	3.20	1.61	7.10	2.22	1.97
Ottumwa, Iowa	29,500	1945	0.61	3.48	1.56	4.92	1.41	1.26
Mason City, Iowa	23,200	1945	0.71	3.47	1.32	6.10	1.76	1.60
Average	......	......	0.60	3.35	1.53	4.64	1.41	1.25
Range	......	......	0.42-0.87	2.77-5.42	1.10-2.14	2.60-8.30	0.78-2.54	0.39-1.97

* Based on all vehicle trips regardless of origin or destination or residence of driver. (All other columns based on trips by internal residents only.)

passenger travel is of great significance in the planning, designing, and operation of a street network. Origin-destination studies show the relative importance of each type of carrier in the consideration of traffic improvement measures. The percentage of passenger trips made by private auto, taxicab, and mass transit for typical cities is set forth in Table 6-5. Mass-transit passengers constitute from 12.7 to 67.9 per cent

TABLE 6-5. TRIP MODE*

City	Approxi-mate population	Date of survey	Per cent of persons using each mode of travel					
			Auto drivers	Auto pass-engers	Taxi pass-engers	Sub-total	Mass transit	Total
St. Louis, Mo........	1,480,000	1945	26.6	5.3	0.2	32.1	67.9	100.0
Baltimore, Md.......	912,900	1945	27.4	14.7	1.3	43.4	56.6	100.0
Seattle, Wash.......	518,600	1946	43.4	21.5	0.5	65.4	34.6	100.0
Portland, Oreg.......	422,700	1947	47.1	24.5	0.4	72.0	28.0	100.0
Indianapolis, Ind.....	411,600	1945	47.0	17.6	0.5	65.1	34.9	100.0
Denver, Colo........	360,000	1944	38.4	23.1	1.2	62.7	37.3	100.0
Sacramento, Calif....	201,300	1947	59.0	25.7	†	84.7	15.3	100.0
Spokane, Wash.......	138,400	1946	46.7	21.5	1.4	69.6	30.4	100.0
Little Rock, Ark......	130,300	1944	36.5	20.3	†	56.8	43.2	100.0
Tuscon, Ariz.........	126,900	1949	57.2	29.8	0.3	87.3	12.7	100.0
Reading, Pa.........	119,900	1946	39.0	16.4	†	55.4	44.6	100.0
South Bend, Ind.....	119,400	1945	63.9	17.9	0.3	81.1	17.9	100.0
Ft. Wayne, Ind......	114,900	1944	49.3	13.9	0.5	63.7	36.3	100.0
Harrisburg, Pa.......	95,000	1946	37.3	17.9	0.5	55.7	42.3	100.0
Lincoln, Nebr........	88,200	1944	47.0	27.4	0.7	75.1	24.9	100.0
Columbus, Ga.......	79,200	1947	39.7	19.8	†	59.5	40.5	100.0
St. Joseph, Mo.......	75,700	1945	39.0	14.9	0.5	54.4	45.6	100.0
Port Huron, Mich....	33,800	1945	51.6	30.0	0.3	81.9	18.1	100.0
Ottumwa, Iowa......	29,500	1945	44.1	20.4	†	64.5	35.5	100.0
Mason City, Iowa....	23,300	1945	55.8	27.3	†	83.1	16.9	100.0
Average.............	....	....	44.8	20.5	0.4	65.7	34.3	
Range								
Minimum........	....	....	26.6	5.3	0.0	32.1	12.7	
Maximum........	....	....	63.9	30.0	1.4	87.3	67.9	

* Internal trips only—excluding truck drivers and passengers and taxi drivers.
† Taxi passengers—unreported or less than 0.1 per cent.

of all passenger trips, and in general the mass carrier is increasingly important as the population size of the city increases.

Trip Purpose. Origin-destination studies recording the purpose of each trip show that from 53.4 to 81.1 per cent of all trips are made in order to reach work or return home. Actual percentages in typical cities are set forth in Table 6-6. It should be noted that social and recreational uses of the highway vehicle amount to only 11.6 per cent in the typical city.

TABLE 6-6. TRIP PURPOSE

City	Approximate population	Date of survey	Per cent of total trips* to objective:										
			Home	Work	Social-recreational	Business	Shopping	Medical-dental	School	Serve passengers	Eat meal	Other	Total
St. Louis, Mo.	1,480,000	1945	48.2	32.9	6.6	3.6	5.7	1.4	0.9	0.5	0.2	....	100.0
Baltimore, Md.	912,900	1945	45.1	25.1	10.2	6.4	6.3	1.1	....	1.1	....	4.7	100.0
Seattle, Wash.	518,600	1946	42.3	22.9	12.9	6.2	7.9	1.1	2.0	3.5	1.2	....	100.0
Portland, Oreg.	422,700	1947	38.2	20.9	15.4	7.4	9.3	1.0	0.7	5.0	2.1	....	100.0
Indianapolis, Ind.	411,600	1945	46.2	22.8	11.2	7.2	8.1	1.5	0.6	1.7	0.7	....	100.0
Sacramento, Calif.	201,300	1947	38.2	25.6	12.0	4.2	8.2	1.0	5.5	3.4	1.4	0.5	100.0
Phoenix, Ariz.	161,600	1949	38.0	16.4	14.6	6.5	10.6	1.3	5.5	4.2	2.0	0.9	100.0
Tacoma, Wash.	138,700	1948	39.9	21.7	13.6	6.3	7.7	1.2	2.2	4.9	1.8	0.7	100.0
Spokane, Wash.	138,400	1946	46.4	19.6	11.5	7.1	7.7	1.5	2.5	2.4	1.3	....	100.0
Tucson, Ariz.	126,900	1949	37.4	16.0	13.9	7.9	10.7	1.2	4.1	6.4	2.1	0.3	100.0
Reading, Pa.	119,800	1946	41.3	28.3	8.3	6.5	5.2	1.2	2.3	3.2	3.1	0.6	100.0
South Bend, Ind.	119,400	1945	45.8	16.6	12.2	9.0	9.0	1.0	0.3	3.7	2.4	....	100.0
Harrisburg, Pa.	95,000	1946	41.7	26.7	13.3	3.8	6.6	1.1	0.2	3.0	3.2	0.4	100.0
Columbus, Ga.	79,200	1947	44.8	21.5	8.0	8.9	7.4	0.9	3.9	2.6	2.0	....	100.0
St. Joseph, Mo.	75,700	1945	46.4	27.9	9.6	....	4.1	....	2.2	1.4	....	8.4	100.0
Average	....	....	4.27	23.0	11.6	6.5	7.6	1.2	2.3	3.1	1.8	2.1	
Range													
Minimum	....	....	37.4	16.4	6.6	3.6	4.1	0.9	0.2	0.5	0.2	0.4	
Maximum	....	....	48.2	32.9	15.4	9.0	10.7	1.5	5.5	6.4	3.2	38.4	

* Includes only those trips made by persons accounted for in the *Internal Home Interview Survey*.

122

CHAPTER 7

STREAM CHARACTERISTICS

7-1. Elements of the Traffic Stream. Highway traffic returns to its origin: the best route for going from A to B is also the best for going from B to A. Thus a traffic stream will ordinarily have flow and counterflow along a common route. However, a traffic stream may be separated into a pair of one-way flows by design or regulation.

The Traffic Lane. The smallest practical transverse dimension of a roadway is one which permits single-file passage of vehicles with enough side clearance for a reasonably safe and satisfactory speed. A strip of roadway intended to accommodate a single file of moving vehicles is termed a *traffic lane* and is the basic unit of width in measuring the traffic stream. It has a lower limit of about 9 ft on low-speed urban routes at grade, but 11 ft or more is required on high-speed thoroughfares.

Types of Traffic Streams. Because of road design and marking, traffic regulation and enforcement, and the training and experience of road users, the traffic stream tends to become laminated into lanes of flow. Traffic streams may be classified by the number of lanes of movement. Figure 7-1 shows a graphical description of the common types of traffic streams classified by number of lanes, indicating the maneuverability of individual vehicles and representative spatial arrangements of vehicles which may develop as traffic movement approaches saturation.

Single-lane routes can develop only one way of flow at a time and do not permit overtaking and passing maneuvers, thus limiting speed of the individual unit to that of a slower-moving vehicle ahead. The two-lane route provides for either one-way or two-way operation. Under one-way operation, overtaking and passing are readily accommodated even up to high densities, at which point queues of vehicles begin to form in the right lane and then in the adjacent lane. Two-lane two-way movement inhibits overtaking and passing to a greater extent, and queues of vehicles form at much lower densities. On three- and four-lane roads "weaving" of vehicles occurs, unbalanced flow is possible, and preempting of the central lane for exclusive movement in the major flow direction may occur.

Because of the inherent range of speeds among the vehicles in the stream, overtaking and passing maneuverability is mandatory if queue

123

formation is to be avoided or reduced to a minimum.   The distribution of vehicles into lanes is of vital significance in reckoning the quality of transmission and the resulting capacity of a given route.

Modifying Factors.   There is a wide range of variability in road users and vehicles, and each driver and his vehicle is a discrete, independent

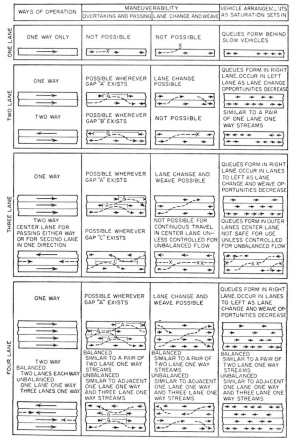

FIG. 7-1. Types of traffic streams.   "Keep to right" rule assumed.

unit with only limited coherence to other such units on the roadway. Traffic-stream performance is vitally affected by width, arrangement, and other geometrical design features of the roadway; condition of the road surface; the width, number, and separation of lanes; gradients; sight distances; frequency and form of intersections; drainage structures; roadway appurtenances; and even the aspect of the route.   Parking, streetcar tracks, and other extraneous factors also cause considerable variation in lane formation and vehicle distribution, especially on urban routes.   In addition, traffic-stream characteristics are affected by signs, signals, markings, traffic rules and regulations, pedestrian and other nonvehicular uses of the road, weather, lighting, and other environmental conditions. The traffic engineer must evaluate these modifying factors in order to develop, through control and design measures, a high degree of safety and efficiency in traffic-stream performance.

7-2. Transverse Distribution of Vehicles. *One Lane.* In a one-lane one-way stream, road users steer their vehicles so that there is a central tendency of their paths about some optimum point within the lane.

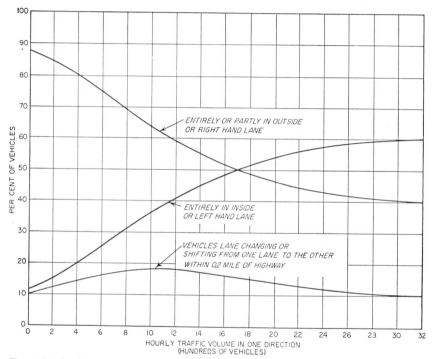

Fig. 7-2. Distribution of vehicles using lanes of a four-lane highway at various volumes. (From O. K. Norman, "Results of Highway Capacity Studies," U.S. Bureau of Public Roads, *Public Roads*, vol. 23, no. 4, June, 1942, p. 65.)

Two Lanes. The distribution of vehicles in the two-lane one-way stream is shown in Fig. 7-2. As total volume of traffic flow increases, the percentage of vehicles in the right-hand lane decreases from 88 to 40 per cent while increasing in the left lane from 12 to 60 per cent. When the total hourly volume is 1,700 vehicles, the distribution in each lane is equalized at 850 vph. Within a 0.2-mile section of highway, the number of vehicles straddling lanes or shifting from one lane to the other ranges from 10 to 20 per cent. At volumes above 1,700 vph, more vehicles pass in the left-hand lane than in the right. The slower vehicles will be found in the right-hand lane, while the faster vehicles will tend toward the left to overtake and pass the slower vehicles. The inside lane accordingly develops higher speeds and volumes, though the outer lane may appear to have the heavier flow.

Distribution in two-lane two-way traffic streams is entirely dependent on origin-destination demands. During a peak hour when the most unbalanced flows occur, there may be as many as two vehicles in one lane and direction to each vehicle in the lane of countermovement. The percentage of vehicles traveling in a given direction which can use the lane of counterflow to overtake and pass (on level tangent sections in rural areas)

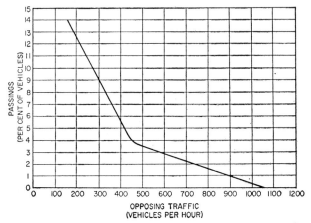

Fig. 7-3. Per cent of vehicles traveling in one direction performing passing maneuvers against various volumes of opposing traffic. Two-lane two-way tangent section. (From O. K. Norman, "Preliminary Results of Highway Capacity Studies," U.S. Bureau of Public Roads, *Public Roads*, vol. 19, no. 12, February, 1939, p. 240.)

decreases as the volume of vehicles per hour in the counterflow increases: at 1,050 vehicles per hour in the direction of counterflow, all vehicles in the given direction will be confined to single-lane operation (Fig. 7-3).

Three-lane Two-way. In the three-lane two-way traffic stream, vehicles are distributed as shown in Fig. 7-4 during those times when two-thirds of the traffic movement is in one direction. The center lane is not an efficient carrier of traffic. At its maximum density it accommodates less than one-sixth (15 per cent) of the total flow, and when the two directions of flow are more equally balanced it carries even lower percentages of the total stream. Since it is used for overtaking and passing, the central lane carries the faster-moving vehicles. While average speed decreases linearly in the outer lanes with an increase in volume, the center-lane average is uniformly high.

Four-lane Two-way. The four-lane two-way stream distributes the vehicles in each direction in accordance with O-D demand, as shown in Fig. 7-2. The severest condition of unbalance usually develops two vehicles in one direction to each vehicle in the counterdirection.

Multilane One-way. In the one-way stream of three or more lanes, the great majority of vehicles travel in the outermost lanes at low volumes. As volume increases, faster-moving vehicles tend toward the central lanes, and consequently the inner lanes carry more volume than the extreme right-hand lanes (Fig. 7-2). At high volumes all inner lanes carry about the same volume.

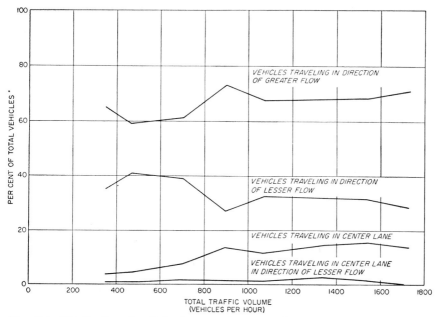

FIG. 7-4. Distribution of vehicles between lanes on a level tangent of a three-lane highway. (From O. K. Norman, "Results of Highway Capacity Studies," U.S. Bureau of Public Roads, *Public Roads*, vol. 23, no. 4, June, 1942, p. 68.)

7-3. Longitudinal Distribution of Vehicles.

The longitudinal arrangement of vehicles in the traffic stream gives the road user a sense of safety with freedom of movement, or a sense of hazard and congestion, and continuously affects his choice of the *speed* and *position* of his vehicle.

Time and Space Gaps in the Traffic Stream. Unoccupied road spaces, or gaps, ahead of each vehicle determine the longitudinal distribution of vehicles in any one-way stream. A given one-way stream may be composed of one or more lanes, each of which develops its own gap lengths. Gaps in the stream as a whole, regardless of transverse placement, also occur.

To compare one-way streams of different types, stream-gap values are employed. The length of gap between any two vehicles, measured from head to head of successive vehicles, may be stated in feet or any other linear unit. Linear measurement of gaps evaluates the number of vehicles along a stretch of road and is a direct measure of traffic density.

As ordinarily measured, each gap includes one vehicle length, and thus 1 mile of roadway includes gaps totaling 1 mile in length.

Since there is a time interval between the passage of vehicles going by a fixed point on the route, gaps may also be measured in seconds or any other time unit, called a *headway*. Headway is also measured from head to head of successive vehicles, so that each includes the time of passage of a vehicle. The number of headways in a given period of time is dependent on the rate of traffic flow and is, therefore, a direct measure of volume. Headway values are especially suited for stream analysis since operation, design, and planning work are based largely on volume data.

Frequency Distribution of Headways. Variation of gap length in a traffic stream ranges from a few feet to relatively great distance and is related to the over-all density at the time of observation. Similarly, there is a wide variation in headway values. The average headway may be computed as follows:

$$\text{Average headway (sec)} = \frac{3,600 \ (\text{sec/hr})}{\text{volume (hr)}}$$

In a one-way traffic stream of 400 vph, the average headway is 3,600/400 or 9 sec. It would seem reasonable to suppose that all headways in the stream would be symmetrically distributed about this 9-sec value, but this central tendency is not found. Figure 7-5 portrays the distribution of headways in a stream at 400 vph. In a two-lane two-way road, where except for overtaking and passing maneuvers, all movement in one direction is confined to one lane, the most frequently recurring headway value is between 1 and 2 sec. In a two-lane one-way stream, such as one direction of a four-lane road, the most frequently recurring headway for the stream as a whole is between 2 and 3 sec at 400 vph.

The cumulative curves in Fig. 7-5 show that 72 per cent of all headways are 9 sec or less for the two-lane road, and 64 per cent of the headways are 9 sec or less for the four-lane road at this volume. On the two-lane road 50 per cent of the drivers have headways less than 3 sec in length (less than a third of the average value). On the four-lane road 50 per cent of the drivers have a headway of 6 sec or less.

When 400 vph are required to travel in one lane, more crowding into smaller stream headway occurs than when they travel in two lanes, and interference from the vehicle ahead is more frequent.

Initial-interference Headways. If a road user is traveling without interference, his speed is not influenced by any other vehicle. In a given lane of travel the faster driver gradually shortens the space gap in front of him, continuing operation at his own speed level until it becomes necessary to reduce speed in order to avoid collision with the vehicle ahead. The rate at which headway is reduced is dependent on the relative speed of the vehicle to the vehicle ahead. At some minimum headway the relative speed must be reduced to zero.

As Fig. 7-6 shows, when the average absolute speed of the rear vehicle is 38 mph, there is an average relative speed of 6 to 7 mph between front and rear vehicles for all headways from 10 to 40 sec in length. At about

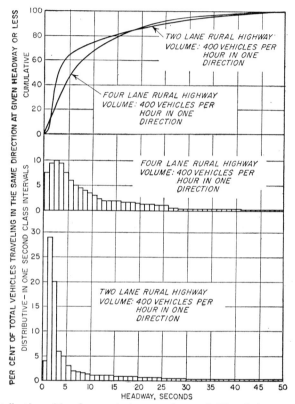

FIG. 7-5. Distribution of headways at stream volume of 400 vph (one way) on typical two-lane and four-lane highways. (Adapted from O. K. Norman, "Results of Highway Capacity Studies," U.S. Bureau of Public Roads, *Public Roads*, vol. 23, No. 4, June, 1942, p. 72, Figs. 17 and 18.)

9 sec headway, however, the absolute speed of the rear vehicle begins to fall rapidly to approach the speed of the vehicle ahead, and there is a marked drop in the relative speed of the first and trailing vehicle. From this it may be concluded that at 9 sec headway, initial interference is given to the average driver on rural roads relatively free of frequent cross traffic at grade and other influences associated with urban areas.

Minimum-lane Headways. As Fig. 7-5 shows, road users do not demand a 9-sec headway at all times. Shorter headways in the traffic stream may be determined (1) solely by the proximity of the vehicles

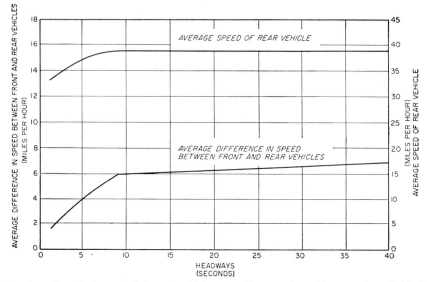

FIG. 7-6. Speed characteristics of vehicles traveling at given time spacings behind preceding vehicles. (From O. K. Norman, "Preliminary Results of Highway Capacity Studies," U.S. Bureau of Public Roads, *Public Roads*, vol. 19, no. 12, February, 1939, p. 228.)

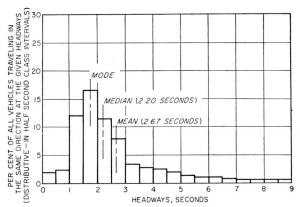

FIG. 7-7. Distribution of short headways (9 sec or less) on a typical two-lane rural highway. Volume is 400 vph in one direction. (Adapted from O. K. Norman, "Results of Highway Capacity Studies," U.S. Bureau of Public Roads, *Public Roads*, vol. 23, no. 4, June, 1942, p. 72, Figs. 17 and 18.)

ahead and the judgment of the driver in the rear vehicle or (2) solely by the ability of the rear vehicle and its driver to keep up with the vehicle ahead. The first case is significant when traffic is moving with a high degree of continuity, as on a rural road, the second when a queue of vehicles which is standing or moving very slowly is suddenly released, as on the departure with a green signal from a signalized intersection.

Case I. Continuous Movement. Figure 7-7 shows the distribution, for a two-lane rural highway carrying 400 vph in each direction, of lane headways shorter than the 9-sec initial-interference headway. For this rate of flow, which is not unusually heavy, the average headway is 9 sec, but 72 per cent of the drivers travel with lesser headways.

Only 4 per cent have headways less than 1 sec. The most commonly found headway is between 1½ and 2 sec and accommodates nearly one driver out of every five. The median value of all those with less than 9 sec headway is between 2 and 2½ sec, and the mean is nearly 3 sec.

Case II. Proceeding from a Stop. Average minimum headways which develop when a queue of vehicles is suddenly released is set forth in Table 7-1. These values are for passenger-car performance only and increase

TABLE 7-1. AVERAGE MINIMUM HEADWAY FROM QUEUE FORMATION

Vehicle position in queue	Average minimum headway, sec
1st	3.8
2d	3.1
3d	2.7
4th	2.4
5th	2.2
6th	2.1

SOURCE: B. D. Greenshields et al., *Traffic Performance at Urban Street Intersections*, Yale University, Bureau of Highway Traffic, Technical Report 1, New Haven, Conn., 1947, p. 27.

by about 50 per cent for the ordinary bus or truck because of sluggishness in starting. Each headway value includes time for the driver to respond to the signal or vehicle ahead as well as to get the vehicle under way. Thus headway values are gradually reduced until about the fifth or sixth vehicle in line, when a constant of 2.1 sec is developed.

Relationship of Minimum Headways and Speed. The relationships of minimum headways and gaps to the speed of the traffic stream are set forth in Fig. 7-8 for both rural and urban conditions. The rural values are the average for all headways less than 4 sec in length for different highways. The urban values illustrate those found on two city streets where it seemed apparent that the speed of the rear vehicle was controlled by the lead vehicle and the relative speed of the pair was less than 1 mph.

Under rural conditions there is a small range of headway values, averaging 2 sec, throughout speeds of 30 to 60 mph. The lowest values are found in the left lane of four-lane roads at speeds of about 38 mph when the average minimum headway is nearly 1.7 sec. The highest values occur at night on two-lane roads and at higher speeds and amount to as much as 2.5 sec.

All headway values tend to increase at speeds below 30 mph, particularly below 20 mph.

Because of the relatively low speeds on urban streets, the range of average minimum headways is inherently greater than on rural roads.

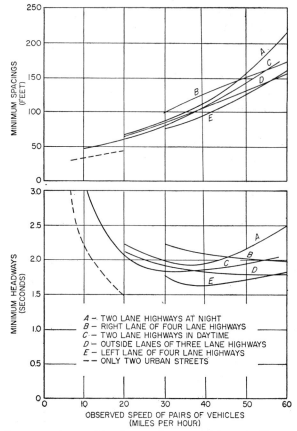

FIG. 7-8. Minimum spacings and headways of average drivers trailing another vehicle, on various types of highways at various speeds. (From O. K. Norman, "Results of Highway Capacity Studies," U.S. Bureau of Public Roads, *Public Roads*, vol. 23, no. 4, June, 1942, p. 58; and B. D. Greenshields et al., "Traffic Performance at Urban Street Intersections," Yale University Bureau of Highway Traffic, Technical Report 1, New Haven, Conn., 1947, p. 31.)

It has been found in urban traffic that the minimum *intervehicular headway* (from the front of the second vehicle to the rear of the lead vehicle) is relatively constant at 1.1 sec for speeds from 6 to 22 mph.[1] To obtain true headway values the time required for the lead vehicle to traverse

[1] B. D. Greenshields et al., *Traffic Performance at Urban Street Intersections*, Yale University, Bureau of Highway Traffic, Technical Report 1, New Haven, Conn., 1947, p. 31.

its own length must be added to this 1.1-sec time gap. If the lead vehicle is 18 ft long and the speed of movement is 12 mph, the headway is about 2.12 sec. On urban streets average minimum headway values range from 1.3 to over 4 sec.

Space gaps are relatively constant at lower speeds but increase rapidly with higher values. For a given headway, space gap varies directly as speed: for a given space gap, headway varies inversely as speed. Thus

$$\text{Headway (sec)} = \frac{\text{space gap (ft)}}{1.47 \times \text{speed (mph)}}$$

7-4. Natural Distribution of the Traffic Stream. Knowledge of the natural longitudinal distribution of vehicles in the traffic stream can be most useful in predicting the characteristics of flow and stream action.

In defining the distribution, two significant observations may be made: (1) Each individual positions his vehicle independently of others except when his headway is very small, and thus his position at any time is ordinarily independent of all other vehicles. (2) At any given volume, within physical limitations, the number of vehicles passing a point in a given length of time is independent of the number passing in any other equal length of time. These two facts are essentially the statistical requirements for "randomness"[1] of a given set of events and suggest that the natural vehicle distribution may be one directly related to the basic laws of probability.

Examination of field data reveals that the distribution of headways does not display a central tendency about the mean. Instead, the majority of drivers travel at lower-than-average headways, while a few extend the range to many times the mean. These phenomena suggest that the distribution may follow the particular expression derived from probability laws known as the Poisson distribution. Applied to traffic this distribution may be expressed as follows:

$$p(x) = \frac{e^{-m}m^x}{x!}$$

where $p(x)$ = probability of the arrival of x vehicles at a point during a given length of time

m = mean number of vehicles arriving in the given length of time = $tv/3,600$

t = given time length of gap (sec)

v = volume, vph

e = base of Naperian logarithms = 2.71828

To relate this distribution of vehicle arrivals to headway distribution the following logic may be followed:

[1] T. E. Fry, *Probability and Its Engineering Uses*, D. Van Nostrand Company, Inc. New York, 1928, pp. 216–220.

(1) A given headway length is equal to a period of time during which no vehicle arrives; (2) therefore for a given volume the Poisson distribution may be solved for the probability that no vehicles arrive in a given length of time:

$$p(o) = \frac{e^{-m}m^o}{o!} = e^{-m}$$

and (3) because no vehicle can arrive in a lesser time, if none arrive in the time t upon which m is based, the calculated probability may be interpreted as the per cent of all headways which will be at least as long as the selected time. Conversely it is the per cent of all headways greater than the selected time.

For example, it has been shown empirically that at 400 vph 36 per cent of all headways on a four-lane road and 28 per cent on a two-lane road are greater than 9 sec. The derived equation would produce the following results:

$$m = \frac{vt}{3,600} = \frac{400 \times 9}{3,600} = 1$$
$$p(o) = e^{-1} = 0.368$$

Thus the equation predicts that 36.8 per cent of all headways will be greater than 9 sec at 400 vph. Close agreement on the free-flowing multilane road, but relatively poor agreement on the two-lane road which has greater intervehicular interference, is noted.

Actual distributions of observed headways at various volumes is shown in Fig. 7-9. The percentage of vehicles traveling at several selected headways as computed by use of the Poisson law are also shown for purposes of comparison. Throughout all ranges of volume, the observed headways for the stream as a whole on the four-lane road are nearly identical with that which may be expected in purely random distribution, the only significant difference being in the 1- and 2-sec-headway classes, where observed numbers are lower than computed and significantly demonstrate that some individual drivers demand, and are able to keep, more than 1- or 2-sec headways.

For all larger headways drivers' choice is not a significant factor, because the observed headway distribution at volumes up to 1,800 vph on four-lane roads occurs in accordance with the laws of chance. Computed values for all headways above 3 sec are practically congruent with observed values, but some 2 per cent more vehicles are crowded into 3-sec or less headways above a volume of 500 vph.

The longitudinal distribution of vehicles on a single lane is the same as for a two-lane one-way flow, when the volume in the single lane of flow is equivalent to the volume in the two lanes of flow. Thus the curves in Fig. 7-9 may be employed for single-lane or stream headways, provided the volume figure used is that which holds for the single lane or the two lanes which form the stream as a whole.

In the two-lane road, however, significant departures from random distribution are found. The refusal of some drivers to accept 1-sec or less

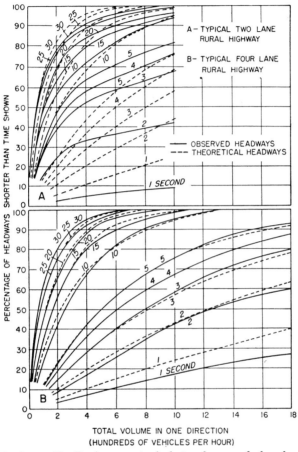

FIG. 7-9. Headway distribution on typical two-lane and four-lane highways. (Adapted from O. K. Norman, "Results of Highway Capacity Studies," U.S. Bureau of Public Roads, *Public Roads*, vol. 23, no. 4, June, 1942, p. 72, Figs. 17 and 18.)

headways is again demonstrated. More drivers are forced into 2-sec or less headways than would be found by chance up to 900 vph. Even greater percentages of drivers are forced into the 3-sec or less headway classes. Throughout all ranges of volume, especially at lower values, two-lane two-way operation causes undue numbers of vehicles to be concentrated into the short-headway classes. There is also an excessive number of lengthy headways in the two-lane two-way stream. The observed numbers of headways longer than 30 sec at all volumes above

150 vph are greater than those which would be expected in a random distribution. Thus the two-lane two-way stream tends to concentrate vehicles into short-headway queues separated by long headways, even at relatively low volumes. Because of inherent differences in speed of individual vehicles and the inadequate opportunity to change lanes or overtake and pass, the two-lane two-way road lacks the ability to produce a fluid traffic stream.

7-5. Lane Changes. The transfer of a vehicle from one lane to the next adjacent lane is defined as a lane change.[1] When the headway in the initial lane of a lane-changing vehicle is rapidly approaching zero because of the proximity of a slower-moving vehicle ahead, the lane change is *forced*. Any other lane change is classed as *optional*.

The gap between the vehicles in the adjacent lane which must accommodate the lane-change vehicle may range from one which is (1) so short that it is not acceptable to the lane-change driver, to (2) an acceptable gap to the lane-change driver but so short that both the lead and trailing vehicles influence his maneuver, to (3) so great that only the lead vehicle or only the rear vehicle or neither vehicle significantly influences the lane-change driver. When the lane change is made under condition (2), it is termed a *gap* lane change. When made under condition (3), if influenced by lead vehicle only it is termed a *retarded* lane change, if influenced by the rear vehicle only a *conflict* lane change, and if influenced by neither vehicle a *free* lane change.

Length and Duration. When average speed is 30 to 40 mph, the average length of roadway required to accomplish lane change varies from about 125 to 260 ft, depending on the type of lane change made. The average optional lane change requires nearly 220 ft, and the average forced lane change about 140 ft, at this speed (Table 7-2). Range limits for both types are 100 to 450 ft.[2]

The average time required for lane change, as shown in Table 7-2, varies from $2\frac{1}{2}$ to $4\frac{1}{2}$ sec. The time for optional-type lane changes is rather constant, about 4 sec, for normal ranges of speed. Lane changes which require greater acceleration require greater time with increase in speed and vary from $2\frac{1}{2}$ to $4\frac{1}{2}$ sec in duration.[3] The distance traveled

----

[1] The term *lane change* as employed by the authors is synonymous with the term *weave* as employed by F. H. Wynn et al., *Studies of Weaving and Merging Traffic: A Symposium*, Yale University, Bureau of Highway Traffic, Technical Bulletin 4, New Haven, Conn., 1948. Weaving is here reserved for use in describing oblique stream crossings which are required, as at traffic circles and other junctions, when (1) two separate streams flowing in nearly the same direction merge into a common stream, (2) vehicles from one street are required to cross completely the path of vehicles from the other stream, (3) this common stream then diverges into separate streams again, and (4) the limits wherein this merging, oblique crossing, and diverging take place are established by the design of the junction and are as short as practicable.

[2] *Ibid.*, p. 18.

[3] T. M. Matson and T. W. Forbes, "Overtaking and Passing Requirements," *Proceedings of Highway Research Board*, Washington, 1938, vol. 18, p. 110.

during a lane change may be taken as six times the velocity in miles per hour.

Lane-change Frequency. The greater the range of speed of vehicles in the traffic stream, the greater the frequency of demand for lane change. As volume increases, frequency of demand for lane change also increases, but the opportunity for lane change decreases at a rate which varies in

TABLE 7-2. AVERAGE LENGTH, DURATION, AND SPEEDS OF VARIOUS TYPES OF LANE-CHANGE MANEUVERS

Type of lane change	Length, ft	Duration, sec	Speed, mph	
			Start	End
Optional				
Free..............	221.9	4.1	35.8	36.0
Retard............	167.4	3.3	33.3	33.3
Conflict..........	260.7	4.3	40.6	41.0
Gap..............	217.7	3.9	37.1	37.1
Average.........	217.5	4.0	36.1	36.2
Forced				
Free..............	131.9	2.7	34.4	31.2
Retard............	123.5	2.6	33.4	32.7
Conflict..........	168.6	2.9	39.3	43.7
Gap..............	145.2	3.0	34.9	32.1
Average.........	136.8	2.8	34.8	33.2

SOURCE: F. H. Wynn et al., *Studies of Weaving and Merging Traffic*, Yale University, Bureau of Highway Traffic, Technical Report 4, New Haven, Conn., 1948, Tables I, II, III.

accordance with the minimum gap size required and actual distribution of gaps in excess of such minimum.

Required Gap Length. Since most drivers require a minimum headway of about $1\frac{1}{2}$ sec, the minimum gap into which a lane-change vehicle should enter would appear to be 3 sec, allowing $1\frac{1}{2}$-sec headways for each vehicle after completion of the maneuver. However, even optional lane changes are made into gaps of as low as 1-sec headway and average about 2 sec. Forced lane changes are made into gaps averaging $1\frac{1}{2}$ sec, so that at the end of the maneuver the lane-change vehicle and that immediately trailing it have headways of less than 1 sec.

In a two-lane one-way stream of traffic, as volume increases, the longitudinal distribution of vehicles in a given lane (A) will vary in accordance with the laws of randomness, and the number and size of gaps will depend on the volume of flow in such lanes. The position of any vehicle in an adjacent lane (B) relative to the vehicles or gaps in lane A is a matter of chance. The probability of any vehicle in lane B being opposite a gap of long-enough headway to permit a lane change is equal to the percentage of total time contained in all headways of equal or greater length in lane A at its existing volume.

Assume that a four-lane road is carrying 1,000 vph in one direction. As Fig. 7-2 shows, 360 vehicles will be found in the inside lane (A) and 640 in the outside lane (B). At a volume of 360 vph, 84 per cent of all headways are greater than 2 sec. If 2 sec is the minimum acceptable gap into which a vehicle from lane B may change into lane A, 84 per cent of all gaps in lane A could accommodate a lane-change vehicle from lane B.

Lane-change Opportunity. Being opposite a gap of sufficient length presents lane-change opportunity only if the lane-change vehicle is in the proper position relative to the gap. If a gap in lane A is just longer than 2 sec and drivers require 1 sec ahead and behind in making a lane change, there is only one single position for the vehicle in lane B with respect to the gap in lane A that provides for a lane change. If the gap is 15 sec, the vehicle in lane B may change if it is at any position relative to this gap from 1 sec behind the front vehicle to 1 sec ahead of the rear vehicle. The per cent of total time during which a vehicle in lane B may lane-change, then, is equal to the amount of time in lane A which consists of gaps 2 sec and longer reduced by a clearance time of 2 sec for each gap.

In a stream of 1,000 vph with 360 vehicles in lane A, the per cent of time during which vehicles in lane B may not change is computed as follows:

$$\text{Gaps 0 to 1 sec} = 7\% \times 360 \qquad\qquad = 25$$
$$\text{Gaps 1 to 2 sec} = (16\% - 7\%) \times 360 = \underline{32}$$
$$57 \text{ gaps}$$

$$\text{Time spent in gaps 0 to 1 sec} = 25 \times 0.5 = 12.5 \text{ sec}$$
$$\text{Time spent in gaps 1 to 2 sec} = \underline{32 \times 1.5 = 48.0 \text{ sec}}$$
$$60.5 \text{ sec}$$

Since there are 57 gaps of less than 2 sec, there are 303 gaps long enough for lane change, but an additional 606 sec (2 sec from each gap) are lost owing to time for clearance. In 1 hr, then, there are 3,600 − 606 − 61, or 2,933 sec (81.5 per cent of the total time) during which a vehicle may lane-change from lane B to lane A. Conversely, lane change is prevented from being initiated 18.5 per cent of the total time.

Similar computations for several selected volumes are shown below:

Total one-way-stream volume	Per cent of total time that lane-change opportunity exists (from outside to inside lane only)
500	94.0
1,000	81.5
1,500	66.0
2,000	53.0

Delay to Lane Change. The length of delay to driver in lane B who cannot immediately make a lane change is dependent on his speed and the speed and volume in lane A. The higher the relative speed and the lower the lane A volume, the sooner he will overtake or be overtaken by a gap of sufficient length to accommodate a lane change.

The average length of headways and portions of headways of insufficient length for a lane change, which may be considered as general blockades moving in the stream, may be computed as follows:

Av. blockade length

$$= \frac{\text{total time in 1 hr of headways too short} \times \text{stream speed}}{\text{number of gaps in 1 hr of sufficient length}}$$

Division of this blockade length by the relative speed (of the lane-change vehicle and the blockade) determines potential total delay time of the blockade. Finally, since a delayed vehicle is as likely to be at the head as at the tail of such a blockade at the moment of desired lane change, the total delay time must be divided by 2 for average delay time. The final equation is as follows:

$$\text{Av. delay} = \frac{1}{2} \times \frac{\text{average blockade length}}{\text{relative speed of lanes A and B}}$$

Assume that blocking vehicles in lane A are traveling 50 mph in a 1,000-vph two-lane one-way stream. A driver in lane B is moving closely behind a vehicle in his lane at 40 mph and wishes to make a lane change. His average waiting time may be computed as follows:

$$D = \frac{1}{2} \frac{(3,600 \times 18.5\%)50}{303 \times (50 - 40)} = 5\frac{1}{2} \text{ sec}$$

Similar calculations for the same speed values are shown below:

Total volume of stream, vph	Volume in inner lane, vph	Average delay, sec
500	115	4.9
1,000	360	5.5
1,500	708	6.3
2,000	1,060	7.3

The above basis of computing delay assumes uniform blocking periods and uniform distribution of delayed vehicles in such blocking periods. Because of the randomness of the traffic stream, it seems likely that blocking periods, as well as the distribution of lane-changing vehicles with respect to such blocking, actually would follow the laws of randomness. If so, far more than half of the delayed vehicles will be delayed for shorter-than-average times, while a very few will be delayed for extremely long periods of time.

Actual Lane-change Frequency.   Figure 7-2 shows the actual percentage of vehicles making lane change in a two-lane one-way stream.   This percentage increases gradually with volume, until the rate of flow is approximately 1,000 vph.   Beyond this volume, the lack of opportunity to lane-change causes a decrease in the percentage of lane-change movement, and queue formation is engendered.

7-6. Overtaking and Passing in the Two-lane Two-way Stream. Overtaking and passing in the two-lane two-way stream is of special significance in the design and operation of two-lane highways because of the sight-distance requirement of the maneuver.

The overtaking and passing maneuver is made up of two lane changes, usually in rapid succession.   In the simplest form of the maneuver, the first lane change places the overtaking vehicle in the lane of counterflow, and the second lane change returns the overtaking vehicle to its initial lane of movement.   The time-space relationship between the *overtaking* (lane-change) vehicle and the *overtaken* vehicle, and generally a third *oncoming* vehicle, must be reckoned with.   These relationships impose complex questions of vehicle operation which must be rapidly and smoothly decided by the driver of the overtaking vehicle.

Types of Overtaking and Passing Maneuvers.   Overtaking and passing maneuvers may be classified as *simple* or *multiple*.   Multiple maneuvers which involve more than one overtaking or overtaken vehicle account for 20 to 40 per cent of all overtaking and passing maneuvers in the two-lane two-way rural highway when volumes are between 200 and 800 vph. Since multiple maneuvers occur in almost endless variety, only the simple types have been measured.   They are classified into four types based on the conditions affecting the overtaking vehicle: (1) flying start—voluntary return, (2) flying start—forced return, (3) accelerative start—voluntary return, (4) accelerative start—forced return.

Length and Duration.   There is a wide range of variation found in the distances and times required to make an overtaking and passing maneuver.   The distance required increases nearly linearly with the speed of the overtaken vehicle.   All simple types of the maneuver range in distance from about 380 to 950 ft when the speed of the overtaken vehicle is 20 and 50 mph respectively.   The average overtaking and passing distance in feet is roughly 19 times the speed of the overtaken vehicle in miles per hour.   Voluntary-return types are longer, whereas forced-return and accelerative types are shorter than average (Fig. 7-10).   Time required ranges from about 8 to nearly 12 sec (Table 7-3).

Headways Required in Counterflow.   Table 7-3 shows the length of time the overtaking vehicle requires free road space in the counterflow lane.

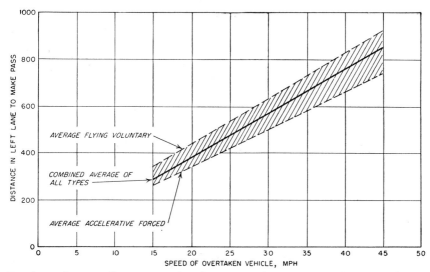

Fɪɢ. 7-10. Average distances required in left lane for passing vehicles traveling at various speeds. (Adapted from data presented by C. W. Prisk, "Passing Practices on Rural Highways," *Proceedings of Highway Research Board*, Washington, 1941, vol. 21, pp. 366–378; and T. M. Matson and T. W. Forbes, "Overtaking and Passing Requirements as Determined from a Moving Vehicle," *Proceedings of Highway Research Board*, Washington, 1938, vol. 18, pp. 100–112.)

Minimum headway in the lane of counterflow must be reckoned on the basis of the relative speed between overtaking and oncoming vehicle. If T sec are allowed for an overtaking vehicle traveling at V_1 mph and the oncoming vehicle is traveling V_2 mph, the minimum headway H_c in the counterflow may be calculated as follows:

$$H_c V_2 = T(V_1 + V_2) \qquad \text{or} \qquad H_c = T\frac{(V_1 + V_2)}{V_2}$$

Tᴀʙʟᴇ 7-3. Dᴜʀᴀᴛɪᴏɴ ᴏꜰ Vᴀʀɪᴏᴜs Tʏᴘᴇs ᴏꜰ Oᴠᴇʀᴛᴀᴋɪɴɢ ᴀɴᴅ
Pᴀssɪɴɢ Mᴀɴᴇᴜᴠᴇʀs

Type of maneuver		Duration of maneuver in seconds for various speeds of passed vehicles, mph					
Start	Return	0–19	20–29	30–39	40–49	50–59	Average all speeds
Flying................	Voluntary	10.0	9.9	11.0	11.8	9.6	10.8
Accelerative	Voluntary	8.7	8.8	9.8	10.9	10.5	9.7
Flying................	Forced	8.1	8.9	9.8	11.8	9.3	9.8
Accelerative...........	Forced	7.7	8.0	8.8	9.4	8.4	8.6
Average, all types	....	8.6	8.8	9.8	10.9	9.9	9.6

Sᴏᴜʀᴄᴇ: Adapted from C. W. Prisk, "Passing Practices on Rural Highways" *Proceedings of Highway Research Board*, Washington, 1941, vol. 21, p. 371.

Assume that an overtaking vehicle in a two-lane two-way road carrying 400 vph in each direction is traveling at 60 mph. A vehicle in the counterflow lane is traveling 40 mph. If the time required to make the overtaking and passing maneuver is 10 sec, the required headway in the counterflow lane may be determined as follows:

$$H_c = 10 \frac{(60 + 40)}{40} = 25 \text{ sec}$$

Frequency of Overtaking and Passing. The greater the range in speed, the greater the need for overtaking and passing. The volume of flow in a given direction determines the frequency of need to overtake and pass, while at the same time the volume of flow in the counterdirection limits the opportunity to make the maneuver.

Effect of Volume on Adequate Headway. The size of gap required and the relative rates of speed are fundamental in determining the per cent of gaps which permit overtaking and passing. The time needed to complete the maneuver is relatively constant, but the minimum gap required in the counterflow varies with the relative speed of overtaking and oncoming vehicles. Figure 7-9 shows the frequency of gaps of sufficient size (25 sec in the example above) occurring in a two-lane two-way stream carrying 400 vph. Note that only 10 per cent of all the gaps in the counterflow are 25 sec or greater in length.

Effect of Volume and Speed on Overtaking and Passing Opportunity. The per cent of time when overtaking and passing is possible may be computed by deducting from the total time (1) all time in the counterflow composed of headways less than 25 sec, (2) all time when the overtaking vehicle is less than 25 sec from any oncoming vehicle in those headways of the oncoming flow which are greater than 25 sec. Using the curves shown in Fig. 7-9 at a volume of 400 vph, an approximation may be made as follows:

Headway range, sec	(Per cent headways in each range)	X Volume, vph	= No. of headways in range	X Average headway, sec	= Duration, sec
0–1	(4–0)	400	16	0.5	8
1–2	(33–4)	400	116	1.5	174
2–3	(52–33)	400	76	2.5	190
3–4	(58–52)	400	24	3.5	84
4–5	(63–58)	400	20	4.5	90
5–10	(74–63)	400	44	7.5	330
10–15	(80–74)	400	24	12.7	300
15–20	(86–80)	400	24	17.5	420
20–25	(89–86)	400	12	22.5	270
Total			356		1,866

Of the total 400 gaps, then, only 44 are greater than 25 sec duration.

Since no maneuver may be started less than 25 sec from the oncoming vehicle, there is a loss of 1,100 additional sec (25 for each gap) during which overtaking and passing is prevented. Accordingly, there remain only 634 sec (3,600 − 1,866 − 1,100), or 17.5 per cent of the hour, when the maneuver may be initiated.

Similar computations for various required headways are indicated below:

PER CENT OF TIME THAT IMMEDIATE OPPORTUNITY TO OVERTAKE AND PASS EXISTS*

Required headway in opposing stream, sec	Volume in opposing stream, vph			
	200	400	600	800
15	55.4	34.0	19.5	8.0
20	46.5	24.5	11.0	2.0
25	39.0	17.5	5.5	1.0
30	33.0	12.5	2.5	0.5

* More accurate, and slightly higher, values can be found by use of smaller increments of observed headway in computation.

Effect of Volume and Speeds on Delays to Overtaking and Passing. The length of time a driver in the two-lane two-way stream must wait to make an overtaking and passing maneuver, if he is opposite headway conditions in the counterflow lane which temporarily prevent it, may be computed as follows:

$$\text{Average delay} = \frac{1}{2} \frac{(3,600 \times \text{per cent of blocking time}) \, S_2}{(\text{no. gaps} > \text{minimum}) \, (S_1 + S_2)}$$
$$= \frac{1}{2} \left[\frac{3,600 \times 82.5\%)40}{44(50 + 40)} \right] = 15.0 \text{ sec}$$

where S_1 = overtaken speed
S_2 = oncoming speed

Similar computations for other headway and volume conditions are shown below:

Assumed speeds			Required headway in opposing stream, sec.	Volume in opposing stream, vph Average delay, sec.			
On-coming	Over-taking	Over-taken		200	400	600	800
60	30	20	15	8.6	11.1	15.1	19.0
50	50	40	20	9.2	13.5	18.6	30.7
40	60	50	25	10.1	15.0	25.1	52.0
30	60	50	30	10.8	18.5	36.7	84.0

Required Frequency of Overtaking and Passing Maneuver. The number of overtaking and passing maneuvers required for each vehicle to maintain its speed may be computed from the free-speed distribution curve found on such a route as follows:

Let N_F = number of fast vehicles going by a point in 1 hr
   N_S = number of slow vehicles going by a point in 1 hr
   V_F = speed of fast vehicles, mph
   V_S = speed of slow vehicles, mph

Then in 1 hr:

Distance traveled by a fast vehicle = V_F miles
Distance traveled by a slow vehicle = F_S miles
         Difference = $(V_F - V_S)$ = additional number of miles fast vehicles go in 1 hr

In this distance there will be on the average $[(V_F - V_S)N_S]/V_s$ slow vehicles, all of which will be passed by a fast vehicle. Since there are N_F fast vehicles, each of which will pass the above number of slow vehicles in this distance, the total number of passings in the hour = $[N_F N_S(V_F - V_S)]/V_S$. Since in an hour these fast vehicles will have gone V_F miles, there will be $[N_F N_S(V_F - V_S)]/V_F V_S$ required overtaking and passing maneuvers per mile in 1 hr. By taking small percentile increments of speed throughout the actual speed-distribution curve, the total number of overtaking and passing maneuvers required per hour may be approximated.

The number of overtaking and passing maneuvers required to maintain the desired speed of each driver in a typical two-lane two-way tangent are set forth in Fig. 7-11.

Actual Frequency of Overtaking and Passing. Figure 7-11 also shows the actual average number of overtaking and passing maneuvers which have been observed in two-lane two-way streams. The total number of maneuvers rises less rapidly than the number required to maintain desired speed and reaches a maximum at about 1,300 vph, but the number per vehicle reaches its maximum at about 800 vph. At about 2,000 vph, overtaking and passing is entirely prevented because of lack of sufficient headway in the counterflow. It is also noted that although theoretically all required maneuvers could be accomplished as volumes approach zero, the data showed only about 90 per cent being made.

7-7. Relative Speeds and Capacities. The capacity of any stream is inescapably limited by the decreased opportunity for lane change or overtaking and passing coupled with an increased need for such maneuvers as volume increases. The retardation of these maneuvers due to absence of sufficient gaps results in modification of the natural free-flowing speed range. The speed of every vehicle which would tend to be faster than other vehicles is decreased. The decrease in speed, while more severe in the higher percentiles of the speed distribution, varies linearly with increase in volume for any given percentile group. The 90-percentile class falls back from about 58 mph when free-moving to about 34 mph when the volume is 1,800 vph; the decrease in speed is uniform at nearly 8 mph for each 600 vph increment of volume. The decrease in

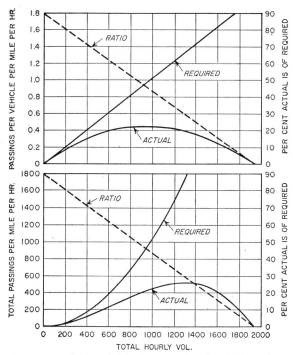

Fig. 7-11. Passings required to maintain free speed and actual passings noted on two-lane highways. (From O. K. Norman, "Results of Highway Capacity Studies," U.S. Bureau of Public Roads, *Public Roads*, vol. 23, no. 4, June, 1942, p. 70, Figs. 15 and 16.)

speed brings the relative speed of all vehicles to lower and lower values until, finally, passing maneuvers are impossible and queue formation is engendered.

The characteristic manner in which the relative speed of vehicles decreases linearly with volume is shown in Fig. 7-12. When relative speed becomes zero, every vehicle is found in a queue with no freedom to lane-change or overtake and pass. This situation occurs in the two-lane two-way stream at 1,500 to 2,000 vph, in the three-lane two-way stream at 2,800 to 3,600 vph, and in the two-lane one-way stream at 3,700 to 4,200 or more vph. Accordingly, the volume of traffic which may be passed increases with speed but reaches a maximum of about 2,000 vehicles per lane per hour at a speed of 35 to 40 mph (Fig. 7-13).

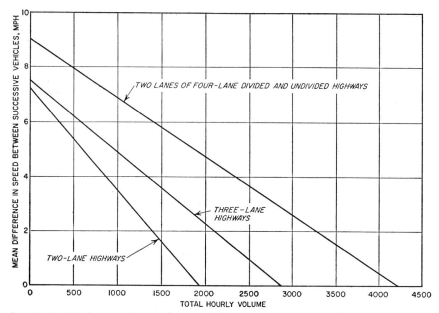

FIG. 7-12. Relative speeds of pairs of vehicles at various hourly volumes. (Adapted from O. K. Norman, "Results of Highway Capacity Studies," U.S. Bureau of Public Roads, *Public Roads*, vol. 23, no. 4, June, 1942, pp. 61–63, Figs. 6 and 8.)

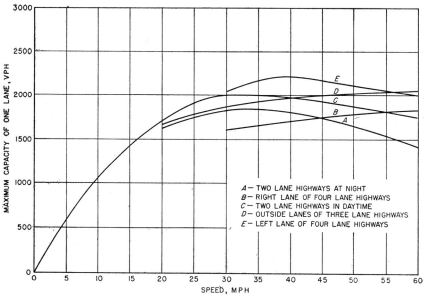

FIG. 7-13. Theoretical maximum capacity of various roads carrying vehicles at uniform speeds and average minimum headways. (From O. K. Norman, "Results of Highway Capacity Studies," U.S. Bureau of Public Roads, *Public Roads*, vol. 23, no. 4, June, 1942, p. 59.)

In order to accommodate reasonable amount of opportunity for overtaking and passing or for lane-change purposes, the "practical" capacities for uninterrupted flows under ideal rural traffic and roadway conditions have been determined as follows:

Two-lane two-way highway (total both lanes)......... 900 vph
Three-lane two-way highway (total all lanes)......... 1,500 vph
Multilane highway (average per lane)................ 1,000 vph

Traffic stream operating at these volumes permits each individual road user to maintain his speed a sufficient percentage of the time so as to be "practicable." Road design, traffic regulation, and other features will affect these values.

CHAPTER 8

INTERSECTION CHARACTERISTICS

Intersections are of profound importance to the traffic engineer because of their influence on the movement and safety of traffic flow. The *place* of intersection is determined by design, and the *act* of intersection is modified by regulation and control. While in the vicinity of an intersection, road users must coadjust their performance by reduced speed or change of path to avoid collision with each other. In the area of intersection the individual road user may (1) transfer from the route on which he has been traveling to another route of different bearing or (2) cross the lanes of any extraneous stream which flows between him and his destination. In the crossing function there is no inherent requirement for interflow relationship; any interference which may be encountered is merely an annoyance or obstacle to the individual road user. In the transfer function, however, interflow relationship is vital, because vehicles in one flow must become associated with vehicles in another flow.

8-1. Elements of Intersection Operation. As a road user transfers from one flow to another or continues in his original flow, he will find it necessary to (1) *diverge* from, (2) *merge* with, or (3) intercept and *cross* the paths of other road users.

Diverging. Diverging to the right or left of the original flow is elemental in every transfer made at an intersection. One variation of the diverging maneuver occurs when there is mutual departure by all road users from the entering direction. Multiple diverging paths may be developed as shown in Fig. 8-1.

Merging. Once a road user diverges from a stream, he may remain in the new flow thus created or merge with another stream. Merging may occur from the right, from the left, mutually, or in multiple as shown in Fig. 8-1.

Crossing. Each road user who passes through an intersection, or transfers by left turn to another flow at an intersection, must cross the paths of vehicles in all intervening flows between his points of entry to and exit from the intersection. Crossing may be made from the right or left of the intervening flow and may follow a diverging maneuver as a portion of a left-turn transfer. If the angle between the path of the road

148

user and the intervening flow is slight (considerably less than 90°), it is an opposed crossing.

Under some conditions the road user making an oblique crossing must bend his path to conform with the path of the intervening flow, in which case the maneuver becomes a weave. This weave might be interpreted as a merging maneuver followed by a diverging maneuver. Typical

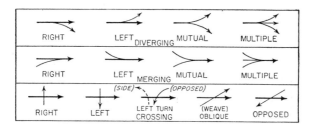

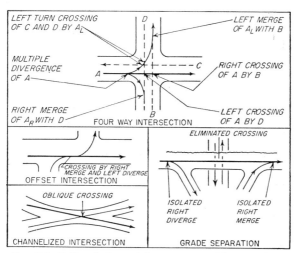

Fig. 8-1. Paths of elemental maneuvers at intersections.

occurrence of elemental crossing maneuvers is set forth graphically in Fig. 8-1.

Conflicts. Whenever diverging, merging, or crossing occurs, potential, if not actual, *conflict* between two or more road users arises. It may concern only the two road users whose paths are joined, crossed, or divided or may be reflected back in each of the flows approaching the area of potential collision. *Conflict areas* may be defined as those areas which include not only the *area of potential or actual collision* but also the *zones of influence surrounding such collision areas* in which approaching road users are delayed by reduction in speed below normal.

All intersection design and control should recognize the nature and importance of the elemental maneuvers which must be accommodated within the intersection areas. Both design and control should strive to minimize and, if possible, eliminate the effects which grow out of conflict between road users in their joint use of the intersectional area.

Modifying Factors. Traffic performance at an intersection is modified by the composition, volume, and speed of the entering traffic streams, the form and arrangement of the intersectional area, environmental conditions of weather, light, and other values, and traffic regulations and controls.

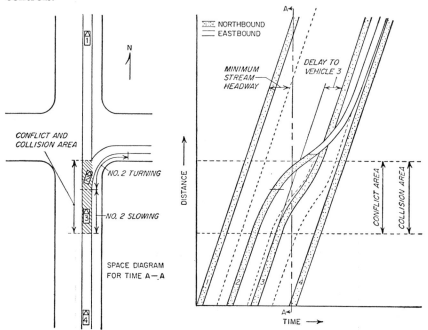

Fig. 8-2. Time-space relationships of diverging maneuver.

8-2. The Diverging Maneuver. Diverging is perhaps the simplest and most readily accommodated of the elemental intersectional maneuvers. The time-space relationships of this maneuver are illustrated in Fig. 8-2. The potential area of conflict begins at the point when the speed of the diverging vehicle 2 becomes nominally lower than that found in its lane of travel and continues on as vehicle 2 moves forward and out of its lane of original travel. In this instance the collision area is nearly congruent with the entire conflict area of the maneuver.

Almost simultaneous with diverging, additional conflicts not inherently a part of the maneuver, such as those of the merging or crossing type, may be incurred. If a left divergence is made, a crossing conflict may arise from either cross flow or counterflow. A merging conflict may arise with cross flow. Crossing or weaving conflicts of the diverging vehicle with other vehicles in the same flow may arise if divergence is initiated from an ill-chosen position in a multilane flow. In any case, the diverging vehicle inherently develops an area of conflict in the lane from which it departs. Figure 8-2 shows vehicle 3 in this conflict area.

The chart of time-distance relationships shows that vehicle 1 has moved through the intersection without conflict or delay. Vehicle 2, the diverging vehicle, reduces its speed at a point some distance removed from the intersection in order to make a comfortable turn, thus establishing the beginning of the conflict area. This conflict area is continuous to the point where vehicle 2 leaves the lane of original travel. Vehicle 3, shown in this conflict area, is delayed because of its conflict with vehicle 2.

Vehicle 4, like vehicle 1, proceeds through the intersection without conflict from the diverging maneuver, but it suffers reduction in headway and continues at nearly the minimum stream headway behind vehicle 3 in its travel through the conflict area. Vehicle 3 enjoys a greater headway after divergence has taken place.

Diverging interference to pure stream flow is dependent on (1) the density of volume of flow in the lane from which the maneuver is made, (2) the relative speed with which divergence will take place, and (3) the relative frequency of this diverging maneuver.

8-3. The Merging Maneuver. Unlike the diverging maneuver, the merging maneuver cannot be made at will but must be deferred until an adequate gap occurs in the lane of flow into which the merging vehicle is to enter. Typical time-space relationships which develop in the merging maneuver are shown in Fig. 8-3. In this case the conflict area begins at a safe distance back of the collision area and extends to a point beyond, where the merging vehicle has achieved approximately normal speed. The collision area extends from the point of entry of the merging vehicle to the forward limits of the conflict area.

The relative special position of the vehicles concerned is shown at the instant A-A. Vehicle 1 has proceeded through the intersection and has moved beyond the conflict area, incurring no delay. Vehicle 2, the merging vehicle, has partially entered the collision area, suffering delay due to the proximity of vehicle 3 (the driver of which took precedence through the collision area). Vehicle 3 reduced speed while in the conflict area until reaching a *point of decision*, when its driver decided that precedence over vehicle 2 was assured. The delay to vehicle 3 was slight since vehicle 2 refused to accept the *time lag* between the time of his arrival and that of vehicle 3 into the collision area. The driver of vehicle 2, after waiting for vehicle 3, accepted the *time gap* between vehicles 3 and 4.[1] In so doing, however, vehicle 2 delayed vehicle 4

[1] A *gap* is the time interval between the arrival at a point by one unit and the arrival at the same point by the next succeeding unit traveling in the same direction. It is equivalent to headway. A *lag* is that interval of time from the instant one unit is in position to accept or reject the opportunity to enter the collision area until the arrival of an approaching conflicting unit into a collision area. It is the length of the lag that is perhaps the critical concern of the conflicting units.

slightly as shown by its time trace.   Vehicle 5, like vehicle 1, proceeded through the intersection without any delay.

In the merging maneuver, the amount of delay is dependent on (1) the density or volume of flow in the entered stream, (2) the relative velocity between merging vehicles and the vehicles in the entered lanes, and (3) the frequency of occurrence of merging vehicles.

Critical Gap (or Lag) for Merging.   In every merging maneuver, a gap or lag must exist in the lane being entered by the merging vehicle.   The

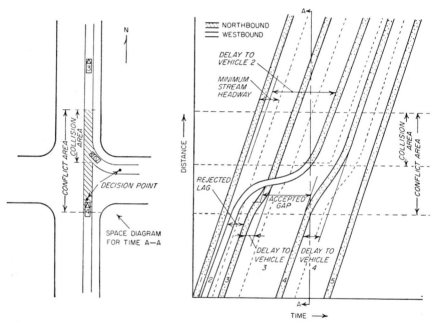

Fig. 8-3. Time-space relationships of merging maneuver.

acceptable size of such gap for drivers of private passenger vehicles ranges from 1.5 to 6.0 sec or greater, the larger gap sizes being required for high relative speed.

Figure 8-4 summarizes the gap acceptance observed in merging maneuvers at urban intersections.   When the merging vehicle was stationary or moving at a low speed (5 mph or less) just prior to the act of merging, and accordingly a high relative speed existed, the great majority of road users rejected gaps of less than 5.5 sec but accepted gaps greater than 6 sec. The relatively few road users who accepted gaps less than 5 sec were usually retarded, the rear vehicle forming the gap.

When the merging vehicle was moving at a medium relative speed (6 to 14 mph), nearly all road users accepted gaps as short as 2.5 sec duration, and a few even entered gaps as small as 1.5 sec.   Nearly all rear vehicles were retarded when the merging maneuver was carried out at this relative speed into gaps of 1.5 to 2.5 sec, and a decreasing proportion was delayed up to 4.5 sec.

At low relative speed, when the merging vehicle was moving at 15 to 28 mph, nearly all road users accepted gaps as low as 1.5 sec. Interference to the rear vehicle occurred at gaps of 1.5 to 2.8 sec, but little or none was experienced at gap acceptances of 2.5 sec or greater.

Delays Due to Merging. Figure 8-4 shows that the bulk of delay incurred from merging is suffered by the merging road user. The amount

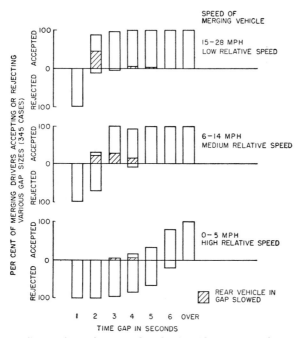

FIG. 8-4. Comparison of accepted and rejected gaps at various speeds.

of delay incurred is not only contingent on the gap size required to make the maneuver, but also on the volume of flow in the lane of merging. The computation for merging opportunity and delay is similar to that for the lane-change maneuver.

Assume that 50 vph are to be merged into one flow of a two-way stream and that the flow in one direction amounts to 500 vph. Assume that no merge will be carried out in gaps 2.5 sec or less in length and that those merging road users who are not given the opportunity of a greater than 2.5-sec gap will then create high relative speeds for the merging maneuver, making a 6-sec gap necessary.

The percentage of headways (2.5 sec or less) when merging opportunity is not present may be read from Fig. 7-9 (two-lane at 500 vph). The number and duration of these short headways is computed as follows:

Gap size	Scaled readings	Per cent of gaps of given size	Total no. of gaps	No. of gaps of given size	Av. size of gap	Time used by given gaps, sec
0–1.0	6–0	6	500	30	0.50	15
1.0–1.5	20–6	14	500	70	1.25	88
1.5–2.0	35–20	15	500	75	1.75	131
2.0–2.5	46–35	11	500	55	2.25	124
Total......	.....	..	...	230	....	358

Thus 230 gaps of 2.5 sec or less, occupying a total of 358 sec of time, still occur in 1 hr of operation. Assuming that no vehicle will merge with less than a 2.5 sec lead over the rear vehicle,[1] there will be a further time loss of 675 sec/hr (2.5 for each of the remaining 270 gaps). Thus for 675 + 358 sec, a total of 1,033 sec/hr, no opportunity to merge exists. Hence 1,033/3,600 × 50 or 14 merging vehicles must markedly reduce speed or stop and will then require a 6-sec gap for merging. On the other hand, 36 of the 50 merging vehicles will be able to merge without delay.

Similar computations for the total number and duration of gaps 6 sec or less in length for the lane carrying 500 vph, into which the remaining 14 vehicles must merge, show that a total of 345 gaps occupying 779 sec is developed. Thus there remain 155 gaps 6 sec or greater in length, each preceded by a blockade of vehicles and short gaps. The duration of each blockade is the sum of the small gaps plus a 6-sec zone of influence in front of the leading vehicle into which no merging movement will be made. In addition to the 779 sec consumed by the short gaps, 155 × 6 or 930 sec are therefore consumed by the zones of influence in front of each blockade. Thus for the merging vehicles which are stopped, the total blocking period is 779 + 930 or a total of 1,709 sec. Each blockade will average 1,709/155 or about 11 sec. If it is assumed that arrivals of delayed vehicles may occur at any instant during a blockade period, the average delay per vehicle is about 5.5 sec, and the total delay is 14 × 5.5 or 77 sec/hr.

While most delay is suffered by the merging vehicles, there is also delay to the vehicles in the lanes of flow into which merging takes place, varying with the relative volumes to be merged and the relative speed at which merging takes place.

[1] Note that this assumption differs from that described in Chapter 7, page 137, concerning required position of lane-change vehicles in regard to gaps being entered. Here it has been assumed that the merging driver is concerned only with the rear vehicle of a gap, and thus the entire zone of influence will precede that vehicle Any requirements that the merging driver has regarding position behind the lead vehicle are assumed to be accounted for *after* the gap has been accepted and the merging maneuver undertaken.

8-4. The Crossing Maneuver.   Typical time-space relationships found in the crossing maneuver are graphically shown in Fig. 8-5.   The conflict area begins at some point which is a safe distance from the collision area and is dependent on speed as well as stopping ability of each of the conflicting units.   In the case shown, the conflict area begins some distance from the intersectional area and extends through the collision area.

The relative position of vehicles involved is shown for instant A-A. Vehicle 1 has proceeded through the intersection without delay.   Vehicle

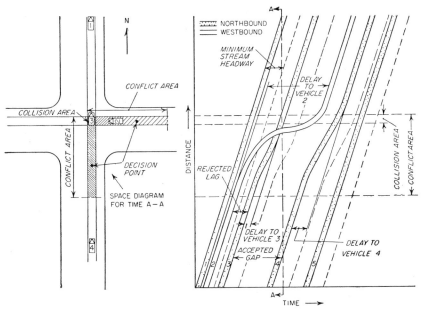

FIG. 8-5. Time-space relationships of crossing maneuver.

2, the crossing vehicle on the E-W route, has entered into the conflict area. Vehicle 3 is following vehicle 1 at nearly minimum headway, too short to be acceptable for a crossing maneuver by vehicle 2.   Nevertheless, the driver of vehicle 3 reduced speed slightly in view of the impending condition when he entered the conflict area.   This deceleration continued up to the *point of decision*, when the driver of vehicle 3 decided that vehicle 2 would yield the right of way and accordingly resumed his normal rate of speed.   Because of the circumstances which developed, the driver of vehicle 2 was required to stop.   When vehicle 3 cleared the collision area, vehicle 2 accepted the next gap and took precedence over vehicle 4, resuming a normal rate of speed in the E-W course.   The driver of vehicle 4 reduced speed slightly in the conflict area to the point of decision when right of way was assured.   Vehicle 5 proceeded through the intersection without delay.

The degree of interference in a crossing maneuver is again dependent on (1) the volume of vehicles in each flow, which determine the frequency of crossing maneuvers, and (2) the normal level of speed of each flow.

Critical Gap (or Lag) for Crossing. The size of gap or lag required differs for direct right-angle crossings, left-turn crossings, oblique crossings, and pedestrian crossings.

Right-angle Crossing (Vehicular). There are two cases to be considered in the measurement of acceptable right-angle-crossing gaps or lags: (1) neither vehicle is stopped, (2) one of the crossing vehicles is at rest just prior to the crossing maneuver.

Limited observations at uncontrolled urban intersections show that when neither vehicle stopped the average minimum time interval between passage of conflicting vehicles through the collision area was 2.85 sec. Thus in nonstop crossing maneuvers conflicting vehicles may be expected to cross over the collision area at the rate of about one each 3 sec.[1]

When one of the conflicting vehicles is at rest just prior to the crossing maneuver, the average driver of the stopped vehicle requires a lag of 4.5 to 8 sec before undertaking the crossing maneuver. Characteristic curves of acceptable lags are shown in Fig. 8-6.

Oblique Crossing (Vehicular). Exploratory measurement of nonstop oblique crossings at a channelized intersection indicates that smaller gaps are more acceptable than those required for right-angle crossing. One study of this type found that the average driver would accept a gap of 1.8 sec.[2]

Left-turn Crossing (Vehicular). A study of opposed left-turn crossings at an urban intersection found that delayed left-turning drivers refused all gaps less than 3.75 sec and accepted all gaps of 4.75 or greater. The median value was 4.25 sec.[3]

Pedestrian Crossing. The median acceptable lag for pedestrian crossing was found to be 5.7 sec for a one-way vehicular stream and 7.3 and 7.7 sec for the near and far flows, respectively, for a two-way vehicular stream.[4]

Delays Due to Crossing. Delay to crossing vehicles may be calculated by methods similar to that shown for merging maneuvers, but with proper analysis this method of arithmetic integration of graphic experimental results may be replaced by direct mathematic computation.

[1] B. D. Greenshields et al., *Traffic Performance at Urban Street Intersections,* Yale University, Bureau of Highway Traffic, Technical Report 1, New Haven, Conn., 1947, p. 70.

[2] R. E. Conner, "Some Time-Space Relationships of Free-flowing, Channelized Cross Traffic," student thesis manuscript, Yale University, Bureau of Highway Traffic, 1951.

[3] F. J. Kaiser, Jr., "Left Turn Gap Acceptance," student thesis manuscript, Yale University, Bureau of Highway Traffic, 1951.

[4] C. C. Robinson, "Pedestrian Interval Acceptance," student thesis manuscript, Yale University, Bureau of Highway Traffic, 1951.

Numbers Delayed. If it is assumed that both conflicting flows are random and that every crossing vehicle is in position to take immediate advantage of any gap, the theoretical proportion of vehicles delayed may be computed. The percentage delayed will be the same as the percentage of time made up of gaps shorter than the critical gap, plus that

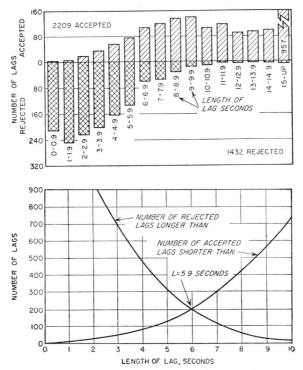

FIG. 8-6. Distribution of accepted and rejected lags.

percentage of time made up of critical lags found in all gaps larger than the critical gap. This percentage equals $100(1 - e^{-NL})$

where e = Naperian base of logs = 2.71828

N = vehicles per second in flow to be crossed

L = critical lag, sec*

The percentage derived by this formula is somewhat lower than that occurring in practice, since vehicles queue up to cross and in many cases a given vehicle is not in position to take advantage of any gap. Limited experimental results show that the percentage of vehicles which will actually be delayed may be computed as follows:[1]

$$P = 100\left[1 - \frac{e^{-2.5N_s} \times e^{-2NL}}{1 - e^{-2.5N_s}(1 - e^{-NL})}\right]$$

---

* M. S. Raff, "Theory of Uncongested Traffic," unpublished manuscript, Yale, 1951.

[1] M. S. Raff, *A Volume Warrant for Urban Stop Signs*, Eno Foundation for Highway Traffic Control, Inc., Saugatuck, Conn., 1950, p. 48.

where P = percentage of side-street cars delayed

N = main-street volume, vehicles per second

N_s = side-street volume, vehicles per second

L = critical lag, sec

e = base of natural logarithms

In the above formula (1) the limit of P, as N_s approaches 0, is $100(1 - e^{-NL})$, which is the theoretical formula: if there are no side-street

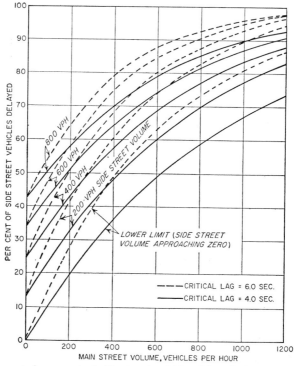

FIG. 8-7. Computed per cent of side-street vehicles delayed at stop-controlled intersections.

cars, there is no sluggishness; (2) P always exceeds $100(1 - e^{-NL})$, except when N_s equals 0: sluggishness delays more cars than would be delayed without this effect; (3) P is always less than 100 per cent for any finite volumes, and (4) the partial derivatives of P with respect to N, N_s, and L are all positive: an increase in either of the two volumes or the critical lag causes an increase in the percentage of cars delayed.

This formula allows for delays to vehicles forced to stop because of the presence of one or more previously stopped vehicles in the same flow. Percentages of vehicles delayed, based on this empirical equation, are shown graphically in Fig. 8-7. Increased volume on either street or increased critical lag causes an increase in the percentage of delayed side-street vehicles. Even at zero main-street volumes, a number of side-street vehicles on the side street stop at least momentarily for reasons of caution before crossing the main street. (The graph does not include the percentage of side-street vehicles stopping only for the sake of caution, as that number is theoretically 100 per cent.)

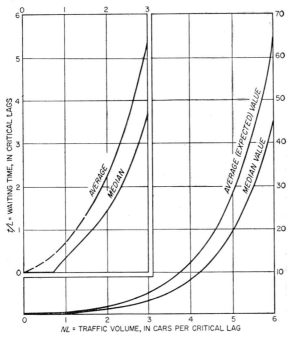

FIG. 8-8. Average and median waiting time as a function of volume.

Delay Duration. Assuming that the traffic flow to be crossed is a random series,[1] the average delay for any vehicle may be computed on a theoretical basis as follows:

$$D = \frac{(1 - e^{-NL} - NLe^{-NL})}{Ne^{-NL}}$$

where D is the average delay per vehicle in seconds.

The average delay per vehicle in terms of the critical lag as a unit of time measure is set forth in Fig. 8-8. Assume that a critical lag of 6 sec is required and the stream being crossed carries 600 vph. The crossed flow then averages one vehicle per critical lag. While not all of the crossing vehicles would be stopped, the average delay would be expected to be about 0.7 of a critical lag or about 4 sec. These values of delay do not

[1] William F. Adams, "Road Traffic Considered as a Random Series," *Institution of Civil Engineers Journal* (England), November, 1936, p. 127.

allow for time lost because of the queueing of crossing vehicles, and therefore are somewhat lower than may be found in practice.

8-5. Intersection Operation.   It is not possible to synthesize accurately intersection performance out of the diverse combinations of the three elemental maneuvers which may be predicted in a given case, as in many

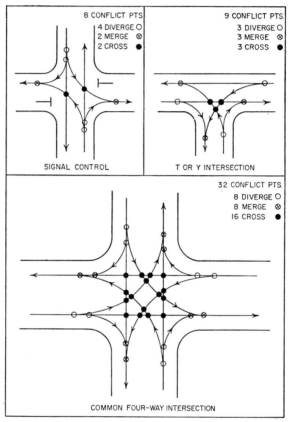

FIG. 8-9. Conflict points at intersections.

instances they are so closely interdependent that they require seriation in their development.

In the actual operation of intersections, safety as well as capacity is a matter of practical importance.   In the interest of safety many stops are made or required by regulations, whether or not they are required by actual conflict arising from the elemental maneuvers.   Such delay thwarts capacity of the entering streams in order to increase safety.

Number and Types of Conflicts.   The number and types of conflict which develop at intersections are shown in Fig. 8-9.   The focal points indicated represent the area in which conflict is most severe and collision most likely. The normal right-angle intersection of two two-way streams develops 32 conflict points, 16 of which are of the more severe crossing type. If one stream is removed by design to make a T or Y intersection, there occur only 9 conflict points, of which only 3 involve crossing maneuvers. If two of the flows are stopped simultaneously by signal control, there remain only 8 conflict points, of which only 2 involve the more severe crossing type of conflict. The number of entering and departing streams which flow at a given time greatly modify the number and types of conflict points which develop in intersection operation.

Frequency of Conflicts. The frequency of interference at conflict points is dependent on the volume of units found in each path of flow. In the standard intersection shown in Fig. 8-9, if each entering stream carries 200 vph, and 10 per cent of each stream turns right and 10 per cent left, there will develop 1,200 potential conflicts per hour. The calculations leading to this conclusion are as follows:

$$\begin{aligned}
\text{Right-turn volume } 10\% \times 200 \text{ vph} \times 4 \text{ approaches} &= 80 \text{ vph} \\
\text{Left-turn volume } 10\% \times 200 \text{ vph} \times 4 \text{ approaches} &= 80 \text{ vph} \\
\text{Through volume } 80\% \times 200 \text{ vph} \times 4 \text{ approaches} &= \underline{640 \text{ vph}} \\
\text{Total} &= 800 \text{ vph}
\end{aligned}$$

From Fig. 8-9:

8 diverging conflicts for 8 turning movements = 1 conflict/turn $(80 + 80) \times 1 =$	160
8 merging conflicts for 8 turning movements = 1 conflict/turn $(80 + 80) \times 1 =$	160
12 crossing maneuvers involving 4 turning movements = 3 conflicts/turn (80×3)	= 240
4 crossing maneuvers involving 4 through flows = 1 conflict/flow through 640×1	= 640
Total conflicts/vehicle-hr =	1,200

This figure represents exactly the number of times per hour under the given conditions that one vehicle departs from, enters in front of, or crosses the path of another vehicle, but does not indicate the proximity of any of the vehicles involved or the degree of hazard or delay encountered.

Added Time Loss Due to Stops. Because of real or potential conflicts, the road user is frequently required to decelerate prior to his passage through the intersection. He may merely slow down until reaching the point of decision or be required to come to a full stop near the collision point. If required to stop, he must wait for the next opportunity to get under way and accelerate to normal speed, thus losing time in deceleration, waiting, starting, and acceleration.

Deceleration Time Losses. While the comfortable rate of deceleration is 8 to 9 ft/sec², actual deceleration rates employed by road users in slowing or stopping for intersections varies from free-wheeling values to as much as 16 ft/sec², varying with speed of travel and distance from the collision area. Observed experimental values are shown in Fig. 8-10.

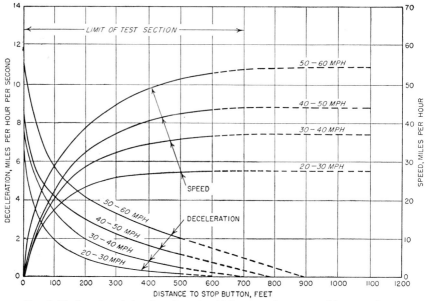

FIG. 8-10. Speed and deceleration of vehicles stopping at rural intersections.

The time loss by passenger vehicles for various rates of approach, if a complete stop is required, is shown in Table 8-1.

TABLE 8-1. TIME REQUIRED FOR DECELERATION TO STOPPED POSITION*
(Rural area and passenger cars)

Approach speed	Time to travel at speed shown, sec	Actual time required, sec	Time loss, sec
20–30	19.1	21.6	2.5
30–40	13.6	17.5	3.9
40–50	10.6	15.4	4.8
50–60	8.7	12.9	4.2

* Based on last 700 ft of travel prior to collision area.

SOURCE: Computed from data presented by John Beaky, "Acceleration and Deceleration Characteristics of Private Passenger Vehicles," *Proceedings of Highway Research Board*, Washington, 1938, vol. 18, pp. 85–88.

Time Loss in PIEV. Once the road user has been brought to a halt, time is consumed in perception, intellection, and volition before he gets under way again. As Chapter 2 showed, over 1 sec is required to shift glances from one side of the road to the other. Depending upon circumstances prevailing at a given instant, the time required for intellection and decision will vary widely. Some observations show that even under familiar conditions, such as in response to a green or go signal, the road user requires about 2 sec to get under way.

Acceleration Time Losses. The time loss in bringing the stopped vehicle back to speed must also be added to the time spent while waiting for opportunity to merge or cross. The rate of acceleration varies with

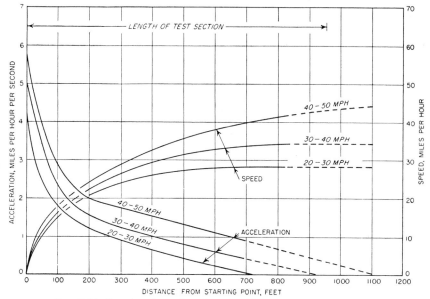

FIG. 8-11. Speed and acceleration of vehicles leaving intersections.

speed class, vehicle type, and other factors. Characteristic curves of acceleration are shown in Fig. 8-11. The time loss for passenger vehicles for various rates of nominal speed is shown in Table 8-2.

TABLE 8-2. TIME REQUIRED TO ACCELERATE FROM STOPPED POSITIONS*

Normal class of speed resumed	Average speed developed	Time to travel at speed shown	Actual time required	Time loss
20–30	27.8	23.3	26.2	2.9
30–40	34.0	19.0	22.6	3.6
40–50	41.5	15.6	20.5	4.9

* Test section, 950 ft in length.

SOURCE: Computed from data presented by John Beaky, "Acceleration and Deceleration Characteristics of Private Passenger Vehicles," *Proceedings of Highway Research Board*, Washington, 1938, vol. 18, pp. 85–88.

Total Additional Time Loss. The average passenger vehicle which is stopped will suffer from 7 to 12 sec delay for slowing, getting under way, and acceleration, in addition to the delay incurred waiting for opportunity to use a collision area. This time value is as great or greater than many of the blockades found in the conflicting stream.

If a flow of 600 vph is to be crossed by road users requiring a 6-sec critical gap, the *average* road user will be delayed about 5 sec waiting for a satisfactory gap. If 100 vph were to cross such flow, a total vehicle delay of 500 vehicle-sec would be incurred. But since only 69 per cent of these crossing vehicles will be stopped, only 69 vehicles would suffer delay due to deceleration, getting under way, and acceleration. The additional loss is therefore 69 × 7 or 483 vehicle-sec/hr if normal speed is 20 to 30 mph. In this case, nearly as much time is lost by stopping as by blocking.

Practical Operating Capacities. Figure 8-7 shows that when a 6-sec lag or gap is allowed for a crossing maneuver, over 50 per cent of the entering vehicles will be required to stop even if total intersection volume is as low as 400 to 500 vph (an average of little more than 100 in each flow). For reasons of safety many intersections carrying even lower volumes are modified by "stop" signs. As a rule, when intersections develop 3 to 4 personal-injury accidents per year, "stop" signs will be installed. If a standard intersection handles only 100 vph on each approach with 10 per cent turning left and 10 per cent turning right, 120 potential merging or crossing conflicts will occur in 1 hr. If the rate of 100 vph is maintained for 12 daylight hours (7 A.M. to 7 P.M.), there will arise over 3 million potential conflicts per year of the serious merging or crossing types. Accordingly, "stop" signs are installed at many intersections long before 50 per cent of the entering vehicles would be required to stop because of blocking traffic movement.

Even if 100 vph per flow is allowed as satisfactory, safe, free-flowing (uncontrolled) traffic performance at intersection, such volume is but a quarter or less of the practical capacity in the two-lane two-way stream. Satisfactory intersection performance without control is obviously limited to very low volumes.

CHAPTER 9

PARKING CHARACTERISTICS

Demand for parking in a given area is strongly influenced by land use and by competing forms of transportation. As a parcel of land is changed from agricultural to residential, to industrial or to commercial use, or as the density or bulk of a given activity increases, parking demand increases. Parking problems are likely to develop in areas devoted to multistoried residences and industrial buildings and are especially acute in commercial or business districts.

Common carriers and pedestrians do not create parking problems. As the population of a community increases there is, in general, a decrease in the percentage of road users who select the private motor vehicle as the means of movement. In smaller cities as many as 80 per cent of the persons entering central business districts do so by private auto. This percentage decreases to an average value of 70 per cent in cities of 100,000 to 500,000 population and is as low as 40 per cent in cities of 500,000 and larger.[1] There is generally a decrease in the number of vehicles parked (in the central business district) per 1,000 population as the size of city increases. However, absolute numbers of private vehicles demanding parking space generally increase with city size, and it is these that generate the parking problem.

9-1. Parking Studies. To determine the relative importance of different forms of transportation and to evaluate amount and character of parking as well as the facilities available, three basic types of field studies are required: (1) the cordon count, (2) space inventory, and (3) parking practices. Because of the outstanding importance of the central business district as a traffic generator and the resultant concentration of parking, most studies of parking are concerned with this district where (1) land values are high, (2) land occupancy is dense, (3) land use is primarily business (frequently multistoried), (4) major routes and their flows converge, and (5) curbs are usually saturated with parked vehicles (or "no parking" regulations).

The Cordon Count. In making a cordon count, observers are placed on all routes leading to or from the selected areas, and they record the

---

[1] W. S. Smith and C. S. LeCraw, *Parking*, Eno Foundation for Highway Traffic Control, Inc., Saugatuck, Conn., 1946, p. 26.

number of vehicles, the number of occupants in each, and the number of
pedestrians entering or leaving the area by elemental time units, usually
half-hour periods. The period of study is usually continuous throughout
the business day (for example, 7 A.M. to 11 P.M.).

Cordon counts yield the relative importance of each route as a traffic
carrier by all forms of transportation. Algebraic summation of these
directional counts develops the amount of traffic, by each mode of transportation, which is accumulated within the district at any instant during
the entire period of the study. Such studies when repeated from year to
year also develop changes in the amounts of accumulation and the modes
of transport.

Space Inventory. All types of parking space in the area under consideration must be listed and classified. There are two general classes
of parking space, curb and off-street. Curb space is subclassified by
location and type of regulation in effect (no parking at any time, no
parking during certain hours, loading zone, taxi stands, bus stops, driveways, time limits, metered space, hydrant, corner clearance, safety zones,
etc.). Off-street space is subclassified by location, type, fees charged,
and other features (lot, garage, public, private, attendant, self-park, fees
charged, etc.).

Parking Practices. Observations are made in the field of the use of
various types of parking space. Type of vehicle, time of arrival, time of
departure, and any time or space violations incurred may be recorded for
each vehicle on prepared forms.   If proper steps are taken, parkers may
also be queried as to origin, destination after leaving the parked vehicle,
trip purpose, frequency of parking in the area, and other desired
information.

9-2. Concentration of Traffic Demand.   *Traffic Flow at the Cordon Line.*
When all flows observed are combined into a single value for a given
period of time, characteristic fluctuation of traffic flow observed on each
route will be reflected in the traffic flow of the cordon district as a whole.
Figure 9-1a shows vehicular traffic movement associated with a central
business district. Note the difference in magnitudes of volume of the
inbound and outbound flows. Inbound movement is peaked in the
morning hours, outbound in the evening.

Cordon Accumulation. The difference between the volume of entering
and leaving traffic in a given period of time is the measure of accumulated
traffic which must either park or remain in motion.

Characteristic accumulation in a central business district is set forth in
Fig. 9-1b. Some base number of vehicles (2,000 in this case) remains
overnight in the district. With the onset of the working day, accumulation above this base begins, rising rapidly with heavy inbound and minor
outbound movement. As the day proceeds, shoppers and those engaged
in other business or errands add to the net accumulation so that the peak

hour is usually from 2 to 3 P.M. Outbound traffic gradually grows larger than inbound, and with the evening rush there is a general exodus, leaving only a third or a fourth as many in the district as at midday. During the late evening hours there is a tendency toward a lesser rate of discharge or an actual increase in accumulation due to recreational activities.

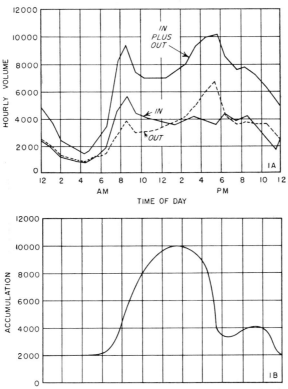

FIG. 9-1. Hourly volumes and vehicle accumulation in a central business district as determined by cordon count.

The general shape of the vehicle-accumulation curve for a given district is highly significant. The peak value is a measure of the capacity required for the accommodation of vehicles within the district. The total area under the curve is a measure of the vehicle-hours accommodated during the period of study. Of all the vehicles which must be accommodated by the central business district at any instant, the great majority are usually parked vehicles throughout the peak periods of accumulation.

9-3. Space Considerations. *The Area of Central Business Districts.* Roughly a third of all trips originate or terminate in the central business district (CBD), which usually occupies a small fraction of the total metropolitan area. The CBD averages 0.1 to 0.6 sq mile in American cities (see Table 9-1).

While population increases from 16,900 to 663,650 (thirty-nine-fold), total area of the CBD increases only from 0.12 to 0.54 sq mile (4.5-fold). Density of development in the largest cities would be 15 times that in the small cities, if all functions were maintained at the same rate per 100,000 population, but the functions actually carried out in the CBD change as city size increases. While the proportion of street space remains about the same, certain types of business functions (filling stations and grocery stores, for example) are reduced or eliminated, while other types (such as specialty stores) are founded and prosper as the city grows.

TABLE 9-1. AREA AND POPULATION RELATIONS IN CITIES IN SIX DIFFERENT POPULATION GROUPS

Population group (000 omitted)	No. of cities	Av. pop., metropolitan area	CBD area, sq miles		CBD area, no. of blocks
			Total	Per 100,000, pop.	
Less than 25	6	16,900	0.12	0.74	27
25–50	3	32,300	0.11	0.36	35
50–100	2	66,550	0.22	0.27	36
100–250	9	131,750	0.44	0.26	76
250–500	6	280,700	0.46	0.12	97
500 and over	2	663,650	0.54	0.05	134
Total.........	28				

SOURCE: *Parking*, Highway Research Board, Bulletin 19, Washington, 1949, p. 39.

As city size increases, there is increased competition for the use of CBD space, and while the demand for parking space increases, the problem of securing adequate parking space becomes more severe.

Use of Space in CBD. Figure 9-2 illustrates the use of land for traffic and parking purposes. A fourth to a third of the total land area in the CBD is usually devoted to streets for the transmission of traffic. In the central section or core, densely occupied by commercial, governmental, or other uses which generate large volumes of traffic, demand for parking space exceeds the supply. Around this core there is generally a *ring* of lower-density commercial use and a certain amount of available parking. The spaces in this area usually must serve both the ring and core demands.

Parking Space Available in CBD. Table 9-2 indicates total space available for parking purposes in CBD, in terms of parking spaces.

Available parking space does not increase nearly so rapidly as population. Curb space does not increase so rapidly as off-street space, and total available space per 1,000 population decreases from about 90 stalls in small cities to only 12 stalls in the largest cities.

Space per Parking Stall. Space required to park a private passenger vehicle ranges from 150 to 300 sq ft. In off-street development of the

TABLE 9-2. CURB AND OFF-STREET SPACES AVAILABLE IN THE CENTRAL
BUSINESS DISTRICTS OF CITIES IN SIX POPULATION GROUPS

Population group (000 omitted)	No. of cities	Number of spaces			Spaces per 1,000 population		
		Curb	Off-street	Total	Curb	Off-street	Total
Less than 25	5	981	668	1,649	54	36	90
25–50	3	1286	775	2,061	41	25	66
50–100	2	1688	2401	4,089	23	34	57
100–250	8	2684	3765	6,449	17	25	42
250–500	6	2961	8132	11,093	7	21	28
500 and over	2	2510	7675	10,185	3	9	12
Total.........	26						

SOURCE: *Parking*, Highway Research Board, Bulletin 19, Washington, July, 1949,
p. 40.

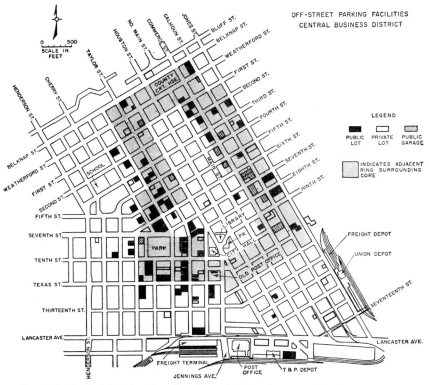

FIG. 9-2. Typical distribution of parking facilities in central business districts.

single-level or parking-lot type, averages of 200 sq ft for attendant parking and 250 sq ft per vehicle for nonattendant or customer parking is found. Good design and operation may require as much as 300 sq ft or more for a given facility.

9-4. Parking Demands. The number of vehicles parked in a specified area at a given instant is one measure of parking demand, referred to as *parking accumulation.* The unit measure of accumulation is the vehicle.

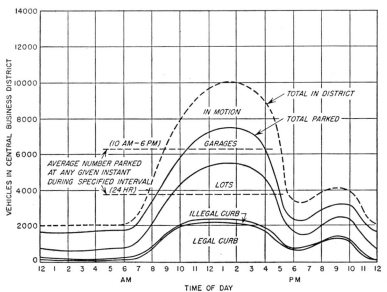

Fig. 9-3. Vehicle accumulation in various types of parking facilities in central business districts.

The unit measure of *parking load,* the integration of accumulations over a specified period of time, is the vehicle-hour. Parking load is ordinarily expressed by a time rate as vehicle-hours per day. The length of time a vehicle remains in a parked condition is referred to as *parking duration.* The actual number of vehicles in a parking load is designated as *parking volume,* expressed as vehicles per day.

Accumulations may be used as a measure of the amount of parking space required to meet demand at a given time. Volume indicates the number of vehicles served. Durations point up the amount of time that a particular road user may be expected to stay. Load is a measure of the over-all usage of parking space. Each of these measures is modified by various factors. Purpose of trip affects times of arrival and departure as well as duration and therefore has much effect on the shape of the accumulation curve. City size affects volume and, to some extent, duration. Factors which grow out of space limitations, such as costs and walking distances, also affect these measures.

Parking Accumulation. Algebraic summation of in- and out-parking movements yields the accumulation of parking at any time throughout the period of study. Characteristic parking accumulation for a CBD is shown in Fig. 9-3. The parking-accumulation curve is very similar to, but always less than, the cordon-accumulation curve shown in Fig. 9-3. The difference between these two accumulation curves is a measure of the number of vehicles in motion within the district at any given time. Ordinarily, 15 to 30 per cent of all vehicles accommodated within the CBD are in motion during peak hours of accumulation, increasing when the rate of change in accumulation is high.

TABLE 9-3. USAGE OF PARKING SPACE

Population group (000 omitted)	No. of cities	Present usage, space-hours		Maximum number parked	
		Number	Per 1,000 pop.	Total	Per 1,000 pop.
Less than 25	4	8,654	511	1,141	62
25–50	3	9,799	303	1,350	43
50–100	2	14,632	220	2,185	30
100–250	5	33,659	255	5,168	28
250–500	4	51,578	184	8,245	21
500 and over	2	65,846	99	9,564	11
Total..........	20				

SOURCE: *Parking,* Highway Research Board, Bulletin 19, Washington, July, 1949, pp. 41, 43.

Parking Loads. Integration of the parking-accumulation curve yields the parking load in vehicle-hours, which establishes the space-hour usage of parking facilities. Peak load occurs at the time of peak accumulation, and this is the time when capacity is used to the fullest extent.

Influence of city size on the magnitude of parking loads developed in the CBD from 10 A.M. to 6 P.M. is shown in Table 9-3.

Total parking load increases with city size, but the load per 1,000 population decreases. Maximum accumulation increases with city size but decreases on the population-rate basis. These measures do not show potential demand but only that portion of such demand as is found under actual conditions. The ratio of peak load to average load generally increases with city size, averaging 5 to 28 per cent in excess of the average space required during the hours of 10 A.M. to 6 P.M. This ratio is much higher when the 24-hr base is employed (Fig. 9-3).

Parking Durations. Length of parking time is dependent on the drivers' purposes and is somewhat greater in large cities. Characteristic curves of parking durations are shown in Fig. 9-4. Of all curb parking 70 to 80 per cent is of 1-hr duration or less, while only 10 to 20 per cent of off-street parking is 1 hr or less. The largest single class of curb parking is 15 min or shorter, while the largest class of lot parking and garage parking is from 1 to 2 hr. The median average duration for curb parking is about 30 min, while the median averages for lots and garages are about 3 and 5 hr respectively. Thus the median average for off-street parking duration is 6 to 10 times that for curbs. The influence of trip purpose and city size on duration of parking is set forth in Table 9-4.

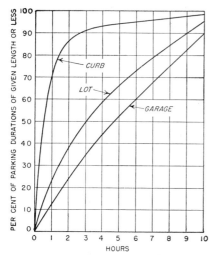

FIG. 9-4. Parking durations in various types of facilities.

The average duration of parking for purpose of work is 3 to 5 hr, compared with less than 1½ hr for any other purpose.

Parking Volume. The total volume of vehicles which park in the CBD is dependent on the size of city, but cyclical and trend variations associated with similar characteristics of traffic volume occur in any given city. Parking volumes in the CBD's of different-sized cities are set forth in Table 9-5.

The general practice of making parking studies only during the heavier portion of the business day necessitates showing values for 10 A.M. to 6 P.M. only.

TABLE 9-4. PARKING DURATIONS
(Average length of time parked for each purpose of trip in cities of six population groups)

Population group (000 omitted)	No. of cities	Average time parked for each trip purpose, hr				
		Work	Shopping	Business	Other	All purposes
Less than 25	5	3.1	0.7	0.7	1.1	1.1
25–50	3	2.9	0.7	0.8	0.9	1.3
50–100	2	3.3	0.8	0.7	0.9	1.3
100–250	5	4.0	0.9	1.0	1.5	1.7
250–500	3	4.5	1.4	1.2	1.5	1.8
500 and over	2	5.1	1.4	1.4	1.2	2.5
Total.........	20					

SOURCE: *Parking*, Highway Research Board, Bulletin 19, Washington, July, 1949, p. 46.

TABLE 9-5. PARKING VOLUME
(Total number of vehicles parked in central business districts of cities,
10 A.M.–6 P.M.)

Population group (000 omitted)	No. of cities	Total	Per 1,000 pop.
Less than 25	5	7,905	432
25–50	3	7,378	239
50–100	2	11,866	164
100–250	7	20,156	112
250–500	5	32,436	83
500 and over	2	29,957	34
Total............	24		

SOURCE: *Parking*, Highway Research Board, Bulletin 19, Washington, July, 1949, p. 41.

TABLE 9-6. EFFECT OF LAND USE ON REQUIRED PARKING SPACE*

Land use	Gross floor area per parking stall, sq ft	Other units of measure per parking stall
Industrial plant.....................	4,223	6.9 employees
Bus terminal..................	1,917	
Retail and mail-order house...........	1,816	
Office building A....................	.....	1,628 sq ft net rentable area
University.........................	1,401	7.6 students
High school........................	1,263	7.5 students
Hotel..............................	1,013	2.7 guest rooms
Hospital (private)...................	934	1.9 beds
Office building B....................	818	
Neighborhood shopping center.........	813	
Department store B.................	686	518 sq ft selling area
Department store A.................	475	263 sq ft selling area
Theater...........................	318	19.1 seats
General market.....................	199	
Department store C.................	180	
Railroad passenger station............	95	

* The data shown here is for a city where 43 per cent of all trips are made by automobile. The parking space required is on the basis of actual demand and not on the basis of actual existing space.

SOURCE: Adapted from J. Thompson and J. Stegmaier, *Effect of Building Space Usage on Parking Demand*, Highway Research Board, Bulletin 19, Washington, 1949.

Total parking volume (10 A.M. to 6 P.M.) increases rapidly with city size, but the volume parked per 1,000 population decreases and in larger cities is only 34 compared with 432 in smaller cities.

Fluctuations in parking volumes are ordinarily assumed to be similar to the weekly and monthly patterns found in traffic volume (see Chapter

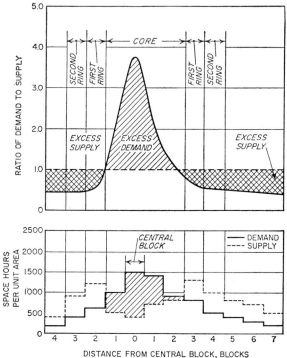

FIG. 9-5. Effect of location on space-demand relationships.

5). During a daily cycle, maximum *accumulation* of parked vehicles usually reaches a peak shortly after noon (Fig. 9-3). Saturday is usually the peak day of parking volume, while minimum volumes occur on Sundays and holidays in business districts. Except for sales days, which vary with local custom, Mondays through Fridays usually develop volumes of similar magnitude. The annual cycle is modified by preholiday shopping, which extends over several weeks before Christmas and Easter and tends to raise parking volumes. Low volumes occur in midsummer and midwinter.

9-5. Space-Demand Relations. Under ideal conditions, adequate space for parking would be provided on, or immediately adjacent to, the section of land used for the purpose which creates the parking demand. Relationships between parking space and demand, as far as location and sufficiency are concerned, are therefore significant factors in the termination of highway traffic.

Effect of Land Use on Demand. Table 9-6 illustrates the effect of land use on parking demand.

Location Factors. In general, the location and amount of parking space in CBD's is not coextensive with the location and amount of demand. Figure 9-5 typifies the arrangement of parking space in the CBD and its lack of congruity with the location of parking demand.

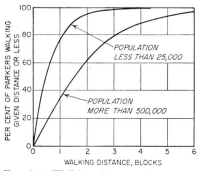

FIG. 9-6. Walking distance of parkers.

The parking demand per unit land area rises from low values at the periphery of the CBD to peak value at the center of the district, where land values are highest. Parking space per unit of land area is lowest at the location of peak demand, but on both sides of this point, and at some distance removed, supply of parking space is highest. Successive rings of area surrounding the central core show a decreasing ratio of demand to supply, approaching the ideal ratio where demand and supply are balanced in the same location.

Walking Distance. Because of the lack of congruity in the location of demand and supply, the road user must park at some point removed from his ultimate destination and complete his trip as a pedestrian. Because of parking costs in areas of high land value, many road users will walk a considerable distance to save expense. However, there is a complex relationship between road-user desires to walk short distances opposed to desires to incur minimum parking costs. Even in the largest cities, 60 per cent of parkers park their vehicles not more than two blocks from their destination (Fig. 9-6). Parking facilities more than two blocks from a particular generator can be expected to accommodate only a minority of road users whose trips are related to that generator.

CHAPTER 10

ACCIDENT CHARACTERISTICS

Accidents are the fourth highest cause of death in the United States, with traffic responsible for 39 per cent of all accident fatalities and 13 per cent of all accident injuries. Traffic accidents result from actual failure of the road user, the vehicle, or the fixed facilities to discharge properly their respective functions in traffic movement. Since traffic movement with safety is the objective of the traffic engineer, study of the facts surrounding traffic accidents is vitally important.

10-1. Accident-record System. Accurate accident reports and records are the foundation for the analysis and prevention of traffic failures. They serve not only in guiding engineering measures but also in shaping traffic law enforcement and traffic educational policies and procedures, and in legislation and administration of motor vehicle law.

The Individual Accident Report. A facsimile of the generally employed form for individual accident reports is set forth in Fig. 10-1. Individual road users involved in accidents are required by law to report facts on prescribed form to public authority (see Chapter 14, General Controls). Police officers who witness, come upon, or are called to the scene of accidents are also required to prepare reports. Supplemental reports of an accident may be made by a coroner, health officer, garage, transit or commercial-fleet operator, or insurance officer. A rough check on completeness of coverage is obtained by the ratio of fatal accidents to the total number reported. In urban areas, for each fatal accident there may be expected 30 to 40 personal-injury accidents and 100 to 150 accidents resulting only in property damage in excess of $25. Similar ratios for rural areas are 1:11:25.

Central Records Agency. Because engineering, enforcement, and education officials are concerned with information furnished by accident reports, it is desirable to route all reports to a central record bureau which is properly managed, staffed, and equipped to serve all agencies. This centralization avoids duplication, provides specialists for analysis and research, fixes responsibility for the preparation of summaries, maintenance of files and records, and in general leads to efficient utilization of accident facts.

Records, Files, and Summaries. The central accident records bureau will maintain (1) driver records, (2) fatality records, (3) location files, (4) spot maps and will carry out other functions necessary for the summarization, analysis, and interpretation of accident data. A typical flow diagram of accident reports and records is shown in Fig. 14-1 (General Controls). An illustrative location file for an urban area (required to determine and keep up to date the points of high accident frequency) is shown in Fig. 10-2.

The standard system of accident records generally employed in the United States develops uniform definitions, classifications, summaries, and files. The National Safety Council, Chicago, publishes and distributes complete details of this system.

10-2. Accident-record Analysis. *Classification and Definition.*[1] The principal classification of traffic accidents is by severity of results: fatal, nonfatal injury, or property damage. Secondary classifications are by location, such as rural or urban, and by collision and noncollision classes. Many other distributions are required and used for analytical and comparison purposes.

Figure 10-1, the individual accident report form, shows that data is collected for each accident concerning (1) location, (2) time, (3) vehicles, (4) persons, (5) road, (6) actions, (7) violations, and (8) the conditions and circumstances of the accident. Each of these may furnish a subclassification pertinent to accident analyses.

Macro and Micro Analysis. The magnitude and general character of hazard for a given state, city, or other subdivision of government for a year, month, or week is measured by the summarization of all accidents into general categories or classes (*macro* or *mass* analysis). Detailed analysis of accidents for a particular intersection, curve, or other location (*micro* or *spot* analysis) is required to bring the accident record and its meaning into sharp focus for such an area. The detailed record of a particular driver of a type of vehicle is also useful in the analysis of accidents.

Accident Rates. Since summaries and totals do not develop the relative degree of hazard for different sets of conditions on a common basis, accident rates are employed based on (1) population, (2) registered vehicles, and (3) vehicle-miles. Annual accidents per 100,000 persons, per 10,000 vehicles, or per 100,000,000 vehicle-miles are rates commonly used.[2] Population is a measure of the number of persons that are

---

[1] A standard system of definitions is provided by the U.S. Public Health Service in their publication *Uniform Definitions of Motor Vehicle Accidents.*

[2] In determining vehicle-miles for state-wide areas, the gasoline gallonage for motor vehicle purposes is obtained from tax records, and each gallon is reckoned as contributing 13.6 vehicle-miles of travel. For a limited area or route, the product of vehicle-volume and route mileage is used. It is obviously difficult to obtain total vehicle-miles from a city-wide area.

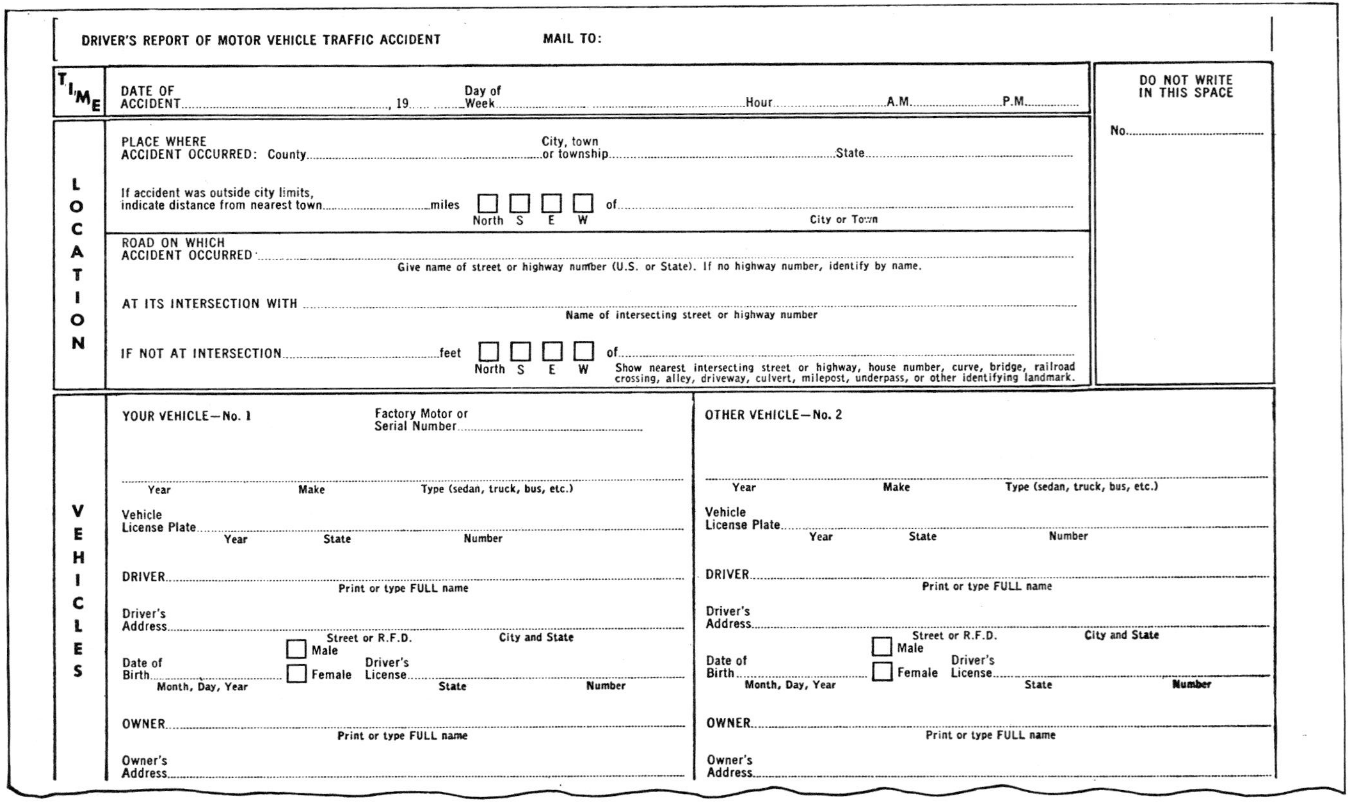

DRIVER'S REPORT OF MOTOR VEHICLE TRAFFIC ACCIDENT   MAIL TO:

TIME

DATE OF ACCIDENT............................, 19...... Day of Week............ Hour............ A.M............ P.M............

DO NOT WRITE IN THIS SPACE

No.............

LOCATION

PLACE WHERE ACCIDENT OCCURRED: County............ City, town or township............ State............

If accident was outside city limits, indicate distance from nearest town............ miles ☐ North ☐ S ☐ E ☐ W of............ City or Town

ROAD ON WHICH ACCIDENT OCCURRED............ Give name of street or highway number (U.S. or State). If no highway number, identify by name.

AT ITS INTERSECTION WITH............ Name of intersecting street or highway number

IF NOT AT INTERSECTION............ feet ☐ North ☐ S ☐ E ☐ W of............ Show nearest intersecting street or highway, house number, curve, bridge, railroad crossing, alley, driveway, culvert, milepost, underpass, or other identifying landmark.

VEHICLES

YOUR VEHICLE—No. 1   Factory Motor or Serial Number............

Year............ Make............ Type (sedan, truck, bus, etc.)............

Vehicle License Plate............ Year............ State............ Number............

DRIVER............ Print or type FULL name

Driver's Address............ Street or R.F.D............ City and State

Date of Birth............ Month, Day, Year   ☐ Male ☐ Female   Driver's License............ State............ Number............

OWNER............ Print or type FULL name

Owner's Address............

OTHER VEHICLE—No. 2

Year............ Make............ Type (sedan, truck, bus, etc.)............

Vehicle License Plate............ Year............ State............ Number............

DRIVER............ Print or type FULL name

Driver's Address............ Street or R.F.D............ City and State

Date of Birth............ Month, Day, Year   ☐ Male ☐ Female   Driver's License............ State............ Number

OWNER............ Print or type FULL name

Owner's Address............

FIG. 10-1A. Standard accident report form, obverse side.

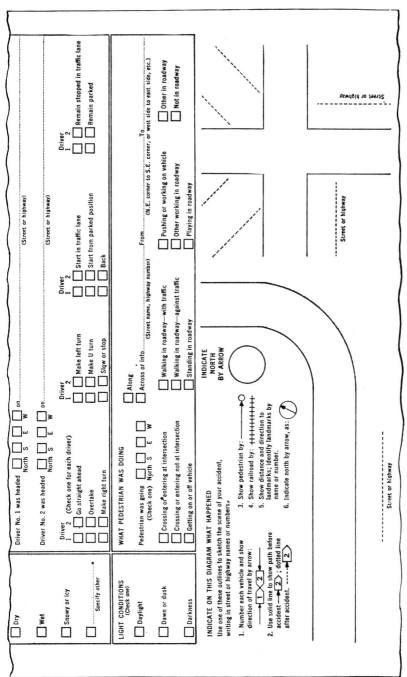

Fig. 10-1B. Standard accident report form, reverse side.

179

exposed to hazard; registered vehicles are a measure of the magnitude of the injuring agency; and vehicle-miles[1] are a measure of the magnitude of the traffic streams in which accidents occur.

Spot Analysis.   Both the location file and the spot map may be used to determine areas, sections of routes, specific intersections, or other *spots*

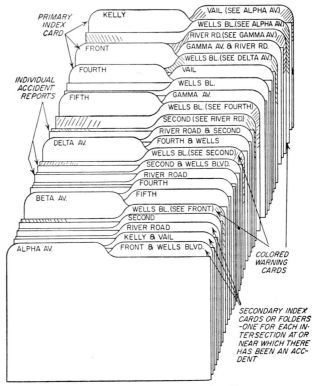

FIG. 10-2. Typical city location file.

of high accident frequency.   From the viewpoint of the hazards presented by a particular location, the collision diagram (Fig. 10-3) is useful. Such a diagram is schematic and is devised to show the paths of vehicles and pedestrians just prior to the accident.   Since the collision diagram may furnish clues as to reasons for traffic failure, it should always be reviewed with a diagram of traffic flow (volume, speed, timing, move-

[1] Vehicle-miles do not indicate number of vehicles per mile or density of traffic. On a given route the opportunity for single-vehicle accidents is directly proportional to the number of vehicles which pass.   On the other hand, it may be argued that the opportunity for two-vehicle collisions varies as the square of the number of vehicles, since two vehicles passing two other vehicles create chances for four accidents, and four vehicles passing four others create chances for 16 accidents.   This law, however, has not been demonstrated, and while some exponential function seems evident, it is not of the pure square form (see inset in Fig. 10-7).

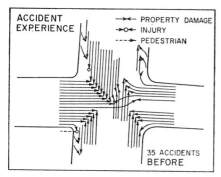

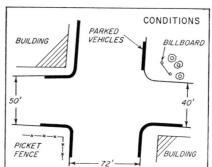

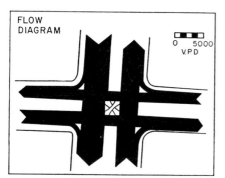

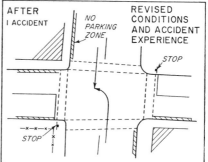

FIG. 10-3. Typical collision diagram analysis.

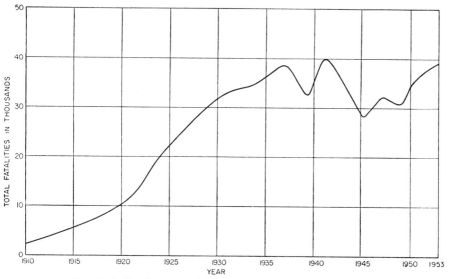

FIG. 10-4. Total motor vehicle fatalities in the United States.

ment) values.   By correlation of the accident history with traffic flow and roadway geometry and conditions, underlying factors of accident causation may become apparent, and remedies for their correction may be suggested.

10-3. Accident Characteristics. *Magnitude and Trends.* The record of fatalities since the advent of the motor vehicle is set forth in Fig. 10-4.   During the year 1952, 38,000 fatalities and 1,350,000 injuries were incurred.   The peak year of accident occurrence was 1941, when nearly 40,000 persons were killed and about 1.4 million were injured.

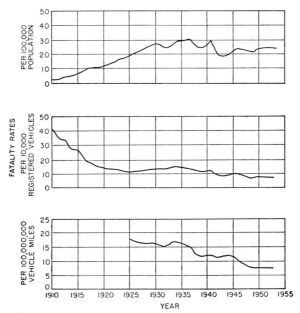

Fig. 10-5. Trends in motor vehicle fatality rates.

During the last 20 years, except for the war years, there have been between 30,000 and 40,000 deaths per year.

Because of growing population, registration, and vehicle usage, the trend in accident rates, as measured by the ratio of fatalities per year to these basic factors, gives a better measure of the trend of traffic accidents (Fig. 10-5).   On a population basis, the rate has stabilized since 1930 at about 25 deaths per 100,000 population.   There has been a steady decline in the rate of deaths per 100 million vehicle-miles.   This rate is currently about 7.3.

Cyclical Variations in Traffic Accident Occurrence.   Cyclical variations in the times of traffic accident occurrence are related to the diurnal, weekly, and seasonal patterns in traffic volume.   As volume of traffic increases, there is a tendency for traffic accidents to increase.

The characteristic seasonal pattern of traffic fatality occurrence is shown in Fig. 10-6. For comparative purposes this graph also shows the index of fatalities per 100 million vehicle-miles, as well as the seasonal variation in mileage. The index of *deaths* is lowest in the winter and spring months, passes through normal in early summer, and remains above average for the remainder of the year. The *death rate* on a mileage basis is highest in the fall and early winter months, passing through normal in February and remaining subnormal until September.

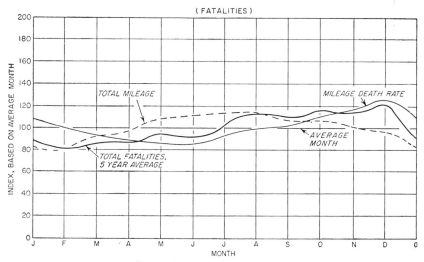

FIG. 10-6. Seasonal variation in accident occurrence.

When accidents are distributed by days of the week the resultant distribution is markedly similar to the variation of traffic volume (Fig. 10-7). Week ends have high frequency, and midweek days develop the lowest ratio of accident occurrence. When accidents are at a normal level, so is traffic volume. As volume departs from normal, accident departure is in the same direction but at a higher rate.

Characteristic daily variation in accident occurrence is set forth in Fig. 10-8, showing clearly that the hours of darkness increase accident occurrence. During hours of daylight, when traffic is above normal, the index of accidents is generally below the index of traffic volume, but in hours of darkness the index of accident occurrence is above that of traffic. Fatigue and alcohol are other factors which may increase hazard in evening and night hours.

Regional Distribution of Accidents. The incidence of traffic accidents throughout the United States on a mileage rate basis shows a range of 7.3 to 8.0 deaths per 100 million vehicle-miles for the 5 years preceding 1953. Their current mileage rate when applied to six regional areas of the nation shows a spread of 5.0 to 9.2, with lowest rates in the North

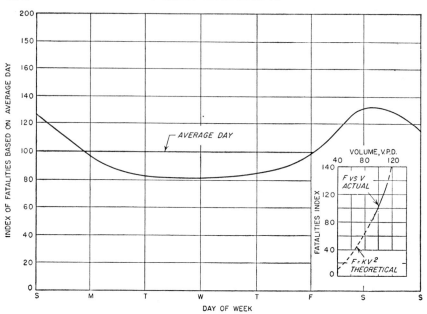

FIG. 10-7. Weekly variation in accident occurrence by day of week.

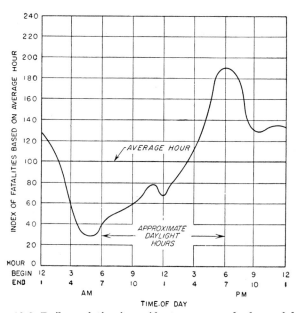

FIG. 10-8. Daily variation in accident occurrence by hour of day.

Atlantic states, next lowest in the North Central states and the Pacific states. The South Atlantic, South Central, and Mountain states develop higher rates, generally above 8.2.

On the population-rate basis for the same period (1948 to 1952), the range of values is from 21.2 to 24.4 deaths per 100,000 population for the nation as a whole. The same six regional areas developed rates from

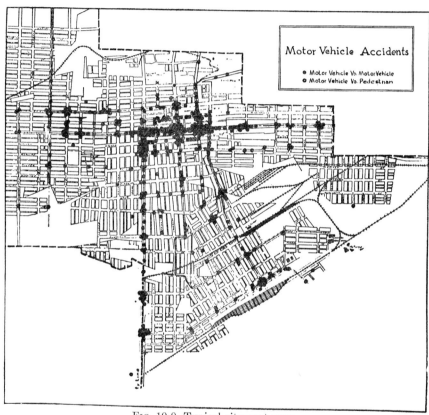

FIG. 10-9. Typical city spot map.

14.0 to 39.7, with the North Atlantic states the lowest and the Pacific and Mountain states highest, with rates above 29.5.

Rural-Urban Distribution of Accidents. For several decades the proportion of total accidents in rural areas has steadily increased in relation to the number in urban places. In the mid-twenties rural and urban locations contributed about the same number of fatalities per year, but by the mid-thirties rural areas were contributing 1½ times as many fatalities as cities. In the first half of the forties the rural rate was twice the urban-area rate, and by 1952 this ratio was about 2.4:1 when reckoned either on total fatalities or mileage rates.

Distribution of Accidents by Routes.   Accidents tend to concentrate in those routes which carry the heaviest volumes of traffic.   Spot maps are usually employed to review the characteristic pattern of accident distribution for a given area (Fig. 10-9), which shows the concentration of accident occurrence along a few routes and at a few intersections.   Similar concentration is shown for a state-wide area in Fig. 10-10.   The principal arteries of travel are clearly indicated by the frequency of accident occurrence.   In this case the average rate for all mileage was about 2 accidents per mile per year, but certain portions of the routes developed as many as 153.

Intersection Concentration of Accidents.   In urban areas nearly half of the fatal accidents occur at intersections, while in rural areas this percentage is only 10 to 15.   Types of hazards and number of intersections per mile are important factors in the distribution of accidents.   The hazard to pedestrians, especially in urban areas, is evident from Table 10-1.   Collision between motor vehicles is also of major importance. However, in rural areas noncollision fatal accidents make up over a quarter of the total.

TABLE 10-1. ELEMENTS INVOLVED IN FATAL TRAFFIC ACCIDENTS

Elements	Urban, per cent	Rural, per cent	Total, per cent
Collision			
Pedestrian...................................	56.4	18.8	29.5
Other motor vehicles........................	19.5	36.5	31.6
Railroad train..............................	4.9	5.0	5.0
Streetcar...................................	1.1	0.0	0.3
Bicycle.....................................	2.2	1.4	1.6
Animal-drawn vehicle........................	0.1	0.2	0.2
Animal......................................	0.0	0.2	0.1
Fixed object................................	7.1	8.6	8.1
Not known..................	2.3	2.6	2.5
Total collisions.........................	93.5	73.4	79.1
Noncollision			
Overturning on roadway......................	1.1	4.2	3.3
Running off roadway.........................	3.2	19.5	14.9
Other noncollision..........................	2.2	2.9	2.7
Total noncollisions.......................	6.5	26.6	20.9
Total collisions and noncollisions...........	100.0	100.0	100.0

SOURCE: "Motor Vehicle Accident Fatalities," U.S. Public Health Service, *Vital Statistics,* vol. 35, no. 15, 1948, Table 8B.

Frequency and Severity.   Since the foregoing data are based on fatalities only, percentages do not reflect frequency of accident occurrence.

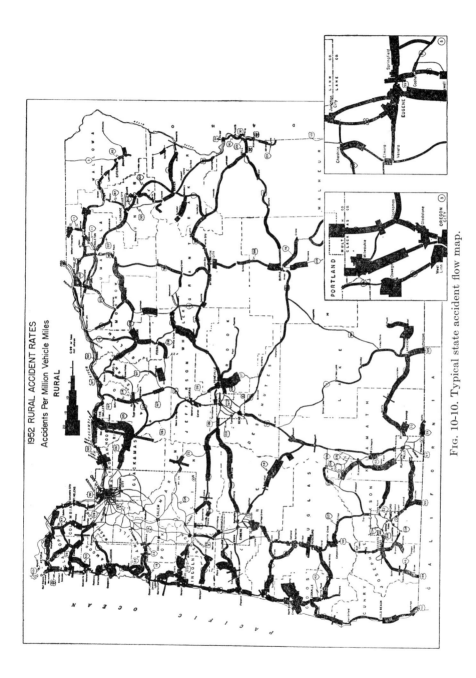

FIG. 10-10. Typical state accident flow map.

187

While over one-half of urban fatalities involve pedestrians, pedestrian accidents account for only 6 to 7 per cent of all urban accidents.

Movements Involved in Accidents. Pedestrian movement is involved in 25 per cent of fatal traffic accidents, and movement of another vehicle in 31 per cent of the cases. In 44 per cent of fatal cases the movement of only a single vehicle is involved. Illustrative distributions are shown in Table 10-2.

TABLE 10-2. MOVEMENTS INVOLVED IN ACCIDENTS
(Per cent of total occurrence in each class)

Movement involved	Urban		Rural		Total
	All	Fatal	All	Fatal	Fatal only
Pedestrian......................	7	56	1.5	15	25
Another vehicle.................	83	23	64.0	37	31
One vehicle only...............	10	21	34.5	48	44
Total.......................	100	100	100.0	100	100

Rural accidents involving pedestrian movement are 10 times as likely to result in fatality as the average accident, but in urban areas pedestrian movement is only about 8 times as likely to result in fatality as the average accident.

Further detailed analysis involving the distribution of accidents in rural and urban areas by types of movement in 1952 is shown in Table 10-3.

Modifying Circumstances. Accident occurrence is modified by the road user involved, vehicle defects, design and condition of maintenance, road conditions, weather, lighting, and illumination. For example, persons under 25 and over 65 years have excessive fatality-accident rates on a mileage-driven basis. The increase in hazard during hours of darkness was clearly shown in Fig. 10-8. It is seldom possible to assign a single cause to a given accident, but the incidence of certain modifying factors is as follows:

Driver under alcoholic influence. 5 to 8 per cent of fatal cases.
Physical defects, drowsiness, fatigue, or sleep. 3 to 6 per cent of fatal cases.
Alleged violation of rules of the road. 50 to 60 per cent of accident cases.
Vehicle defects. 15 to 20 per cent of fatal cases. Defective brakes, tires, and lights present in half of such cases.
Obstructions to vision due to weather, unclean windshields, low sight distances, glare, and other factors. 20 per cent of fatal cases.
Abnormal road surface conditions due to sleet, ice, mud, and other factors causing low friction values. 20 per cent of fatal cases.
Highway defects. 10 per cent of accident cases.

10-4. Interpretation and Use of Accident Facts. Perhaps the only class of traffic accidents which is fully known and reported involves

TABLE 10-3. DIRECTIONAL ANALYSIS, MOTOR VEHICLE TRAFFIC ACCIDENTS, 1952, IN PER CENT

Location and movement	State-wide fatal (18 states)	Urban		Rural	
		All accidents (346 cities)	Fatal (346 cities)	All accidents (13 states)	Fatal (13 states)
Total accidents.....................	100.0	100.0	100.0	100.0	100.0
Pedestrian intersection accidents.........	4.0	3.0	25.0	0.3	1.0
Car going straight, entering intersection.	1.7	0.9	8.5	0.1	0.4
Within intersection................	0.6	0.3	3.7	0.1	0.2
Leaving intersection..............	1.2	0.6	7.9	0.1	0.3
Car turning right, entering intersection.	0.1	0.1	0.3	*	0.1
Within intersection................	*	0.1	0.4	*	*
Leaving intersection..............	0.1	0.2	1.1	*	*
Car turning left, entering intersection..	*	0.1	0.3	*	*
Within intersection................	0.1	0.1	0.4	*	*
Leaving intersection.	0.2	0.5	2.0	*	*
All others........................	*	0.1	0.4	*	*
Pedestrian nonintersection accidents.....	16.0	3.0	28.0	1.3	12.0
Car going straight..................	15.0	2.7	24.9	1.2	11.1
Car backing.......................	0.4	0.1	0.7	*	0.2
All others........................	0.6	0.2	2.4	0.1	0.7
Two motor vehicle intersection accidents.	11.0	43.0	18.0	19.4	10.0
Entering at angle, both going straight..	7.2	18.5	13.1	5.5	5.7
One right, one straight............	0.1	1.6	0.3	0.6	0.2
One left, one straight..............	0.6	3.2	0.9	1.6	0.6
All others........................	0.2	0.7	0.1	0.6	0.7
Entering from same direction, both going straight..................	0.3	3.7	0.4	1.1	0.2
One right, one straight............	0.1	1.6	0.1	0.8	0.2
One left, one straight..............	0.5	2.4	0.2	4.0	0.5
One stopped......................	0.2	5.2	0.2	2.2	0.1
All others........................	*	1.0	0.2	0.4	*
Entering from opposite direction, both going straight..................	0.5	1.1	0.6	0.6	0.6
One left, one straight..............	1.2	3.5	1.8	1.5	1.1
All others........................	0.1	0.5	0.1	0.5	0.1
Two motor vehicle nonintersection accidents........................	23.0	42.0	8.0	44.9	28.0
Opposite directions, head-on collision...	7.6	1.1	3.0	2.9	10.6
Sideswipe collision................	7.1	2.2	0.9	10.8	7 6
Same direction, rear-end collision......	2.6	7.1	0.6	8.2	3.3
Sideswipe collision................	1.3	5.6	0.5	5.0	1.4
One car parked, proper location.......	0.6	10.6	1.4	1.4	0.3
Improper location.................	0.8	0.7	0.2	1.2	0.9
One car stopped in traffic............	0.9	5.4	0.3	5.8	0.9
One car leaving parked position........	0.1	4.7	0.2	0.7	0.1

TABLE 10-3. DIRECTIONAL ANALYSIS, MOTOR VEHICLE TRAFFIC ACCIDENTS, 1952, IN PER CENT (*Continued*)

Location and movement	State-wide fatal (18 states)	Urban		Rural	
		All accidents (346 cities)	Fatal (346 cities)	All accidents (13 states)	Fatal (13 states)
One car entering alley	0.1	0.2	*	0.3	*
One car leaving alley	*	0.3	*	*	*
One car entering driveway	0.9	1.1	0.1	4.9	1.3
One car leaving driveway	0.3	1.7	0.3	1.1	0.4
All others	0.7	1.3	0.5	2.6	1.2
Other accidents	46.0	9.0	21.0	34.1	49.0
Collision with nonmotor vehicle, inter-section	1.2	1.0	2.6	0.2	0.7
Not at intersection	4.1	0.9	2.2	1.8	3.0
Collision with fixed objective in road, inter-section	0.5	0.5	0.9	0.7	0.9
Not at intersection	4.1	1.1	1.5	2.9	4.4
Overturned in road, intersection	0.4	0.1	0.2	0.2	0.2
Not at intersection	3.2	0.2	0.6	1.9	3.3
Left road at intersection, then over-turned	0.8	0.1	0.3	0.9	0.9
Then struck fixed object	0.7	1.0	1.9	0.9	0.8
Then struck other vehicle	0.1	*	*	0.2	*
Then struck pedestrian	0.1	*	*	*	*
Left road at curve, then overturned	8.6	0.1	0.7	8.3	13.4
Then struck fixed object	5.2	0.5	2.6	2.5	4.5
Then struck other vehicle	0.1	*	*	0.2	0.3
Then struck pedestrian	0.1	*	*	*	0.1
Left road on straight road, then over-turned	8.5	0.2	0.9	7.1	8.9
Then struck fixed object	5.0	1.9	3.5	3.8	4.1
Then struck other vehicle	0.3	*	0.1	0.4	0.5
Then struck pedestrian	0.1	*	0.3	*	0.1
Fell from vehicle, boarding, alighting in traffic	0.3	*	0.2	*	0.3
Not boarding or alighting	1.8	0.2	0.8	0.2	1.6
Injured within vehicle	0.1	0.1	0.1	*	*
Driverless moving vehicle	*	0.3	0.1	0.1	*
Fire (no other event)	*	*	*	0.2	*
All others	0.7	0.8	1.5	1.6	1.0

* Less than 0.05 per cent.

SOURCE: Reports of state and city traffic authorities. Urban based on reports from cities with more than 10,000 population. *Accident Facts*, National Safety Council, Chicago, 1953, p. 64.

fatality. In cities where reporting of injury accidents is required by law, there is an average of 20 to 60 injury accidents reported for each fatal accident. However, many accidents are never reported, especially those involving property damage only.

If complete reporting of accidents could be accomplished, any analysis designed to determine the correlation of accident occurrence with a particular factor would require merely the elimination of influence by other factors. Unfortunately, when accidents are distributed by time, space, light conditions, and other fundamental factors, the number remaining for comparison under a controlled set of conditions is frequently too small to have statistical or practical significance. If the scope of *cause* is enlarged, certain categories of causative factors may be discerned. For example, it seems clear that darkness, with such related factors as fatigue, poor night vision, and effects of alcohol, creates a condition of extra hazard. As categories of causative factors are enlarged, the dependability of analysis is increased, but specific, useful qualities are obscured.

Legal versus Natural Causes. In the search for accident causation there is a tendency to charge road users with violation of some preconceived notion of moral or statutory law and thus to establish the cause of accident. While the traffic engineer is vitally concerned with the system of traffic regulation and accepted conventions of society, it is his responsibility to search for the scientific facts which surround accidents and if possible find the laws of nature which influence or govern accident causation.

In one case, for example, right-angle collisions at a signalized intersection on a high-speed road were numerous. In the attempt to reduce accidents, many persons were charged with violation of signals. It was later found that the mere lengthening of the amber of clearance period practically eliminated all right-angle collisions and numerous rear-end collisions. Here it is clear that violation of the natural laws of inertia, momentum, and human PIEV time, rather than intended violation of legal statute governing the meaning of signal legend, was the cause of accident.

SECTION 2

REGULATIONS

CHAPTER 11

BASIC CONSIDERATIONS IN
TRAFFIC REGULATION

The job of regulating the public is difficult, since many persons feel that traffic controls are an encroachment on their individual driving rights. It becomes necessary to show that controls which might be interpreted as restrictions on the part of a particular individual are necessary for the general welfare. Where it can be demonstrated that regulations do not curtail the rights or actions of the majority, the fact should be well publicized. Legislative bodies and traffic authorities must keep in mind that unreasonable restrictions or regulations, such as overrestrictive speed limits, are not likely to last very long.

To develop the most reasonable and effective traffic regulations, facts must be sought through traffic surveys, accident studies, driver records, and other sources of data so that bias, political influences, and other undesirable approaches can be avoided. In too many cases traffic regulations are developed because a legislative body is finally "driven to action" or because traffic conditions have become so chaotic that legislative groups are "willing to try anything." Often regulations are not adequate because they lag far behind needs and changing conditions. Regulations must apply to all phases of traffic. One component should not be overemphasized in relation to those of all groups and agencies having a responsibility in the total problem.

Why Traffic Regulations. The regulation of highway traffic covers all aspects of the control of both vehicle and driver. The vehicle must be controlled as to registration, ownership, mechanical fitness, accessories, size, and weight. Drivers must be regulated as to age, ability to operate specific types of vehicles, and financial responsibility. Even with basic controls on the vehicles and drivers, regulations must be prescribed as to how they will operate in the traffic stream on public ways. For legality and effectiveness, all highway-traffic regulations are dependent upon the laws of states and local governments, especially the traffic ordinances of cities. In addition, some control is prescribed by Federal laws and regulations, specifically in the operation of vehicles in interstate commerce. The wide adaptability and universal ownership of the automobile make rigid regulations essential.

It has always been demonstrated that education as to what is good, or correct, is not enough; in matters of public concern, the government must regulate. It must be recognized that in traffic, regulations come as near to affecting the actions of individuals as in any other category of social organization, because practically every person is a road user—if not a driver, a pedestrian.

Regulations Should Be Rational. Irrational regulations cannot be enforced except by tremendous effort and expense, and then usually not for long. There are social problems, economic problems, and human habits which must be considered. If the habits of a community are greatly at variance with the regulations, success cannot be attained for any substantial period of time.

Regulations should be developed progressively. Changes of too great a magnitude appear radical and thereby irrational. Like all other phases of traffic and highway transportation, regulations must be planned over a long period of time, and the effects must be carefully observed so that alterations can be made as experience dictates.

Regulations Alone Often Not Enough. Regulations constitute but one approach to the over-all traffic problem. Alone they cannot correct any serious traffic condition. When public acceptance is poor and enforcement is lax, regulations may be totally ineffective. However, they are necessary to define the rights of individuals and to specify correct actions by individuals. They may, in themselves, reduce severity of the problem or provide a long-lasting cure. They must be used in combination with control devices, over-all highway planning and design, and administrative policies for a comprehensive traffic engineering job.

11-1. Authority to Regulate. The authority to fix practices and to make changes in existing practices can come only from official regulations, even though in some cases broad powers to promulgate regulations may be provided by basic legislation. All enforcement and court practices are directly affected by the various regulations prescribed by legislative and administrative bodies. Local authorities must be careful to ascertain that necessary enabling legislation has been provided by the state to allow the establishment of local traffic ordinances.

Many traffic regulations are effected under the general police powers provided for the general welfare and safety of the public. It has been held, however, that police power must be restricted to the regulation of the use and enjoyment of private property.[1] The extent to which regulations can be developed under police power varies widely within the different states. It is largely a matter of court opinions and interpretations and, in some cases, relates to the charter or constitution of the state and cities involved. For example, in an Ohio case,[2] the court held: "It lies

---

[1] *Noble State Bank v. Hasket,* 31 Sup. Ct. 168 (1911).

[2] *Ragland v. Wallace,* 80 Ohio App. 210, 70 N.E.2d 118 (1946).

within the power of the legislature to control the operation of motor vehicles on the public highways, prescribing who may operate such motor vehicles and under what conditions such right shall be denied. The legislature is not restricted in the exercise of this power unless it violates some specific constitutional provision or the legislation is found to be an unreasonable exercise of such power." Similarly, it was found in a Michigan case[1] that "regulation of traffic on the public highways of the state is a primary responsibility of the state to the public. The use of vehicles and travel by foot is subject to regulation under police power, and a large discretion is vested in the legislature in its exercise."

The legal extent of control which cities may exercise has also been the subject of many cases. "The regulation of motor vehicles on particular streets even to the complete exclusion therefrom, when deemed necessary in the public interest, is within the police power delegated to municipalities, and while such regulation may be considered drastic in its operation, a court is not at liberty to substitute its judgment for that of the municipality as to the best methods of relieving traffic congestion in a specified area in the interest of public welfare," according to a New Jersey court.[2] In contrast, in a Connecticut case[3] the Town of Darien was found to have exceeded its police powers in prohibiting trucks from using an important bypass route. The finding was, in effect, that the state had delegated the power to make traffic rules to the town but had retained the power to regulate vehicles. Since this local ordinance was discriminatory to trucks, it was held a vehicle regulation and thereby beyond the powers of the town's police commission. The need for checking the extent of the local authority is discussed in more detail in Chapters 34–36.

11-2. Development of Traffic Laws. The five Uniform Traffic Acts and the Model Municipal Ordinance developed by the National Conference on Street and Highway Safety serve today as a general guide for uniform traffic legislation throughout the United States.

As motor vehicle registrations increased rapidly following World War I and as accident and congestion problems developed, it became increasingly evident that traffic laws were in many instances inadequate to meet the conditions and that there was a great disagreement between the laws of different states and different localities within the same state.

In 1924, the sovereign power of the individual states to regulate and control traffic activities was recognized at the First National Conference on Street and Highway Safety in Washington. This conference was sponsored almost solely by business and commercial groups and by others having a direct interest in highway transportation. The conference was pointed up as a "life-and-death" conference, at which drastic actions

[1] *Jacobson v. Carlson*, 302 Mich. 448, 4 N.W.2d 721 (1942).
[2] *People's Rapid Transit Co. et al. v. Atlantic City*, 105 N.Y.L. 286, 144 A. 630 (1928).
[3] *Adley Express Co. v. Town of Darien*, 125 Conn. 501, 7 A.2d 446 (1939).

were needed to develop programs and means for reducing accidents. There was no uniformity in traffic laws at that time. (Accidents were blamed on a small group of "vicious" or "ignorant" drivers. Records of police and motor vehicle departments did not permit accurate determination of accident repeaters.) It is important to note that at the first conference, the sovereign power of the individual state to regulate and control traffic activities was recognized. The principal accomplishments of the first conference were the adoption of recommendations that (1) states adopt driver-license laws, (2) states establish motor vehicle departments, (3) certificate of title be issued for automobiles, (4) accident reporting be made compulsory, (5) compulsory safety education be adopted in public schools, (6) more state funds be made available for traffic improvement, (7) uniform speed limits be passed in all states (15 mph minimum, 35 mph maximum prima facie limit), (8) basic road-construction standards covering lane widths, grades, and curvatures be adopted, and (9) uniform colors for signs and signals be agreed upon.

At the Second National Conference in 1926, every attempt was made to have official representatives from each of the 48 states. The first Uniform Motor Vehicle Codes were adopted. At that time the codes consisted of (1) Uniform Motor Vehicle Registration Act, (2) Uniform Motor Vehicle Anti-theft Act, (3) Uniform Vehicle Operators' and Chauffeurs' License Acts, and (4) Uniform Act Regulating Operation of Vehicles. In addition, the first edition of the Model Municipal Ordinance was approved, and the use of traffic engineers was formally suggested and recommended for the first time by an important public body.

The one outstanding achievement of the Third Conference, in 1930, was the approval of the *sign manual* prepared by the American Engineering Council, thereby making it the first edition of the *Manual on Uniform Traffic Control Devices*.

The Fourth Conference in 1934 emphasized the alarming increase in accidents during the early thirties, particularly 1933. Low accident rates reported by states that had adopted substantially the recommendations of previous conferences were cited as reason enough for a nationwide adoption of uniform traffic laws covering all important aspects of driver and vehicle control. Out of this conference came the basic codes which are still in use: (1) Uniform Motor Vehicle Administration, Registration, Certificate of Title, and Anti-theft Act, (2) Uniform Motor Vehicle Operators' and Chauffeurs' License Act, (3) Uniform Motor Vehicle Civil Liability Act, (4) Uniform Motor Vehicle Safety Responsibility Act, and (5) Uniform Act Regulating Traffic on Highways. This Conference also adopted a revised Model Municipal Traffic Ordinance and approved a revised *Manual on Uniform Traffic Control Devices* as one of three sponsoring agencies (the other two were the American Association of State Highway Officials and the U.S. Bureau of Public Roads).

In 1947 the National Conference was officially disbanded, but the work of revising and keeping up to date the Uniform Codes and Model Ordinance was taken over by the National Committee on Uniform Traffic Laws and Ordinances. Seventy-two national agencies constitute the membership, and each agency designates a representative. Through the work of this committee, the Codes and Ordinance are kept current with revisions and changes. At least one formal meeting of the National Committee is held each year.

The first President's Highway Safety Conference, taking up some of the activities of the National Conference, was held in May, 1946. Since then there have been additional President's Highway Safety Conferences. The functions of the President's Conference are carried on through eight committees: (1) Committee on Accident Records, (2) Committee on Education, (3) Committee on Enforcement, (4) Committee on Engineering, (5) Committee on Laws and Ordinances, (6) Committee on Motor Vehicle Administration, (7) Committee on Organized Public Support, and (8) Committee on Public Information. These committees analyze and discuss the problems of each subject and make recommendations for increasing the effectiveness of the program, or the phase of highway safety under consideration.

The first President's Conference, made up of over 2,000 delegates from all of the 48 states, worked out an "action program" which is recognized as the most comprehensive collection of highway-safety measures ever assembled. Following conferences improved and expanded the action program and reported progress made as a result of the adoption of the action recommended by the preceding conferences. The President's Highway Safety Conference is now called at about two-year intervals.

The Uniform Motor Vehicle Code. Act I of the Uniform Vehicle Code, entitled Uniform Motor Vehicle Administration, Registration, Certificate of Title and Anti-theft Act, provides for a separate motor vehicle department in the state government, to be controlled by a motor vehicle commissioner. Details connected with motor vehicle registration, certificate, and transfer of title, recording of liens and other financial encumbrances on the vehicle, special registration requirements of dealers, manufacturers, and transporters, licensing of dealers and wreckers, and motor vehicle theft are also covered.

Act II, Uniform Motor Vehicle Operators' and Chauffeurs' License Act, deals with minimum age and other requirements for licensee; special requirements for chauffeurs; cancellation, suspension, or revocation of licenses; constitution of license-provision violations; and penalties which may be imposed for committing such violations.

Act III, Uniform Motor Vehicle Civil Liability Act, treats of the liability of states, counties, municipalities, and other public corporations in motor vehicle accidents; the imputing of negligence or willful misconduct of operator to owner; liability for injury or death of a guest (otherwise known as the hitchhiker law); method of service of process on a non-resident; and control of the owners of for-rent vehicles.

Act IV, Uniform Motor Vehicle Safety Responsibility Act, covers financial responsibility and security as a result of automobile accidents. Main points include report of accidents, details and conditions governing the filing of security, the mechanics of future proof of responsibility in case of accident, penalties for violation of provisions of the act, and general provisions such as exceptions and constitutionality. It has been found that more states have adopted this act or an acceptable substitute than any other.

In Act V, Uniform Act Regulating Traffic on Highways, the traffic engineer will find his tools—the authority to set up such control measures as one-way streets, speed zones, or parking regulations. Included in this act are accidents and accident reports; traffic signs, signals, and markings; rules of the road such as driving on the right, overtaking and passing, turning and starting, and right of way; speed restrictions; pedestrian rights and duties; equipment and inspection of vehicles; and regulation of size, weight, and load. Such technicalities as penalties and fines, obedience to traffic laws, and criminal procedure are also provided for.

The Model Traffic Ordinance, for use in municipal traffic control, contains many of the provisions found in Act V. Traffic administration within the city government is adequately provided for and the authority of the traffic engineer is set forth. Traffic control devices are dealt with in much the same manner as in Act V, as are speed restrictions, pedestrians, rules of the road, and parking regulations. Additional provisions cover loading zones, parking limitations, and other local-control measures, and their enforcement, violations, and penalties.

The five acts of the Uniform Vehicle Code and the Model Traffic Ordinance, as described above, make up a comprehensive set of laws to serve as a standard to be sought for motor vehicle regulations throughout the country.

11-3. Need for Uniformity in Traffic Laws. One of the most urgent legislative needs in this country is the enactment of modern, comprehensive, and uniform motor vehicle laws. The establishment and development of the Uniform Vehicle Code has done much to bring this about, but a considerable amount of work remains in many areas.

In motor vehicle legislation, as in any other legislation, uniformity is virtually unattainable if important terms and wordings are defined differently by various groups. Too often regulatory terms vary in meaning from state to state and sometimes even from city to city within a single state. When this condition exists it is unreasonable for motor vehicle and traffic authorities to expect obedience to traffic laws. In the present era of high speed and long-distance travel, it is not unusual for a driver to pass through a number of legislative jurisdictions in each of which the traffic and motor vehicle laws are different and terms and words used vary. If uniformity is to be achieved, it is important that definitions be considered and arrived at sensibly, so that authorities throughout the country will have the same interpretation of words and phrases and confusion will be kept to a minimum.

Law-enforcement agencies and authorities would benefit from the adoption of uniform laws. Uniformity would reduce the need for arrests and promote greater understanding between vehicle operators and enforcement officials. It would help to standardize safety equipment and reduce the cost of operation of larger and heavier vehicles. Vehicles built and equipped according to the laws of one state would be able to travel legally in all other states. Many present-day highway "barriers" such as differing size and weight restrictions would be removed, and transportation costs would be decreased as a result of the reduced amount of handling necessary. Uniformity would improve traffic conditions throughout the country and would simplify and standardize engineering problems. Accidents would be reduced and efficiency of streets and highways would be increased, with higher speeds and less congestion as a result of the uniform rules. Provisions could cover extreme local conditions, so that extensive use of uniform laws should not be a hindrance to authorities in a solution of local problems.

The numerous advantages to be gained through full-scale adoption of uniform traffic laws could possibly be offset by the stagnation which would result in legislative groups in regard to further rule changes. Perhaps the only advantage of having varied laws throughout the country is that the experience gained by a certain state in the use of one type of law may be of value to another state in setting up a better law regulating the same subject.

Degree of Uniformity Achieved. Since the establishment of the Uniform Vehicle Code in 1926, major strides have been taken by most states in an effort to make their laws conform with the Code. Complete acts of the Code have been placed into the statutes of many states, and others have adopted verbatim parts of the various acts. A recent check showed that all states but one (Louisiana) had at least one act of the Uniform Code in their laws.[1] Table 11-1 indicates the states which may be considered to have adopted the various acts of the Uniform Vehicle Code.

Since its inception in 1946 the President's Highway Safety Conference through its Committee on Laws and Ordinances has recommended that each state make a detailed analysis of its motor vehicle laws and compare them with the Uniform Vehicle Code. Some states have completed such studies, and others have projects under way. Marked legislative changes

[1] President's Highway Safety Conference, *Report of Committee on Laws and Ordinances*, June, 1949, p. 19.

TABLE 11-1. UNIFORMITY IN STATE MOTOR VEHICLE LAWS
(States considered to have adopted Uniform Vehicle Code acts or
reasonable substitutes*)

Act I Registration, certificate of title, etc.	Act II Driver licensing	Act IV Safety responsibility	Act V Traffic regulation
Arkansas	Arkansas	Alabama	Alabama
California	California	Arizona	Arkansas
Delaware	Colorado	Arkansas	California
Idaho	Delaware	California	Colorado
Maryland	Florida	Colorado	Delaware
Michigan	Idaho	Connecticut	Florida
New Mexico	Kansas	Delaware	Idaho
North Carolina	Maryland	Florida	Illinois
North Dakota	Michigan	Georgia	Iowa
Ohio	Minnesota	Idaho	Kansas
Pennsylvania	Nevada	Illinois	Maryland
Tennessee	New Mexico	Indiana	Michigan
Utah	North Dakota	Iowa	Mississippi
Virginia	Ohio	Kansas	New Mexico
Washington	Oregon	Kentucky	North Dakota
West Virginia	Pennsylvania	Maine	Ohio
	South Carolina	Maryland	Oklahoma
	Tennessee	Massachusetts	Oregon
	Texas	Michigan	Pennsylvania
	Utah	Minnesota	South Carolina
	Virginia	Missouri	Tennessee
	Washington	Montana	Texas
	West Virginia	Nebraska	Utah
	Wisconsin	New Hampshire	Washington
	Wyoming	New Jersey	Wisconsin
		New Mexico	
		New York	
		North Carolina	
		Ohio	
		Oklahoma	
		Oregon	
		Pennsylvania	
		Rhode Island	
		South Dakota	
		Tennessee	
		Utah	
		Vermont	
		Virginia	
		Washington	
		West Virginia	
		Wisconsin	
		Wyoming	

* Provisions of Act III were not analyzed.
SOURCE: President's Highway Safety Conference, *Report of Committee on Laws and Ordinances*, June, 1949, p. 18.

result from most of the studies because they bring into focus the legislative needs in relation to present traffic and highway conditions—conditions which can change rapidly.

Cities are urged to take action of the same sort as states and compare their ordinances with the Model Traffic Ordinance. Some cities have completed studies of this sort, and their findings support the idea and show the values that can be derived. It is quite probable that general uniformity in city ordinances is further from achievement than uniformity in state vehicle codes.

In recognition of the work accomplished in completing such comparative studies as have just been discussed and of the continued and added interest of civic-minded organizations throughout the country, it is apparent that progress is being made in the attainment of complete uniformity of motor vehicle laws. The continued support of responsible individuals, committees, pressure groups, and other organizations is bound to have its effect upon legislative bodies and officials and gives promise of greater uniformity of traffic laws in the future.

CHAPTER 12

DRIVER CONTROLS

There are three elements with which the traffic engineer must deal—the road, the vehicle, and the driver. The road and the vehicle are subject to constant change and improvement. Actually, however, the traffic engineer considers them inflexible, since once they are built their characteristics are inherent and can be changed only with great expense. A major portion of existing regulations are aimed at the driver. Minimum age limits (varying from 14 to 18), vision tests, law tests, and a road test have all been instituted since 1900, and steady improvement has been noted. However, the only drivers examined are those who learn to drive after the law is put into effect. Hence a large portion of present-day drivers have never taken an examination.

It is primarily through driver licensing that improvement in driving habits and practices can be achieved. When it can be reliably assumed that all drivers in a state have a reasonable minimum ability, the problems of traffic design and regulation are considerably reduced. Furthermore, since the engineer may be called upon to suggest changes or complete revisions of state license laws, he should be thoroughly familiar with the problems involved.

Present license laws are designed to exclude few.[1] No state requires a physical examination; visual acuity is the only eye test given, with no special test for night driving; the road test usually takes only 15 to 20 min. License laws have not kept pace with growth in vehicle and roadway characteristics, particularly speed and volume. Maximum speed has increased from 58 mph in 1926 to 95 mph in 1948[2] and now to more than 100 mph; volume of traffic in vehicle-miles increased from 55 billion in 1921 to 398 billion in 1948 to about 500 billion in 1954.

There are wide variances in existing license laws. Minimum age

[1] Forty-nine per cent of population 16 or over are drivers. See *Automobile Facts and Figures*, 28th ed., Automobile Manufacturers Association, Detroit, 1948, p. 38.

[2] Henry K. Evans (ed.), *Traffic Engineering Handbook*, 2d ed., Institute of Traffic Engineers, New Haven, Conn., 1950, p. 60.

requirements range from 14 to 18.[1] Visual-acuity requirements range from 20/30 both eyes without glasses to 20/70 with glasses.[2]

12-1. Driver Licensing. A license has been defined as "a privilege granted by competent authority, to do that which would be unlawful without such privilege."[3] A driver's license fits this definition, and in almost every state with a driver's license law there have been court decisions upholding that law on the grounds that driving a motor vehicle (on public roads) is a privilege and that no person has a God-given right to operate a vehicle on a public roadway.[4]

One of the major purposes of licensing is to impress drivers with the fact that they are being granted a privilege and that misuse may lead to suspension or revocation. The average law disqualifies only those totally unfit: totally blind, epileptics, and persons addicted to narcotics or alcohol. However, drivers with physical deficiencies are made to realize that they must exercise extra care for safe driving. A license law also forces drivers to become familiar with state laws and regulatory signs[5] and in most cases to become reasonably skillful before they can expect to secure licenses.

The need for driver regulation and control was first met by towns and cities. In 1899 New York City first required drivers' licenses under its authority to regulate business or trade within its limits. Only commercial drivers were required to have a license, and requirements were more related to the applicant's character than his ability to drive. As vehicle accidents increased in number, cities and towns began to require every driver to obtain a license. Soon, however, states passed license laws, and cities were prohibited from requiring licenses.

By 1908, eight Eastern states had license laws.[6] The age limit varied from 16 to 18, and the license was issued for 1 year for a fee which was usually $2. As the need for examination of applicants became more apparent, an eye test and an oral or written test on laws and signs, and a road test when time and money permitted, were given. As applicants

---

[1] Summary of information supplied by various states, *Digest of Motor Laws*, 17th ed., American Automobile Association, Washington, 1950.

[2] Milton D. Kramer (ed.), *Safety Supervision in Motor Vehicle Fleets*, National Conservation Bureau, New York, 1947, p. 125.

[3] *Home Insurance Co. v. Augusta*, 50 Ga. 530 (1874).

[4] *Cusak v. Laube & Co.*, 104 Conn. 487, 133 A. 584 (1926). A typical case. The court said, "An operator's license is purely a personal privilege granted by the state on account of fitness."

[5] "In 21 states more applicants were rejected on written test than on driving skill." President's Highway Safety Conference, *Inventory and Guide for Action*, 1948, p. 42.

[6] Maine, New Hampshire, Vermont, Massachusetts, Connecticut, New Jersey, Pennsylvania, New York (chauffeurs only). See A. C. Wyman, *Automobile Laws of the New England States, New York, New Jersey and Pennsylvania*, Rhode Island State Library, Legislative Reference Bureau, Bulletin 2, Providence, 1908, p. 29.

became more numerous, the license period was increased to 2, 3, 4 years or even "until revoked."

In 1939[1] the Interstate Commerce Commission, under the authority granted it by the Motor Carrier Act of 1935, set up licensing requirements for drivers engaged in interstate commerce. The licensing requirements under this act are stricter than those of any state. Minimum age of 21 and a year's experience, plus a physical examination,[2] are required. Physical handicap disqualifies any applicant.

Uniform Motor Vehicle Code Requirements. Since the first form of the Motor Vehicle Code was adopted in 1926, it has provided that all persons operating a motor vehicle on public ways must be licensed. Licenses are not recommended for persons under 16 years of age or chauffeurs under 18 years. By 1934 school buses were in such common use that the National Conference on Street and Highway Safety added special revisions to the model driver-license act requiring school-bus operators to be 21 years of age or older. Recent editions of the Code require the following tests for licenses: (1) eyesight, (2) ability to read and understand highway signs, (3) knowledge of traffic laws, (4) road test, (5) physical and mental examination as deemed desirable.

Early drafts of the Code provided that when states first passed license laws drivers with a year's experience could get a license without examination by applying within a 3-month period.

The Code recommends that authority to revoke and suspend licenses for traffic violations rest with a state department, usually with the Motor Vehicle Commissioner. More serious traffic violations have always been weighed by the courts. Mandatory suspensions may be recommended in case of convictions for manslaughter, drunken driving, perjury, repeat convictions for reckless driving, felonies, and hit-and-run driving. The Code also provides that licenses be reissued every 2 or 3 years but makes no recommendation as to the amount of fee.

Under the Model Act of the National Conference on Street and Highway Safety, the Motor Vehicle Commissioner has broad authority, including the right to suspend and revoke licenses, to prescribe forms and procedures to be used in various aspects of license applications and issuances, to work out details concerning examinations and methods of conducting them, to determine when restricted licenses should be issued, and to decide what records are required of various officials and agencies of the government relative to driver licensing.

Present Practices. Every state but South Dakota has a license law. Some laws are as old as 1907,[3] and some are as new as 1947[4]; More than

---

[1] 4 *Fed. Reg.* 2295 (1939)—Order Ex Parte No. MC-4, adopted by Interstate Commerce Commission for addition to Code of Federal Regulations.

[2] *Ibid.*, secs. 2604–2615.

[3] Connecticut.

[4] Louisiana and Wyoming.

half the states have laws the same or equal to the recommendations of the Uniform Act.

It is difficult to determine the effect of license laws on accidents, but a study made during the 1930s concluded that a properly administered license law could be expected to cut fatalities 20 per cent. Detailed studies[1] have been made to determine whether the mere existence of licensing laws had any appreciable effect on safety on the highways. It appears that distribution and density of population are more important factors in fatalities than license laws, but it can be conceded that a license law with proper administration can have a very marked effect on fatalities. State accident reports indicate that operators are responsible for 65 per cent of all fatal accidents;[2] and an analysis of 1948 records shows that 57 per cent of drivers involved in fatal accidents (they were at least partially responsible for 71 per cent of the total fatalities) were violating some law.[3]

Drivers' licenses of a restricted nature are issued by many states. A learner's permit[4] may be issued to persons one or two years below the minimum age for an operator's permit, and they must be accompanied by a licensed driver when operating a vehicle. Other restricted licenses may permit the operator to drive only while wearing his glasses, only during the daytime, or only if his vehicle is equipped with special attachments.

Persons to Be Licensed. Nonresidents over 16 who are licensed in their home states are usually permitted to drive for at least 30 days, or for as long as 6 months in "tourist" states. Residents who operate farm vehicles, road machinery, or official Army or Navy vehicles are usually exempted from the license requirement.

The trend is toward higher minimum-age requirements, as Fig. 12-1 shows. Minors' applications usually must be signed by a parent, guardian, employer, or some other person willing to bear the liability for negligence.

Fees and Duration of Licenses. Yearly fees for licenses range from 12½¢ to $3 per driver.[5] About one-third of all states require annual renewal of licenses, while others, like Maryland, provide a license which remains good until revoked. In Delaware, a driver who has completed 3 consecutive years of safe driving may obtain a permanent driver's

---

[1] William J. Cox, "The Drivers' License Law Reappraised," paper presented at joint session of the Institute of Traffic Engineers and the National Safety Congress, Louisville, Ky., 1935, p. 12.

[2] Amos E. Neyhart, *Good Driving Course*, 1935, p. 1. (Mimeographed.)

[3] *Accident Facts*, National Safety Council, Chicago, 1949, p. 55.

[4] Issued by 35 states. See *Motor Law Digest*, American Automobile Association, Washington, 1949.

[5] *Ibid.*

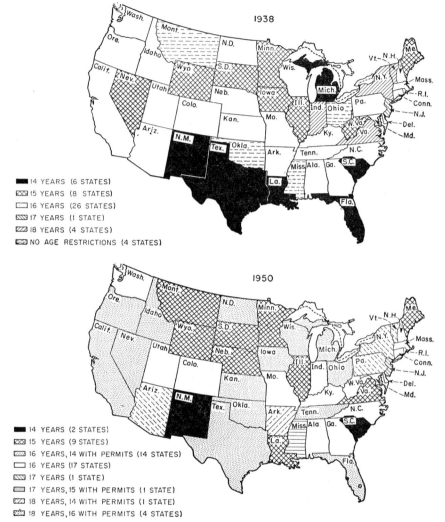

■■ 14 YEARS (6 STATES)
▨▨ 15 YEARS (8 STATES)
☐ 16 YEARS (26 STATES)
▧▧ 17 YEARS (1 STATE)
▨▨ 18 YEARS (4 STATES)
▨▨ NO AGE RESTRICTIONS (4 STATES)

■ 14 YEARS (2 STATES)
▨▨ 15 YEARS (9 STATES)
▤▤ 16 YEARS, 14 WITH PERMITS (14 STATES)
☐ 16 YEARS (17 STATES)
▧▧ 17 YEARS (1 STATE)
▤ 17 YEARS, 15 WITH PERMITS (1 STATE)
▨▨ 18 YEARS, 14 WITH PERMITS (1 STATE)
▨▨ 18 YEARS, 16 WITH PERMITS (4 STATES)

FIG. 12-1. Minimum age for unrestricted driver licenses. (From *Motor Vehicle Traffic Conditions in the United States, Non-Uniformity of Traffic Laws,* U.S. Department of Agriculture, 1938, part I, p. 45; and *Digest of Motor Laws,* American Automobile Association, Washington, 1950.)

license through the payment of a \$5 fee. Some states, like Delaware,[1] require a reexamination of drivers over a certain age. Licensing fees should be high enough to support a good licensing program,[2] including examinations, clerical work, and driver-improvement plans. At present

[1] T. E. Transeau, "The Suspension and Revocation of Drivers' Licenses," paper presented before Eastern Conference of Motor Vehicle Administrators, Hartford, Conn., 1941, p. 6.

[2] *Highway Safety—Motor Truck Regulation,* The Council of State Governments, Chicago, 1950, p. 24.

North Carolina is spending 31 cents per driver per year for examinations only, more than any other state, while Pennsylvania and Massachusetts are high with 20 cents per driver per year for driver improvement. If a minimum of 45 cents per driver per year were spent for all aspects of the driver-licensing program, adequate funds would be available for improvements in licensing and control.

Requirements and Examination Methods. Almost every examination includes a test of visual acuity.[1] The American Association of Motor Vehicle Administrators recommends 20/40 as a screening minimum,[2] with persons having less acuity to be examined by a carefully trained examiner, who decides whether or not the applicant should be allowed to drive.

Color-vision tests (discrimination between red, yellow, and green) are used by several states, but few if any rejections are based on this test. It is generally used only for its educational value:[3] if a driver is color-blind, he is made to realize that fact and usually compensates for it.

Field-of-vision, depth-perception, and night-vision tests have also been recommended, but few states have adopted them.

Law tests usually consist of a number of questions relating to right of way, turns, etc. Applicants who are rejected for failing this test usually pass after a little study of the drivers' manual.

The signs test consists of identification and explanation of signs by their size and shape and knowledge of directional hand signals.

The road test is given by an examiner who notes the applicant's ability to handle the vehicle safely, ability to park, speed at which he drives, and observance of traffic laws and regulations, but as it is ordinarily administered today it does not accurately determine the ability of the driver being tested. Ideally, it should include some of the difficult situations commonly encountered on the road, so that the driver's ability to handle himself and his vehicle under critical circumstances could be determined.

Clinical Tests. Certain clinical tests, including reaction time, steering, speed estimation, glare, traffic-light color, and depth perception, have been developed. There seems to be very little correlation between any individual test and accident proneness, but a low average score is usual for "repeaters."[4] The main value of these tests is educational.[5] If

[1] More than 34 states require such a test. Only 4 states require an eye test for renewal. Source: M. D. Dramer (ed.), *Safety Supervision in Motor Vehicle Fleets*, National Conservation Bureau, New York, 1947, p. 124.

[2] *Manual on Drivers' Vision Test*, American Association of Motor Vehicle Administrators and the Motorists Vision Committee of the American Optometric Association, Pittsburgh, 1949, p. 23.

[3] *Ibid.*, p. 26.

[4] Earl Allgaier, *Interrelationships of Driver Test Scores*, American Automobile Association, Research Report 11, Washington, 1939, p. 7.

[5] H. R. DeSilva and T. W. Forbes, *Driver Testing Results*, Harvard University, Harvard Traffic Bureau, and Works Progress Administration, Cambridge, Mass., 1937, p. 49.

drivers are made aware of physical limitations, such as weak eyes or slow reflexes, they will usually correct the fault or compensate for it. Cab companies and fleet operators have found such clinical tests, when far stricter than those given to the general public, to be helpful in selecting safe drivers.[1] They are generally administered with the *moving road scene* driver-testing apparatus,[2] which tests eye-hand coordination, braking reaction time, and ability to do several things in quick succession.[3] Also used to check glare blindness is a device with controlled illumination and a gauge to measure the amount of light it takes to "blind" the person being tested.

Enforcement of License Laws. Under present conditions most suspensions and revocations are mandatory under various provisions of the license laws. In 1947 nearly 430,000 licenses were suspended, an increase of more than 25 per cent over 1946. Four out of every ten were barred from the roads as a result of drunk-driving convictions.[4] A brief summary of suspensions and revocations in New York State for 1946 and 1947 is shown in Table 12-1.

12-2. Financial Responsibility. Since financially irresponsible persons usually carry no insurance because they have no property to protect, financial-responsibility laws came into being in a few states in 1926 and 1927 to provide compensation for injured parties in automobile accidents.[5]

Massachusetts passed a compulsory-insurance law in 1927 which prevented vehicle registration unless the applicant furnished proof of financial responsibility by insurance policy, bond, or cash deposit. This plan made the automobile owner financially responsible for damage, provided there was no contributory negligence on the part of the injured party, but it had the disadvantages of high insurance costs and frequent fraudulent claims.

Plans have been presented for compulsory compensation,[6] under which every injured party would be compensated regardless of accident cause. The cost of such systems, however, is disproportionate to the benefits, and compensation is not in accordance with economic need or actual responsibility for the accident.

---

[1] Amos E. Neyhart, "Driver Training for the Fleets," *Traffic Quarterly*, vol. I, no. 3, New Haven, Conn., July, 1947, p. 238.

[2] H. R. DeSilva, *Why We Have Automobile Accidents*, John Wiley & Sons, Inc., New York, 1942, p. 63.

[3] *Ibid.*, p. 64.

[4] American Association of Motor Vehicle Administrators, "Licensing Authorities Bear Down on Unsafe Drivers," *Highway Research Abstract*, vol. 18, no. 7, July, 1948, p. 8.

[5] State of New York, *Report of the Joint Legislative Committee to Investigate Automobile Insurance*, Albany, 1937, pp. 30, 198.

[6] Robert S. Marx, former Judge of the Superior Court of Cincinnati, was the first advocate of such a plan.

The most widely accepted solution is the financial-responsibility law requiring that a motorist's license and all registration certificates be canceled if he does not meet a judgment or furnish an insurance certificate or bond or put up cash to cover any future accident. Minimum judgment is usually $25 to $100. Only drivers who are not financially responsible are affected by this plan, and insurance companies are not forced to insure bad risks.

TABLE 12-1. SUSPENSIONS AND REVOCATIONS IN THE STATE OF NEW YORK, 1946-1947

	1946	1947
Revocations—causes		
Operating while intoxicated..............	885	885
Reckless driving or speed................	520	750
Leaving scene of an accident.............	436	478
Unlicensed operator.....................	306	362
Other...................................	535	810
Total...............................	2,682	3,285
Suspensions—causes		
Fatal accident.........................	1,814	1,803
Serious accident.......................	1,221	1,503
Reckless driving or speeding............	1,294	1,835
Failure to notify of change of address.......	596	791
Improper registration...................	1,799	1,747
Physical or mental disability.............	1,205	1,161
Habitual or persistent violator...........	515	942
Other...................................	1,862	1,775
Total...............................	9,756	11,557
Total suspensions and revocations......	12,438	14,842

SOURCE: State of New York, Department of Taxation and Finance, *Report of Bureau of Motor Vehicles*, Albany, 1948.

The Uniform Vehicle Code[1] recommends the filing of security in connection with recent accidents. Licenses should be suspended when accident reports are not filed within 10 days of the accident and when security is not posted within 60 days of an accident which causes death, personal injury, or property damage in excess of $50. The Code also provides that responsibility should be required for possible future liability to the extent of $10,000 when personal injury or death to two persons is involved, $5,000 for the personal injury or death of one person, and $1,000 for property damage. Proof is to be made through a motor vehicle policy, money, or security bonds.

[1] National Committee on Uniform Traffic Laws and Ordinances, Subcommittee on Financial Responsibility, *Annotation of Redraft of Act IV, Uniform Vehicle Code*, Washington, 1948.

Type of Law. There are two predominant types of responsibility laws in use today, the "old type" and the "new" or security type.

Old-type law. This law affects the driver after he has had an accident and a judgment against him. If the judgment is not satisfied up to statutory limits of $5,000-$10,000-$1,000, then the license and registration in the offender's name is suspended. To remove suspensions he must prove future responsibility up to statutory limits.

New-type law. This law provides that within a stated period of time after an accident, but before any judgment, proof of financial responsibility can be required by the Motor Vehicle Commissioner. Failure to provide proof will result in suspension, which may be lifted when the driver (1) furnishes proof of future financial responsibility, (2) shows a release from liability, (3) obtains a judgment in his favor, or (4) satisfies the judgment against him. If legal procedures are not instigated within a period of 1 year after an accident, the coverage requirements are generally allowed to lapse, but once a judgment has been made, requirements are continued until payment is made and proof established. (The Uniform Vehicle Code even provides that discharge in bankruptcy following a judgment does not relieve the judgment debtor from these requirements.) The period of "future proof" is sometimes extended by the Motor Vehicle Commissioner for a period of several years.

Under either law, the motorist is given "one chance" to cause damages and not be forced to pay. Under the new law, responsibility may be required for negligence or suspicious actions even without an accident.

Of the financial responsibility requirements, some 16 states have *old-type* laws and about 25 have *security laws.* Several states still do not have laws.[1]

12-3. Civil Liability in Automobile Use. Special provisions must be made for certain cases, such as governmental liability in auto accidents, stolen or borrowed vehicles, gratuitous guests, nonresidents, and for-rent vehicles. Each of these contains peculiar problems, and it is important that the cases should be clearly and logically stated in the motor vehicle laws of each state if driver or road-user controls are to be fully utilized.

Governmental Liability for Accidents. It is inevitable that situations arise in which Federal, state, and local governments find themselves defendants in litigations involving automobile accidents. The theory has always been that the sovereign nature of the Federal and state governments made them irresponsible for torts unless they specified that they could be held liable. This provision still is the basis on which Federal and state units function; however, the municipal governments are set up on a somewhat different basis. In tort liability cases against municipal

[1] See President's Highway Safety Conference, *Report of Committee on Laws and Ordinances,* 1949, pp. 54, 55; and *Motor Law Digest,* 16th ed., American Automobile Association, Washington, 1950.

organizations, the distinction lies in whether the acts were committed in the performance of governmental or proprietary functions. States have used a number of different methods to achieve the same result. Some have adopted legislation which specifically provides that the states themselves, or some of the government units under them, will be civilly liable for damages as a result of negligent operation of a vehicle by employees or agents acting within the scope of their employment. This is the type of legislation which is regarded as the most acceptable solution to the problem. Claims courts and boards are used in some states, and others use special legislative acts to allow the state to be sued.

In 1944 the Federal Tort Claims Act was passed, providing that claim against the various Federal agencies "on account of damage to or loss of property or on account of personal injury or death, caused by a wrongful or negligent act or omission of any employee of the government while acting within the scope of his office or employment, under circumstances where the government would be subject, if a private citizen, to the claims of such persons" would be legal.

Borrowed and Stolen Vehicles. Responsibility for a claim frequently rests on the difference between a borrowed and a stolen vehicle. The main point of contention is the phrase "express or implied permission." At least a third of the states have laws which place the responsibility squarely with the owner if he has expressed or implied the loan of his car. Almost all states make the owner liable for negligence of an agent or servant or member of the family.

Gratuitous Guests. At present 28 states have laws providing that the vehicle owner or driver should not be held responsible for any gratuitous guest or hitchhiker. The ideal law states that gross negligence or intoxication on the part of the driver must be proven if recovery is to be made by gratuitous guests.

Legal Action against Out-of-state Drivers. Section 9 of Act III of the Uniform Vehicle Code provides that any out-of-state driver who uses the roads is subject to the same legal action as he would be were he similarly involved in his own state. Service of process should be made through notification of the Commissioner of Motor Vehicles, and payment of a bond made to ensure continuation of the complaint. Specification of method of mailing and the time limit to the due date of the return of the process are also included in the Code. The Motor Vehicle Commissioner can act for the driver in regard to legal actions brought against him outside his home state, and a $500 bond must be posted by the plaintiff to cover fraudulent complaints in an attempt to coerce the defendant.

Control of For-rent Vehicles. Legislation controlling for-rent vehicles is centered on the question of whether liability should lie with the owner or the renter. According to the Uniform Code, owners of for-rent vehicles must be registered with the motor vehicle department and show proof of a public liability insurance policy to the extent of the $5,000-$10,000-$1,000 basis (for each vehicle rented) set forth in the financial-responsibility laws. The owner is jointly and severally liable with any person operating his vehicles, and usually no action can be had by a passenger against the owner.

Civil liability as a road-user control is not being used. Every provision under the Civil Liability Act is important as a driver or road-user control While these provisions are not as effective controls as good driver-license or financial-responsibility laws, it is agreed that without such provisions control of the road user cannot be sufficiently complete.

CHAPTER 13

VEHICLE CONTROLS

Extensive regulation of motor vehicles was not necessary in the United States prior to 1900. There were not enough automobiles in use to require regulations. However, with the turn of the century the auto became more commonplace throughout the country. In 1901, New York State passed a registration law, and within the few years following many more states adopted vehicle-control measures. As the development of vehicles increased rapidly, all states recognized the need for regulating vehicle ownership and use and enacted registration laws. Since the first laws were enacted, many developments in the vehicle and the traffic conditions have necessitated changes in basic requirements and in practices. The appearance of commercial vehicles following World War I and the rapid acceptance of such vehicles brought about the need for many changes in registration requirements.

While registrations are basic to all vehicle controls, there are other significant areas of control. These include requirements for equipment and accessories, sizes and weights of commercial vehicles, and inspections. Before discussing each of these controls, a brief review of some of the significant developments in motor vehicles is in order.

Improvements in Motor Vehicles. The period 1920 to 1930 saw many significant developments in the automobile. During that period the all-steel bodies, four-wheel brakes, low-pressure tires, and closed-style bodies first appeared. Significant developments occurred during the same period with relation to such things as antifreeze solutions, chains, ethyl gasoline, oil-cooling systems, and reliable batteries. It was not until 1925 that the number of closed-type car bodies surpassed the open type. Much of the equipment (bumpers, mirrors, safety glass, etc.) which had been previously considered "special" became standard. In addition to the marked advances in styles and types of vehicles and in accessories, there have been significant technological improvements. The self-starter, the lowered centers of gravity, improved transmissions, more efficient motors, etc., have come about naturally. It is significant that many of the improvements and developments have had a profound influence on traffic operations.

Almost all data collected in traffic surveys and in accident analysis must be correlated with basic facts obtainable from registration records and other files of the state motor vehicle department. The limitations imposed on vehicles by legislation and administration by the motor vehicle departments are important considerations for the engineer in planning, traffic controls, geometric design of routes, and in the application of many traffic regulations.

Motor Vehicle Administration. Functions of the department of motor vehicles in most states are extremely broad. In addition to supervising the major controls of operators, including driver licensing, certificate of title, financial responsibility, and civil liability, as discussed in the preceding chapter, the department must also maintain and direct the work of vehicle registration. In some states the motor vehicle departments also include a strong enforcement arm, either the highway patrol or inspectors. In most states, the responsibility for maintaining and circulating information on stolen vehicles is assigned to the department of motor vehicles. In a few cases, the departments are charged with the collection of gasoline and other highway-user taxes. The motor vehicle division is a key agency of state government and an essential element in the over-all traffic program.

Many states have now developed motor vehicle departments as separate divisions of state government, although in some states the functions are still attached to an older governmental unit.

The Uniform Motor Vehicle Code recommends the establishment of the motor vehicle department as a separate state agency to absorb all duties and functions attached to the registration and control of vehicles as well as the licensing of operators.[1] The Code further recommends the creation within the department of motor vehicles of a state safety patrol,[2] but there is considerable disagreement on this point among motor vehicle administrators and police officials. The trend is toward the establishment of state police organizations as separate departments of state governments on a par with motor vehicle departments.

13-1. Vehicle Registration. Basic registration records include registration numbers, alphabetical listings of owners, numerical files of serial numbers, vehicle classification, information on vehicle use and additional information for enforcement, taxation, highway planning, and other statistical uses. When a vehicle is registered, the motor vehicle department issues a registration card, license plates, and, if the law provides, a certificate of title. (In those states without certificate of title laws, a bill-of-sale receipt and registration card are usually accepted as evidence of ownership and transfer.)

In 1901, when New York State enacted the first registration law, fewer

---

[1] *Registration, Certificate of Title, and Anti-Theft Act,* Uniform Vehicle Code, act I, art. II, sec. 20, Uniform Motor Vehicle Administration, 1945, p. 4.

[2] *Ibid.,* sec. 19, p. 4.

than 1,000 vehicles were registered in the state. Fees were $1 and licenses were issued bearing the owners' initials. Within 10 years practically all states enacted registration laws as a first component of motor vehicle department functions. Registration increased at a tremendous rate; in Pennsylvania only 5,000 vehicles were registered in 1903 when registrations were initiated,[1] producing an income of about $32,500, but in 1948 more than 2,000,000 vehicles were registered at an income of $40,000,000. Trends in registrations by vehicle types are given in Table 13-1.

Revenue from motor vehicle registration is now generally earmarked for highway improvement.

One of the first cases testing the legality of motor vehicle registrations was *Unwen v. State of New Jersey*,[2] in which the court held that provisions of a law passed in 1903 requiring a statement by the owner of a motor vehicle of his name and address, maker of the machine, and registration and license of the machine were within the exercise of the state police power to secure public safety in the use of the highways. This ruling is typical of the many which have followed.

The right of a state to require vehicles to display license tags was the subject of early court cases. In Massachusetts the court ruled that a statute passed in 1903 requiring owners of automobiles to have same registered was a proper exercise of the state's police power. Since that time the license plate has come to be regarded as standard motor vehicle equipment.

Fees. In 1921 there were 17 different methods of determining vehicle registration fees, with most of the states using either weight or horsepower rating as a basis. In 1931 certain types of commercial vehicles were taxed on the basis of as many as five measures in Texas, Wyoming, and Wisconsin.[3] Table 13-2 shows the variations in present-day passenger-car registration fees, which range from $1.50 to $87.50 per year.

Registration fees for trucks are even more varied than for passenger cars. Gross weight is used by 17 states as a basis for determining fees, 9 use empty weight, 11 use rated capacity, and 12 use some combination of these methods.

In addition to fees required for registration of commercial vehicles with the state motor vehicle department, there are special registrations and fees required for certain types of commercial vehicles at the city level. In addition, fees are imposed by the Interstate Commerce Commission on trucks engaged in interstate commerce.

Other variations in registration fees develop because of different

---

[1] *Highway Statistics, Summary to 1945*, U.S. Bureau of Public Roads, 1947, p. 25.

[2] 73 N.J.L. 529, 64 A. 163 (1906).

[3] *Taxation of Motor Vehicle Transportation*, National Industrial Conference Board, Inc., New York, 1932.

TABLE 13-1. MOTOR VEHICLE REGISTRATIONS IN THE UNITED STATES
(Does not include publicly owned vehicles)

Year	Passenger cars	Buses	Trucks	Total
1895	4	....	....	4
1896	16	....	....	16
1897	90	....	....	90
1898	800	....	....	800
1899	3,200	....	....	3,200
1900	8,000	....	....	8,000
1901	14,800	....	....	14,800
1902	23,000	....	....	23,000
1903	32,920	....	....	32,920
1904	54,590	....	700	55,290
1905	77,400	....	1,400	78,800
1906	105,900	....	2,200	108,100
1907	140,300	....	2,900	143,200
1908	194,400	....	4,000	198,400
1909	305,950	....	6,050	312,000
1910	458,377	....	10,123	468,500
1911	618,727	....	20,773	639,500
1912	901,596	....	42,404	944,000
1913	1,190,393	....	67,667	1,258,060
1914	1,664,003	....	99,015	1,763,018
1915	2,332,426	....	158,506	2,490,932
1916	3,367,889	....	250,048	3,617,937
1917	4,727,468	....	391,057	5,118,525
1918	5,554,952	....	605,496	6,160,448
1919	6,679,133	....	897,755	7,576,888
1920	8,131,522	....	1,107,639	9,239,161
1921	9,212,158	....	1,281,508	10,493,666
1922	10,704,076	....	1,569,523	12,273,599
1923	13,253,019	....	1,849,086	15,102,105
1924	15,436,102	....	2,176,838	17,612,940
1925	17,439,701	17,808	2,483,215	19,940,724
1926	19,220,885	24,320	2,807,354	22,052,559
1927	20,142,120	27,659	2,969,780	23,139,559
1928	21,308,159	31,982	3,171,542	24,511,683
1929	23,060,421	33,999	3,408,088	26,502,508
1930	22,972,745	40,507	3,518,747	26,531,999
1931	22,330,402	41,880	3,489,756	25,862,038
1932	20,832,357	43,476	3,256,776	24,132,609
1933	20,586,284	44,918	3,245,505	23,876,707
1934	21,472,078	51,530	3,430,396	24,954,004

TABLE 13-1. MOTOR VEHICLE REGISTRATIONS IN THE UNITED STATES. (*Continued*)

Year	Passenger cars	Buses	Trucks	Total
1935	22,494,884	58,994	3,675,865	26,229,743
1936	24,108,326	62,618	4,001,464	28,172,318
1937	25,390,773	66,166	4,249,219	29,706,158
1938	25,167,030	65,198	4,210,477	29,442,705
1939	26,139,526	68,859	4,406,702	30,615,087
1940	27,372,397	72,641	4,590,386	32,035,424
1941	29,524,101	88,800	4,859,244	34,472,145
1942	27,868,746	102,093	4,608,086	32,578,925
1943	25,912,730	106,702	4,480,176	30,499,608
1944	25,466,331	106,518	4,513,340	30,086,189
1945	25,691,434	112,253	4,834,742	30,638,429
1946	28,100,188	119,937	5,725,692	33,945,817
1947	30,718,852	128,983	6,512,628	37,360,463
1948	33,213,905	132,603	7,209,961	40,556,469
1949	36,312,380	135,002	7,692,569	44,139,951
1950	40,185,146	143,206	8,272,153	48,600,505
1951	42,525,217	143,290	8,657,931	51,326,438
1952	43,653,545	145,227	8,817,140	52,615,912

SOURCE: *Highway Statistics, Summary to 1945* (also *1946–1952*), U.S. Bureau of Public Roads, 1947, p. 18.

classifications. In some states taxicabs are classed as passenger vehicles, in others as commercial vehicles. School buses are also classified differently in various states.

Certificate of Title. The Uniform Motor Vehicle Code[1] recommends that certificates of title be made a part of the standard registration procedure of every state. Some 30 states have certificate of title laws, but the administration and coverage of the laws vary widely. In theory, certificate of title laws provide the maintenance of a complete record of liens and encumbrances on motor vehicle registrations in a given state. These comprehensive records make possible careful and detailed investigations for the sale and transfer of used cars. However, a majority of the states' laws do not specifically require recording of liens with the motor vehicle agency, and often chattel mortgages and other liens which are recorded locally are not admitted by the owner in applying for certificate of title even if required to do so by law. If properly written and executed, certificate of title laws can be very effective in curtailing sales of stolen vehicles and the transfer of vehicles on which mortgages or liens have been issued.

[1] *Registration, Certificate of Title, and Anti-Theft Act,* Uniform Vehicle Code, act I, arts. III–V, secs. 41–65, pp. 10–18.

TABLE 13-2. PASSENGER CAR REGISTRATION FEES
(Fee bases and range in rates for various sizes and conditions of vehicles)

Weight	Flat fee	Horsepower	Combination
Alabama.......... $5.50-$18.50	Arizona......... $3.50	Illinois.......... $6.50-$22.00	Arkansas (hp and wt) . $7.00-$26.00
Colorado........ 5.00- 11.00	California....... 6.00	Maine........... 10.00- 16.00	Indiana (hp and wt)... 11.00- 12.00
Connecticut..... 7.00- 11.00	Idaho.......... 5.00	Massachusetts... 3.00- 7.50	Iowa (wt and % list price)........... 14.00- 38.00
Delaware 8.00- 12.00	Kentucky 4.50	Missouri 5.00- 37.50	Kansas (flat and wt). 6.00- 15.00
Dist. of Columbia.. 5.00- 12.00	Louisiana....... 3.00	New Jersey...... 4.00- 24.00	Mississippi (hp and wt and tag)......... 9.00- 20.00
Florida.......... 5.00- 25.00	Nevada......... 5.00		Oklahoma (mfg. delivered price)...... 5.50- 70.00
Georgia.......... 1.50- 6.00	Ohio........... 10.00		Vermont (year of model)........... 14.00- 22.00
Maryland......... 10.00- 18.00	Oregon......... 10.00		
Michigan......... 7.00- 16.00	Pennsylvania.... 10.00		
Minnesota....... 5.00- 75.00	Utah........... 5.00		
Montana......... 5.00- 10.00	Washington...... 5.00		
Nebraska........ 3.00- 5.00	Wisconsin....... 16.00		
New Hampshire.. 10.00- 17.00	Wyoming........ 5.00		
New Mexico..... 15.00- 62.00			
New York....... 10.00- 17.50			
North Carolina.. 10.00- 15.00			
North Dakota... 5.00- 87.50			
Rhode Island.... 8.00- 18.00			
South Carolina.. 1.60- 7.60			
South Dakota... 13.00- 55.00			
Tennessee........ 7.50- 10.00			
Texas........... 2.80- 25.00			
Virginia......... 6.00- 14.00			
West Virginia.... 11.00- 27.00			

SOURCE: *Highway Statistics, Summary to 1945* (also *1946-1952*), U.S. Bureau of Public Roads, 1947, p. 18.

220

Anti-theft Laws. In 1919 Congress enacted a National Motor Vehicle Theft Act giving jurisdiction in automobile thefts to Federal enforcement agencies.[1] The Uniform Vehicle Code recommends that owners of stolen vehicles and all law-enforcement agencies of the state be required to report thefts to a centralized state agency, usually the motor vehicle department. This department maintains a complete record of stolen vehicles and issues periodical lists, usually weekly, giving information available, such as registration, registration numbers, motor numbers, serial numbers, etc., on vehicles which have been reported stolen. Official registration of the vehicle is automatically canceled when it is reported stolen.

13-2. Equipment and Accessories. It has generally been held that states and municipalities have the authority to pass laws regulating motor vehicle equipment as long as such laws or regulations meet the requirements for validity as exercises of police power, under the requirements of certainty.

Trends in Regulations. When the first Uniform Motor Vehicle Codes were drafted in 1926, it was recommended that there be two separate means of braking and that horns should be audible at a distance of 200 ft. It was suggested that head lamps should clearly show a person walking along the road 200 ft ahead, that a red rear lamp should be visible at a distance of 500 ft, and that two clearance lamps should be placed on the left side of all commercial vehicles. These equipment requirements have been expanded as technological improvements in the vehicle have been made and more rigid controls have been required. Four-wheel brakes were introduced on passenger cars about 1924, and safety glass in 1926, and both became generally required by state motor vehicle laws on new vehicles during the thirties. Requirement of certain equipment in a few states usually makes it standard equipment in the manufacture of motor vehicles. For example, when a few states required that all vehicles sold therein must be fully equipped with safety glass, the manufacturers soon started making all their cars to meet these standards.

In 1934, brake-performance standards and new lighting equipment requirements were added to Act IV of the Uniform Vehicle Code of 1926. It was recommended that all new vehicles be equipped with four-wheel brakes capable of stopping in 30 ft, that the hand brake be capable of stopping the vehicle in 55 ft, and that any older vehicle equipped with only two-wheel brakes be capable of stopping in 40 ft from a speed of 20 mph. Stricter control was also imposed on type, placement, and ability of lighting equipment.

The 1938 edition of the Code adopted still stricter limits upon the performance ability of brakes, which are still regarded as the standard. Changes in lighting provisions, chiefly in regard to clearance and tail-

[1] National Motor Vehicle Theft Act, 66th Cong., 1919.

lights, were made so that the truck-light code would conform to Interstate Commerce Commission (ICC) regulations. Further changes in recommended-equipment provisions have been made during postwar years.

Present Standards. Modern standards for equipment and accessories are typified by the provisions of Act V of the Uniform Vehicle Code.[1]

Brakes. It is generally required that standard passenger cars be equipped with two separate means of applying brakes, each affecting at

TABLE 13-3. REQUIRED MOTOR VEHICLE BRAKING DISTANCES
(Summary of state requirements)

Uniform vehicle code recommendations, 30 ft at 20 mph	Shorter stopping distance, less than 30 ft at 20 mph	Longer stopping distance, more than 30 ft at 20 mph	No stopping-distance requirement
Delaware	Arkansas	California	Alabama
Illinois	New Hampshire	Colorado	Arizona
Kansas	North Carolina	Dist. of Columbia	Connecticut
Kentucky	Rhode Island	Iowa	Florida
Maryland	Virginia	Louisiana	Georgia
Michigan		Nebraska	Idaho
Minnesota		New Jersey	Indiana
Mississippi		Oregon	Maine
Nevada		South Dakota	Massachusetts
New York		Texas	Missouri
Ohio		Washington	Montana
Oklahoma		Wisconsin	New Mexico
Pennsylvania			North Dakota
South Carolina			Vermont
Tennessee			
Utah			
West Virginia			
Wyoming			

SOURCE: Various existing state motor vehicle codes.

least two wheels, and that in the failure of one of the brakes the other will be automatically engaged. Passenger vehicles must be equipped with service brakes on all wheels. Tractors and semitrailers, except the lightest types, must have brake mechanisms which permit the operation of the brakes on the trailer in conjunction with actuation of the brakes on the tractor. In addition, the motor vehicle commissioners of the various states are usually empowered with a broad right to require braking equipment which they consider necessary for individual types of vehicles and special vehicle uses.

Where brakes are provided on all wheels, the vehicle should be capable of being stopped at a distance of 30 ft from speed of 20 mph, developing a

[1] *Uniform Act Regulating Traffic on Highways,* Uniform Vehicle Code, act V, art. XVI, secs. 123–159, 1948, pp. 32–46.

deceleration rate of 14 ft/sec². When the vehicle does not have brakes on all wheels, a stopping distance of 40 ft at 20 mph is prescribed (a deceleration rate of 10.7 ft/sec²). General brake-performance standards of different states are shown in Table 13-3. There are 15 states which do not specify any performance abilities for brakes but require merely that they be "in good repair," able to stop the vehicle in a distance to be specified.

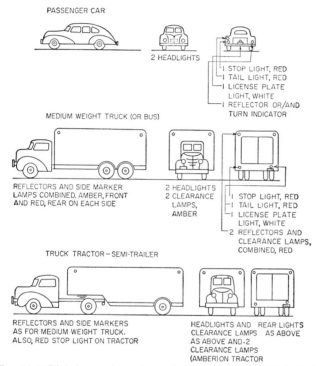

FIG. 13-1. Lighting equipment required by Uniform Vehicle Code.

Lights. Lights and locations prescribed by Act V of the Uniform Code for the more common types of vehicles are shown in Fig. 13-1. Interstate Commerce Commission standards for commercial vehicles, shown in Fig. 13-2, vary little from those prescribed by the Uniform Code,[1] except in the basis of classing the vehicle. While the Code classes vehicles according to gross weight (greater or less than 3,000 lb), the ICC groups them by width (more or less than 80 in.).

In general, standard passenger cars must be equipped with two headlights, a tail lamp, and a license-plate lamp. Autos sold after a specified time must also have two red reflectors on the rear. Commercial vehicles

[1] *Motor Carrier Safety Regulations,* U.S. Interstate Commerce Commission, 1947, part III, p. 2.

are required to have additional lighting equipment. The Uniform Code recommends that buses and trucks have two reflectors and a stop light on the rear, in addition to the lamps required for standard automobiles.

Since the effectiveness of lighting equipment is dependent upon its location, it is necessary to provide rigid standards for locating each lighting and reflector device, so that maximum visibility will be provided without undue glare conditions.

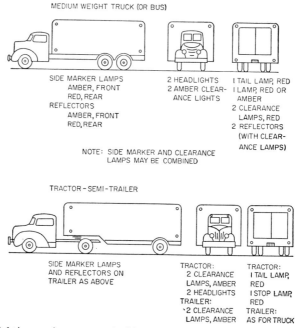

Fig. 13-2. Lighting equipment required by Interstate Commerce Commission. (From *Motor Carrier Safety Regulations*, U.S. Interstate Commerce Commission, 1947, part III, p. 2.)

Light intensities must also be specified. Headlights are generally considered adequate if, with the upper beam on a level roadway, persons and objects are visible for at least 350 ft under normal atmospheric conditions. In order to avoid serious glare, restrictions are imposed on the upward aiming of the beams. In an unloaded vehicle the maximum apparent candlepower must not exceed 8,000 at 1° above the horizontal, and maximum intensity of the light beam is not to exceed 75,000 candlepower at any point, according to Act V of the Uniform Code.

The lower light beam, use of which is required when approaching within 500 ft of an oncoming car, should be regulated so that the high-intensity portion to the left of the vehicle is 8 in. or more below the center of the lamp at a distance of 25 ft when the vehicle is unloaded. To the right of the lamp the high-intensity portion of the light beam must be 3 in. or more below the center of the lamp at 25 ft. In no instance is the high-intensity portion to be more than 32 in. above the ground at 75 ft. It is necessary, however, that there be sufficient light to reveal persons or objects at 100 ft on a level roadway.

The number of vehicles in use today equipped with single-beam head-lights is so few that the regulations regarding them are relatively unimpor-tant. For safety reasons, particularly when glare is caused, single-beam regulations must approach closely those for the lower beams of multiple-beam lights.

At least 25 states have adopted the majority of the lighting-equipment provisions of the Uniform Code, and many other states have adopted some of the suggested provisions. The sealed-beam headlight, which appeared in 1940, has done much to improve the lighting standards of motor vehicles, since focal properties are fixed by the manufacturer. The sealed-beam unit assures a much greater sustained brilliance than the old-type reflector elements, provides about 50 per cent more light and relief from glare, and its efficiency is not materially lowered with age.

Motor vehicle codes usually provide special lighting provisions for certain types of vehicles, such as motor-scooter, emergency cars, horse-drawn vehicles, and on certain vehicles operated at very low speeds. Most states permit auxiliary equipment, including spotlights, fog lamps, backup lights, auxiliary driving lamps, turn signals, fender or side cowl lamps, and running-board lights. The Uniform Code allows one spot-light, two fog lamps, one auxiliary passing lamp, one auxiliary driving lamp, two backup lights, turn signals for front and rear, two fender or side cowl lamps, and one running-board lamp on each side of the vehicle. Spotlights may be legally used in all states, but the number allowed varies. Some states allow one light as the Code recommends, others allow two, and others specify no limiting number. Fog-lamp limitations vary from one to four, but most states allow two, as does the Uniform Code.

Laws requiring the dimming of headlights when meeting oncoming traffic have been passed by at least 43 states. In a study of the observ-ance of the headlight-dimming law in Connecticut in December, 1947, out of 152,097 cars observed 9,822 were listed as violators, and 350 more as possible violators (a total of 6.2 per cent of the drivers checked).[1]

Laws must include provisions as to the time when different types of lamps are to be employed, actions required by the driver under different conditions (such as in approaching and passing other vehicles), operations in conjunction with lighted and unlighted road sections, colors and methods of illuminating all lamps, and special operating conditions, such as the illumination of projecting loads, parking, and backing.

[1] Burton W. Marsh, "Report of Committee on Night Visibility," *Proceedings of Highway Research Board*, Washington, 1948, vol. 28, p. 508.

Other Equipment. Most states require vehicles to be equipped with horns that are audible for at least 200 ft but are not unreasonably loud or obnoxious. Sirens, whistles, and bells are prohibited except on emergency vehicles. Many cities impose heavy fines on the unnecessary use of horns. Several new-model automobiles have horns which will operate only when the ignition is on, and some have been announced wherein the horns will operate only when the vehicle is in motion.

All states now require that vehicles be equipped with mufflers. Mirrors are required on all vehicles where the driver's view to the rear is obstructed. Provisions are generally worded to the effect that wherever a driver's view to the rear is obstructed because of the construction or loading of the vehicle, a mirror should be provided so that a view to the rear for a distance of 200 ft may be obtained.

Mirrors are now provided as standard equipment on passenger vehicles and commercial vehicles alike, and some vehicles are equipped with more than one mirror.

Most states prohibit the operation of vehicles with obstructed windshield or those not equipped with windshield wipers in operating condition on the driver's side.

Safety glass is now so generally required that practically all vehicles, including school buses and trucks, are fully equipped by the manufacturer.[1]

Trucks and buses operating at night outside of urban areas are required to carry flares and warning devices which can be used in case of breakdown or accident to give proper warning to other motorists if the vehicle is parked on the roadway or causes obstruction in the roadway. Some states have approved the use of red emergency reflectors by buses and trucks in lieu of the previously required flares and fuses.

New-equipment Developments. It is likely that new-equipment developments will lead to regulations requiring turn signals on all vehicles, defrosting devices, crash pads, windshield sprayers, etc. Laws have already been passed in some states[2] requiring all vehicles manufactured after a certain date to be equipped with mechanical directional signals.

Considerable study is being given to polarized head lamps. Before any polarized headlights may be used, however, state laws in all states will have to be changed in order to permit their use. Polarized headlights are considerably brighter than the maximum intensity now allowed and can apparently offer advantages over the lights currently in use. They will increase down-the-road illumination, decrease the risk involved in passing cars traveling in the opposite direction, provide complete protection against glare, and eliminate the regulation-and-enforcement problem of beam depression. However, distance judgment may at first be difficult

---

[1] *Report of Committee on Laws and Ordinances*, President's Highway Safety Conference, 1950, pp. 54–55.

[2] Minnesota, effective July 1, 1949, and New York, effective Jan. 1, 1952.

because of the dimness of oncoming lights as seen through the polaroid viewer, and because the glare to the naked eye is much greater, the pedestrian may be greatly bothered by them. The polaroid viewer will cut down some glare visible over dips and rises and on curves, caused by the headlights of approaching vehicles, which is helpful on hilly, curved roads.

13-3. Sizes and Weights of Commercial Vehicles. The earliest size and weight limitations were put into effect in 1913,[1] when Maine, Massachusetts, Pennsylvania, and Washington passed laws limiting the weights of trucks using their highways. By 1919, 21 states had some type of size limitation.

The serious effect of heavy freight trucks on the roads and highways became so obvious in the 1920s that by 1930, 45 states had imposed gross-weight limitations, 31 states had axle-weight limits, and 40 states specifically limited weight per inch width of tire.[2] Much of the truck damage in those days was due to the use of solid rubber tires instead of the heavier pneumatic tires of today. Since oversize trucks and carriers were also creating traffic hazards, by 1930, 42 states had introduced width limits, 29 had height limits, and 28 had length limits.[3]

The matter of sizes and weights of commercial vehicles has caused continual conflict between commercial-vehicle interests and other public and private groups. There have been steady demands for increased sizes and weights, and accusations that states were creating barriers to interstate commerce and allowing lobbying to result in unfair legislative actions. On the other hand, questions have been raised as to hazards created by larger and heavier trucks, whether or not commercial vehicles are "paying their way," and the economic effects of heavy loads on highway structures and surfaces.

Claims and counterclaims continue, and increasing efforts and finances are going into the battle on both sides. The trucking groups claim that more width will make possible improved vehicle designs which will produce very definite safety factors in truck operation. They feel that increased load limits will permit them to handle commerce in a more economical way, which will be reflected in rates to the public, and that the larger trucks will make it unnecessary to use as many trucks, thereby reducing the volume of commercial vehicles in the traffic streams; and they further claim that road pavements are not unduly damaged by heavy loads, provided they are properly distributed. The opposition groups point to the fact that 40 per cent of the primary system of rural

---

[1] *State Limitations on Sizes and Weights of Motor Vehicles*, U.S. Interstate Commerce Commission, 1940, p. 173.
[2] *Ibid.*
[3] *Ibid.*

state highways were, in 1947, less than 20 ft in width[1] and are, thereby, physically incapable of accommodating wider vehicles with safety. They point to the effect of slow and cumbersome movements of trucks, particularly on grades, as a factor in congestion and delays to all highway users, and they claim that major damages to highways are being caused by heavily loaded vehicles. The claims, more than anything else, reflect the need for extensive research into the question of the effect of different sizes and loads on various types of pavements and structures.

The need for more accurate economic analysis of highway revenues allocated to the different highway-user groups is also indicated. It cannot be expected that the desires of commercial groups can ever be completely satisfied by factual reports, but much sounder judgment can be rendered when the results of fundamental researches are available in this field.

Legislative bodies are placed in a difficult position since powerful lobbyists, local highway agencies, the trucking group, and other factions are all involved in the question of sizes and loads on various types of pavements and structures.

When transportation was at a premium during World War II, comprehensive truck-bus inventories were undertaken at the request of the Federal agencies in most of the states. The data collected have been useful in many traffic engineering studies. The inventory was based upon the registration records of the motor vehicle departments of each state. Among the types of information collected were average annual mileage, year, model, and body styles, manufacturers' rated capacity, and the type of operation in which the vehicle is engaged. Information on buses also included seating capacity, passenger service, and mileage figures based upon the capacity and service.

Present Regulations. Act V of the Uniform Motor Vehicle Code recommends a maximum vehicle width of 96 in. (motor buses, tractors, and trolleys can be 102 in. wide), a maximum height of 12 ft 6 in., and a maximum length of 35 ft for single-unit vehicles with two axles, 40 ft for single-unit vehicles with three axles, and 50 ft for combinations. Maximum axle loads are fixed at 18,000 lb per axle. Maximum gross loads recommended by the American Association of State Highway Officials (AASHO) are shown in Table 13-4. These agree with the recommendations contained in the Uniform Code, except that combinations employing full trailers are allowed a maximum length of 60 ft. However, the AASHO recommended a maximum length of 50 ft for truck tractor and semitrailer combinations.

In 1948 the American Trucking Association recommended a maximum width of 102 in., a maximum height of 13 ft 6 in., a maximum length of 40 ft for single vehicles, 50 ft for tractor, semitrailer combinations, and

[1] *Highway Statistics, 1947,* 1948, p. 91.

65 ft for other combinations. The truckers recommended that axles spaced 4 ft apart should be allowed 36,000 lb on the groupings, and axles spaced 18 ft apart should be allowed a maximum load of 46,000 lb.

Numerous changes in size and weight regulations occur during legislative sessions each year.[1] Most changes liberalize the allowances. The

TABLE 13-4. AMERICAN ASSOCIATION OF STATE HIGHWAY OFFICIALS LOAD TABLE

Distance in feet between first and last axles in group	Maximum load in pounds on group	Distance	Load	Distance	Load
4	32,000	22	45,700	40	60,800
5	32,000	23	46,590	41	61,580
6	32,000	24	47,470	42	62,360
7	32,000	25	48,350	43	63,130
8	32,610	26	49,220	44	63,890
9	33,580	27	50,090	45	64,650
10	34,550	28	50,950	46*	65,400
11	35,510	29	51,800	47	66,150
12	36,470	30	52,650	48	66,890
13	37,420	31	53,490	49	67,620
14	38,360	32	54,330	50	68,350
15	39,300	33	55,160	51	69,070
16	40,230	34	55,980	52	69,790
17	41,160	35	56,800	53	70,500
18	42,080	36	57,610	54	71,200
19	42,990	37	58,420	55	71,900
20	43,900	38	59,220	56	72,590
21	44,800	39	60,010	57	73,280

* 50 ft being maximum length limit (UVC recommendations), all values in table in excess of 45 ft are for cases in which special permits are granted for oversize vehicles.

SOURCE: *Policy Concerning Maximum Dimensions, Weights, and Speeds of Motor Vehicles to be Operated over the Highways of the United States*, American Association of State Highway Officials, Washington, 1946, pp. 6–7.

wide variances in regulations for commercial vehicles in the different states are shown in Table 13-5.

Enforcement. There is perhaps no area of traffic operations causing greater difficulties than the enforcement of size and weight regulations, particularly weight. In some cases effective police enforcement is difficult because of policies and limited forces. In almost every state, tremendous problems are created because some commercial fleets disregard

[1] Up-to-date information on sizes and weights is available from the U.S. Bureau of Public Roads, and National Highway Users Conference, National Press Building, Washington.

TABLE 13-5. STATE SIZE AND WEIGHT LIMITS
(Maximum legal dimensions and practical weights)

State	Width, in.	Height, ft	Length, ft	Gross weight, 1,000 lb	State	Width, in.	Height, ft	Length, ft	Gross weight, 1,000 lb
Alabama	96	12½	45*	53.9	Nebraska	96	12½	50	64.6
Arizona	102	13½	65	77.6	Nevada	NR	NR	NR	76.8
Arkansas	96	12½	60	73.2	New Hampshire	96	13½	45	50
California	96	13½	60	76.8	New Jersey	96	12½	50	60
Colorado	96	12½	60	73.6	New Mexico	96	12½	65	72.7
Connecticut	102	12½	45*	50	New York	96	13	50	61.5
Delaware	96	12½	60	60	North Carolina	96	12½	48	58.8
Dist. of Columbia	96	12½	50	65.4	North Dakota	96	12½	45	57.7
Florida	96	12½	50	64.6	Ohio	96	12½	60	78
Georgia	96	13½	45	53.9	Oklahoma	96	12½	50	60
Idaho	96	14	65	72	Oregon	96	12½	60	72
Illinois	96	13½	45	72	Pennsylvania	96	12½	50	62
Indiana	96	12½	50	72	Rhode Island	102	12½	45	80
Iowa	96	12½	45*	60.8	South Carolina	96	12½	50	68.3
Kansas	96	12½	50	63.8	South Dakota	96	13	50	64.6
Kentucky	96	12½	45*	42	Tennessee	96	12½	45	42
Louisiana	96	12½	60	68	Texas	96	12½	45	48
Maine	96	12½	45	50	Utah	96	14	60	79.9
Maryland	96	NR	55	65.2	Vermont	96	12½	50	50
Massachusetts	96	NR	45	51	Virginia	96	12½	45	50
Michigan	96	13½	50	120-W	Washington	96	12½	60	72
Minnesota	96	12½	45	57.7	West Virginia	96	12½	45	102.4
Mississippi	96	12½	45	52.6	Wisconsin	96	12½	45	63
Missouri	96	12½	45	53.9	Wyoming	96	12½	60	73.9
Montana	96	13½	60	73.2					

* Full trailers prohibited.

W = Maximum possible on maximum axle load basis.

SOURCE: State Motor Vehicle Size and Weight Laws, National Highway Users Conference, Washington, October, 1949.

the prevailing regulations and because of the "grapevine" interchange of information among the truckers concerning enforcement practices. In some states, police simply do not have adequate manpower and equipment to do a satisfactory job.

Act V of the Uniform Code recommends that officers be allowed to weigh or measure any vehicle which they suspect of violating the size and weight laws and that if the weight is excessive, the load must be reduced to a legal amount before the vehicle can be moved. Penalty for failure to comply with the rules is set at a maximum $100 fine or 10 days imprisonment for the first offense and up to $500 or 6 months in prison for a third or subsequent conviction. Owners or operators are held liable for damage to any highway as a result of use by any vehicle in illegal or overweight operation.

Some states have made special plans and provision for the enforcement of truck weights. As early as 1941 the state of Oregon established a special Division of Weigh Masters, which now operates more than 50 fixed pit scales as well as about 10 portable axle-weighing loadometer crews. Results achieved during a 5-year period are shown in Table 13-6.

TABLE 13-6. ANALYSIS OF WEIGHT CHECKS IN OREGON, 1944–1948
Average weight of legal loads = 58,467
Average weight of illegal loads = 68,406

	1944	1945	1946	1947	1948
Loads weighed............	187,471	136,638	146,097	181,537	166,584
Number of violations......	4,441	4,606	6,934	9,212	8,681
Per cent violations........	2.37	3.37	4.75	5.07	5.21
Number of operators......	1,337	1,505	2,105	2,548	3,104
Number of trucks.........	2,291	2,483	3,202	....	4,737
Fines and court costs......	$74,961	$97,061	$171,875	$227,231	$236,185
Average cost per violation.	$16.88	$21.07	$24.79	$24.67	$27.21

SOURCE: W. W. Stiffler, "Control of Overloads on Oregon Highways," *American Highways*, vol. 28, no. 3, July, 1949, p. 5.

Other successes in enforcing restrictions are available. For example, Indiana established a comprehensive truck-weight law in 1949, and the Indiana State Highway Commission has installed permanent weighing stations at strategic points and made available portable scales for spot checks by state police officers. In 1949 violations were reported in 25 to 30 per cent of the trucks checked, but by spring 1950 the figure was down to 3.7 per cent. Copies of violation tickets are sent to the state truckers' organization, and a representative is assigned to follow up all violations under the Indiana system.

The Highway Research Board, in collaboration with state, trade, and Federal agencies, is conducting tests on different types of highways in an effort to determine as accurately as possible the extent to which the highways are damaged by heavy wheel loads. Single-axle loads and tandem loads of different magnitudes are being used to determine the amount of damage which may be expected under each set of conditions.

13-4. Vehicle Inspection. In favoring inspections, it has been argued that they improve the general standard of automobile conditions, develop better garage workmanship in repairs, and inform drivers of the conditions of their vehicles, thereby educating them as to the need for periodic mechanical examinations. In opposition, it may be contended that compulsory inspections are too expensive, that they develop undue inconveniences to drivers, that relatively few vehicles are in poor condition, that most of the faults discovered are of a petty character, and that undue emphasis has been given to the part of defective vehicles in accidents. Motor vehicle inspections grew out of early Save-A-Life campaigns. In the period 1927 to 1929, six[1] Eastern states instituted such programs by proclamations of governors. Owners of vehicles were required to present their vehicles to certain garages for inspection and repair of all safety equipment. Campaigns usually lasted for only a few weeks. In 1929 Pennsylvania, Maryland, and Massachusetts became the first states to enact laws calling for periodic inspection of all vehicles.[2] In 1930, six New England states embarked on a similar program calling for periodic inspections. At the start of World War II, five states[3] and the District of Columbia had state-owned and -operated stations in service, and nine states[4] had state-appointed stations.

Numerous cities became inspection-conscious. Memphis, Tenn., was the first city to take definite action in 1934.[5]

Inspection legislation of some type is in effect in about one-half of the states and the District of Columbia. The effect of compulsory state-wide motor vehicle inspection on the accident rate is demonstrated by the fact that states which had some type of inspection program had rates of 7.7 in 1948 and 6.5 in 1949 (per 100 million vehicle-miles), while the remaining states had rates of 8.4 and 7.4 respectively.[6] Delaware, District of Columbia, New Jersey, and Washington, which were conducting compulsory inspection programs through state-owned stations in 1947 and 1948, had an average fatality rate of 5.9 and 5.5.

---

[1] Delaware, Maryland, Massachusetts, New Jersey, New York, and Pennsylvania.

[2] T. F. Creedon, *Study of Motor Vehicle Inspection in Various States*, Automobile Manufacturers Association, Detroit, 1948, sec. I, p. 1.

[3] Connecticut, Delaware, District of Columbia, New Jersey, South Carolina, and Washington.

[4] Colorado, Maine, Maryland, Massachusetts, New Hampshire, New Mexico, Pennsylvania, Vermont, and Virginia.

[5] Creedon, *op. cit.*, p. 2.

[6] 1949 death rates from National Safety Council, provisional as of February, 1950, Chicago.

Fitness of Vehicles. The ICC has been especially active in research on the condition of commercial vehicles in the traffic stream. In 1940 tests were carried out in 18 states to determine the effectiveness of brakes on trucks engaged in interstate commerce.[1] Of the vehicles tested, 51 per cent of the buses, 42 per cent of the two-axled trucks, 21 per cent of tractor semitrailer combinations, and 12 per cent of tractor full-trailer combinations had brakes which would meet the ICC requirements which call for any vehicle having brakes on all wheels to stop in 30 ft from 20 mph and any vehicle not having brakes on all wheels to stop in 45 ft from the same speed. The tests showed a close correlation between gross vehicle weight and stopping distance and indicated that in quite a few cases poor performance was due to overloading or inadequate maintenance.

Accident records of the ICC indicate that the most commonly reported defect is brakes, but usually only 1.5 to 2.5 per cent of all accidents involving commercial interstate vehicles can be traced to defective brakes.

Results of a check by the International Association of Chiefs of Police of over 1 million vehicles during May and June, 1946, on United States highways, showed that a total of 31.6 per cent were defective, 29.3 per cent had faulty rear and stop lights, 17.7 per cent had bad headlights, and 12.0 per cent were driving with defective brakes.[2] A series of inspections conducted in Connecticut in 1943[3] found that 35.3 per cent of the vehicles checked had defects in the lighting system, brakes, tires, steering equipment, or horns. Lighting accounted for over half of the defects, possibly because of wartime dim-out restrictions. During July, 1949, a check of 854,917 vehicles by the Ohio State Police found one of every six to be defective.[4]

Studies of the National Safety Council show that vehicular defects, principally unsafe brakes, are reported as contributing causes in from 10 to 15 per cent of fatal accidents and in from 8 to 10 per cent of all accidents. The extent to which vehicular defects have a part in the accident picture varies widely between the different states.

Records of vehicle inspections also reveal the conditions of vehicles. Data released by the District of Columbia show that 18.5 per cent of the vehicles reporting for initial inspections in 1949 had defective brakes, 10.6 per cent had bad steering, and 27.6 per cent had lighting defects. Further interesting figures show that throughout the 1949 inspection in

---

[1] *Brake Performance on Commercial Vehicles and Combinations,* U.S. Interstate Commerce Commission, Washington, 1941, p. 17.

[2] *International Association of Chiefs of Police Traffic Safety Check,* Highway Research Board, Highway Research Abstracts, No. 132, July, 1946, p. 1.

[3] *Checking Traffic on the Spot,* Department of Motor Vehicles, Hartford, Conn., 1943.

[4] "Car Safety Check," Ohio Department of Highways, *Traffic Safety,* vol. 12, nos. 2 and 4, March–April, 1950, p. 3.

the District of Columbia only 51 per cent of the vehicles were approved on their first inspection. This is favorable when compared with results in previous years. In 1945, for instance, only 23 per cent of the vehicles were approved on their first inspection.

It is significant that checks at inspection stations and at random reveal a much higher percentage of defective vehicles in use than is indicated by defects reported in accidents.

Type of Inspection Systems. Motor vehicle inspections may be carried out by (1) state-owned and -operated stations, (2) officially designated garages, and (3) on-the-road spot inspections by enforcement officers. The Uniform Vehicle Code recommends that state-owned stations or officially designated stations should be supplemented by on-the-road inspections; that periodic inspections be required once or twice per year; that appointed stations be required to post a bond and keep records and be examined periodically by state officials; and that inspection fees should not exceed 50 cents each.

State-owned and -operated Stations. Though most generally favored by motor vehicle administrators and other public officials, the system of state-owned and -operated stations is currently in use only in Delaware, New Jersey, Washington, and the District of Columbia. This system involves a combination of fixed stations near the centroids of heavily populated areas (roughly speaking, where there are 20,000 registered vehicles in a 20-mile radius) and portable inspection lanes for coverage of more sparsely populated sections. Stations are usually administered by the motor vehicle department, but sometimes by the state police.

Automobile owners are generally notified of the time they are expected to bring their vehicles to the stations for the periodic inspections. If the vehicle satisfactorily meets inspection standards, a certificate of approval is issued upon the payment of the required fee. (In some instances the fees are collected as a part of the registration fee.) If the vehicle is found defective, a rejection sticker is placed on the windshield and it must be returned for reexamination within a specified period of time. In some stations vehicles with major defects are condemned and must be towed away, repaired, and towed back for a second inspection.

State-owned inspections assure uniform inspection techniques and are usually self-supporting from a small inspection fee. The level of work performed by garages in the locality is audited, and inspections are made by unbiased persons. However, objections to the system have been made on the grounds of competition of government with private garages, unnecessary expense to government (where no fees or inadequate fees to cover expenses are involved), and inadequate distribution of stations to permit sufficient service convenience to motorists in terms of waiting time and travel distance.

Appointed Stations. When private garages are authorized to conduct inspections required by state laws, as in Massachusetts and Pennsylvania, items to be covered and standards to be met are prescribed by the state. In most instances, the state authorities prescribe the minimum amount and the type of equipment which the inspection garages must have, and they frequently review the qualifications of garage personnel.[1] In most states, inspection garages receive nominal fees, but the repair business derived from the inspection work is usually of considerable magnitude.

The system of appointed stations does not require a heavy outlay for equipment on the part of the state, and it provides an easily organized extensive network of garages. However, it is sometimes difficult to administer, and appointments may be influenced by political considerations. In addition, garages have been accused of suggesting unneeded repairs to increase business and of approving defective vehicles for financial or friendly considerations.

Spot Inspections. In the spot-inspection method, law-enforcement officers stop vehicles at random or stop those which they suspect need mechanical attention. Since such tests must usually be made without benefit of equipment, they cover only such things as brakes, lights, horns, windshield wipers, and mirrors.

Municipal Inspections. In some states without state-wide inspection laws, cities have obtained enabling legislation under which they have established and operated municipal inspection stations.

Inspection Coverage. Ideal inspection programs cover the following equipment:

1. Lighting systems: (*a*) operation of bulbs, (*b*) electrical systems, (*c*) lenses, (*d*) colors, (*e*) beams—focus, aim, and intensity
2. Brakes—service and emergency: (*a*) minimum stopping requirements, (*b*) equalization, (*c*) pedal reserve, (*d*) parts—hoses, levers, cylinders, etc.
3. Steering: (*a*) play in wheel, (*b*) assembly mechanical fitness
4. Tires, wheels, alignment: (*a*) excessive tire wear, (*b*) exposed fabric in tires, (*c*) tire bulges, scuffs, etc., (*d*) bad rims, bolts, flanges, etc., (*e*) alignment
5. Accessories: (*a*) horn, (*b*) glass—required type, ability to raise and lower on driver's side, obstructions, cracks and discoloration, (*c*) wipers, (*d*) defrosters, (*e*) mirrors, (*f*) wiring and switches, (*g*) directional signals, (*h*) exhaust systems—mufflers, (*i*) registration plates, (*j*) fuel tank and lines, (*k*) body fittings, (*l*) spare parts

Since inspection of the above items requires considerable equipment and trained mechanics, they are not all covered under the average inspection program. Inspection of the lighting system often covers only the operation of bulbs and compliance with color requirements; brakes may be checked only as to stopping abilities and pedal reserves; steering as to the amount of play in the wheel; tires as to general fitness (alignments not

[1] *Motor Vehicle Inspection,* Association of Casualty and Surety Companies, Bulletin 44, New York, July, 1950.

ascertained); the examination of accessories may be limited to the horn, glass, windshield wipers, directional signs, and the placement of mirrors.

Ideally, inspection garages are equipped with headlight-test apparatus, wheel-alignment indicators, brake testers, and pits of lifts for quick examination of brake connections and front-end assembly. In the design of an inspection station, adequate reservoir space must be provided so that large accumulations of vehicles in peak periods will not unduly interfere with passing traffic. Care must also be exercised that the structures housing inspection lanes are well ventilated and properly lighted.

CHAPTER 14

GENERAL CONTROLS

There are certain basic rules which every motorist must know and which also affect pedestrians, bicyclists, and other road users. These general controls, intended for state-wide application, are so fundamental that their incorporation in city ordinances is usually not necessary, but a few (like those affecting bicycle riders) are almost always included in local ordinances rather than in state codes. Few specific rules of road use can be made operative without prior enactment of basic general controls essential to the orderly regulation of highway transportation.

State laws usually provide that all vehicles and persons on public ways are to be governed by the motor vehicle codes, but certain exceptions may be conferred upon emergency vehicles, enforcement authorities in pursuit of violators, and state maintenance or construction crews. There is a trend toward eliminating exceptions so that all vehicles are subject to the same regulations. In the Uniform Code, and in many states, special privileges and exemptions for certain Federal and state employees have been discontinued. Act V of the Uniform Code permits emergency vehicles to violate "stop" signs, traffic signals, and other regulations when responding to emergency calls only if approved audible signals are given and the specified special lighting equipment is in use.

In the interest of uniformity and elimination of overly restrictive local regulations, there is an increasing tendency to curtail the power of local authorities to deviate from general motor vehicle laws adopted by the state. However, state laws generally give local authorities broad powers over traffic situations which prevail only in cities and which vary in character from city to city. These include the establishment of parking-time limits, manual direction and control of traffic, regulation of processions, establishment of stop streets (excluding state highways), and detailed bicycle controls.

For uniformity in the application and interpretation of traffic control devices, it is important to state clearly in the general codes the actions to be taken by the road user. State laws usually require that all traffic regulatory devices erected and maintained within the state must conform with standards prescribed by a central state agency, usually the highway department. For example, Connecticut provides that a state traffic commission, made up of the commissioners of highways, motor vehicles, and state police, shall establish a manual of uniform traffic control devices to serve as a guide in the placement of any control device in the state. In Vermont, on the other hand, selection and placement of signs and signals is left to local officials. Reference to a uniform manual appears in the laws of 27 states, and in almost all states the state highway agency is authorized to place and maintain traffic signs and other control devices. It is common practice to grant state agencies absolute control over devices placed on state highways and streets used as state highways.

Some general regulations affecting control devices are taken for granted by both public officials and road users. When traffic signals were first used, a few traffic codes specifically defined the meaning of the colored lenses, and to ensure legal effectiveness it was common to superimpose "stop" and "go" on the lenses. (Under modern laws, word messages on the lenses are no longer necessary.) On the other hand, little thought was given to pedestrian observance of traffic signals in the early days of vehicle use, but now state laws specifically state the action to be taken by pedestrians for each signal indication.

Proper placement, design, and maintenance of control devices should be required before they are legally effective. It is not reasonable to expect observance of inadequate, improperly located, or poorly maintained control devices.

The failure of a state to provide adequate traffic control devices has been the subject of many legal cases, among them the case of *Zeim v. State of New York*,[1] in which it was held that "the traffic signs involved in this action were confusing, improperly worded, improperly located, insufficient in height, insufficient in number, not reflectorized or improperly reflectorized, misleading, and an invitation to disaster at night time. . . . The State and the State Traffic Commission were negligent in failing to order the removal of the signs in question and in failing to replace them with signs which would conform with nationally accepted standards. . . . The testimony is undisputed that nationally accepted standards required that the dead-end stop sign should have been replaced long prior to the accident with a reflectorized double pointed arrow to warn the motorist that he was approaching a main highway."

In addition, many cases have been brought before the courts on the question of the failure of traffic signals. When the traffic signals are operated by cities, recovery for injuries received as a result of light failure is invariably denied because it is held that the operation of the traffic signals is a governmental rather than a corporate function, as, for example, in the case of *Avey v. City of West Palm Beach*.[2]

However, a dissenting opinion was set forth in this case stating, "As the

[1] 270 App. Div. 876 (1946).
[2] 152 Fla. 117, 12 So.2d 881 (1943).

city is charged with keeping its streets in safe condition as a corporate function, I think the duty to keep its traffic lights in good and safe condition is also a corporate rather than a governmental function."

On the other hand, it has been held that the state may be held liable if negligent maintenance by the state can be proven, in the event of an accident resulting from signal failure.[1]

General laws usually stipulate that control devices must not be destroyed, obstructed from view, or otherwise affected to give an inadequate or improper indication to road users. The erection of signs or devices which have the appearance of official devices is likewise prohibited.[2] The actual enforcement of such provisions, however, is often difficult.

14-1. Accident Reporting. There are only two states in the United States which do not require by law that drivers report accidents. Reports of accidents involving personal injury or property damage in excess of $50 are required in 25 states, while 8 states require reports regardless of the extent of damage. Since reports on minor accidents are as valuable as those for serious accidents in prevention work, there is a marked trend toward requiring reports on a greater percentage of accidents. Act V of the Uniform Motor Vehicle Code requires drivers involved in accidents to notify the police immediately and to file a written report on forms prescribed by a designated state agency within 5 days of any accident involving personal injury or property damage to the extent of $25. If the operator is incapable of making the report, passengers, owners, or witnesses can be required to submit reports. Police are required to file written reports within 24 hr after an accident investigation, and coroners must report to the central accident-record bureau whenever a death results from a motor vehicle accident. Garages are also required to report any vehicles they repair which have been damaged in accidents, but it is difficult to enforce such a provision unless garages are licensed by a state agency.

The National Committee on Uniform Traffic Laws and Ordinances and other agencies recommend that operators' accident reports should be for the confidential use of state officials and that they be filed without prejudice. In practice, accident bureaus operating under confidential reporting laws are free to give the names of operators and factual information which does not divulge general statements or opinions of operators.

Figure 14-1 indicates the procedure generally followed in accident analysis. At the state level some form of mechanical tabulation is essential; after accident information has been collected, supplemental records such as fatality cards and driver record cards are prepared, and reports numbered serially, and the pertinent information is coded. Before placing original reports in a location or serial file, spot-map data

---

[1] *Foley v. State of New York*, 43 N.Y.S.2d 587 (1943).

[2] *Alabama State vs. Mobile*, 30 Am. Dec. 564 June, 1937.

and other special study facts may be derived. Many states file the bulky original reports in serial files after making special card reports for the location file.

14-2. Vehicle Maneuvers. *Passing.* Many states have special requirements to restrict overtaking and passing. In 18 states laws

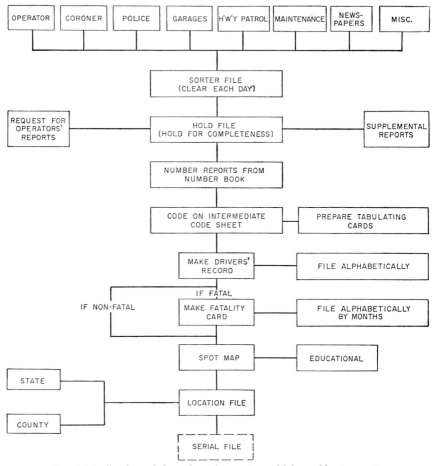

FIG. 14-1. Condensed flow chart for motor vehicle accident reports.

require a clear-sight distance of some specified value, ranging from 150 ft in Arizona and Texas to 1,000 ft in Wisconsin for curve locations, before overtaking and passing can be performed. In 34 states overtaking and passing is unlawful within 100 ft of a bridge, tunnel, intersection, or railroad crossing. Passing on the right is legally permitted in 23 states on multilane facilities and when another vehicle is preparing to make a left turn.

When no-passing zones are established and properly marked, they alter the basic rules for overtaking and passing in the same manner as a "stop" sign alters the basic right of way at an intersection. On multilane roads, movements from one lane to another must usually be made at the responsibility of the driver of the vehicle in motion.

Turning. Traffic codes should specify the exact manner in which turns are to be made, particularly at intersections. As common sense

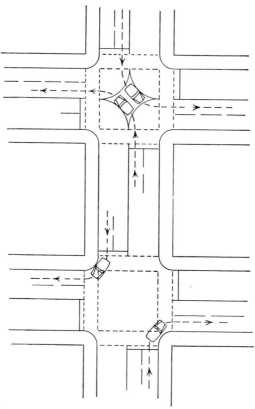

Fig. 14-2. Recommended procedure for negotiation of left and right turns.

dictates, right turns should always be made from the lane nearest the right-hand curb and into the right-hand lane of the street entered. The proper position for vehicles making left and right turns is shown in Fig. 14-2. Inadequate turning radii, improperly regulated parking, and unusually long vehicles sometimes make it impossible to negotiate turns in precisely the manner desired.

There is some controversy about the most desirable manner of making left turns. In the early days of motor vehicle use, cities commonly required that left turns be made around the center point of the intersection in order to reduce speed and eliminate corner cutting, and 18 states still have laws requiring this type of turn. As traffic volumes and intersection congestion increased, however, this type of turn became unsatisfactory, as Fig. 14-3 shows. Act V of the Uniform Motor Vehicle Code and most states require that a left turn on two-way streets be made from the lane nearest to the center line of the street and be completed in the lane nearest the center line of the entered street. Many state laws provide that left turns must not only be made in the lanes nearest to the center line, but that "whenever possible such turns shall be made to the

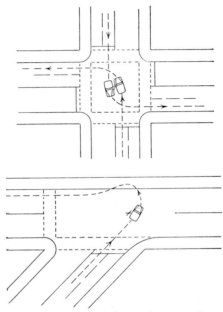

Fig. 14-3. Situations in which old-type left-turn law complicates intersection movements.

left of the center of the intersection." This method of turning avoids conflicts between left turns from opposite directions, permits the negotiation of a turn with less delay, and makes it possible to complete the turn in a legal manner at all types of intersections.

The left-turn rule must be altered to fit one-way streets, but the turn should still initiate from the left-hand lane and should enter the left-hand lane of traffic moving in the same direction. Because methods of turning must be modified at irregular intersections, it is important that state laws grant local authorities the right to vary basic practices by markers or appropriate signs.

There is increasing doubt as to the safety of turning signals, which seem to be of far less importance than the position of the vehicle. Most state laws provide for three types of hand signals, two of which relate to turns. Three-position hand signals, as recommended in the Uniform Motor

Vehicle Code, are now required by 27 states (Fig. 14-4). The chief difference in hand signals relates to right turns: in one type there is a forward rotary motion of the hand and arm, in the other an upbending of the forearm at the elbow. Left-turn signals are always the same: a straight horizontal extension of the arm and palm with the fingers pointing outward.

Mechanical signaling devices on motor vehicles are decreasing the necessity for hand signals. Certain commercial vehicles, designed or loaded so that hand signals are not readily visible from the rear, have long been required to have mechanical signaling devices, and New York State

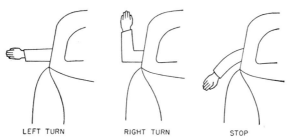

LEFT TURN   RIGHT TURN   STOP

Fig. 14-4. Three-position hand signals as recommended by Uniform Vehicle Code.

passed a law in 1949 prohibiting the operation of motor vehicles manufactured after Jan. 1, 1952, which are not equipped with mechanical directional signals. It is usually required that a hand signal be given.

Stopping. When a stop is to be made or speed is to be markedly reduced, the arm is extended in a downward position with the palm to the rear (Fig. 14-4). However, the importance of hand stop signals has been decreased by equipping motor vehicles with stop lights. Also, the increasing number of closed-type car bodies, requiring the lowering of glass in order to make a hand signal, decreases the observance of the regulation. Some further mechanical improvements, such as buttons which automatically lower the glass, may encourage use of hand signals.

Right of Way. State traffic codes usually give the right of way to vehicles approaching from the right at an uncontrolled intersection. The logic of this rule has been questioned by some on the contention that if the left vehicle had right of way, crossing position for the preferred vehicle would be transferred to the far side of the intersection. More time and opportunity would be available for the driver to ascertain whether his right of way would be respected. While this argument appears to have merit, the consensus seems to be that the necessary reeducation of drivers would make the change prohibitive.

The right-of-way rule must be modified to fit peculiar conditions, such as when one vehicle has already entered the intersection, and at intersections having "stop" signs or signals. Vehicles turning left at intersections are usually given the right of way over vehicles approaching from the opposite direction, after they have first yielded and ascertained that it is safe to negotiate the turn. Right of way for turning vehicles requires the proper hand or mechanical signal.

Where the basic right-of-way rule is altered by the erection of "stop" signs, a motorist is allowed to proceed through the intersection after he has first stopped, yielded the right of way to any vehicles approaching on the artery, and ascertained that it is safe to enter or cross.

Right-of-way rules in most states require persons approaching from driveways or private alleys to yield the right of way and to go to the right of the road and stop.

14-3. General Pedestrian Rules. In early traffic laws no mention was made of pedestrians' responsibilities at signalized intersections, but traffic codes now generally require pedestrians to obey traffic signals in the same manner as motorists. While it is difficult in most areas to enforce pedestrian observance of traffic signals, there are many examples of effective pedestrian-control programs in the downtown sections of cities. Most pedestrian fatalities occur in outlying sections of the cities, where it is almost impossible to effect a satisfactory degree of pedestrian signal observance because of the light traffic volumes and the small number of pedestrians desiring to cross at each location.

At unsignalized intersections pedestrians are usually given the right of way when they are in crosswalks and have already begun a crossing of the traveled portion of the roadway. When not crossing at an intersection or an officially designated crosswalk, they are required by law in most states to yield the right of way to motorists. Pedestrians may be required to cross only at crosswalks where signals are provided at adjacent intersections. Motorists are required to exercise "due caution" and are not entirely relieved from blame in case of an accident involving a pedestrian crossing at some place other than a designated crosswalk.

In rural areas and other locations where sidewalks are not available, traffic codes usually provide that the pedestrian must walk on the left-hand side of the roadway facing traffic.

Regulations prohibiting the solicitation of rides from the roadway, commonly known as *anti-hitchhiking laws*, are largely unenforceable because of public sentiment and pressure brought by college groups and civic organizations.

Bicycle Controls. Modern traffic laws make bicycle operations an integral part of the over-all codes. While general rules of bicycle operation and the rights and duties of bicyclists in relation to motorists and pedestrians are generally prescribed by state laws, control by licensing and other specific regulations are usually left to local authorities. State codes usually require parents to assume responsibility for children, prohibit double riding or clinging to other vehicles, require bicycles to use the right side of the roadway, prohibit the operation of bicycles two abreast except on special bicycle pathways, require bicycles to use paths if provided, restrict riding without hands on handle bars, require lights if bicycles are operated at night, specify brakes, and stipulate that bells, horns, or adequate audible signals must be provided on each bicycle. City ordinances may provide for periodic inspection, licensing, and registration, may establish special speed limits for bicyclists, designate additional required equipment, regulate bicycle parking in public places, and prescribe regulations controlling the activities of bicycle dealers.

It is estimated by the Bicycle Institute of America that there are now about 14 million bicycles in use in the United States, a ratio of one for every 2.7 automobiles and every 11 persons.

Bicycle Accidents. About 2 per cent of automobile fatalities involve bicycle collision. During 1952 about 500 persons lost their lives and 27,000 were injured as a result of collisions between motor vehicles and bicycles.[1]

TABLE 14-1. SUMMARY OF BICYCLE ACCIDENTS, 1952 TOTALS

	All ages	0–4 yr	5–14 yr	15–24 yr	25–44 yr	45–64 yr	65 yr and over
Deaths							
Urban...............	240	...	130	50	10	10	40
Rural...............	260	...	220	10	20	10	
Injuries							
Total urban and rural...	27,000	600	21,000	3,500	900	600	400

Studies show that two out of three bicyclists are violating one or more traffic laws when they become involved in collision with an automobile. In one out of three accidents involving motor vehicles and bicycles, the primary fault is placed with the motorist. About one out of every five bicycles involved in motor vehicle accidents is found to have a mechanical defect, most frequently the lack of proper lighting equipment.

In establishing a municipal control program for bicycles, there are three important elements to be considered: (1) registration, (2) passage and enforcement of bicycle-traffic regulations, and (3) instruction of cyclists in regard to the regulations. Approximately 250 cities in the United States have bicycle ordinances of one type or another.[2]

Since many bicycle riders are juveniles and because of the attitudes of the public and the courts, police face a difficult situation in strict enforcement of bicycle laws. In most cases warnings are given rather than

[1] *Accident Facts*, National Safety Council, Chicago, 1953, pp. 45, 60, 61.

[2] *Bicycle Registrations, Thefts and Accidents*, American Automobile Association, Washington, 1943.

formal citations. Sometimes ordinances are written which empower the police to impound bicycles, suspend registrations, and to require special instructions when violations of bicycle laws are observed.

14-4. Streetcars and Safety Zones. Streetcars are in operation today only in the largest cities, and even there the number is diminishing rapidly. General regulations controlling the actions of motorists in relation to streetcars are therefore becoming less important.

Passing streetcars on the left is usually prohibited except where motorists are directed otherwise by police officers, on one-way streets, and on streets where the streetcar tracks may be located on the extreme right-hand portion. Motorists must usually stop before passing on the right of a streetcar which is in the process of taking on or discharging passengers, except at points where official safety zones are available for use by streetcar passengers.

14-5. Miscellaneous Rules of Road Use. Numerous miscellaneous regulations are necessary to provide for many minor details concerning vehicle use. Miscellaneous traffic rules provide that (1) unattended vehicles must be locked and properly parked, (2) parking is allowed on a public way only when it can be achieved with safety and without interfering with other traffic, (3) the number of passengers and parcels in the front seat must be limited so as not to obstruct the driver's view, (4) the riders of motorcycles must be limited to the number that can occupy permanent seats on the vehicle, (5) special care must be exercised in driving on mountain roads, (6) coasting is prohibited, (7) following fire trucks and emergency vehicles is not allowed, (8) motorists are prohibited from crossing fire hose except when directed by signs or officials, (9) glass and other hazardous objects are not to be placed on public roadways, (10) special rules must be followed in the overtaking and passing of school buses, and (11) detailed regulations must be adhered to in the operation of school buses.

14-6. Disposition of Traffic Violation Cases. Through actions of the higher courts and through legislation, uniform penalties are developing in many states for traffic violations. Several cities in Michigan have recently installed a uniform court policy, suggested by heads of local enforcement agencies, which accurately indicates the nature and seriousness of the violation in question and sets a uniform fine based upon the violation. This system eliminates political factors such as ticket fixing, creates a warning system for first offenders, provides an exchange of records between cities, automatically increases or decreases fines at the level of cases normally handled by the violations bureau in accordance with the degree of violation and the seriousness of attendant conditions. Table 14-2 shows the uniform traffic violation and the seriousness of attendant conditions. Table 14-2 also shows the uniform traffic violation notice with the penalties inserted. In cases above the level of the violations bureau, such as accident cases, repeater cases, cases of driving 20 miles or more above the speed limit, and reckless driving, a uniform level for fines is suggested.

In 1949 New Jersey adopted a set of "Rules of Practice for Local Criminal Courts," providing for the inclusion of all requisite information

TABLE 14-2. MICHIGAN UNIFORM FINE SCHEDULE

Six principal violations causing most accidents	$2 fine	$3 fine	$4 fine
Speed (over limit)	__5 miles	__6–9 miles	__Over 10*
Improper left turn	__No signal	__Cut corner	__From wrong lane
Improper right turn	__No signal	__Into wrong lane	__From wrong lane
Disobeyed traffic signal (when light turned red)	__Past middle of intersection	__Middle of intersection	__Not reached intersection
Disobeyed "stop" sign	__Wrong place	__Walk speed	__Higher speed
Improper passing	__At intersection	__Cut in	__Wrong side of pavement

Other violations and conditions that increased seriousness of violation	
Slippery pavement	Caused person to dodge‡
__Rain__Snow__Ice	__Pedestrian__Driver
Darkness	Just missed accident‡
__Night__Fog__Snow	__1 ft
Heavy traffic†	Actually hit object¶
__Pedestrian__Vehicles	__Pedestrian__Vehicle

District: __Business__Industrial__School__Residential

* $5 additional fine for each 5 miles above 10 miles over limit.
† $1 additional fine each.
‡ $2 additional fine each.
¶ $3 additional fine each.
NOTE: Above amounts are for first offense. Add 50 per cent for second and 100 per cent for third offense.
SOURCE: Maxwell Halsey, "Michigan Courts Develop Uniform Fine Schedule for Violations Bureau," *Michigan Municipal Review*, vol. 21, no. 5, May, 1948, p. 70.

on the original summons; the issuance of warrants to violators who ignore summonses, specifying a definite period of 5 days between date of offense and trial; calling for traffic cases to be handled separately for greater court efficiency; and requiring that except for parking and out-of-town violators, the defendant must appear to face the charges.

In 1940 Colorado adopted a form ticket for traffic violation cases, known as the penalty assessment ticket, providing a uniform penalty schedule for cases involving minor offenses and no accidents.

The establishment of collateral schedules by the courts and the fixing of penalties for serious traffic offenses are other ways of developing uniform penalties. Many laws provide that persons are to be taken before a judicial agent immediately after a serious traffic violation such as negligent homicide, driving under the influence of intoxicants, hit-and-run accident, refusal of a motorist to give proper promise to appear in court, or refusal to give proper identity. In less serious cases (misdemeanors), motorists are usually given up to 5 days in which to appear in court.

Many state laws require enforcement agencies at all levels of government within the state to use triplicate or quadruplicate ticket systems, providing a careful and continuing audit of traffic charges and eliminating or reducing ticket fixing.

14-7. Intoxication Tests in Traffic Cases. The difficulty of obtaining convictions in drunken-driving cases, delinquency in prosecutions of such cases, and the charging of innocent persons have led to legislative consideration of laws authorizing scientific and chemical tests in drunken-driving cases.

Tests made by the National Safety Council indicate that the normal chance of accident increases about 55 times when a driver has a blood concentration of alcohol of 0.15 per cent. Some tests have shown that even very small amounts of alcohol in the body affect the neuromuscular coordination of drivers to such an extent that they are unfit for driving. Chemical tests to determine the concentration of alcohol in the blood can be made from specimens of urine, blood, saliva, breath, spinal fluids, or brain tissues. The latter two are used in fatal accidents.

As an adjunct or substitute for chemical tests, various types of questions may be used to secure basic information about the driver which can be related to his actions and to results of chemical tests. Some police departments have developed forms for general observations of drivers suspected of being under the influence of alcohol or drugs. In some cases, clinical examinations, conducted by city physicians, may be required.

Some courts have held that a driver cannot be required to submit to clinical or chemical tests, while others have found that police can require the accused to submit. One Iowa court[1] held that blood tests taken from an unconscious driver (without his permission) were not acceptable evidence. Results of a blood test made after a driver had died were admissible evidence, since the analysis showed 0.28 per cent ethanol.[2] The United States Court of Appeals found that the prohibition of compelling a man to testify against himself does not exclude his body as evidence: "His body may be examined either in or out of court, with or without his consent."[3]

Local Traffic Ordinances. It has been pointed out previously that many of the foregoing regulations are prescribed specifically in state codes,

---

[1] *State v. Weltha*, 228 Iowa 519, 292 N.W. 148 (1940).

[2] *Lawrence v. City of Los Angeles*, 53 Calif. App. 2d 6, 127 P.2d 931 (1942).

[3] Opinion 8959, *Carl J. McFarland, appellant v. United States of America. appellee.* 1945.

while others are treated more generally by the state with specific application granted to local traffic authorities. There has been developed a model traffic ordinance useful as an aid in setting up adequate local ordinances.[1]

Local traffic ordinances usually prescribe in detail those regulations which state laws grant the right to be prescribed locally. They may also cover the various requirements for administration of traffic engineering and enforcement agencies.

In some states local authorities are encouraged to adopt in total all applicable state laws as local ordinances to provide for proper legal basis in enforcement. This apparent duplication of laws has been found helpful in classification of proper jurisdiction and application.

[1] National Committee on Uniform Traffic Laws and Ordinances, *Model Traffic Ordinance*, 1946, 1952.

CHAPTER 15

SPEED CONTROLS

Since the early days of the steam engine, increase in speed has been a major objective in the technical advancement of all types of vehicles. The automobile and the highways have been improved to the extent that average 20-mph speeds of 1910 have increased to an average of 50 mph at present.

Paralleling growth of speed potential, there has been growing recognition of the necessity of regulating speeds. As early as 1678, Newport, R. I. passed an ordinance that no one should operate any conveyance in a hazardous or reckless manner. Throughout the 1700s and 1800s a number of communities dealt with the problem, although it was not until 1890 that types of vehicles began to appear which presented serious problems.

State legislation governing motor vehicles was adopted by Connecticut in 1901, setting forth maximum limits of 15 mph in rural areas and 12 mph in cities.[1] Today every state in the country has some type of speed law.

A commonly accepted criterion of the need for and importance of speed control is the relationship of speed and accidents. A summary of 1952 accidents shows that in one of every three fatal accidents a speed violation was involved: exceeding a speed limit or driving too fast for prevailing conditions.[2] Table 15-1 shows the effect of speed on accident severity

TABLE 15-1. SPEED AND ACCIDENT SEVERITY

Reported speed, mph	Severity index No. of fatal-accident drivers per 1,000 injury-accident drivers
20 or less	12
21–30	21
31–40	36
41–50	48
51 and over	92

SOURCE: Special reports from seven states: Minnesota, Mississippi, Nebraska, Tennessee, South Carolina, Virginia, 1940, and Indiana, 1939, Statistical Bureau, National Safety Council, Chicago. See "Speed Control," *Report of Joint Committee on Post War Speed Control*.

[1] Joseph N. Kane, *More First Facts*, H. W. Wilson Co., New York, 1935, p. 39.
[2] *Accident Facts*, National Safety Council, Chicago, 1953.

250

as measured by the number of fatal-accident drivers per 1,000 injury-accident drivers at various speeds. It is evident that high speeds do generally accompany high accident-severity ratios, and speed control at a reasonable level appears to be justified.

Experience on controlled-access roads and modern turnpikes indicates that very good accident rates can be achieved with relatively high speeds. Rates of about three fatalities per 100 million vehicle-miles are common on high-type roadways, while the nationwide rate is about seven. However, since it is impossible to design and build new roads in the quantity and quality needed for extensive high-speed travel, the only solution lies in controlling speeds on our existing road systems.

15-1. Methods of Control. Three types of speed legislation and control are in use today: (1) the basic speed rule, which states in effect that all motorists should travel at a "reasonable and proper" speed, (2) numerical speed limits, generally in the form of maximum posted limits in certain districts or under special road conditions, and (3) the authority, given by law, for speed zoning, thus providing for the establishment of special numerical limits suited to physical and traffic conditions at certain locations.

Numerical speed limits are of two types, absolute and prima facie. The absolute limit fixes maximum allowable speed for a particular location, any excess of the limit being sufficient to warrant arrest. The prima facie limit places the burden of proof of safe driving upon the driver if he exceeds the stated maximum.

At present about one-quarter of the states use the basic rule on all roads outside of urban areas; another quarter of the states specify an absolute maximum limit, ranging from 40 to 60 mph in the various states; one state has a 45-mph prima facie limit and a 60-mph absolute limit, and the remaining states have prima facie limits ranging from 50 to 60 mph.[1] Prima facie limits were adopted by the National Committee for Street and Highway Safety in the Uniform Vehicle Code. Most of the states still favoring the basic rule are Middle Western states.

A number of less familiar legislative speed controls are in common use in some states. Four states employ a lower speed limit for after-dark driving[2] and a number of cities also use changing limits, generally increasing the maximum speed for night travel in business and residential zones. Special truck speed limits, usually less than passenger-car limits, are in effect in a number of states and cities. Most states have a general minimum speed rule, although this provision is seldom enforced.

Speed legislation can be only as effective as its enforcement, but the problems connected with adequate speed enforcement are numerous and a

[1] President's Highway Safety Conference, *Laws and Ordinances*, 1949, p. 29.

[2] *Traffic Speed Enforcement Policies*, Eno Foundation for Highway Traffic Control, Inc., Saugatuck, Conn., 1948.

well-staffed and well-equipped enforcement agency is necessary if the desired results are to be obtained. Effective enforcement policies can do much to curb excessive speeds and reduce accident frequency. Table 15-2 shows the results of a state-wide enforcement program, based on a 50-mph absolute limit, carried out in Pennsylvania in 1938.

TABLE 15-2. COMPARATIVE ACCIDENT EXPERIENCE IN PENNSYLVANIA, 1937, 1938, 1939

	In rural areas			In urban areas		
	Total accidents	Persons injured	Persons killed	Total accidents	Persons injured	Persons killed
1937	35,934	30,506	1,516	37,532	30,939	1,048
1938	27,136	21,543	1,095	35,846	29,055	801
1939	27,709	22,626	1,062	36,323	29,254	809
Per cent change 1937–1938	−24.5	−29.4	−27.8	−4.4	−6.1	−23.6
Per cent change 1938–1939	+ 2.1	+ 5.0	− 3.0	+1.3	+0.7	+ 1.0

SOURCE: Pennsylvania Department of Revenue.

Proper publicity through educational newspaper and magazine articles, radio broadcasts, and posters is a valuable aid in attacking the speed problem. Driver education can be a very effective tool in promoting safe driving and observance of speed regulations. Since regulations are enforceable only if they receive the support of a majority of the people, legislation, enforcement, and education must go hand in hand.

15-2. Control through Devices. In many problem locations where speed is a factor in high accident rates and corrective design measures are not economically justified, effectively placed control devices may temporarily afford adequate protection. They may even reduce the problem to such an extent that design changes are not needed.

Signs. Blanket numerical limits to indicate general speed laws are posted in all states. In most cases signs are erected wherever the speed limit changes, as determined by the type and density of roadside culture or by corporate boundaries.

Speed-limit signs are frequently used at curves and intersections. The most positive speed control at intersections is the "stop" sign, but the "yield right of way" sign has been developed for locations where a stop may not be necessary. Studies made with these signs indicate that they are effective in reducing accidents because they develop alertness and care not present in "stop" sign violations. The value of "yield right of way" signs has not been conclusively proved, but they are becoming commonly used, and a standard design has been approved by the Joint Committee on Uniform Traffic Control Devices.

Because of the seriousness of intersection accidents, especially in rural areas, other types of signs have been employed in an attempt to reduce approach speeds. "Slow" signs, in combination with symbols and directional signs, were standard treatment a few years ago, but studies of speeds and accidents have shown that such "standard" treatments were totally ineffective. Signs giving safe-approach speeds have also failed, because calculations involving possible actions of drivers on two intersecting roads usually give speed values that are far too low for the average driver.[1]

Missouri was the first state to indicate extensively safe speeds on curves, in 1937, and many states have since adopted this practice.[2] In general, speed signs on critical curves are simply informative and do not represent a legal maximum limit.[3] Table 15-3 shows the results of safe-speed indications at curves in Illinois.

The Indiana State Highway Commission reports very favorable results in accident reduction through the use of stated-speed signs on curves. In 1939 a 95-mile section of Route 37, a winding hilly road between Indianapolis and Paoli, was posted with safe-speed signs on every curve. Despite a 15 per cent increase in traffic, there were 36 fewer accidents on curves, 33 fewer injuries, 7 fewer fatalities, and a reduction of 9 million dollars in property damage.[4]

Speed-control signs may also be effectively used at low-visibility intersections, especially in isolated rural areas where high average speeds are found on the open roads approaching the intersection. At problem locations of this type where the accident rate is excessive, the use of oversize or novel signs may be helpful. Barriers, overhead and flashing signs, and special signals have also been used, chiefly for their value as attention-getters. However, when the novelty of the control wears off its effectiveness diminishes. Enlarged standard signs used in South Carolina showed that the number of vehicles traveling above the safe-approach speed was materially reduced, and accidents decreased from a 5-year average to a total of two for the first 18 months after the change.[5] An unusual type of speed control is that in which striped barricades create a funneling effect as seen by the approaching driver. Barriers suspended overhead may be

[1] For methods of calculating safe-approach speeds at intersections, see National Safety Council, Public Safety Memo 73, Chicago; *Normal Safe Approach Speeds at Intersections*, American Automobile Association, Washington; *Policy on Intersections at Grade*, American Association of State Highway Officials, Washington.

[2] R. G. Paustian, *Speed Regulation and Control on Rural Highways*, Highway Research Board, Washington, 1940.

[3] R. A. Moyer and D. S. Berry, *Proceedings of Highway Research Board*, Washington, 1940, vol. 20, p. 425.

[4] *Ibid.*

[5] W. S. Smith, "Control of Speed with Signs and Markings," *Proceedings of Institute of Traffic Engineers*, New Haven, Conn., 1939, p. 34.

most effectively applied at locations where the approach to the intersection is on an upgrade and the intersection is not visible over a top of the grade or over a knoll.

Any type of regulatory device can be useful only if it is reasonably and logically placed and regulates traffic in a practical manner. Enforcement

TABLE 15-3. COMPARATIVE SPEEDS BEFORE AND AFTER ERECTION OF
STATED-SPEED SIGNS AT SELECTED CURVES ON U.S. 40 IN STATE OF ILLINOIS

Class of speed indication	Speed		Per cent of vehicles exceeding—		No. of vehicles observed
	20%,* mph	85%,† mph	Posted speed	5 mph above posted speed	
50 mph					
Before	31.7	49	12.6	8.1	604
After	32.1	46	9.3	5.5	889
45 mph					
Before	29.0	44	12.5	5.2	596
After	30.0	43	6.9	3.4	824
40 mph					
Before	32.5	42	43.6	3.6	55
After	33.0	44	44.1	13.5	111
30 mph					
Before	23.3	34	29.9	12.3	228
After	23.5	33	33.2	10.0	361
25 mph					
Before	22.0	32	57.7	20.7	329
After	23.0	31	55.2	18.6	382

Comparison between foreign vehicles and Illinois registered vehicles

45 mph, foreign cars					
Before	33.3	47.0	20.8	8.2	159
After	34.1	43.8	10.7	5.8	326
45 mph, Illinois cars					
Before	29.1	45.4	13.3	5.4	279
After	28.7	41.3	6.3	2.7	333

* Speed at or below which 20 per cent of vehicles travel.
† Speed at or below which 85 per cent of vehicles travel.

and court action are doubtful when unusual signs are concerned. In some instances, such signs have proven confusing to motorists and have produced actions that are hazardous, rather than improving traffic conditions. Impractical use of only one device will build up strong sentiment against all similar devices, and observance of the controls will be reduced.[1]

[1] W. S. Smith and C. S. LeCraw, "Travel Speeds and Posted Speeds in Three States," *Traffic Quarterly*, January, 1948, p. 101, and C. C. Wiley et al., *Effect of Speed Limit Signs on Vehicle Speeds*, University of Illinois, Urbana, 1949.

Markings and Signals. Pavement markings are sometimes employed in attempts to control speeds. Stenciled speed limits on the pavement are regularly maintained in some cities, but usually they are used principally for short periods of intensive educational or enforcement efforts. Wigwag flashers, in combination with signs, are effective in speed control where abrupt physical changes in the roadway occur.

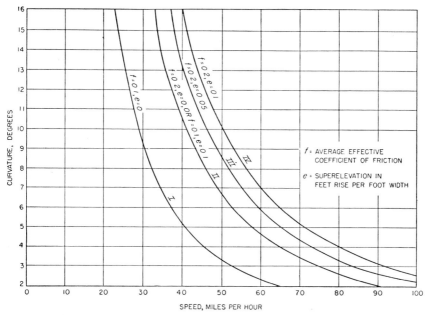

FIG. 15-1. Speeds on curves for different superelevations and coefficients of friction.

There is some experimentation with signals to control speeds along multilaned principal roadways in rural areas, and block signal systems are being considered for limited-access roadways. It seems likely, however, that rear-end accidents, unnecessary stops, and delays would offset the gains achieved. At sharp curves and narrow structures, vehicle-actuated signals may be installed to reduce approach speeds.

15-3. Speed Zoning. Speed zoning, or the application of special posted speed limits to sections of roadway, is the area of speed control that most frequently involves the traffic engineer. Properly applied, zoning becomes a guide to safe and proper driving rather than an arbitrary restriction. The most obvious advantages are that zoning (1) aids motorists in adjusting speeds to conditions, (2) furnishes officers a guide as to what is excessive speed under favorable conditions, (3) develops more uniform speed (lowers high speeds and raises low speeds), (4) reduces accident frequency and severity, and (5) provides temporary corrective for outmoded roadways pending reconstruction.

Speed zoning is commonly employed on entrances to cities. The posting of graduated zones, so that motorists are induced progressively to lower speed from high rural speeds to safe urban speeds, has proven effective in reducing accidents.

A thorough speed-zoning project might include the preparation of strip maps showing (1) curves (degree and superelevation), (2) intersections (angles, road types, nature of traffic), (3) grades (vertical curve

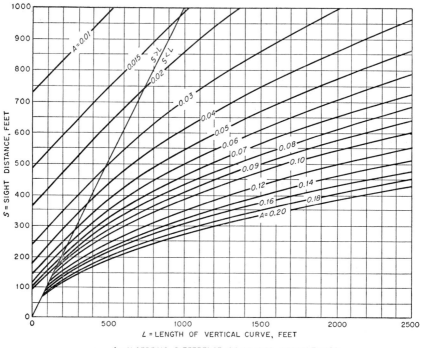

L = LENGTH OF VERTICAL CURVE, FEET

A = ALGEBRAIC DIFFERENCE OF GRADES, PERCENT ÷ 100

WHEN $S > L$, $S = \dfrac{7.28}{A} + \dfrac{L}{2}$ WHEN $S < L$, $S = 3.82\sqrt{\dfrac{L}{A}}$

HEIGHT OF EYE 4.5 FEET HEIGHT OF OBJECT 4 INCHES

FIG. 15-2. Relation between sight distances, grades, and vertical curve lengths. (From *A Policy On Sight Distance for Highways*, American Association of State Highway Officials, 1940, Fig. 3, p. 18.)

data), (4) pertinent physical factors (streams, railroads, roadside culture), and (5) existing traffic controls and speed limits. Accident and traffic volume data can be superimposed on such maps.

Zoning Methods. The proper limits for zoning can be determined by field measurements of prevailing speeds and by calculations based on critical physical factors.

Spot-speed studies provide the most reliable data for zoning. At selected locations (usually at points where some reduction over normal route speed appears necessary for safety, i.e., curves, bridges, or intersections), speeds of vehicles are determined under favorable driving conditions. When the speed distribution is thus determined, the value to be used for zoning is usually taken as the 85 or 90 percentile (see Chapter 4). This method assumes that most motorists are capable of selecting a safe speed at a given location and that only a few motorists normally drive at excessive speeds.

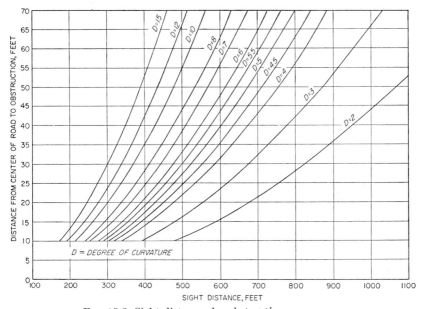

FIG. 15-3. Sight distances by obstructions on curves.

Calculating Speeds. The initial factor for the calculation of safe speed may be the side-friction force which develops between the tires and the roadway on curves, or the sight distance required for safe stopping for the given speed.

Critical speeds for curves under various conditions are shown in Fig. 15-1. The equation for computing critical speed must take into account curve radius, superelevation, and coefficient of friction as shown below:

$$v = \sqrt{15R(e + f)}$$

where v = speed, miles/hr

R = radius of curvature, ft

e = superelevation, ft/ft

f = coefficient of friction for side skidding (for most surfaces a value of about 0.2 is used)

Values for zoning of curves can be determined and checked by speed surveys. The vehicles used in such surveys can be equipped with ball-bank indicators, specially conceived instruments employing pendulum principles, or mercury-filled U tubes. Standard ball-bank indicators give speed values on curves. Outside cities, a deflection of 10° is commonly used to obtain the limit for posting (this corresponds to a frictional coefficient of about 0.15). At low speeds, such as in cities, higher coefficients are developed on curves and turns, and higher deflection angles are used to give reasonable limits. For speeds of from 20 to 30 mph, a ball-bank-indicator deflection of 12° gives a reasonable speed for zoning. Where speeds are under 20 mph, deflections of 14° are reasonable.

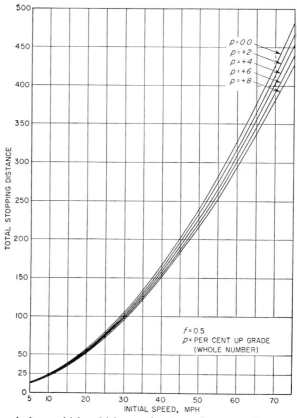

Fig. 15-4. Speeds from which vehicles can be stopped on upgrades, assuming a good dry road surface, average brakes, and a 1-sec reaction time.

Safe-stopping distance is usually taken as 50 to 75 per cent of the total sight distance. Figure 15-2 shows sight-distance values over vertical curves for typical grade conditions, and Fig. 15-3 shows sight distances around obstructions on curves. Stopping distances, for both uphill and downhill, are represented by the family of curves shown in Figs. 15-4 and 15-5. These distances are computed from the equations shown below.

Let S = stopping distance, ft
   V = speed, mph
   T = driver's reaction time, sec
   f = coefficient of friction (braking)

Solving the equation,

$$S_g = \frac{V^2}{30f \pm 0.3p} + 1.467\,VT$$

where p = per cent grade expressed as a whole number. The sight distances existent under given physical conditions may be used to determine

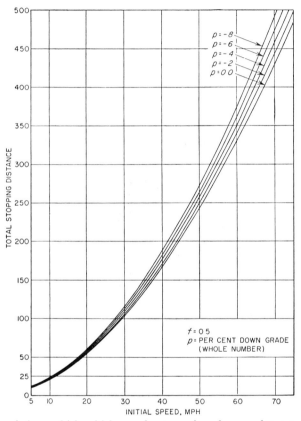

FIG. 15-5. Speeds from which vehicles can be stopped on downgrades, assuming a good dry road surface, average brakes, and a 1-sec reaction time.

safe-stopping distance, which in turn will provide safe-speed values by direct reading of Figs. 15-4 and 15-5 for the given physical conditions and driver-reaction times.

Factors in Zoning. A thorough zoning project will include the speed-distribution, stopping-distance, and side-friction methods of study. Further, field speed checks which use special instruments are desirable.

After studies are of value in determining effectiveness. These require information on speeds, accidents, enforcement, and public opinion. They will result in increasing effectiveness, in indication of needed changes, and they will provide valuable data for records and educational purposes.

CHAPTER 16

ONE-WAY STREETS

Expressways, bypasses, street widening, and one-way street systems are among the aids suggested for increasing traffic access to and flow within congested districts. One-way streets generally afford the most immediate and least expensive method of increasing traffic flow within a city, costing little more than the expense of signs and markings necessary to indicate the direction of flow. Dating from the days of Pompeii[1] (prior to 79 A.D.), when very narrow streets forced such restrictions, it appears that one-way regulations first went into effect in this country on Philadelphia's Chestnut Street in 1906.[2] New York City made several streets one-way in the Park Row section in 1907,[3] and in 1908 Boston followed with a number of one-way streets. Full use of a one-way system covering pairs of streets, or an area of the city, to provide greater access to the central business district was not common until the early 1920s, when Cleveland, Los Angeles, St. Louis, Providence, and Kansas City all made use of the principle in controlling traffic in the alleys of their downtown districts. A typical regulation stated that "In all east and west alleys, traffic shall move west; and in all north and south alleys, traffic shall move south."[4] Application of one-way principles to wide streets followed at a later date. During the 1930s a total of 1,750 miles of streets in New York were converted to one-way operation,[5] and since 1940 many cities have adopted extensive one-way systems. It is certain that the one-way street will become increasingly common, in cities of all sizes, as a means of facilitating traffic movements.

Financially, the conversion to one-way operation is far preferable to the installation of expressways or street-widening programs. Four streets were made one-way in Baltimore, Md., at a total cost of $70,000, including signal improvements, while an expressway built to carry a corresponding

[1] Miller McClintock, *Street Traffic Control*, McGraw-Hill Book Company, Inc., New York, 1925, p. 101.

[2] *One-way Streets*, Highway Research Board, Bulletin 32, Washington, 1950, p. 1.

[3] T. W. Rochester, "Relieving Congested Districts by a System of One-way Streets," *Proceedings of Institute of Traffic Engineers*, New Haven, Conn., 1939, p. 59.

[4] McClintock, *op. cit.*, p. 103. Quoted from *St. Louis Traffic Laws and Regulations*, 1921, p. 14.

[5] Rochester, *op. cit.*, p. 58.

number of vehicles would have cost an estimated $16,000,000.[1] In Sacramento, Calif., the cost of conversion of two 48-ft streets to three-lane one-way streets was $23,000, while to create one four-lane two-way street by widening would have cost over $1,300,000.[2] In addition, one-way streets usually reduce accidents, compared with two-way streets carrying similar traffic volumes, and are ideal for the use of progressive-signal systems which practically guarantee volume and speed increases and delay reductions. Transit operations are usually greatly facilitated, and satisfactory traffic movement may be maintained without applying stringent parking and loading restrictions. Confusion at complex intersections may be reduced and traffic diverted from residential areas. The use of one-way streets or one-way street systems fits well a basic traffic engineering aim, "make the best use of existing facilities." A good one-way system usually affords so many advantages to motorists that public approval develops as soon as a short trial period has provided the opportunity to overcome previous habits.

One-way streets are effective because of the changes they produce in traffic characteristics: (1) intersectional conflicts reduced—those between vehicles and other vehicles and those between vehicles and pedestrians; (2) head-on and sideswipe conflicts eliminated; (3) medial conflicts (in same direction) minimized; (4) parking maneuvers made less dangerous; (5) headlight glare eliminated; (6) field of concern of drivers and pedestrians reduced; (7) speed differentials reduced; and (8) odd lanes of roadway efficiently utilized.

One-way streets may be of three general types. The most common type limits traffic operation to one direction at all times and is found principally in cities making use of extensive grid systems for the handling of traffic in congested districts. Another type is the reversible one-way street, on which the direction of flow is reversed at predetermined times. A particular street may operate one-way inbound during all hours of the day except during the outbound rush period, when it will operate outbound. The third type is operated as a two-way facility during off-peak hours and is made one-way in the direction of predominant flow during peak hours. Washington, D. C., has had since 1935 several streets of this type which operate inbound during the morning rush period and outbound in the evening rush.

One-way streets may be necessary where the roadway is too narrow (less than 20 ft if no parking is allowed, less than 35 ft if curb parking is allowed) for two-way traffic operations, where fixed transit facilities (streetcars) do not allow passing on narrow streets, where the street pattern develops circles or squares at focal points, or where traffic

[1] *One-way Streets*, p. 5.

[2] D. J. Faustman, *Improving Traffic Access into Central Business District*, Sacramento, Calif., 1950, pp. 13, 22.

demands cannot be met in other ways. They should also be considered in cases involving (1) intersection simplification, as an aid to routing (reducing interlocks) and signalizing complex intersections, (2) steep grade, to reduce danger, especially where commercial traffic is heavy, (3) control of street use, to remove commercial traffic from residential streets where it has developed to achieve a cut-off or to avoid signals, (4) routing of highway traffic through cities and towns, (5) handling special events which generate heavy traffic, and (6) military movements.

16-1. Potential Achievements. One-way operations may be applied to city streets to effect basic changes in the traffic pattern and thereby to

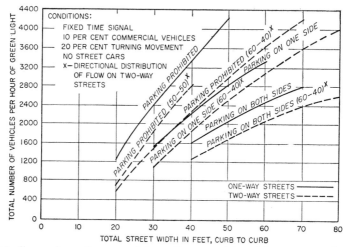

FIG. 16-1. Comparison of average intersection capacities with one-way and two-way operation in downtown areas. (Adapted from *Highway Capacity Manual*, Highway Research Board, Committee on Highway Capacity, and U.S. Bureau of Public Roads, 1950, Figs. 26 and 27.)

produce many benefits. One-way operations bring about fundamental traffic changes.

Increased Street Capacity. The removal of traffic in one direction, reduction of intersectional conflicts, and utilization of odd lanes produce a great increase in the number of vehicles which can move in the same direction. Figure 16-1 indicates the relative capacities of one-way and two-way streets for measured peak-hour traffic volumes. Note that a 40-ft downtown street on which parking is prohibited is capable of passing 3,400 vph with one-way traffic, while with two-way traffic evenly divided (directionwise), only 2,800 vph can be accommodated. For similar conditions in intermediate areas, the one-way street will handle 2,950 vph while the two-way street handles 2,650.[1]

[1] *Highway Capacity Manual*, Highway Research Board, Committee on Highway Capacity, and U.S. Bureau of Public Roads, Washington, 1950, p. 86.

It may be noticed that where parking is prohibited, the downtown area will have greater capacities for given street widths than the intermediate area. However, when parking is permitted on either one or both sides of the street, capacities will be larger in the intermediate area. Parking demands and extent of police enforcement are the likely reasons for this.

In Detroit, Mich., when one street was changed to one-way operation in 1930, there was a volume increase for that street of 44 per cent.[1] The conversion of two boulevards in Chicago to one-way traffic resulted in a 64 per cent increase in volume.[2] (A factor in the increase on these streets, however, was the prohibition of curb parking which went into effect at the same time.) Installations in Baltimore in 1947 produced an average volume increase of 100 per cent.[3] In Oklahoma City in 1948 two one-way and two two-way streets showed gains ranging from 9 to 43 per cent, while one two-way street registered a 4 per cent decrease. In Philadelphia, volumes have increased as much as 80 and 200 per cent over a five-year period.

Traffic volumes are decreased by a change to one-way operation in a few instances, generally in a network of one-way streets where one or two streets may be better suited physically to handle more traffic in one direction than another. The wider streets will therefore attract a portion of the traffic which would otherwise use the narrow street.

Increases in volume on one-way streets are related to the ease of movement: given a wide street carrying traffic in one direction only, most drivers choose such a route provided it takes them reasonably close to their destination. Overtaking and passing are much easier on a one-way facility, and odd traffic lanes which cannot be utilized under two-way operations are readily used under one-way traffic. Since ease of movement is apparent under one-way operation, it is reasonable to expect increased traffic volumes on such facilities.

Elimination of Need for Parking Restrictions. In solving congestion problems, the choice frequently lies between the installation of a one-way system or the application of some type of parking restrictions. Although in some instances other types of control may be dictated (turn regulations, transit or general traffic rerouting, channelization, or signalization), the choice between one-way streets and curb-parking restrictions offers greater possibilities for congestion reduction.

The choice between parking restrictions and one-way operation must be based upon a number of considerations. Parking demand is the most

[1] W. S. Canning, "Report of Committee on One-way Streets," *Proceedings of Highway Research Board*, Washington, 1938, vol. 18, part I, p. 334.

[2] H. F. Hammond and L. J. Sorenson, *Traffic Engineering Handbook*, Institute of Traffic Engineers, New Haven, Conn., 1941, p. 175.

[3] Charles J. Murphy, "Baltimore's One-way Streets," *Traffic Quarterly*, vol. 4, no. 3, Saugatuck, Conn., July, 1950, p. 283.

important consideration, with the choice primarily dependent upon the availability of off-street facilities to handle cars displaced from the curb. As Fig. 16-1 indicates, a 40-ft street with one-way operation and parking permitted on both sides can accommodate 1,600 vph of green-signal time. The same street with cars operating in both directions in an equal distribution is capable of passing 2,800 vph of green signal if parking is prohibited. Prohibition of parking on the heavily traveled side of a 40-ft two-way street carrying 60 per cent of the cars in the direction of heavy flow will result in a capacity of 1,800 vph of green. When parking is permitted on both sides and traffic is split 60-40 in direction, the street-intersection capacity per hour of green is only 1,250. A change from two-way to one-way operation without a change in parking limitations, then, may be expected to increase traffic flow from 15 to 30 per cent, depending on street width. Parking prohibition on both sides of the street, strictly enforced, should result in an increase of 75 to 80 per cent.

Street capacity, therefore, may not be as greatly increased by one-way operation as by the adoption of parking prohibitions, but a good system of one-way streets will provide other more favorable results. In most areas fewer objections are raised to one-way streets than to the elimination of curb parking; and parking cannot be prohibited on a wholesale basis in congested downtown districts, with the possible exception of peak-hour periods, because adequate off-street facilities have rarely been provided. In addition, one-way regulations are largely self-enforcing, while rigid enforcement of no-parking bans required much police manpower and equipment and frequently results in bad public relations. Revenue from parking meters may also be an important factor in favor of the adoption of one-way regulations as opposed to curb-parking prohibitions.

Increased Speed and Reduced Traffic Delays. Closely associated with volume and capacity increases for one-way streets are vehicle speeds and delays. As volumes increase, speeds are found to decrease on a typical two-way city street. The restriction of traffic flow to one direction will result in speed increases as well as volume increases. As Table 16-1 shows, the introduction of one-way streets reduces speed differentials and increases over-all speeds without necessarily increasing top speeds. Many traffic improvements on one-way streets are traceable to the even flow of vehicles. Early installation in Chicago[1] increased average speeds from 14 to 28 mph or 100 per cent. Portland, Oreg., showed average-speed increases of 10 to 20 per cent on its initial pair of one-way streets in 1946, while the volume increased 37 per cent.[2] Since 10 to 20 per cent increases in average vehicle speeds may be insignificant in view of the low speeds found in congested areas, speed increases may be better indicated

[1] Hammond and Sorenson, *loc. cit.*

[2] F. B. Crandall, "One-way Streets," *Traffic Engineering*, vol. 28, no. 5, 1948, p. 203.

in terms of reductions in vehicle delays.[1] Figure 16-2 shows travel and delay times before and after one-way installation on two streets in New Haven, Conn.   It will be noted that while the travel time or actual moving time varied only slightly, the length of delay time was materially reduced so that the trip time was shortened.   Table 16-2 shows the effect of one-way streets on delays and vehicle speeds in Houston, Tex.,

TABLE 16-1. VOLUME AND SPEED BEFORE AND AFTER INTRODUCTION OF ONE-WAY STREETS

Location	Volume, per cent changes		Speed	
	Before	After	Before	After
Detroit, Mich., 1930				
1st installation (1 street)	680/max. hr	980/max. hr	15 mph	22 mph
2d installation (4 streets)	....	+10–40%	9–18 mph	21–35 mph
Chicago, Ill.				
Warren and Washington Blvds	29,000/24 hr	47,500/24 hr	14 mph	28 mph
Philadelphia, Pa.				
Walnut St	4,052/12 hr	12,098/12 hr	17 mph	25 mph
Chestnut St	7,137/12 hr	12,938/12 hr	17.6 mph	22 mph
Syracuse, N. Y.				
Warren and Clinton Sts	....	+20–40%	10 mph	23 mph
Portland, Oreg., 1946				
Columbia and Jefferson Sts	....	+60%	....	+10–20%
Kansas City, Mo.				
Thirteenth St	500/hr	525/hr	13.2 mph	19.6 mph
Fourteenth St	460/hr	510/hr		
Fort Wayne, Ind.				
Jefferson St	1,774/max. hr.	2,741/max. hr.	9.0 mph	11.9 mph
New Haven, Conn.				
Chapel St	6,820/12 hr	8,500/12 hr	8.2 mph	11.2 mph
York St	5,512/12 hr	8,276/12 hr	13.9 mph	13.3 mph
Orange St	6,337/12 hr	6,039/12 hr	10.9 mph	13.1 mph

where a volume increase of 18.7 per cent on one-way streets and a decrease of 13 per cent on two-way streets resulted.

The reduction of the number and duration of traffic delays and the increase in travel speed through the use of one-way streets can be primarily attributed to the increased lateral movement which is possible. Vehicle speeds are not limited to those of the slowest cars, and weaving and overtaking and passing are facilitated.   Where the slow-moving driver, the cruising parker, or the maneuvering parker may delay a long line of vehicles on a two-way street, such individuals may be easily bypassed on a one-way street.   Thus with more freedom of movement

[1] ". . . the average driver is more concerned with the number of stops he has to make. . . ." He is not "too much concerned if a trip takes nine or eleven minutes or if his average speed is 27 or 28 mph; but he is concerned if during the course of travel he has to stop five or six times instead of two or three."   Murphy, *op. cit.*, p. 280.

the efficiency of the traffic facility is increased through the application of one-way regulations.

Reduction of Accidents. It has been found that streets on which one-way traffic is introduced experience a reduction in virtually all types of accidents. As already indicated, and as shown in Fig. 16-3, while there are 16 possible vehicular conflicts and 16 vehicle-pedestrian conflicts at

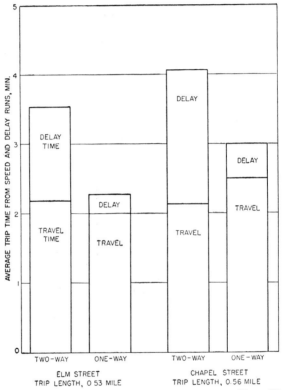

Fig. 16-2. Travel time versus delay time on one-way streets. (From W. S. Smith and J. Hart, "A Case Study of One-way Streets," *Traffic Quarterly*, vol. 3, no. 4, October, 1949, p. 378.)

the intersection of two two-way streets (one lane in each direction), the intersection of two one-way streets provides only 4 vehicular conflicts and 10 vehicle-pedestrian conflicts on two-lane streets. Conflict points vary widely with the number of traffic lanes, but one-way operation greatly reduces conflicts, regardless of street widths. The chance of collision is obviously reduced.

Medial conflicts (causing head-on and sideswipe accidents) are non-existent on a one-way street, since there is no possibility of colliding with a vehicle moving in the opposite direction unless the driver of one vehicle violates the regulation. The absence of headlight glare of opposing

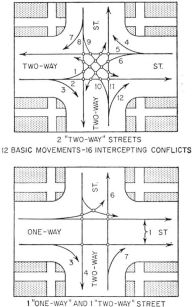

2 "TWO-WAY" STREETS
12 BASIC MOVEMENTS-16 INTERCEPTING CONFLICTS

1 "ONE-WAY" AND 1 "TWO-WAY" STREET
7 BASIC MOVEMENTS-7 INTERCEPTING CONFLICTS

2 "ONE-WAY" STREETS
4 BASIC MOVEMENTS-4 INTERCEPTING CONFLICTS

FIG. 16-3. One-way versus two-way intersection movements.

vehicles makes night driving easier and gives the driver a better chance to see pedestrians and marginal vehicle movements. Marginal conflicts are greatly reduced on one-way streets because of the added space which enables the driver to avoid dangerous street-side locations. Parked vehicles are less hazardous on one-way streets. The reduction in speed differentials is also a factor in reducing the accident potential.

Before-and-after studies of accident records for one-way traffic have shown a surprising consistency of results. In Detroit, accidents were

TABLE 16-2. CHANGES IN TRAFFIC DELAYS AND VEHICLE SPEEDS AFTER ADOPTION OF ONE-WAY REGULATIONS IN HOUSTON, IN PER CENT

	Vehicle delays	Average running speed	Over-all speed
Off-peak traffic movements			
All streets..................	−14	+ 7	+13
Two-way streets...........	+ 1	+ 5	+ 4
One-way streets...........	−35	+10	+26
Peak traffic movements			
All streets..................	− 6	+ 5	+ 8
Two-way streets...........	− 3	+ 2	+ 4
One-way streets...........	−13	+ 9	+20

SOURCE: State Traffic Engineering Department, Houston, Tex.

decreased from 54 to 38 after 1 year of one-way operations.   In Erie, Pa.,. a system in which a wide thoroughfare was operated one way between two narrow streets handling traffic in the opposite direction resulted in a 75 per cent accident reduction on the wide street but produced increases in accidents on the two narrower streets, chiefly because volumes on these streets were increased by 58 and 107 per cent.[1]   In Medford and Eugene, Oreg.,[2] the total number of accidents was reduced 55 per cent

TABLE 16-3. ACCIDENT EXPERIENCE AT INTERSECTIONS IN CENTRAL BUSINESS DISTRICT

Conditions	Injury accidents		Property damage
	Pedestrian	Auto	
Two-way–two-way intersections changed* to one-way–one-way intersections (16)			
Before†	3	5	38
After‡	1	3	21
Per cent change	−67%	−40%	−46%
Two-way–two-way intersections changed to two-way–one-way intersections (24)			
Before	7	5	65
After	5	4	45
Per cent change	−29%	−20%	−31%
Entire city—all accidents			
Before	501	227	4,271
After	502	219	3,707
Per cent change	0%	−3%	−13%

* One-way street system installed Feb. 8, 1948.
† "Before" period—Mar. 1–Aug. 31, 1947.
‡ "After" period—Mar. 1–Aug. 31, 1948.
SOURCE: *Getting Results through Traffic Engineering*, National Conservation Bureau, Appropriate Case Histories, New York.

after one-way installation.   Pedestrian accidents were reduced 7 per cent, rear-end collisions 73 per cent, and parking accidents 53 per cent. Table 16-3 shows the reduction in intersection accidents produced by one-way streets.   Note that injury and property-damage accidents were reduced more at intersections on the one-way streets than in the whole city.   Pedestrian accidents showed no improvement on a city-wide basis, but were reduced by 67 per cent and 29 per cent at the points of one-way installation.

In some cases certain types of accidents become prevalent for a short

[1] D. W. McCracken, "Warrants for One-way Streets and Their Values," *Proceedings of Institute of Traffic Engineers*, New Haven, Conn. 1941, p. 79.

[2] *An Accident Analysis of Oregon's Urban One-way Streets*, Oregon State Highway Department, Salem, 1948.

time after the change to one-way operation, until drivers and pedestrians are acquainted with the new regulations.

Pedestrian accidents are usually reduced by one-way operation because of (1) the reduction in potential conflicts between vehicles and pedestrians

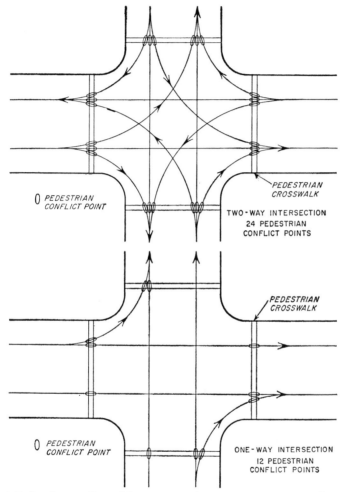

FIG. 16-4. Pedestrian conflicts with one-way and two-way movements at intersections.

at intersections (Fig. 16-4), (2) increased safety at mid-block locations, since there is no possibility of being trapped between opposing streams of traffic, and (3) the progressive-signal system which develops vehicle platoons and gives pedestrians a "break" in traffic for crossing.

Accident reductions are more significant when related to traffic volumes, as travel almost always increases substantially under one-way operation. Accident reductions of 50 per cent are not uncommon with volumes doubled. Sometimes reductions are recorded on one-way streets when increases develop for the city as a whole. Total accident decreases of 30 to 55 per cent have been most frequently reported. Traffic authorities of several large cities, where one-way streets have been in use for many years, make such statements as "one-way streets are twice as safe as comparable two-way streets."

Routing and Diverting Traffic. The one-way principle is frequently used to divert traffic into desirable streets away from problem areas and complex traffic locations. Typical of such an installation is the use of a one-way street to reduce the conflicts and congestion at a complex intersection. Effectiveness of this type of installation depends upon a number of factors. When the intersection is operating with all streets carrying traffic in two directions, the problem of controlling movements within the intersection is very complex. At a five-pronged intersection, the problem is reduced to that of a four-way intersection if the odd street is one-way away from the intersection. This can usually be arranged without the aid of an alternate one-way street unless all five of the approach streets are indispensable as approach facilities.

One-way operation may also be applied to locations where traffic uses residential streets as a means of bypassing traffic signals or congested points. Limiting travel on such streets to the direction opposite to that most frequently used will eliminate most of the through traffic. Such a scheme has been used effectively in Boston.[1]

Some cities and states have effectively employed one-way streets to expedite the movement of through highways. The Oregon State Highway Department[2] frequently uses a pair (complex) of one-way streets to carry highway traffic through an urban area, with much success in relieving congestion and in reducing accidents. Other states, including New Jersey, Virginia, and Pennsylvania, are also adopting the one-way-street plan.

Public parks and squares frequently require the application of one-way principles. If the central square of a city is bounded by narrow streets, or angle parking is allowed, there are certain hours of the day when extreme congestion develops. A rotary traffic plan may provide additional curb-parking facilities or allow retention of existing parking. This use of the one-way rule is one of the oldest, having been in effect in many communities almost since their founding.

In any consideration of routing by one-way streets, the question of trip length should be considered. While actual trip routes cannot be dictated, the layout of a system of one-way streets can have a considerable effect on the choice of route. In the simple grid system shown in Fig. 16-5, the increased distance of some trips and the decreased distance

[1] McCracken, *op. cit.*, p. 87.
[2] See *Traffic Quarterly*, vol. 4, no. 2, April, 1950, p. 136.

of others is obvious. The distance from D to the destination point at X on a two-way system is 0.35 mile while under the one-way plan shown the trip is 0.45 mile. This trip with two-direction traffic requires only one left turn; with one-way traffic, the left turn and two right turns are required. The greatest inconvenience caused by the system illustrated is to the driver desiring to reach point X from location C: trip length is

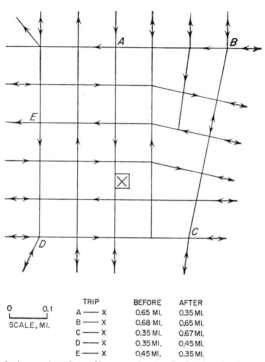

TRIP	BEFORE	AFTER
A —— X	0.65 MI.	0.35 MI.
B —— X	0.68 MI.	0.65 MI.
C —— X	0.35 MI.	0.67 MI.
D —— X	0.35 MI.	0.45 MI.
E —— X	0.45 MI.	0.35 MI.

0   0.1
SCALE, MI.

FIG. 16-5. Travel time related to distance for typical trips in one-way grid system.

nearly double, and four right turns and one left turn, compared to only a right turn before installation, are necessary.

Progressive-signal Operation. The use of one-way streets either in pairs or in block systems offers the signal engineer an unusual opportunity. While one-way operation increases traffic flow without extensive help from other controls or devices, a progressive-signal system can further reduce delays and improve the rate of flow. Without the installation of such a system the performance of the one-way streets may not be much superior to the old two-way streets.

A study of a one-way system in Charleston, W. Va., shows an interesting relation of delays to signals. Before the one-way installation, delays caused by traffic signals accounted for 50 per cent of the total: general congestion was responsible for 35 per cent, and the remainder was charged to slow-moving vehicles, parking and bus maneuvers, cross traffic, and a railroad crossing. One-way operation, without progressive signals, resulted in the disappearance of practically all delays except those caused by signals. The importance of a progressive signal to the one-way streets is indicated. In Baltimore, Md., where a one-way system is in operation,[1] peak-hour stops on an arterial average only about three for a 3.5-mile route (one per mile), with a progressive-signal operation.

An unusual type of signal control has been applied in Detroit, where two pairs of major one-way streets have been equipped with mid-block progressive signals.[2] Since the original reason for the mid-block installation was to control traffic on the main street and at the same time not encourage traffic on any residential side street because of the signal placement, the side streets were all stop-signed and none were indicated as through streets. This type of signal control on two-way streets permits signals to be placed at any location suitable for achieving desired progression with signal equipment available, while if the placement is limited to intersections such progressions may not be possible.

Where needed, pedestrian intervals can usually be tied in with the conventional signal system more readily with one-way streets. Injection of pedestrian intervals will interfere less with signal progression on one-way streets.

Improved Transit Services. Of the many groups interested in one-way streets perhaps none is more directly affected than the local transit company. Anything which changes the street-traffic plan of a city may change transit routings; it is important that such changes should not adversely affect transit operations. With the adoption of one-way regulations, transit-route changes may be needed. Transit lines are routed according to passenger demands. Minimum passenger walking distances are important, but in actual practice it is virtually impossible to plan the route so that all passengers are within the desired two or three blocks of the line. When a pair of one-way streets is installed, some of the transit patrons may have to walk an extra block.

Careful planning with transit officials should result in a minimum of extra walking for transit patrons. Unwise route changes may considerably reduce the transit demand.

The effect of a one-way system on transit is much more serious when trolley buses and street cars are involved. In such cases, expensive track and electrical-distribution changes may be required. The traffic engineer must be sure of the adequacy of his plan before it is tried, or gas buses must be substituted for streetcars and trolley buses during an experimental period.

[1] Murphy, *op. cit.*, p. 274.
[2] *One-way Streets*, Highway Research Board, Bulletin 32, Washington, 1950, p. 8.

In some of the largest cities, where there are numerous streetcar lines, one-way operations may be especially needed because of the congestion created by the fixed-rail operations. In Philadelphia, traffic problems involving streetcar operations have been solved by installation of one-way streets in some locations, with the cooperation of the transit company. Baltimore, Md., and Portland, Oreg., established extensive one-way systems as soon as free-wheel transit vehicles replaced streetcars.

Bus stops have a far less restrictive effect on other traffic on one-way than on two-way streets, as opportunity to pull out and pass is afforded even on narrow streets. Bus turns on narrow streets and at intersections where adequate turning radii cannot be provided are safer and less congesting on one-way streets.

TABLE 16-4. COMPARISON OF BUSINESS TRENDS ON ONE-WAY STREETS AND TRENDS IN ENTIRE CITY OF NEW HAVEN, IN PER CENT
(Six-month periods before and after one-way regulations)

Business group	New Haven	One-way area
Food	− 3.9	+ 6.2
Apparel	−16.6	+10.8
General merchandise	−14.3	+ 8.3
Automotive*	−14.5	−26.0
Furniture and fixture	−41.7	+18.4

* Figures do not include sales of gasoline or other motor fuels, as these are not subject to the state sales tax.
SOURCE: Aart L. Roscam Abbing, "The Effect of One-way Streets on Business," thesis, Yale University, Bureau of Highway Traffic, 1950, p. 30.

Franchises of transit companies must always be considered, as some transit operations are extensively limited to specific streets or areas.

Benefits to Business. In most cases businesses in retail districts are benefited by the installation of one-way operation, though results may not be apparent for several weeks. Some time, usually less than 2 weeks, is required for motorists to become accustomed to the one-way plan. When previous driving habits are overcome, traffic volumes usually exceed those which prevailed under two-way operation, thus bringing more potential customers to the street and area. This should produce benefits to business, and available data reflect such benefits. When an area becomes more accessible, it should increase in value to retail businesses, as it can better compete with other areas for business of motorists and users of mass transportation.

A study of sales-tax data in New Haven, Conn., revealed that one-way regulations did not injure business. As Table 16-4 shows, only automotive business suffered a greater decline on the one-way streets than in the area as a whole. This might be explained by the very small number (inadequate sample) of such businesses in the retail area served by one-way streets. (The decrease is not due to service stations being "on the wrong side of a one-way street.") In general, the competitive position of merchants in the one-way area was improved.

Table 16-5 shows the results of a similar study in Sacramento. It would appear from these and other similar studies that most businesses are not damaged by one-way traffic regulations.

Studies which have been made of the relationship of one-way regulations to land values fail to reveal detrimental results. Increases in assessments are much more numerous than decreases after one-way rules are applied.

TABLE 16-5. COMPARISON OF BUSINESS TRENDS ON ONE-WAY STREETS AND TRENDS IN ENTIRE COUNTY OF SACRAMENTO, IN PER CENT

Type business	County	One-way streets
Auto supply stores, garages, auto dealers, service stations, tire shops...	−4.67	−2.9
Eating and drinking places...........................	−7.18	−3.98
All other businesses................................	−0.43	+8.6
All businesses....................................	−1.30	+2.09

SOURCE: D. J. Faustman, "Improving the Traffic Access to Sacramento's Business District," *Traffic Quarterly*, July, 1950, p. 252.

Other Benefits. There are other direct benefits of one-way streets and many indirect benefits. They are inexpensive; enforcement is easy—they become practically self-enforcing; public approval usually is immediate; and movements of emergency vehicles are expedited. (Some fire officials direct emergency calls into one-way streets wherever practical. In some cases they report favorable results from having fire equipment travel counter to the one-way traffic.) Arguments in favor of one-way streets are numerous and conclusive—where the regulations are properly applied in keeping with local conditions and problems.

16-2. Disadvantages of One-way Operation. One-way regulations provide many advantages, but like all traffic regulations there are some disadvantages which may develop. In some locations, such as residential areas, one-way regulations may encourage excessive speeds.

Some difficulty may be experienced in directing out-of-town motorists through an extensive network of one-way streets, but with modern signing and marking practices such problems may be readily overcome.

If side streets are made one-way a problem may develop if pedestrians cross multiple lanes of traffic without a refuge point. Other hazards may result from pedestrians stepping into the travel portion of the roadway after looking only to the left on streets where traffic is approaching in all lanes from the right.

When one-way streets are first applied a higher accident rate often develops, but it usually persists only until motorists become familiar with the new regulations.

Though most claims that one-way streets adversely affect business are ill-founded, some types of businesses, such as filling stations, are likely to suffer a decline when the regulation is first applied if they are located " on the wrong side of the street." In most instances the initial losses are regained and previous business volumes exceeded after motorists become accustomed to the system.

When basic changes in a city's traffic pattern are necessitated by one-way regulations, many motorists are forced to revise their driving habits and, thereby, develop psychological resistances. Such resistances are invariably overcome in a short period of time.

Some trips may be slightly longer after the application of one-way streets, but motorists are generally more concerned with travel time than travel distance. The favorable effect of one-way movements on traffic operations usually results in a shorter travel time even between points where travel distances are greater.

16-3. Requirements for Satisfactory One-way Operation. Certain requirements must be met when a one-way system is installed:

1. Traffic must be equally accommodated in each direction by adjacent parallel streets, usually not more than 500 ft apart and never more than 700 ft apart.
2. There should be frequent entrance and exit points (cross streets), 300 to 500 ft apart.
3. Wherever possible, the one-way movement should be applied to the predominant direction-of-traffic flow.
4. Where congested areas are concerned, one-way streets should be designed to drain traffic away.
5. Transit routing and loading points as well as transit franchises must be carefully studied.
6. Standard signs readily visible both day and night should be provided.
7. A carefully planned program of public education should precede the installation of the one-way system, and police must be provided to assist in the direction of motorists until they become accustomed to the new system.
8. One-way regulations must have the wholehearted support and backing of public officials.
9. One-way streets which do not meet these requirements may be effective in temporary traffic routing plans such as for large parades, military movements, and other special events.

Legal Complications. It is necessary to ascertain whether one-way streets can be effected under general police powers or whether legislation is required. Sometimes special legal considerations are involved. A franchise of a transit company might upset plans for a one-way street

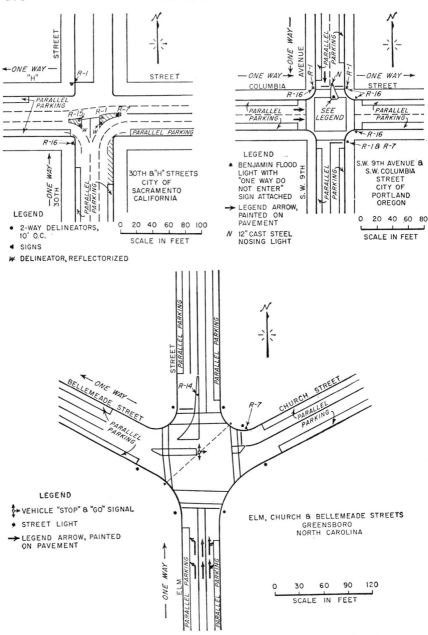

Fig. 16-6. Use of islands, pavements, signals, and signs at extremities of one-way streets.

system,[1] especially in cities where there are competing mass-transportation facilities.

Markings. Adequate signs and markings, visible day and night, are essential to a successful one-way operation. A simple aid is the substitution of arrow lenses on green-signal faces, so that only the proper turns are indicated at intersections of one-way streets. Figure 16-6 shows the use of islands, pavement markings, signals, and signs to direct traffic properly at the extremities of a one-way street.

[1] *Eighth Avenue Coach Co. v. City of New York*, 170 Misc. 243, 10 N.Y.S.2d 170 (1939). The coach company claimed that a franchise gave the right to operate in two directions on Eighth Avenue for a period of 10 years and could not be broken. The court held that one-way installation would reduce the coach company's revenue considerably and was therefore illegal.

CHAPTER 17

CURB-PARKING REGULATIONS

Accumulations of vehicles in the retail districts of large towns and cities have developed serious parking and congestion problems. Sky-scrapers have caused concentrations of persons in small areas which cannot be adequately served by existing street and transportation facilities. Many cities which first grew vertically, because of the ease and economy of concentrating masses of people at focal points by mass transportation, have gradually given way to a horizontal type of development which can be better served by the automobile. Parking demands, however, are still developing more rapidly than downtown urban redevelopments are making additional spaces available. (Traffic studies show that only about 10 per cent of the vehicles in downtown districts are moving during daylight hours, while the remainder are parked or stored.[1]) As a result, suburban shopping centers have sprung up during the past 20 years in all sections of the country.

Parking problems are complicated by the interests of many special groups. Retail merchants desire attractive parking facilities near their businesses but are often unwilling to invest the necessary funds to provide the facilities. Often they resist changes and improvements proposed by public officials for the general improvement of traffic and parking conditions in the downtown area as they think primarily of the availability of curb spaces, though the average business frontage in a retail district has only two or three curb spaces to serve it.

Building and property owners are likely to be even more opposed to certain types of off-street parking developments if the costs are to be paid by taxation or general-benefit assessments. In general, they take the view that parking is a public responsibility which should be solved by public agencies with public funds.

Transit companies are constantly recommending additional curb-parking bans in order to expedite traffic movements and are generally opposed to the public development of off-street parking facilities. Many fleet groups such as taxi companies, delivery vehicles, and intercity truckers place major demands on curb spaces and are usually unwilling to do much

[1] *Controls of Curb Land Use on City Streets*, Vehicular Parking, Newark, N. J., 1942, p. 1.

to help the over-all parking problem by changing their own habits or by meeting their own terminal requirements off the public ways. Many cities have passed restrictive ordinances requiring cabs to use off-street spaces when not hired and requiring commercial vehicles to concentrate their deliveries in off-peak traffic hours.

Individual road users are unwilling to walk great distances after parking, will not use off-street facilities which are inconvenient, and will not

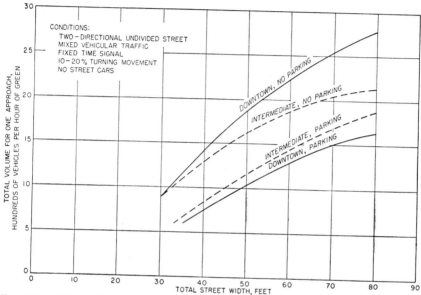

CONDITIONS:
TWO – DIRECTIONAL UNDIVIDED STREET
MIXED VEHICULAR TRAFFIC
FIXED TIME SIGNAL
10 – 20% TURNING MOVEMENT
NO STREET CARS

FIG. 17-1. Effect of curb parking on street capacity. (From *Highway Capacity Manual*, Highway Research Board, Committee on Highway Capacity, and U.S. Bureau of Public Roads, 1950, p. 79.)

patronize lots and garages which are overly expensive. Since these demands are inherent and are imposed unconsciously, parking programs must take them into account.

17-1. Problems Created by Curb Parking. *Street Capacity.* Curb parking seriously affects street capacity through the narrowing of the roadway available for moving traffic, the placement of vehicles at the curb, and the frictions created by cars moving in and out of parking spaces. Downtown street capacities, measured at intersections, are about 45 per cent lower with parking permitted. Figure 17-1 shows the relation between parking and street capacity for streets in downtown and intermediate areas. In a downtown area with parking permitted, a 50-ft street can accommodate a maximum of about 1,000 vehicles on one approach to an intersection, per hour of green signal. The same-width street in an intermediate area with parking can accommodate about 1,200 vph of green on each approach to the intersection. When parking is prohibited, the same 50-ft street can accommodate higher volumes per hour of green at intersections in downtown areas than in intermediate areas (1,900 to 1,700 vehicles), probably because of poorer enforcement of parking prohibitions in the intermediate area. A significant factor in the influence of parking on traffic flow is the amount of traffic moving in search of parking places. A survey made in Chicago[1] showed 7 per cent of the cars on a particular street to be cruising for just this purpose.

Curb parking also has a serious effect on the movement of transit vehicles. The number of transit passengers moved in a given period may be reduced as much as 35 per cent where curb parking is permitted in the downtown area.[2]

Since angle parking has a far greater effect on capacity than parallel parking, many cities have enacted ordinances which ban angle parking within the entire corporate limits. It should be remembered that parkers on a given block constitute only a fraction of the total street users and that undue parking interference penalizes the vast majority to benefit the small minority.

Accidents. A high percentage of accidents on downtown city streets (16 per cent in 1949[3]) involves cars which are parking and unparking. Parking maneuvers also cause many collisions in which the parking car is not directly involved. In rural areas, less than 4 per cent of accidents involve parking. Table 17-1 shows accident records for 1948. Note that pedestrians figured prominently in parked car accidents.

In Chicago, a 63 per cent accident reduction resulted when parallel replaced angle parking.[4] Double-parked vehicles were at least partially responsible for 17 per cent of all bus accidents in one city.[4] In New Orleans,[4] 10 per cent of streetcar accidents and 19 per cent of bus accidents during 8 months of 1946 involved parked cars.

Economic Factors. Parking on important streets consumes expensive areas for storage of vehicles. In most instances, comparable spaces could be provided off-street at far less cost. On the basis of conservative land and maintenance values, the cost for the provision of on-street parking facilities would be $418,863 capital investment (about $1,200 per car space) plus $19,000 per year maintenance per mile of street with parking on both sides.[5] A street-widening program in Pittsburgh to provide

[1] Ladislas Segoe, *Local Planning Administration,* Institute for Training in Municipal Administration, Chicago, June, 1941.

[2] *Parking—Cincinnati Metropolitan Master Plan Study,* Cincinnati City Planning Commission, 1947, p. 24.

[3] *Accident Facts,* National Safety Council, Chicago, 1950, p. 64.

[4] Data from miscellaneous parking reports and local accident reports.

[5] *The Parking Problem,* Eno Foundation for Highway Traffic Control, Inc., Saugatuck, Conn., 1942, p. 52. These figures are derived from the assumption that street right of way can be valued at $5 per sq ft, pavement costs at $6 per sq yd, street cleaning and snow removal at $2 per sq yd per year, a pavement life of 20 years. Cost of pavement repairs and interest charges have not been included.

more parking spaces cost $12,930 per space,[1] while the cost of a parking garage might have ranged from $1,600 to $2,600 per car space.

Shopping studies have shown that people spend more money if they are allowed to park off the streets. One study made in Anaheim, Calif., indicated that people parking in lots spend about four times as much in adjacent or nearby stores as people parked at curbs in the same area. A survey conducted in Bakersfield, Calif.,[2] showed that people purchased almost twice as much in the downtown area when they traveled by car as they did when using transit.

TABLE 17-1. PARKING AND ACCIDENT EXPERIENCE IN SELECTED STATES IN 1948

State	Total number of accidents	Per cent of accidents involving parked cars	Per cent of accidents involving maneuvering cars	Total number of pedestrian involvements	Per cent of pedestrians walking from between parked cars
Alabama..........	5,594	7	1.4	289	12.4
Arizona..........	5,864	3.8	2.6	252	11
Massachusetts....	26,591	1.7	...	7,855	20
Nebraska.........	20,915	10	8.4	627	12
New Hampshire...	19,239	....	...	400	12
New Jersey.......	60,022	12.8	5.7	4,994	22
Oregon...........	66,298	11.4	7	1,791	7
Rhode Island.....	2,673	....	...	641	33
South Carolina....	9,346	3.2	2.6	508	13.8
Tennessee........	2,812	5.2	...	276	6.5
Utah.............	9,961	7.7	4.8	580	11.2

SOURCE: State accident records.

Since rates for public transportation must be adjusted on the basis of costs, expenses developed through parking congestion impose an added cost to the public. Many city officials and traffic authorities contend that the effects of curb parking, especially downtown congestion, are contributing substantially to urban-decentralization trends, which in turn are related to realty and business values (Table 17-2). The decreasing importance of retail stores and sales transactions in city downtown districts is shown in Table 17-3.

Fire Hazards. In some cities, parking regulations have been effected because of the fire hazards created by improper parking, which congests the movement of fire apparatus and blocks access to hydrants, driveways, and key buildings.

[1] *Ibid.*, p. 54.

[2] *Downtown Parking: A Survey of the Parking Needs of the Bakersfield and E. Bakersfield Central Business Districts*, Bakersfield City Planning Commission, 1947, p. 71, Table XIX.

Types of Parkers. Parkers are generally classified in three groups: (1) office workers and other long-time parkers who enter the business area in the morning and leave their cars all day, (2) shoppers and other intermediate parkers who desire to park for 1 to 3 or 4 hr, (3) short-time parkers, such as persons on business trips. As Table 17-4 shows, 54 per

TABLE 17-2. TREND IN CENTRAL BUSINESS DISTRICT VALUATION

Population range and city	Valuation	Year	Valuation	Year	Per cent decline
Over 500,000					
Baltimore, Md........	$ 175,000,000	1931	$ 115,000,000	1945	34.3
Boston, Mass........	540,644,100	1935	408,604,100	1944	24.4
Milwaukee, Wis......	228,745,110	1930	140,119,300	1944	38.7
New York City.......	9,593,395,609	1930	7,435,092,050	1945	22.5
Philadelphia, Pa......	511,893,706	1936	363,734,500	1946	28.9
250,000–500,000					
Columbus, Ohio......	445,000,000	1930	287,000,000	1940	35.5
Louisville, Ky........	105,347,455	1930	74,569,225	1945	29.2
Portland, Oreg.......	38,765,720	1935	23,490,350	1945	39.4
Rochester, N. Y......	151,695,018	1930	108,834,784	1946	28.2
Seattle, Wash........	37,100,000	1928	20,700,000	1944	44.2
100,000–250,000					
Erie, Pa.............	25,314,000	1935	18,651,300	1946	26.3
Fort Worth, Tex.....	39,203,480	1930	30,850,850	1940	21.3
Trenton, N. J........	145,702,080	1938	123,312,975	1945	15.3
Yonkers, N. Y.......	13,820,268	1936	10,228,550	1946	26.0
Worcester, Mass......	310,000,000	1930	247,000,000	1945	20.3
50,000–100,000					
Binghamton, N. Y....	40,445,370	1930	27,320,400	1945	32.4
New Rochelle, N. Y...	37,386,555	1941	31,778,360	1946	15.0
St. Petersburg, Fla....	89,585,870	1930	63,446,853	1945	29.2
Under 25,000					
Anaheim, Calif.......	....	1930	....	1944	18.4

SOURCE: *Parking Manual*, American Automobile Association, Washington, 1946, p. 169.

cent of persons parking at the curb in central business districts generally want to stay 30 min or less, 74 per cent 1 hr or less.

Space is also required by trucks for loading and unloading and by taxis and buses. In New Haven, Conn.,[1] 56 per cent of all commercial vehicles parking in the central business district for business purposes stayed less than 15 min, 80 per cent less than 30 min, according to one study. In Spokane,[2] Wash., only 36 per cent of the truckers using

[1] *New Haven Parking Study*, State Highway Department, Hartford, Conn., 1947, Appendix, Table 10.

[2] *Central Business District Parking Study of Spokane*, State Department of Highways, Spokane, Wash., 1947–1948, p. 128, Table 22.

marked loading zones were found to complete their business in 15 min, while 74 per cent needed to park longer than 30 min.

Characteristics of parkers are generally consistent from one town to another of the same size. Motorists usually will not walk more than 500 ft from where they are able to park, but under certain conditions they will walk 1,000 ft or more (rarely more than 1,200 ft, however). They will walk farther in large cities and when they plan to park longer. The driver's unwillingness to walk far is evident in the development of outlying shopping centers and special business services which permit motorists to transact business without leaving their automobiles.

TABLE 17-3. TRENDS IN DOWNTOWN BUSINESS DISTRICTS, 1939–1948

Population range	No. of cities	Per cent of stores in CBD		Per cent of sales in CBD	
		1939	1948	1939	1948
1,000,000 and over	5	64	61	67	64
500,000–1,000,000	9	63	59	69	65
250,000–500,000	18	76	72	86	83
Total or average	32	66	63	72	68

SOURCE: C. T. McGavin, "Teamwork Can Solve the Downtown Parking Problem," Urban Land Institute, *Urban Land*, Washington, July–August, 1950.

In order properly to develop time-and-space restrictions for curb parking, certain simple types of surveys should be considered. Vehicle-volume and turning-movement counts at intersections within the survey area, during both peak and off-peak hours, determine the ability of the street to handle traffic and may indicate a need for parking prohibitions. Intersection congestion can often be relieved or eliminated by making the curb lane available to turning cars. Vehicle volumes by types are important to a determination of the relative amount of curb space needed for trucks, buses, taxis, and passenger cars.

Speed-and-delay values indicate delay caused by parking maneuvers and are often useful in "selling" a program of parking regulations.

Accident facts indicate the number of both vehicular and pedestrian accidents which may be reduced or eliminated if curb parking is more effectively regulated.

Curb usage, demand, and supply (number of available spaces) are also important. A complete prohibition of curb parking cannot be effected unless there is available sufficient off-street space to handle the displaced cars, therefore data on off-street space must be included.

Parking habits, such as time of arrival and departure, purpose of trip, and distance walked to destination, should also be determined.

Knowledge of business trends for a considerable period of time prior to the proposed change is helpful in determining the need for parking changes, as are physical and structural features of the city or district in question. Street width, layout or pattern, general weather conditions, and the legal authority to set up parking restrictions must be given detailed consideration.

TABLE 17-4. LENGTH OF TIME VEHICLES PARKED AT CURB IN CENTRAL BUSINESS DISTRICTS OF SELECTED CITIES
(Per Cent of Total Number of Curb-parked Vehicles)

City and population	0–15 min	15–30 min	0–30 min	30–60 min	0–60 min	1–2 hr	0–2 hr
Bakersfield, Calif. 30,000	....	....	67	16.4	83.4	10.5	93.9
Denver, Colo. 325,000	41	21	62	20.8	82.8	12	94.8
New Haven, Conn. 175,000	39.5	26	65.5	21.2	86.7	9.2	95.9
Wichita, Kans. 115,000	51.6	10.4	62	15.7	77.7		
Portland, Oreg. 305,000	28.1	20.4	48.5	24.2	72.7	17.3	90.0
Harrisburg, Pa. 90,000	....	....	52.5	15.3	67.8	9.6	77.4
Reading, Pa. 110,000	26	25.8	51.8	23.2	75.0	13.4	88.4
Providence, R. I. 250,000	31.5	25	56.5	22.6	79.1	13.1	92.2
Richmond, Va. 200,000	36.1	71.1	53.2	18.2	71.4	12.7	84.1
Spokane, Wash. 150,000	17.6	30	47.6	20.4	68.0	14.4	82.4
Averages 191,000 population	33.7	20.3	54	20	74.0	13.3	87.3

SOURCE: Parking surveys of various cities.

17-2. Space Controls. Curb-parking controls can be classified as (1) space controls or (2) time controls. In the first, physical arrangement of parking at the curb should be determined both in relation to parking demands and to demands for movement. In the second, space available at the curb for parking or loading should be allocated on a time basis to serve the majority demands.

Legal bases for parking regulations are *public safety* and *public convenience*. State motor vehicle codes generally cover legal authority for space control, including the prohibition of parking on sidewalks, in driveways, within intersections, near fire hydrants, on the immediate approach to traffic control devices, opposite pedestrian safety zones, near railroad grade crossings, and in similar places. Municipal governments may be granted the authority for more specific space restrictions, such as the prohibition or control of parking in the vicinity of schools, on congested streets, night bans, and so forth. Nearly all local ordinances specify the distance of wheels of parked cars from the curb, require parallel parking unless other types are specifically designated, provide for exclusive use of bus and taxi stands by such vehicles, ban the display of vehicles for sale on public ways, and indicate curb areas to be set aside for special uses.

Certain parking restrictions, such as those prohibiting parking on sidewalks and near fire hydrants and driveways, are usually not posted. It is customary to post the regulation where corner clearances are necessary, special prohibitions are established, or where special uses such as bus stops or taxi zones prevail. Most traffic codes now provide that time restrictions are unenforceable unless clearly marked.

Loading Zones (Commercial). Because of inadequate off-street spaces for loading and unloading of commercial vehicles, it is necessary in many downtown districts to set aside curb spaces exclusively for this purpose. A questionnaire survey[1] covering cities ranging in size from 2,000,000 to 50,000 inhabitants found that 85 per cent provided curb-loading space.

Establishment of loading zones requires consideration of availability of off-street or alley-loading space, frequency of loading or unloading operations, general curb-parking conditions, existing curb-loading zones, and available curb space. In some cities zones are established only upon request of property owners or building occupants, while in others a limited number are established per block. About one-third of the cities providing curb-loading space levy a charge, sometimes designated for maintenance of the markings and traffic signs required by the zones. Only a few cities base the fee upon the cost of sign installation. Some base the fee on length of curb provided (dollars per linear foot per year), while others charge a flat fee ranging up to more than $25 on establishment of the zone. A very few cities levy a flat annual charge, which may be higher the first year than succeeding years.

Of 549 cities reporting in another survey,[2] 18 charge application fees ranging from $4 to $70, and 45 charge annual loading-zone fees varying from $1 to $180. Some cities using parking meters base their fees (annual or monthly) for curb-loading space on the amount of money the meters might be expected to collect. In Memphis, Tenn., this system is used, but zones that were in use before parking meters were installed are not charged. In outlying areas, zones are provided for no charge where

[1] *Truck Loading Zones Study, Central Business District, Wichita,* Division of Traffic Engineering, Wichita, Kans., September, 1948.

[2] *Municipal Year Book,* The International City Managers' Association, Chicago, 1949, p. 443.

traffic will be aided by a reduction in the numbers of double parkers, or for a \$10 to \$30 annual charge (depending upon the signs required) where an establishment will use the zone primarily for loading or unloading its own truck.[1]

Signs may be provided on a permanent basis, or they may be portable so that during the hours when loading and unloading is not taking place other vehicles may use the zones. In Philadelphia, signs are placed in sockets buried in the curb, with keys placed in the socket so as to face the signs in the desired direction.

Some cities allow passenger vehicles to use loading zones when not occupied by commercial vehicles. Others do not allow passenger vehicles to enter the zones under any conditions. Nearly 44 per cent of the cities reporting in one survey[2] restricted the use of curb-loading zones to certain hours of the day, usually from 8 A.M. to 6 P.M., but three cities limited use of the zones to night and early morning hours. Some cities require property owners to place portable signs on the curb to designate loading zones at hours that have been authorized. Where meters are in use, they may be required to be hooded or otherwise signed to show that the space is to be used only for loading-zone purposes at certain times of the day.

Progressive property owners and businessmen are doing everything possible to develop off-street loading spaces. Some fleet owners are also cooperating by transferring heavy merchandise at terminals to small-type vehicles for delivery. Others are scheduling pick-up and delivery services to avoid periods of maximum traffic.

In Providence, R. I., police regulations prohibiting tractor-trailer trucks from entering a specified downtown area between 8 A.M. and 6:15 P.M. on week days were adopted early in 1948. Complaints of businessmen resulted in a change of restricted hours to 12 noon to 6:15 P.M.[3] New York City has adopted a regulation banning any vehicle or combination in excess of 33 ft in length from the "garment district" from 8 A.M. to 6 P.M. Mondays through Fridays.[4] Numerous difficulties arose, however, and the regulation was not effectively applied.

In 1952 at least 107 local government units in 30 states had adopted zoning or other types of ordinances requiring off-street truck-loading facilities in connection with various types of building uses,[5] generally within the same block as the structure to be served. Space requirements vary considerably, depending upon the type of business served. A typical ordinance requires one loading space for each 25,000 sq ft and

[1] Data from E. C. Sosman, Traffic Engineer, Memphis, Tenn.

[2] *Municipal Year Book*, 1949, p. 442.

[3] Information received from Dwight T. Myers, Traffic Engineer, Providence, R. I.

[4] Information received from T. T. Wiley, Deputy Traffic Commissioner, New York.

[5] David R. Levin, *Zoning for Truck-loading Facilities*, Highway Research Board, Bulletin 59, Washington, 1952.

fraction thereof in excess of 5,000 sq ft of aggregate gross floor area designed or used for storage, goods display, or department-store purposes.

Taxi Zones. In larger cities taxis are numerous and their place in transportation and in the traffic pattern is important. Some cities set aside curb spaces for the exclusive use of cabs, and a few even set aside spaces for specific companies licensed to operate within the city. Omaha, Nebr., provides permanent stands for exclusive use of one company, while Hartford, Conn., has both exclusive stands and unlimited stands.[1] Generally, however, taxi zones are designated for use by any of the regularly licensed companies within the city. In cities which have refused to provide space at the curb, taxi companies have been forced to develop off-street facilities.[2] The general use of radio by taxi operators makes off-street facilities feasible and encourages constant cruising. To keep down congestion, however, many traffic authorities encourage ordinances prohibiting taxis from cruising. One survey[3] found that approximately 65 per cent of the cities with populations of 50,000 and over prohibited cruising of cabs.

Most cities[4] do not set time limits for parked cabs. Compliance with parking regulations in general is the most common requirement. Toledo, Ohio, provides some stands from 7 A.M. to 6 P.M., and Seattle, Wash., has one stand, with a high night demand, in operation only from 6 P.M. to 6 A.M.

A few cities levy charges for the use of curb space for taxi stands. Some cities include the fee with a taxi- or hack-license fee, while others make separate charges. Omaha, Nebr., collects $2 per foot, San Jose, Calif., $10 per space per month, and Portland, Oreg., levies an initial fee of $20 and subsequent fees of $15 per year.

Bus Stops. On streets with bus routes, it is necessary to set aside adequate space at the curb so that the transit vehicle can pull out of the traffic stream for the loading and discharging of passengers. Zones must be long enough to allow a freedom of movement in adjacent traffic lanes. Requirements for modern vehicles are shown in Table 17-5.

In establishing bus stops, consideration must be given to the headways of transit vehicles and to the careful maintenance of schedules so that a minimum number of vehicles will accumulate at a given point at the same time. The proper location of the bus stop is also important, particularly

---

[1] Luis Gonzales-Solar, *A Study of Taxicab Stands*, Yale University, Bureau of Highway Traffic, 1948.

[2] Ft. Worth, Tex., required that cabs "must stand and remain, except while in the immediate act of discharging or taking passengers, upon and within private depots and grounds upon private premises." A Charlotte, N. C., ordinance states that "no taxicab shall operate on the streets of the city of Charlotte unless the same shall have a depot or terminal on private property."

[3] Gonzales-Soler, *op. cit.*

[4] *Ibid.*

where the bus route turns. Far-side or mid-block stops are advisable at complicated intersections, transfer points, and such other conditions as inadequate sidewalks and inconvenient business or factory entrances.

TABLE 17-5. MINIMUM DESIRABLE BUS-STOP LENGTHS, IN FEET

Approximate bus seating capacity	Approximate bus length	One-bus stop			Two-bus stop		
		Near side	Far side	Mid-block	Near side	Far side	Mid-block
25 passengers or fewer..	25	60	50	85	90	80	115
30 passengers..........	25	70	50	95	100	80	125
35 passengers..........	30	75	55	100	110	90	135
40 to 45 passengers....	35	80	60	105	120	100	145

NOTE: To find the space necessary for additional buses, add the bus length plus 5-ft clearance between buses. For example, a three-bus stop, near side, for a 30-passenger bus is 100 ft + 25 ft + 5 ft = 130 ft.

SOURCE: *Traffic Engineering Handbook*, Institute of Traffic Engineers, New Haven, Conn., 1950, chap. 15, p. 432.

The cooperation of bus drivers in the full utilization of designated stops is a very important factor in maintaining adequate traffic movements. In some states it is necessary to work closely with the public utilities commission.

Other Restrictions. Traffic codes generally specify restricted distances for parking from control devices such as signals, "stop" signs, and crosswalks. The Uniform Vehicle Code, Act V, provides that "no person shall stop, stand, or park a vehicle except when necessary to avoid conflict with other traffic or in compliance with the law or the directions of the police officer or traffic control device . . . within 30 ft of the approach to any flashing beacon, stop sign or traffic signal located at the side of the roadway . . . on or within 20 ft of a crosswalk at an intersection."

It also suggests restrictions "in front of a public or private driveway," "within 15 ft of a fire hydrant," "within 20 ft of the driveway entrance to any fire station and on the side of the street opposite the entrance within 75 ft of said entrance," "between a safety zone and the adjacent curb or within 30 ft of the points on the curb immediately opposite the ends of a safety zone," "within 50 ft of the nearest rail of a railroad crossing," "upon any bridge or other elevated structure upon a highway or within a highway tunnel," and "on the roadway side of any vehicle stopped or parked at the edge or curb of a street."

Type Parking. Angle parking is more convenient for the motorist than parallel parking, but it invariably produces a much higher accident rate than parallel parking at the same location. As a result, the Uniform Vehicle Code and most states require that all parking of motor vehicles shall be parallel to the curb unless otherwise specifically designated by pavement markings or signs, or both. Angle parking is not adaptable to trucks and other types of commercial vehicles; it encourages U turns on streets of certain widths, and the large overhang of front fenders and bumpers on many types of motor vehicles creates a serious sidewalk obstruction where sidewalks are narrow and pedestrian movements are heavy.

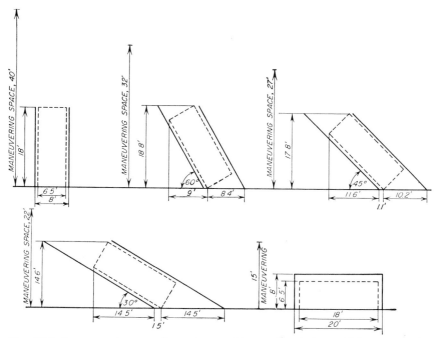

FIG. 17-2. Space requirements for various parking angles, design vehicle 6.5 × 18 ft.

It is often difficult, however, to establish parallel parking on downtown retail streets where angle parking has prevailed. Merchants and property owners are likely to protest because the number of spaces for potential customers is reduced, and motorists frequently complain that parallel parking is too difficult, particularly for women drivers. Figure 17-2 shows the space required at the curb for the average passenger vehicle with different types of parking, taking into account spacing between vehicles as well as absolute vehicle dimensions. Parallel parking requires about 22 ft per vehicle, while 45° angle parking requires 14.2 ft per vehicle where 10-ft stalls are used, a difference of 55 per cent. However, parking breaks necessitated by driveways, fire hydrants, bus zones, and other curb restrictions cause the loss of an appreciable curb frontage (13 ft) for angle parking.

17-3. Time Controls.   Parking-time limits, to be assigned to any given block or portion of a block, should be determined by careful study of parking demands and traffic volumes.   Since time characteristics will be influenced by regulations already in effect and the degree to which they are enforced, it is common practice to evaluate curb-turnover data along with characteristics of demand of nearby generators.   A post office or bank develops a very short time-parking demand, a factor which should be taken into account so that the influence of established time limits and

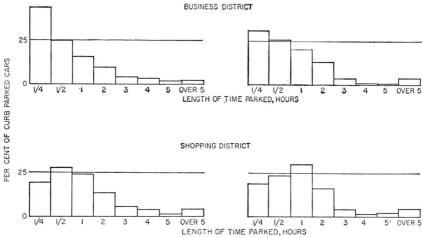

Fig. 17-3. Curb-parking time-demand distribution (typical business- and shopping-district blocks).

of other types of parking generators in the immediate vicinity will not unduly influence turnover data observed at the curb.

Some traffic authorities feel that time limits can be more properly established on the basis of generators than from observed curb data.   In areas where time limits are of most value, it is obvious that the curb should be used primarily for short-time parkers.   In such cases the existence of a 1-hr or longer parking limit in effect at the time surveys are made would not incorrectly influence the results.   Typical time-demand distribution charts are shown in Fig. 17-3.   Maximum limits of 30 min are usually proper in business districts (see Chapter 9), while 1-hr limits are more suitable to downtown retail shopping areas, and periods of 10 to 15 min serve adequately in the vicinity of banks and post offices.   On the periphery of the retail district, 2-hr limits are common.

Period of Prohibitions.   Periods when parking prohibitions should be applied on certain streets are dependent upon the demands for street capacity for traffic movements.   The analysis of traffic volumes by short time intervals will indicate hours during which curb parking should be eliminated in order to develop additional street capacity.   As Fig. 17-4 shows, afternoon peak-hour volumes begin to rise rapidly at 3:45 P.M. The prohibition of curb parking should go into effect in time to clear the streets before volumes exceed the practical working capacity of the street, in most instances from 15 to 30 min before the time when full street capacity is needed.

Since peak-hour parking prohibitions may be largely controlled by increases in transit traffic, it is desirable to classify vehicles by type in

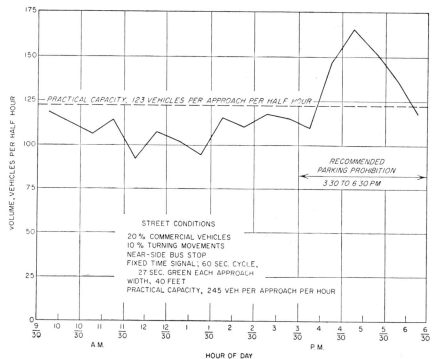

FIG. 17-4. Determination of parking restrictions through the use of volume-count data.

collecting volume information. Where special street uses are a factor in the need for greater capacity, it may be necessary to observe the delays of vehicles at key points along the street during the periods when volume data are collected. One method of determining delay is to record the average number of vehicles accumulated during the red-signal interval at key intersections on the street under consideration. The period of heavy delays can be taken as the time when parking prohibitions should be put into effect.

Rush-hour parking in Chicago[1] has recently been prohibited on the side of the street bearing the predominant traffic flow. Parking is prohibited on the inbound side between 7 A.M. and 9 A.M. and on the out-

[1] *Effect of Rush Hour Parking Control on Street Traffic and Transit Operation,* Chicago Transit Authority, 1949.

bound side between 4 P.M. and 6 P.M. (see Fig. 17-5). The streetcars and trolley buses which used the streets showed increased speeds of 3 to 16 per cent, and there was an increase in parking turnover resulting from the elimination of many long-time parkers from their favorite all-day positions. There was also a reduction in the number of accidents involving transit vehicles.

Traffic conditions may be such that the entire block length need not be closed to curb parking, but peak-hour volumes may warrant a peak-hour

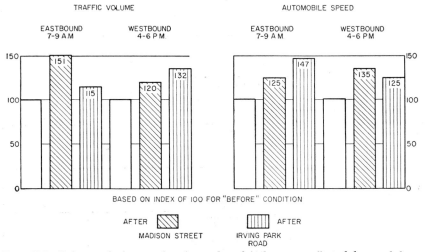

FIG. 17-5. Before-and-after study of speed and volume as affected by peak-hour parking prohibitions. (From *Effect of Rush Hour Parking Control on Street Traffic and Transit Operation*, Chicago Transit Authority, 1949.)

parking prohibition along the curb within some set distance of the intersection. Act V of the Uniform Code provides that "parking shall be prohibited within 20 ft of a crosswalk at an intersection" and "within 30 ft of a traffic control signal or device." However, at peak periods, when traffic backs up considerably at the intersection, prohibition of parking for greater curb lengths permits the formation of a second moving lane and reduces the time required to clear the intersection.

All-night Parking. To facilitate street cleaning, snow removal, the storage of commercial vehicles in residential areas, and to reduce vehicle thefts and fire hazards, most cities have regulations controlling all-night parking. The Model Traffic Ordinance[1] suggests that "No person shall park a vehicle on any street for a period of time longer than 30 minutes between the hours of 2 A.M. and 5 A.M. of any day, except physicians on emergency calls."

The enforcement of all-night prohibitions may vary in different districts of the city. Such regulations can properly be effected at a later

[1] *Model Traffic Ordinance*, Art. XIV, sec. 130, U.S. Bureau of Public Roads, 1946.

hour near nighttime recreational facilities than in business and residential districts. Most all-night bans apply from 2 A.M. to 6 A.M., with 30 min the maximum parking time during this period. In some cities a number of residential streets have no garages or driveways, so that it is very difficult to enforce all-night parking bans.

Some cities have ordinances preventing the long-time storage or abandonment of vehicles at the curb, aimed not at drivers forced to keep cars in the street overnight, but at the practice of leaving old and dilapidated cars at the curb for much longer periods of time. Twenty-four hours is the time limit generally used in such an ordinance, although the figure varies from six[1] to forty-eight hours or more.

Enforcement of Time Limits. Meters. Parking meters, first used in Oklahoma City in 1935, were employed in 1950 in more than 16,000 municipalities throughout the United States. In general, courts have held that parking meters can be used when their principal function is an aid to the enforcement of time regulations.

Revenue is usually limited to the amount necessary for the maintenance and inspection of the meters. Revenue in excess of maintenance costs is usually earmarked for the expansion of off-street parking facilities or for the general use of the traffic, police, or public works departments. On the average, meters can be expected to produce from $50 to $100 each per year.

The rights of the abutting property owner have caused questions, but courts have ruled[2] that "the abutter has no rights superior to the public in the use of the streets except the rights of ingress and egress."

In general, parking meters have the advantage of developing rapid turnover in curb use, of aiding police officers in enforcement of time limits (one of the chief reasons for the success of parking meters is that they become largely self-enforcing), and of producing sizable revenues. On the other hand, they do not effectively reduce police manpower, are not tinkerproof, induce graft, and constitute fire hazards in some congested districts.

[1] A Miami, Fla., ordinance specifies 6-hr limit to parking in any one place, and Detroit and Omaha specify 48 hr.

[2] *Maxwell and Quinn Realty Co. v. City of Columbia,* 193 S.C. 260, 8 S.E.2d 339 (1940).

CHAPTER 18

MISCELLANEOUS REGULATIONS

The daily work of many traffic engineers entails the use of regulations other than those which have been previously discussed, i.e., one-way traffic, speed, and parking. Here, again, both safety and rate of movement are involved in each of the regulations. Some may be applied singly, or they may be used in various combinations. A full study of an urban traffic area often requires consideration of all these regulations before the traffic engineer can be sure that full use is being made of existing facilities and that a properly integrated program of control has been achieved.

18-1. Stop Rule and Through Streets. Under some conditions of traffic, it is desirable to alter the basic right-of-way rule (section 14-2) by erection of "stop" signs on one or more of the intersecting roadways. When properly erected, such signs supercede the basic rule and provide a positive control over traffic operations. They are usually erected only on the minor legs of an intersection to give the heavy traffic flows the right of way over lesser flows. By erecting the signs on successive cross streets along a traffic artery, a *through* street is established. One of the most recent uses of the stop rule, commonly referred to as the *four-way stop*, involves erection of "stop" signs on *all* approaches to an intersection.

Simple Stop Control. The simplest application of the stop rule is at an isolated or independent intersection. Where intersection accident rates are high, where there is a great imbalance in volumes, and where approach sight distances are restricted, "stop" signs are commonly used on the *side* street. They provide an important driving guide and a specific driving instruction. The facility of movement on the main street or road is increased by controlling the cross movements. The delays incurred by cross flow are usually not serious because it is chiefly the minor movement that is affected.

Proper positioning, uniformity, and supplemental pavement markings are essential to good observance of "stop" signs. Enforcement is also a key factor. Motorists are inclined to become lax in observing "stop" signs if the level of enforcement is low, especially where "stop" signs are widely used. In general, it has been found that if the observance ratio drops to 95 per cent or below, serious accident conditions will develop.

Warrants. Factual warrants for "stop" signs are almost totally lacking. Currently recommended placement criteria are largely based upon experience. A "stop" sign is considered warranted wherever at least one of the intersecting roads is a state highway or has been designated as a through street, or where high speeds, restricted sight distance, or high accident records indicate need for stop control. Other criteria, such as granting priority to streets with streetcar tracks or to multilane roads intersecting two-lane roads, or to improved roads intersecting unimproved roads, have also been adopted.

Some attempt has been made to establish warrants upon the basis of interference between crossing vehicles; for example, at any intersection where distribution of volumes is such that without "stop" signs at least half the side-street cars can be expected to interfere with main-street cars. In application, this warrant becomes complex and difficult to apply. The *critical lag*, or car-free time interval required by drivers to enter from the side street and cross the intersection, must be determined. This time interval has been found to be relatively constant for a given intersection, but varies for different intersections from approximately 4.5 to 6.0 sec. It is also necessary to know the volumes on the streets involved. A series of warrant curves can be developed[1] for different lags and volume relationships, but the traffic engineer must still decide the acceptable interference ratio to be used in applying the warrant. This key variable keeps the results from being fixed.

Through Streets. The stop rule is frequently used to create through streets or through highways. Since such application is usually limited to principal streets, the amount of enforcement supervision required is limited and a much higher voluntary observance develops than when the rule is applied profusely. Key radial streets, circumferential routes, and important traffic arteries are often well suited to through-street principles, as are principal roadways between two business areas, between a business area and a residential area, intercity arteries, and urban bypasses. Except under very special conditions, through streets are not extended through business districts, where speeds are relatively low and many of the streets are of the same relative importance.

Warrants. Normally, through streets are established when a street is a natural traffic artery, into which large volumes of through traffic can be routed. Sometimes the decision to establish a through street is based on accident experience. Usually, intersection accidents can be expected to decrease, but accidents on side streets in the area sometimes increase. Increased speeds and pedestrian hazard on wide streets are the principal reasons for accidents on through streets.

[1] Mr. Morton Raff conducted extensive research on "stop" sign warrants and developed the "gap" and probability reasoning in the publication *A Volume Warrant for Urban Stop Signs,* Eno Foundation for Highway Traffic Control, Inc., Saugatuck, Conn., 1950.

Some traffic engineers establish through streets if from one-third to one-half of the intersecting streets require "stop" signs (under one or more of the "stop" sign warrants already mentioned, or under local practices).

State statutes often prescribe that streets and roadways used as state highways shall be through ways, and that "stop" signs must be erected on intersecting roadways.

Factors in Establishing Through Streets. In most jurisdictions, enactment of an ordinance establishing a through way, or express avenue, or any form of through street does not alleviate the necessity for erection of "stop" signs on all cross streets. It is also important to mark properly the extremities of through streets. Motorists should be advised when they are on a section of through street and should be conditioned by means of proper marking signs when leaving an extensive section of a through street.

When two through streets intersect, signal control is usually warranted. If not, the intersection may be controlled with "stop" signs on the street with smallest volumes. In some cases, the four-way stop control has been employed. The treatment of such intersections is most important because of the high-speed conditioning that prevails with a majority of motorists.

The four-way Stop. The *four-way stop* at conventional intersections of two streets is a relatively new concept of traffic control, about which there are divergent views. The principle is usually applied at an intersection of streets of approximately similar characteristics (especially traffic volumes), sometimes as a transitory treatment between simple stop control and signalization.

The four-way stop can be expected to reduce serious accidents at most locations, since two motorists are not likely to violate a "stop" sign simultaneously. Serious angle-type collisions caused by running "stop" signs are greatly reduced or they occur at such slow speeds as to cause little damage.

Like all applications of the stop rule, this regulation can be so widely applied as to greatly decrease the voluntary observance of "stop" signs and thereby reduce basic effectiveness.

The one principal objection to the four-way stop is that it produces much delay, thereby reducing the intersection capacity and creating the ill will of motorists.

In many communities, enabling laws and ordinances do not allow the use of four-way stop controls, or they do not fix legal responsibility by prescribing the actions required of motorists. Provision for the regulation is now made in the Uniform Motor Vehicle Codes.

Yield-right-of-way Concept. At lightly traveled intersections and at points where certain physical conditions exist, it sometimes appears more reasonable to establish by signs the right of way without attempting to evoke a strict stop rule. "Stop" signs are difficult to enforce under conditions of light traffic volumes and at acute and obtuse angle intersections.

The principal difficulty in the use of "yield right of way" signs is the legal right to enforce them. Legislation should provide that the signs alter the basic right-of-way rule and should fix the responsibilities of motorists in observance of the regulation.

To some extent the yield-right-of-way concept is an outgrowth of *stated-speed signs*, which were intended to provide information that would permit motorists to regulate speeds so that they would have time to ascertain whether or not other motorists were approaching before they entered the intersection. Experiences with the stated-speed signs were poor, and they have been almost completely abandoned. The "yield-right-of-way" sign is more flexible. Rather than try to dictate speeds, it simply fixes the responsibility of the motorist with regard to right of way. It is always placed on the minor-volume street.

18-2. Turn Controls. Conflicts at intersections involve vehicle-vehicle conflicts and vehicle-pedestrian conflicts. The elimination of certain conflicts by prohibiting turns has broad possibilities because it requires no important physical changes and can be easily rescinded or altered if necessary.

Turn control may be necessary to reduce bad accident situations or, most frequently, to increase capacity of an intersection. Where through-traffic movements are delayed by vehicles waiting to negotiate a turn, it is obvious that the capacity of the street is greatly impaired, and under heavy traffic conditions considerable delay may result.

Control of turns must be considered along with the progressive application of "stop" signs, yield-right-of-way regulations, and signals.

The traffic engineer is usually concerned with the restriction of left turns, as they generally involve the greatest degree of hazard and conflict. There are certain conditions, however, in which it is desirable to restrict or prohibit right turns. U turns must also be considered in over-all traffic engineering analyses.

Application of Turn Controls. Turn controls can be established on a continuous basis or they can be made effective only during peak hours. In some instances, the turning regulations apply only in emergency traffic situations. Most turning regulations are found along heavy traffic arteries which may or may not be signalized, at interchanges and at other places where one-way movements must be effected, and at irregular intersections which can be simplified through the use of turn controls.

A recent study[1] shows that almost all cities use one or more types of

[1] *Turn Controls in Urban Traffic*, Eno Foundation for Highway Traffic Control, Inc., Saugatuck, Conn., 1951.

turning regulations. Only 7 per cent of the cities of 50,000 or more population reported that they have no turn controls. About one-fifth of the cities use only full-time turn controls, while the remainder use both full-time and part-time controls. Left-turn controls are applied much more frequently than right-turn controls. Cities of 200,000 population and over average about 55 locations with left controls and only 15 with right-turn controls. Experience shows that the use of turn controls primarily benefits traffic flow, but also aids enforcement, aids pedestrians, and reduces accidents. On some streets the prohibition of left turns alone will enable the retiming of signals to increase progressive speeds by 20 to 25 per cent. Substantial reductions in the running time for vehicles and buses may be evident.

There are many cases in which turns can be adequately controlled through special turn-signal indications. This means that time separation can be effected in keeping with the capacity demands of the particular intersection. Street lanes must be available to store vehicles desiring to make certain turns which are permitted by special signal indications.

Channelization to aid in the regulation of turns is most important. Where adequate street width affords it, a properly designed island, supplemented by signs and markings, frequently makes it possible to permit turns that could not be allowed otherwise. The design requirements for such islands and turning protection at channelized locations are discussed in Chapters 29, 30, and 31.

Mass-transit routes must be considered in turn control. Sometimes routes cannot be changed because of franchises, concentrations of transit riders, or because of fixed rails or overhead electrical-distribution systems. In some cases, the pressures of business groups and the public make it difficult to reroute transit lines. Some cities have applied turning regulations to all classes of vehicles except transit. In a few cases transit vehicles are given special indications allowing them to effect turns under special conditions.

Warrants. The establishment of turn control is a question of how much conflict can be permitted. In one study, intersection counts were made of pedestrian volumes in each crosswalk, the volume of vehicular turns across the crosswalk, and the number of delayed vehicles making turns. Empirical formulas were then derived to show the percentage of turn delays which might be expected under different combinations of turning and pedestrian volumes. However, arbitrary decisions still are necessary as to the number of delays or the amount of delay which can be tolerated before turning regulations are applied. A proper warrant for turn prohibition would have to take into account the volumes of non-turning cars, the number of such cars delayed, and the number of lanes on the roadways involved. This has not been carried through to a satisfactory conclusion, and no substantial warrants are now available.

Some attempts which have been reported for establishment of turning controls are (1) when left turns constitute 10 per cent of the total movements, (2) when left-turn volume exceeds 20 per cent of the total traffic, (3) when left-turn movements interfere with straight-through movements of 15,000 vehicles in 24 hr, regardless of number of lanes and at signalized intersections, (4) where a left-turn or a right-turn movement interferes with pedestrian crosswalk volumes in excess of 2,000 persons per hour, (5) when 600 vehicles conflict with 1,000 or more pedestrians per hour, (7) when turning demands average 7 vehicles per green interval for several successive signal changes, (8) where more than three intersection accidents are reported involving turning vehicles in a 12-month period.

Factors in Prohibiting Turns. It must be carefully determined how the movements which are to be affected by the turning regulation can be achieved. Sometimes installations simply shift the problem to another intersection.

The question of added travel must be investigated. Inconveniences to motorists and volumes which will be forced onto certain streets are important in turn-control plans. Forcing additional vehicle mileage, particularly in the critical areas of the central business district, must be evaluated.

Left turns are generally prohibited where there are heavy traffic streams in which through movements are retarded by left turns. In most cases the warrants for right-turn prohibition are dependent upon pedestrian crossings. The time separation of pedestrian vehicle streams with new-type pedestrian signals eliminates some of the necessity for right-turn control. Heavy traffic movements at high speed and on narrow streets are the principal bases for the application of no-U-turns regulations. Wherever the application of the no-turn rule can be prevented, it is usually best to do so.

Another important consideration in the establishment of turning regulations is the routing of highways. Where numbered routes are involved, it is important to work out ways of effecting the routing when the controls are established. In some cases the highway department is reluctant to have turn controls applied because it means adding blocks or mileage to the state highway system. In other cases, it is difficult to make a change without confusion in routing through motorists. Sometimes the highway systems and the extent of highway mileage within an urban area may be so fixed that the proposed controls cannot be approved.

18-3. Pedestrian Regulations. With the advent of serious urban congestion, more thought has been given to the control of the pedestrian, not only to increase safety but also to reduce delays to motorists. Pedestrian aids principally involve the separation of pedestrian and vehicular movements on a time or space basis.

One of the simplest pedestrian aids is the crosswalk. By properly locating crosswalks in relation to natural paths of pedestrian movement and by clearly defining these walks with paint, or various types of plastic and metallic markers, or by varicolored concretes, the proper places for pedestrians to enter and cross roadways are indicated. The principal factor controlling the location of these walks is the predominant natural flow of pedestrian movements. Supplementary signs which direct persons to the proper crossings and which indicate the legal requirements for using defined crossings may also be erected. Again, however, it is essential that the reasonableness of the regulations be noted. For example, there are marked differences between the attitudes and actions of motorists with regard to pedestrians in different sections of the country. In most West Coast areas, motorists faithfully observe signs which indicate the right of way granted pedestrians properly using defined crossings. The observance of such signs in the East and other sections of the country is not reliable, and considerable hazards are risked by persons who take the signs at their face value. These differences in precedent and attitudes must not be overlooked.

Locating pedestrian islands or vehicular-channelization islands which can also be employed by pedestrians in street crossings is also common. In some cases the islands are designed and located for the express purpose of providing pedestrian refuge points and also pedestrian protection. These include streetcar and bus-loading platforms, median islands for the express use of pedestrians, and general refuge islands at complex intersections. In designing and locating these islands (see Chapter 30), particular attention must be given to the safety and requirements of motorists as well as the requirements of pedestrians. Improper design, improper placement, or improper lighting may create more hazards than the islands are capable of correcting. In some cases good results have been reported with painted islands, but generally for effectiveness pedestrian islands must provide some physical separation between the vehicles and persons afoot. This does not necessarily mean that immovable barricades must be a part of each pedestrian island.

Where there are heavy traffic volumes, complex intersections, or unusually heavy pedestrian movements, it may be necessary to erect various types of pedestrian barricades in order to ensure proper observance and usage of other aids which are afforded. Designated pedestrian-crossing facilities can often be made much more effective when augmented by barricades which force their proper usage by preventing persons from crossing at mid-block locations.

The traffic engineer may also consider the physical separation of pedestrian and vehicular movements through the use of pedestrian bridges and tunnels. Relatively, such facilities are expensive and it is difficult to get a high voluntary usage. Persons afoot have a basic resistance to changing grades in crossing roadways and are not prone to go either up or down in order to make use of a special pedestrian facility. Problems are also created by those transporting baby carriages or hand vehicles. There are some locations, however, where the natural topography and other conditions made pedestrian underpass or overpass facilities convenient and attractive. On heavily traveled routes, gaps in the traffic stream occur so infrequently that pedestrians may welcome the opportunity to use a bridge or tunnel. Factors of design, lighting, and maintenance play a key role in the extent to which such facilities are accepted and used.

Pedestrian accidents at night are greatly reduced with satisfactory street lighting. Since the traffic engineer is increasingly assuming the responsibility for street lighting, this fact should be considered in investigations and studies for establishing priorities and standards for lighting installations.

Where heavy conflicts prevail between vehicular and pedestrian flows, it is often desirable to install special pedestrian signals, "walk—don't walk" or "walk—wait." Such pedestrian signals are used in several ways. They may be employed simply to indicate the time during which the pedestrians can cross a given street and comply with the green intervals on the standard traffic signals. The only advantage that such a system affords is that conspicuous and separate indications are given and that a clearance interval can be injected into the signal cycle requiring pedestrians to stop sufficiently in advance of the stoppage of vehicles to ensure clearance of crosswalks before a change in the vehicular right of way. Another type of pedestrian signal involves the use of an exclusive pedestrian interval in the over-all signal cycle, when all vehicles approaching the intersection are stopped and the area is given over exclusively for use by pedestrians. It is common practice under this system to allow pedestrians to cross diagonally as well as in crosswalks. Still another type of signal involves the segregation of pedestrian and vehicular movements without the injection of an exclusive pedestrian interval into the signal cycle. Such segregation normally involves the application of turn restrictions and one-way streets so that pedestrians can be given indications in certain crosswalks without fear of interference from turning vehicles.

The application of pedestrian signals requires very careful examination of the legal authority, particularly in the case of such treatments as that involving diagonal crossings at the intersection during an all-pedestrian interval. Most state codes and city traffic ordinances do not have a provision which makes such movements legal, and if they are to be allowed, the laws should be amended or modified accordingly.

Other forms of pedestrian control involve the use of pedestrian-actuated signals or special types of signal indication which are known locally, to provide special right of way for pedestrians.

Under some conditions of lane capacities, vehicular-turning volumes, and pedestrian volumes, the injection of separate pedestrian intervals does not necessarily reduce the over-all capacity of the intersection. Placement of pedestrian signals should entail the conducting of objective studies to determine factually the extent to which there are pedestrian-vehicular conflicts at given intersections.   The broad-scale application of pedestrian controls is not recommended and should be carefully guarded against because of the tendency to employ such regulations indiscriminately.   Overuse creates undue delays and serious hazards and sometimes produces a restriction in capacity on heavily traveled downtown streets.

18-4. Unbalanced Traffic Flow.   Unbalanced traffic flows may result from unusual street patterns, from peak-hour demands, or from location of unusual types of generators and special events.   Multilane streets are sometimes used to meet the capacity requirements to accommodate such flows.   In some cases, more lanes may be permanently designated in one direction than in the other.   In other cases, unbalanced designations prevail only during special times of the day, commonly being reversed for morning and afternoon conditions.

One of the earliest and best-known applications of the principle is on the Outer Drive in Chicago.   On this roadway, hydraulically controlled islands are used on some sections to provide different numbers of lanes for movements inbound and outbound, depending upon the time and character of traffic demands.   On an eight-lane section of the roadway, it is possible to provide six lanes in one direction and two in the counterdirection.

On key bridge and tunnel facilities the principle of unbalanced flow is commonly applied to accommodate the heavy directional movements which occur during peak hours.

Methods.   The most common method of achieving unbalanced flow is to place portable barricades or traffic cones in the roadway to designate the number of lanes which are available for movement in each direction. This method is quite common at barrier-toll stations for efficient utilization of collection booths, but has the disadvantage of requiring a crew of workmen to place and remove the markers at predetermined times.

Another method is to designate movements in each lane on a given street by lane signals.   Standard-type traffic signal lenses can be mounted over each roadway lane and can be illuminated to indicate whether or not the lane is available in a particular direction.   On very wide streets it is common to find that one or more central lanes are reversed in the morning and afternoon, but as an added safety feature during off-peak periods traffic is prohibited from using them. This is achieved with signals by having the signal show red in each direction.

Movable curbs such as those used on Chicago's Outer Drive can also be utilized, but they are quite expensive and involve high maintenance cost. They are effective, however, and where traffic conditions warrant the expense, they are one of the best means of getting maximum usage from the unbalanced-flow principles.

Through the basic design and construction of the roadway, sectionalized lanes can be provided so that controls at extremity points are sufficient to achieve unbalanced movements.

In some cases, unbalanced flow is made possible by the removal of curb parking during peak traffic hours. In such cases it is common to remove the parking from the side of the street which gives an extra lane for the direction of heavy movement. If *all rolling* regulations can be effectively enforced, this is an acceptable means of getting an additional lane for the heavy flow and thereby making unbalanced flow possible without special barricades or signal designations.

Application. It is possible to make detailed studies of the hourly flow characteristics of traffic flow on a given street or on a given net of streets and to determine when unusual capacity is needed in a particular direction.

At certain locations where heavy turns are necessary and where channelization has been applied, it may be desirable to effect unbalanced flow to *feed* or utilize properly the other traffic control treatments. The development of storage lanes for unusually heavy turning movement is one of the simplest applications of the principle.

It should be carefully determined that application of unbalanced-flow techniques does not bring about hardships and hazards due to absence of a natural unbalance in traffic demands. This can be measured from simple volume counts over an area if not along a particular roadway.

The number of lanes which can be utilized on a particular street is a critical factor. Under most conditions, more than one lane is desirable to accommodate traffic in each direction. Temporary blockage of a single lane will create bad congestion and induce violation of the regulation, which can be extremely hazardous. Thus a minimum of five lanes for moving traffic must be present. One or more of these lanes may be obtained by curb-parking restrictions.

Experience with unbalanced flow has been very favorable where it has been properly applied and where it is understood by enforcement officials and motorists. It is a simple and inexpensive way of getting maximum usage from existing roadways and usually meets with the favor of motorists and downtown interests.

CONTROL DEVICES AND AIDS

CHAPTER 19

TRAFFIC CONTROL DEVICES

Traffic control devices are the means by which the road user is advised as to detailed requirements or conditions affecting road use at specific places and times in order that proper action may be taken and accident or delay avoided. Control devices are even demanded by regular users of a given route and are highly necessary for those who are strangers to the locale.

Traffic control devices fall into three distinct functional groups. *Regulatory* devices have the authority of law and impose precise requirements upon the actions of the road user. *Warning* devices, used to inform road users of potentially hazardous roadway conditions or unusual traffic movements which are not readily apparent to passing traffic, impose responsibility upon the individual to employ added caution as he approaches and proceeds through the danger area. *Guiding* devices are employed simply to inform the road user of route, destination, and other pertinent information.

The management of traffic control devices, including their standardization, use, warrants, placement, design, maintenance, and related matters, is discussed in Chapter 14. Responsibility and authority for traffic control devices should be placed under the specified public agency which has legal authority to compel observance.

There are four elementary requirements which every traffic control device should meet: (1) it should compel the *attention;* (2) it should convey a simple clear *meaning* at a glance; (3) it should allow adequate *time* for easy response; and (4) it should command the *respect* of the road users for whom it is intended (Table 19-1).

These four fundamental traits develop in a logical sequence. In the first place, if the attention of the road user is not compelled, any device becomes valueless regardless of its other qualities. Once the attention is compelled, it is essential that a simple clear meaning be imparted at a glance, for if the meaning is confused or lost to the road user, the purpose of the device is lost. Still further, if the device is noticed and its message is clear, a comfortable interval of time must be provided so that the road user can respond easily in the space provided for; if this is not possible, the device is rendered useless, if not hazardous. Finally, no control device is of value unless it commands the respect of the road user. To the degree that any of these four basic qualities is deficient, to that degree any device is destroyed or rendered ineffective.

The four fundamental traits of a control device are effected through (1) the *design* and outward aspect of the device, (2) the *position* or placement with respect to the road user's normal line of vision, (3) the *maintenance* of the condition, appearance, and visibility, and (4) the uniformity of application and uses (Fig. 19-1).

Fig. 19-1. Regulatory signs. (Adapted from various United Nations and United States sign standards.)

The design of the device is a major factor in its *target* value and encompasses values of size, contrast values through color and brilliance, movement, and sound. It is the design of legends and messages and their familiarity which establish legibility. These aspects of design also determine legibility distance and glance values, and it is design which yields the impression of the "true and official" character of the device.

The position of the device with respect to the normal line of vision and its distance from the eye of the road user is of great importance in the accomplishment of purpose.

The position of a traffic control device in the normal plane of vision is a factor of great importance in compelling attention. As the angle between the axis of vision and a line drawn from the control device to the eye of the road user decreases, the accuracy of identification of the control

TABLE 19.1. THE CRITERIA OF TRAFFIC CONTROL DEVICES
No traffic control device will achieve its purpose unless it fulfills these four basic requirements:

These basic requirements are achieved by	Compels attention	Makes its meaning clear at a glance	Permits time for easy response	Commands respect
Design factors	Target values including size, color, and brilliance Movement Sound	Simplicity of legends and messages used Legibility values (pure and glance) Familiarity of legend	Sufficient legibility distance Glance legibility Value	"True and official" aspect
Position or placement	Within cone of vision for normal road-user attention, 10–12° in horizontal, 5–8° in vertical	Location with respect to point, lane, or curb to which device applies	Accurately sited with respect to legibility, speed, and travel Distance to point of required response	Logically placed so as to carry its message to road users for whom it is intended
Maintenance of condition and visibility	Maintained in fresh, clean condition, free from sight obstructions	Legends, messages, and legibility kept clean and bright	Maintained to secure high legibility values	Maintained to avoid appearance of being abandoned or forgotten Removed when no longer required
Uniformity of application and use	Good standards to assure highest attention-compelling values	The same device for the same purpose at the same relative position	Standards of design, placement, and use to assure adequate time if adhered to	Standardized official devices, logically used for warranted traffic conditions

device increases (see Chapter 2). For optimum attention and identification, control devices should fall within a visual cone of 10 to 12° on the horizontal axis of the cone base and 5 to 8° on the vertical axis, throughout the intended range of effectiveness of the device.

Because of experience and habit, control devices placed in the right half of the cone of vision are readily recognized by the road user, but in some cases devices are placed to fall within an even smaller cone of vision or approach more closely the axis of vision of the road user for whom it is intended.

Control devices which are not close to the axis of vision may gain attention but will require shifting of eyes or head and thereby impose added time for identification and response.

The longitudinal position of control devices is highly important in timing. The time relationships between legibility distance, longitudinal position, point of response, and speed are evident. By custom, legal requirement, or convention, regulatory devices are usually placed in the immediate vicinity of the location where the regulation is to be respected. Warning and guide devices, however, are frequently placed at or before the area of hazard or response. Balance must be provided in the design of such devices so that legibility distance plus longitudinal placement will provide adequate time for response.

Longitudinal placement of control devices must clearly indicate the features of the road system or area to which they apply. Confusion will result if the directed path, rule, or signal is not clearly associated with the roadway and traffic stream for which it is intended. Regulatory control devices placed at points which do not seem reasonable to road users generally lose respect; a stop line too far back from the intersection is of little value, for example.

The importance of maintenance of traffic control devices in fresh, clear, condition, free from vegetation, extraneous poles, signs, and other sight obstructions, is evident. Legends, messages, reflecting materials, and lenses, when kept clean and bright, will secure high legibility values which are important in the conveyance of the message at a glance and in providing the necessary time for response. Finally, proper maintenance will avoid the appearance of a device being "forgotten" or abandoned.

Prompt removal of those devices which are no longer required must be provided because, if left in place, they tend to encourage disrespect for control devices in general.

The importance of uniformity of traffic control laws and devices is of great significance in the application and use of a given device. Good standards will assure high attention value. The use of the same device for the same purpose and at the same relative position is a valuable tool in assuring clarity of meaning at a glance and standardizing time for response. Ideally, a given device should always connote identical meaning, so that road users develop a habitual proper response. The effective standardization of traffic control devices requires study of traffic movement and parking, road-user psychology, vehicle performance, physical conditions of the fixed facilities, and an intimate knowledge of the operations of traffic within each jurisdiction. Uniformity facilitates control, since police, road users, and courts all attach the same interpretation to a given device, and it facilitates economy and accuracy in manufacturing, installation, maintenance, and administration.

Uniform standards for United States traffic control devices have been revised from time to time by the U.S. Bureau of Public Roads and published in a *Manual of Traffic Control Devices for Streets and Highways.* Efforts are being made toward the establishment of international standards through the United Nations. Figure 19-1 gives illustrative examples of comparative United States and international standards recommended by the United Nations.

CHAPTER 20

TRAFFIC SIGNS AND MARKINGS

Signs are employed more frequently than any other device to regulate, warn, or guide road users. The effectiveness of any sign depends upon its attention, meaning, time, and respect values. Typical standardized signs are shown in Fig. 20-1.

20-1. Attention. The characteristics of a sign, which make it stand out from its background and surrounding objects, also give it attention-compelling value and priority value over other signs in the same group. Such values are dependent on size, contrast, shape, and in some cases, sound and motion (apparent or real).

Size. Early attempts to standardize the size of warning signs resulted in 1929 in the adoption of 18 in. as the minimum size of a sign, regardless of place of use. Six years later this minimum dimension was increased to 24 in. By 1948 it was recognized that higher speeds demanded larger signs, and the minimum standard size of signs for multilane rural routes was increased to 30 in. Oversized signs are currently suggested where speed or degree of hazard is high, time to see is limited by sight distance, or extraneous stimuli, such as complicated road layout, advertising signs, lighting, or displays, compete for attention. Where little or no hazard is involved, as in the case of parking or guide signs, present standards permit smaller signs of 12 to 18 in.

Contrast. Contrast depends on color brightness. Two basic colors are currently used for the ground color of traffic signs, yellow for all warning signs and for the "stop" regulation signs, white for all other standard signs. Yellow provides strong contrast against the green of summer and the white of winter. Both yellow and white allow a high degree of contrast with text or symbols, have relatively high light-reflecting properties, and provide a high degree of contrast with the usual backgrounds. Black, green, red, and other colors are employed on occasion, but it is generally agreed that yellow and white have inherently higher attention-compelling value, and they have been adopted as preferable in American standard practice.

Brightness, the intensity of light reflected or emanating from a surface, is measured in foot-lamberts. Relative brightness of a sign and its background determine attention value. The standard ground colors of yellow

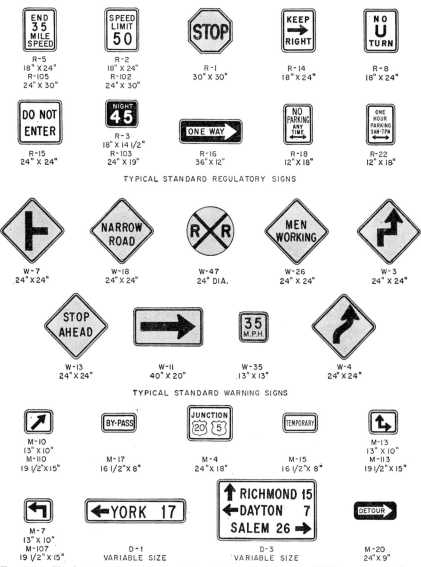

FIG. 20-1. Typical standard signs. (Adapted from *Manual on Uniform Traffic Control Devices*, U.S. Bureau of Public Roads, 1948.)

and white provide a high relative brightness, but reflectorizing materials and interior illumination or floodlighted signs are also used to increase the brightness of the sign over that of its background.

Since background of traffic signs varies from practically 0 ft-L in darkness (outside the beam of the headlights) to the brilliance of the setting sun on a clear day, under extreme conditions it is sometimes necessary to provide a special background affording brightness contrast.

Shape. Attention is naturally directed to simple, regular shapes because of the contrast afforded to the irregular pattern of natural backgrounds in form or outline. Irregular cut-out shapes, such as arrows, are to be avoided as the basic shape of signs.

Sound. Vibratory bells, gongs, and other noise-making devices tend to compel attention, but because of the disturbance, created sound is seldom employed in traffic control devices except at railroad crossings, drawbridges, and similar hazardous locations.

Motion. Apparent movement, commonly achieved in signs by flashing lights, is usually reserved for warning signs at points of very high hazard. It is also achieved by mechanical means at railroad crossings.

Novelty. Any new or strange shape, color, or other feature is prone to attract attention, but becomes ineffective as its familiarity increases. Since novelty is the antithesis of standardization, it is rarely used.

Relative Position. Relative position and its importance in compelling attention have been discussed previously (Chapter 19). In a group of traffic signs within the field of vision, when all other factors are equal, prior attention is directed to the sign which is (1) nearest the road user, (2) proximate to another attention-compelling object, or (3) usually read first because of normal reading habits (top to bottom, left to right).

Use Factors. If road users know through experience that a given type of sign can be relied upon, they tend to give such signs full attention. Route signs are anxiously sought by the strayer, who has learned to rely on them, but are generally overlooked by the user who knows the route. Excessive misuse of a particular sign type destroys its attention values at warranted locations.

Maintenance. It is obvious that attention is highest when signs are maintained as designed.

20-2. Meaning. Meaning is conveyed by design features of shape, color, and message. Familiarity with legend and brevity and clarity of message give traffic signs a *readability* which affects the speed with which meaning is comprehended.

Shape. Eight combinations of shape and mounting arrangements are employed in current United States standards to convey meaning in traffic signs (Fig. 20-2): (1) the octagon shape, for the "stop" regulation, (2) the circular shape, as an advance warning of a railway crossing at grade, (3) the crossbuck, at point of railway crossing, (4) the square, mounted with diagonal vertical, for warning of road or traffic hazard, (5) the rectangle, with major axis on the vertical, to convey a regulatory message, (6) the rectangle, mounted with the major axis horizontal, to inform or guide road users, (7) the shield, for marking U.S. routes, and (8) a variety of characteristic forms for marking state or other routes. In addition, the trapezoid has been suggested for marking relatively high cross-traffic volume not warranting the "stop" sign. The pentagon has been proposed for school purposes. Both solid and hollow triangles mounted with apex up or down have special meaning in the International Standard.

Color. Yellow signs are reserved to mean warning alone except for the mandatory stop signal, which is also yellow. However, strong consideration is now being given to the use of red for "stop" signs. White is used for all other signs except large, oversized signs employed for information or guidance, for which black is permitted. Green, blue, yellow, red, and other colors are sometimes employed to signify a particular route or direction of route.

Message. In the United States, few symbols are employed. Arrows with straight, bent, or crooked shafts are used to indicate course of road or permitted traffic direction, the direction of the arrowhead showing degree of curvature or change in path toward the left or right. Intersection "maps" and simulated crossbuck are the only other symbols found in present United States standards. However, silhouette forms, red cross, and other symbolic legends are sometimes employed in the United States and are heavily used in the International standards. Simple symbols are a valuable method of conveying meaning, especially in the case of language differences among road users, but motorists must be trained to recognize their meaning.

Word messages, including numbers and letters, should be kept as short as practicable; not more than three or four familiar words can be conveyed at a glance.[1] Strange and unfamiliar words or symbols require greater time for identification.

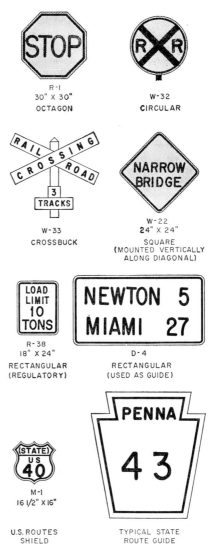

FIG. 20-2. Sign shapes. (Adapted from Manual on Uniform Traffic Control Devices, U.S. Bureau of Public Roads, 1948.)

[1] Cf. T. W. Forbes, "A Method of Analysis of the Effectiveness of Highway Signs," *Journal of Applied Psychology*, vol. 23, no. 6, December, 1939, pp. 669–684.

The distance at which a message can be read or recognized is known as *pure legibility* or legibility distance. The distance at which a sign message can be read at a glance is known as *glance legibility*. The time required for a glance, ranging from 0.5 to 1.4 sec, is composed of (1) eye movement, (2) binocular focus, and (3) fixational pause.

Since it is unlikely that a road user can read in a single fixational pause any part of the message outside of the 3° cone of acute vision, the glance area is about 5 ft in diameter at a distance of 100 ft. It is essential to have a clearly legible message within that time and space limitation of a glance, for there is no time to study the message.

ABCDEGHJKMORS234568
SERIES A

ABCDEGKMORS23568
SERIES B

ABCGMORS23568
SERIES C

ABDEGQRS389
SERIES D

ABDEGOS389
SERIES E

FIG. 20-3. Standard lettering for highway signs. (From U.S. Bureau of Public Roads.)

Legibility distance is influenced by (1) alphabet design, (2) letter height, (3) letter width, (4) letter stroke, (5) letter spacing, (6) sign margin, (7) word and line spacing, and (8) contrast in color and brightness.

Alphabet design should be as simple in style as possible in order to increase legibility. Rounded-block-style letters yield about 5 per cent greater legibility distance than pure-block capital letters, because of purely parallel adjacent strokes which develop in pure block-letter sequence. Recent experimentation has been carried out with combination capital and lower-case letters, which seem to yield a slight increase in legibility distance of comparable single-word signs. However, the need for added space between lines to maintain legibility when lower-case letters are used tends to limit their use. Standard alphabets have been developed in the United States (Fig. 20-3), and specifications may be procured from the U.S. Bureau of Public Roads.

There are currently employed six series of letter dimensions in the United States standards: Series A, B, C, D, E, and F. Height-stroke ratios of the series are approximately 11:1, 8:1, 7:1, 6.5:1, 6:1, and 5:1, respectively. The height-width ratio varies with individual letters, but ranges roughly from 11:1 for letter "I" of Series A to 1:1.2 for the letter "W" of Series F. Legibility distance varies considerably among observers owing to difference in seeing ability and, for example, may range from 300 to 700 ft for a given letter series and height. Figure 20-4 shows average distances for road users with 20/20 vision. Legibility distance increases curvilinearly with letter height for a given series in the smaller letters and linearly in larger letters. Daylight legibility distance of various letter series may be considered as 50 ft/in. of height for Series D, 42 ft/in. of height for Series C, and 33 ft/in. of height for Series B.

In general, spacing between letters of the same word ranges from 1.2 times the stroke width for parallel strokes to almost 0 for strokes of opposing slope. Generous spacing is advisable for high legibility and is mandatory in numerals. A sign margin increases attention and meaning and serves as a point of reference in reading the message. Spacing

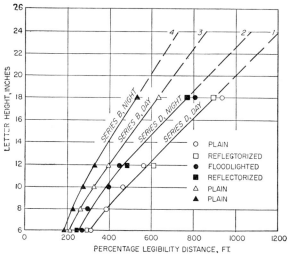

FIG. 20-4. Practical legibility values corresponding to 20/20 vision, 80 percentile figures. (From T. W. Forbes and R. S. Holmes, "Legibility Distances of Highway Destination Signs in Relation to Letter Height, Letter Width and Reflectorization," *Proceedings of Highway Research Board*, Washington, 1939, vol. 19, p. 331.)

between words and lines is usually satisfactory when three-fourths of the letter height is allowed. Interlinear spacing may be reduced to one-half of the letter height when capital letters only are employed.

The color of the message must provide high contrast with the color of the sign background. Since the background is usually yellow or white, black is usually employed for message; when black or other dark colors are employed for background, white messages are used. Red and green are sometimes employed for the message, as in parking-regulation signs. Contrast between message and sign background is also achieved by reflectorizing the message or the sign ground (see section 20-5). In cases of excessive contrast, however, the meaning of message may be lost at night because of halation of light destroying detail of design.

Position. By convention, signs usually are placed laterally on the side of the road nearest the road user for whom their meaning is intended. Overhead signs are placed directly over the lane or lanes to which they apply (Fig. 20-5). Longitudinal placement is also important: the "stop" sign is positioned at the point where the forward motion of a vehicle must be temporarily halted. Great care must be exercised so that no confusion of meaning results from improper placement. A "No parking here to corner" sign should leave no question as to which corner is meant.

Use. Uniformity of use is highly important in conveying meaning to the road user. Standard signs for similar conditions are quickly understood by the public.

FIG. 20-5. Overhead directional signs.

Maintenance. For optimum meaning, street maintenance of the sign to its original design, position, and use specification is required.

20-3. Response Time. In providing adequate time for response there are three significant factors: (1) the speed of the road user and his vehicle as they approach the sign, (2) the legibility distance of the sign and its message, and (3) the longitudinal site of the sign. These factors must be in a relationship which provides adequate PIEV time plus time to maneuver the vehicle into its required lateral position in the roadway. Time is required for the road user to change speed, change lanes, or come to a stop. Simpler messages such as "stop" will require perhaps only a second for PIEV, but multiple choice (as in destination or guide sign) may require 3 or 4 sec. The higher the speed or speed change, the greater the time which should be provided. Lane-change-delay time will increase with traffic volume, (see Chapter 7).

20-4. Respect. While design, position, and maintenance have significant effect, uniformity and reasonableness of use provide the foundation upon which respect for traffic signs is founded.

Design features of size and brightness may be employed to command respect at particularly hazardous locations. Position allowing time for response but not time for forgetting or having the attention diverted provides respect. Shabby, ill-kept signs cannot command respect.

In any event the regulation, warning, or guidance which the sign is intended to convey must appear reasonable, rational, and advantageous to the road user. Experience of the road user in regard to the dependability of meaning and purpose as conveyed by the sign contributes most greatly to the respect it receives. Overuse, misapplication, and even underuse at necessary locations can destroy respect for traffic signs. The pitfalls of misuse are best avoided by employing standardized warrants for placement and use. Those warrants which thus far have been adopted are set forth in United States standards in the *Manual on Uniform Traffic Control Devices for Streets and Highways*.

20-5. Traffic Markings. Traffic markings (lines, patterns, words, symbols, delineations, or reflectors, applied to pavement, curb, roadside, or fixed objects on or near the roadway) are specialized types of traffic signs in which the message is in contrast with the color and brightness of the pavement or other background. Since markings convey a perspective of depth, they are used to signify the delineation of path, its lateral clearance, and the proximity of obstruction. Because of their position, markings are in the normal field of attention and range of visual focus.

Attention. To assure attention, standards indicate a minimum dimension of 4 in. in any longitudinal line. Transverse lines are of necessity wider because of perspective. White and yellow are generally used to provide contrast, but black is sometimes used on light-colored pavements. Contrast is maintained at night through reflectorization. Bells, "jiggle bars," or other devices which give noise or "feel" when crossed by a vehicle are sometimes used to compel added attention.

Meaning. Legends and continuous or broken lines used to form patterns or outlines are employed to convey a message. Illustrative markings and their meanings are set forth in Fig. 20-6. Color and position are also used to give meaning: a continuous yellow line conveys regulatory meaning and must not be crossed by the road user, but a white dash line is used for guidance and may be crossed. Meaning is also conveyed by position of line. A line placed adjacent and parallel to another conveys the regulation of "no crossing." A transverse line used in connection with a "stop" sign indicates limit of forward progress before making the complete stop. Care must be taken in the design to correct for the road user's perspective, as Fig. 20-7 shows.

Time. Markings, like signs, should become meaningful far enough in advance to give road users adequate time for response. A guide line around an obstruction should be placed far enough ahead of the obstruction to effect a comfortable lateral transition at the nominal speed for which it is designed. Because of the low legibility distance of word markings, they must occur considerably in advance of the point of observance. Word markings are usually employed only as auxiliary to traffic signs.

As with signs, respect grows mainly from proper use. Because of the natural tendency of a line to guide the road user along whatever path it follows, proper and intended usage is required at all times.

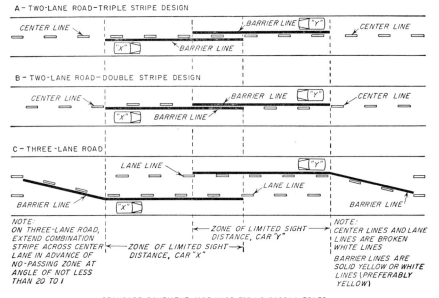

A - TWO-LANE ROAD-TRIPLE STRIPE DESIGN

B - TWO-LANE ROAD-DOUBLE STRIPE DESIGN

C - THREE-LANE ROAD

NOTE:
ON THREE-LANE ROAD, EXTEND COMBINATION STRIPE ACROSS CENTER LANE IN ADVANCE OF NO-PASSING ZONE AT ANGLE OF NOT LESS THAN 20 TO 1

←— ZONE OF LIMITED SIGHT DISTANCE, CAR "Y" —→

←— ZONE OF LIMITED SIGHT —→ DISTANCE, CAR "X"

NOTE:
CENTER LINES AND LANE LINES ARE BROKEN WHITE LINES

BARRIER LINES ARE SOLID YELLOW OR WHITE LINES (PREFERABLY YELLOW)

STANDARD PAVEMENT MARKINGS FOR NO-PASSING ZONES

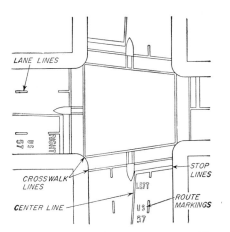

FIG. 20-6. Typical urban pavement markings.

20-6. Reflectorization and Illumination. Because of low levels and limited fields of illumination, the road user must rely greatly on traffic signs and markings when traveling at night. Current United States standards require reflectorization or illumination of all signs carrying messages which regulate or warn vehicular movement or guide such movements on important thoroughfares. Ideally, signs and markings should have the same visibility at night as during the day.

Illumination. Illumination may be achieved by luminous tube lighting (such as neon), light shining through a translucent sign face, or by direct floodlight. Markings generally depend on exterior illumination from street lights or vehicle headlights. A few markers, such as mushroom buttons, use interior sources of light.

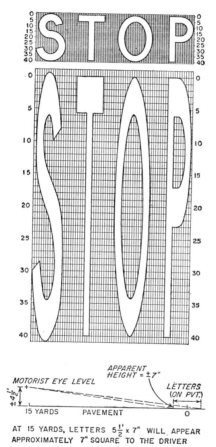

Fig. 20-7. Method of developing elongated letters for pavement markings (Adapted from *Manual of Uniform Traffic Control Devices* and from *Pavement Signs*, Eno Foundation for Highway Traffic Control, Inc., Saugatuck, Conn., 1931.)

Neon tubing is used to illuminate pedestrian signals, "one-way" signs, and the like. It has also been used for special hazard-warning signs and for some large signs. Interior illumination through translucent faces should provide a brightness of about 5 ft-L in order to give visibility performance comparable with daylight illumination. Direct flood illumination of signs through incandescent or other light source should provide an incident intensity of about 10 to 15 ft-c.

The color of light generally used in illuminating is white, but red is often used in neon-tubing signs.   Flashing amber or flashing red is employed in signs for attention-compelling purposes.

Reflectorization.   The basic principle of reflectorization is to return to the eyes of the road user some portion of the light from his vehicle head lamps, thereby producing a brightness which attracts attention and

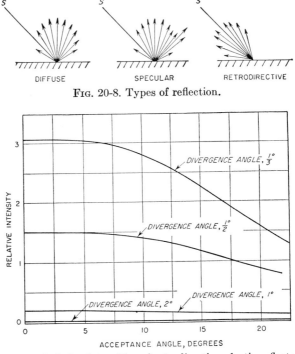

Fig. 20-8. Types of reflection.

Fig. 20-9. Relative intensities of retrodirective plastic reflectors.

imparts meaning at sufficient distance to provide time for the required response.   The three types of reflectors are shown in Fig. 20-8.

Diffuse reflectors (such as flat paint) reflect incident-light rays in all directions, but the reflected light in any given direction is not great and brightness is relatively low.   *Specular* reflectors (such as glossy paint) act in the manner of a mirror and reflect light away from the source at the same angle as that between the source and a perpendicular to the reflecting surface.   Such a surface will appear even darker than the diffuse surface except for the brilliance produced at the given angle of reflection. *Retrodirective* reflectors return light toward the source and show their maximum brilliance when viewed from a point at or near the light source. This form of reflection is employed in reflectorized signs and markings. In signs, either ground or message may be reflectorized to provide brightness and needed contrast. In markings, the marking or delineation is reflectorized.

The effectiveness of reflectorization is measured by the light returned, usually expressed as the ratio of reflected-light intensity to incident-light intensity at specified angles of incidence and reflection. The angle of incidence of light is referred to as the entrance or acceptance angle, measured in degrees between the axis of the sign, which is normal to its surface, and the direction of light, which is incident. The angle of reflection is referred to as the divergence angle and is measured in degrees between the direction of incident light on the sign and the observed line of sight. The divergence angle is usually small, since the distance from vehicle head lamps to road user's eye is small. Entrance angles are relatively large, varying with the longitudinal distance between the head lamp and the sign or marker itself (increasing as the vehicle approaches a sign and decreasing as the vehicle approaches a pavement marking). Characteristic curves of intensity are shown in Fig. 20-9. Intensity decays with acceptance angle and falls off very rapidly with divergence angle.

Pavement-marking reflectorization requires material of quite a different acceptance angle than is required for signs. The traffic engineer may have greater concern for attention-compelling value than for legibility, the former requiring high intensity, the latter requiring optimum contrast to avoid excessive irradiation or halation. Diverse specifications and manufacturers of retrodirective materials are found in practice.

CHAPTER 21

TRAFFIC SIGNALS

At intersections, where the demand for movement with safety heavily taxes the supply of road space—especially during certain periods of the day—the right of way is frequently assigned alternately to the diverse traffic flows. Any control device used for this purpose must not only fulfill the four basic requirements (attention, meaning, time for response, and respect), but in addition must apportion available time among road users. Traffic signals are widely used in the assignment of right of way at intersections.

First applied in London in 1868, signal devices (beginning with manually operated semaphores) have passed through many stages of development. To furnish adequate visibility at night, manually operated semaphores were first supplemented and then gradually replaced with lights. Whistles, gongs, and vibratory bells were employed in early developments for attention-compelling purposes. Manual operation was gradually replaced with automatic mechanisms, which gave way to present standardized stationary lighted lenses of changing aspect.

21-1. Basic Requirements. *Attention.* The design of the signal head and face, its mounting, and relative position are prime factors in compelling attention. Present standards call for 8-in. lenses lighted by 40- to 100-watt incandescent lamps with reflectors, special 60- and 67-watt lamps being generally used to concentrate light into the most effective pattern of distribution. Experience has shown that significant benefits are gained from the use of more than a single face on high-speed, especially wide thoroughfares. Two or more faces are required for urban traffic in present United States standards.

For contrast and attention value, dark backgrounds are required for lighted lenses. On urban thoroughfares, especially at night, the glare of red-neon advertising signs may destroy contrast. Dark-colored baffles are sometimes used under these conditions. Under daylight conditions, however, the contrast of a lighted lens with its background is greatly decreased owing to the high illumination of the background by strong sunlight or actual glare from the rising or setting sun. While dark (green or black) color is generally employed for signal heads and their supports,

324

highway yellow, sometimes diagonally striped with black, is occasionally used because of its added attention value.

While the relative position of the lighted lens in the plane of sight is of paramount importance for reasons of attention, the relative longitudinal position of the lens in the plane of sight changes with the relative longitudinal position of the road user, and an optimum position in the plane of vision throughout the intended range of effectiveness must be maintained.

Because of the diversity of actual conditions, it is extremely difficult to adopt a single standard of signal-head position to compel attention. The form and arrangement of intersections, the width and gradient of approaching roadways, the speed and composition of traffic movement, and other factors, including roadside development, affect each location so that positioning of the signal head is a unique problem at each installation point.

In the design of the signal head and optical system, caution must be taken that a road user in any given stream of traffic will not have his attention diverted to an indication not intended for his use. Louvers and shields are sometimes attached to indicate clearly the traffic stream for which a particular signal head is intended.

Meaning. Generally speaking, red "stop" lights prohibit entry into the intersection. Yellow allows entry but requires clearance of the intersection before the yellow expires, and the green "go" signal permits entry. The green-arrow symbol permits vehicle entry for movement in the direction indicated only. Word messages—"walk," "wait," and "don't walk"—are employed for the benefit of pedestrians.

The relative vertical or horizontal position of light messages in a signal assembly has also been standardized to convey meaning, as a particular aid to the color-blind. United States standards specify the following relative positions (top to bottom or left to right): (1) red, (2) yellow, and (3) green, and the optional addition of (4) straight-through arrow, (5) left-turn arrow, (6) right-turn arrow, (7) wait (don't walk), and (8) walk. The display of two or more signal lights at the same time will change the basic meaning of any single light and give a new meaning to such light when shown in combination with another lighted indication.

In standard United States practice, the red, yellow, and green lights have general applicability to all road users and are not allowed to be shown in combination with each other. However, subclasses of road users may be given special permissive or modifying messages: green-arrow messages, applying only to vehicle operators, and word messages, applying only to pedestrians, are permitted in combination with the red light. These special modifying indications take precedence over the general indications.

Meaning of signal messages is also modified by the relatively rapid flashing of a given light. A steady red light prohibits intersection entry, but a flashing red light merely requires a momentary halt before entry. Similarly, flashing yellow is used for warning only. Flashing green is also employed but is not a recognized standard.

Lane-direction signaling is employed to limit signal meaning for unbalanced flow control by positioning a signal over each lane at suitable spacings (1,000 ft or less) and showing a continuous steady light.

Time to Respond. Since it takes time to stop a moving vehicle, an instantaneous change from green to red ignores natural physical laws. Again, once a road user has been permitted entry into the intersection under signal control, safety demands that he be allowed comfortable journey time to traverse the intersection before cross traffic is released. The yellow light provides for this change in the assignment of right of way in stop-and-go operation. Two criteria are employed in computing the duration of the yellow light, (1) stopping time and (2) clearance time. The time required to stop is PIEV time plus braking time (see Chapter 2). Required clearance time varies directly as the sum of the comfortable stopping distance plus the intersection width, and inversely as the speed of travel of a given vehicle. Clearance time may be reduced in some instances, especially at intersections with irregular layouts, where significant approach time is required by a released cross flow in its acceleration to the collision point.

As speed of travel increases, stopping time increases, and clearance time increases at a lesser rate. Where t_1 is time to stop, t_2 is time to clear, P is PIEV time, S_1 is distance to brake, S_2 is intersection width, and V is speed of travel, then:

$$t_1 = P + \frac{S_1}{V/2} = P + \frac{2S_1}{V} = \left(P + \frac{S_1}{V}\right) + \frac{S_1}{V}$$

or

$$\left(t_1 - \frac{S_1}{V}\right) = \left(P + \frac{S_1}{V}\right)$$

but

$$t_2 = \left(P + \frac{S_1}{V}\right) + \frac{S_2}{V}$$

hence

$$t_2 = t_1 - \frac{S_1}{V} + \frac{S_2}{V}$$

When distance to brake exceeds the intersection width S_2, stopping time t_1 is therefore greater than clearance time. Since the distance to brake increases as the square of the speed while intersection width remains constant, time to stop becomes the critical value in determination of yellow light at higher speeds, though time to clear may be the critical value at lower speeds.

In urban areas where speeds are relatively low, yellow lights of about 3-sec duration are satisfactory at most locations. At rural, high-speed locations where stopping time may have a duration of 5 to 8 sec, road users tend to attempt to clear the intersection rather than stop. Five seconds is probably a practical maximum yellow duration in such location. Time for response in starting must be included in any green "go" interval (Table 7-1).

Where sight distance is low or obstructed, auxiliary equipment must be used to give advance warning time to road users.

21-2. Time Apportionment. When signals are used in stop-and-go operation, it is necessary to determine how much of the total time available at the intersection will be apportioned to each *flow* of traffic movement. At a given intersection the total traffic is broken up into *phases* of movement in which one or more *flows* will take place. Certain flows are ascribed in the right of entry to the intersection and are given a green, yellow, or "walk" signal aspect, while all other flows are stopped and given a red or "wait" ("don't walk") signal aspect. The selection and arrangement of simultaneous flows of movement is known as "phasing."

The objective of phasing is to accommodate all traffic movement with increased safety and minimum delay. Safety urges a phasing which will reduce or eliminate all potential conflict. Consideration for minimum delay impels a phasing which will accept as many simultaneous flows as practicable, to achieve a high volume of accommodation. The use of *leading* or *split-phase* green to accommodate turning or other irregular movements is an example of ingenious phasing.

The sequence of phase occurrence in multiphase operation may become an important factor in irregular intersections in which consideration is given to potential interphase conflicts and the time-space relationships between sequential flows of separate phases. Illustrative examples of phasing are shown in Fig. 21-1.

Generally speaking, the number of distinct phases employed should be kept at a minimum consistent with safety and facilitation. The selection of flows in each phase should develop the minimum frequency and severity of conflict, and the sequence of phases should minimize waste of time.

The shortest complete sequence of phases constitutes a *cycle of operation*. Ordinarily each phase will appear only once in each pattern, but under some circumstances a given phase may be repeated within the cyclic course of operation (as in A, B, C, B; A, B, C, B). Each phase duration has its "go" interval and usually, though not always, its "clearance" (yellow) interval. The total time required for the complete sequence of phases is known as *cycle length*. Phase durations, intervals, and cycles are ordinarily expressed in seconds (Fig. 21-2).

Delay. When a red signal interrupts a flow of traffic, the vehicles stopped by the signal will require time to get started and under way again, so that an additional number of vehicles may be stopped because of the starting performance of the queue which has been accumulated on the red signal. The number of vehicles which will be stopped or delayed

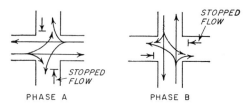

PHASE A   PHASE B
USUAL PHASING OF TRAFFIC FLOW AT ORDINARY
"RIGHT - ANGLE" INTERSECTION. TWO, TWO-WAY TRAFFIC STREAMS

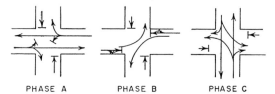

PHASE A   PHASE B   PHASE C
USE OF THREE PHASES AT ORDINARY, "RIGHT-ANGLE"
INTERSECTION. TWO, TWO-WAY TRAFFIC STREAMS WITH
SEPARATE PHASE PROVIDED FOR LEFT TURNS

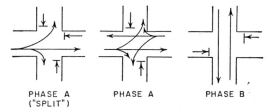

PHASE A   PHASE A   PHASE B
("SPLIT")

(TURNS IN ONE   "TWO" PHASES, UTILIZING "SPLIT"
DIRECTION RECEIVE   FOR HEAVIER TURNING MOVEMENT
PRIORITY GREEN)

FIG. 21-1. Phasing of traffic flows.

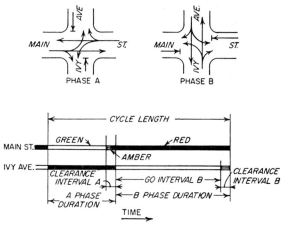

FIG. 21-2. Signal aspect.

and the duration of their delay are dependent on the red interval, the arrival headways in the flow, and the starting performance of the queue.

Let R = interval of stop signal aspect, sec

  n = number of vehicles stopped in R

  i = any selected one of the n vehicles

  A = average headway of vehicles on arrival, sec

  D = headway of departure at intersection entry

  (NOTE: D is known from experimental data to vary for successive vehicles up to $n = 6$ (see Chapter 8) and then remain constant for $n \geqq 6$)

  d_i = delay, sec, for vehicle i

  T = total delay, sec, for all vehicles

On the average, the first vehicle will be stopped at the point of interruption for duration $R - (A/2)$. The second vehicle will arrive A sec later and will be delayed accordingly $R - (A/2) - A$ or $R - (3A/2)$ sec. The third vehicle will arrive an additional headway later and will thus be delayed $R - (3A/2) - A$ or $R - (5A/2)$. Therefore, the ith vehicle will be delayed by the red light $R - [A(2i - 1)/2]$ sec.

But after the red light changes to green, there is an additional delay for each vehicle, dependent on the starting performance of the entire queue of vehicles which has been stopped by the red light, including the given vehicle. Since D is a variable, this additional delay for a given vehicle amounts to $\sum\limits_{x=1}^{i} D_x$, so that the delay to any vehicle becomes

$$d_i = R - \frac{A(2i - 1)}{2} + \sum_{x=1}^{i} D_x$$

NOTE: D_x is the departure headway of the xth vehicle.

The sum of the individual delays then becomes the total delay for n vehicles:

$$T = nR - \frac{n^2 A}{2} + \sum_{x=1}^{n} \sum_{x=1}^{i} D_x$$

NOTE: The sum of $\sum\limits_{i=1}^{n} \frac{A(2i - 1)}{2}$ is the sum of $(1 + 3 + 5 + 7 + \cdots + n) \frac{A}{Z}$

which is $\dfrac{n^2 A}{2}$

Now if D is assumed to be a constant:

$$T = nR - \frac{n^2A}{2} + \frac{n(n + 1)D}{2}$$

Note: The expression $\frac{n(n + 1)}{2}$ is the sum of $(1 + 2 + 3 + 4 + \cdots n)$ which is the coefficient of D.

But D in the expression above is a variable. Empirical values for median values of D found in practice for straight-through passenger automobiles are as follows:

When $x =$	$D_x =$	$\sum\limits_{x=1}^{i} D_x =$	or	$\sum\limits_{x=1}^{i} D_x =$
1	$3.8 = 2.1 + 1.7$	2.1	$+ 1.7 = 2.1$	$+ 3.7 - 2.0$
2	$3.1 = 2.1 + 1.0$	4.2	$+ 2.7 = 2.1i$	$+ 3.7 - 1.0$
3	$2.7 = 2.1 + 0.6$	6.3	$+ 3.3 = 2.1i$	$+ 3.7 - 0.4$
4	$2.4 = 2.1 + 0.3$	8.4	$+ 3.6 = 2.1i$	$+ 3.7 - 0.1$
5	$2.2 = 2.1 + 0.1$	10.5	$+ 3.7 = 2.1i$	$+ 3.7 - 0.0$
6	$2.1 = 2.1 + 0.0$	12.6	$+ 3.7 = 2.1i$	$+ 3.7 - 0.0$
7	$2.1 = 2.1 + 0.0$	14.7	$+ 3.7 = 2.1i$	$+ 3.7 - 0.0$
n	$2.1 = 2.1 + 0.0$	$2.1n$	$+ 3.7 = 2.1i$	$+ 3.7 - 0.0$

Accordingly, for passenger-car queues:

$$\sum_{x=1}^{n}\sum_{x=1}^{i} D_x = \frac{2.1n(n + 1)}{2} + 3.7n - Q$$

in which Q is always 3.5 for $n \geqq 4$ and

When $n =$	$Q =$
1	2.0
2	3.0
3	3.4
4	3.5

NOTE: These empirical values will vary due to composition of flow, turning vehicles, pedestrian interferences, and other factors.

The total delay becomes

$$T = nR - \frac{n^2A}{2} + \frac{2.1n(n + 1)}{2} + 3.7n - Q$$

An example of the delay caused by a signal interruption is set forth graphically in Fig. 21-3, which assumes the following conditions: $R = 40$ sec, $A = 10$ sec or 360 passenger vehicles per hour, $D =$ passenger-car performance—empirical (with no undue delay caused by left turns, pedestrians, heavy vehicles, etc.). It will be seen from Fig. 21-3 that four vehicles were stopped by the red light alone. Two additional vehicles were stopped by the queue formation and its clearance. Thus, six vehicles were delayed for a maximum of 38.8 sec for the first vehicle to a minimum of 1.3 sec for the last vehicle. The seventh vehicle, though not forced to stop, had its headway reduced from 10 to 6.6 sec. Total delay T increased to a maximum of approximately 123 sec when the sixth vehicle entered the intersection.

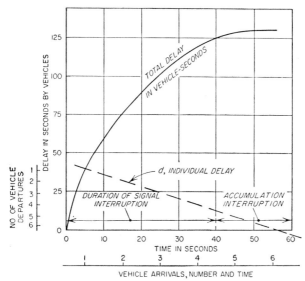

Fig. 21-3. Total delay—vehicle departures to vehicle arrivals.

In any case, there is a unique number of vehicles n which will be delayed. This number occurs when total delay T is at its maximum. The point of maximum delay may be found by taking the first derivative of T with respect to n and equating to zero as follows:

$$T = nR - \frac{n^2A}{2} + 2.1n(n + 1) + 3.7n - Q$$

$$\frac{dT}{dn} = R - nA + 2.1n + 1.05 + 3.7 = 0$$

Hence,

$$n = \frac{R + 4.75}{A - 2.1}$$

In the above example shown in Fig. 21-3:

$$n = \frac{40 + 4.75}{10 - 2.1} = 5.6 \text{ or } 6 \text{ vehicles}$$

Thus the number of vehicles n which will be delayed varies directly with the sum of the duration of the red signal plus a constant composed of the time losses in starting of the queue of accumulated vehicles, and inversely as the difference between arrival headways and departure headways.

As R increases, the number of delayed vehicles increases, and as the sluggishness of departure increases, delay increases, as evidenced by the appearance of the factor of sluggishness in the numerator and departure

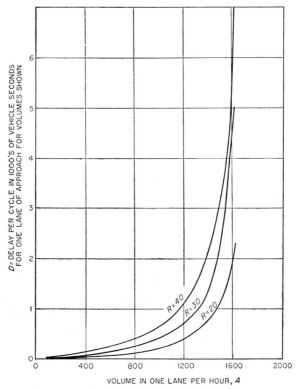

FIG. 21-4. Relationship of delay to volumes (ideal passenger-car performance).

headway in the denominator. The influence of volume is also apparent, remembering that A decreases with increased volume. While the influences of R and A are not so readily apparent in the equation for delay, the effect of sluggishness on departure is evident. Figure 21-4 shows the relationship of delay to volume for ideal passenger-car performance and a given value of R. Delay increases asymptotically as volume approaches a maximum of $3,600/D$, the knee of the curves occurring at about 1,200 vph, or A values of about 3 sec.

The increase in delay for a given volume caused by increasing R is shown in Fig. 21-5. The length of queues which will form for an average interruption of $R = 30$ is shown in Fig. 21-6. These queue lengths

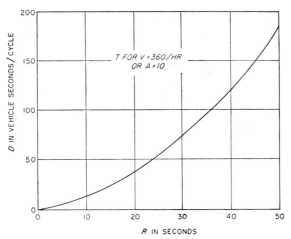

FIG. 21-5. Increase in delay for a given volume.

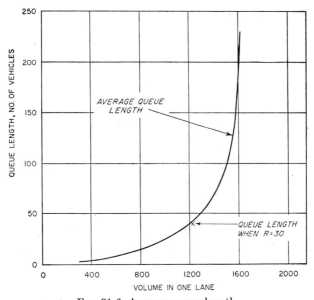

FIG. 21-6. Average queue length.

become very large (over 15) at 800 vph, and they are excessive (over 40) at 1,200 vph.

If the derived value for the number of vehicles delayed per signal interruption is inserted in the equation for delay, and it is then divided by the number of delayed vehicles, the average delay per vehicle becomes

$$\frac{T}{n} = \frac{R + 4.75}{2} - \frac{Q(A - 2.1)}{R + 4.75}$$

It is only at extremely low volumes that the negative term of this equation becomes significant in value. For nearly all volumes, therefore, the average delay per delayed vehicle is approximately $(R + 4.75)/2$. This assumes that every delayed vehicle is allowed to pass on the next "go" interval, as Fig. 21-7 indicates.

21-3. Cycle Length. A primary objective in the signalized operation of an intersection is to achieve the required capacity for each flow with a minimum of delay. From the previous discussion it is to be noted that

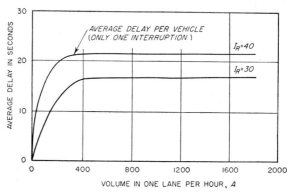

Fig. 21-7. Average delay per delayed vehicle:

$$\frac{T}{n} = \frac{R + 4.75}{2} - \frac{Q(A - 2.1)}{R + 4.75}$$

the least delay is incurred by a given flow when R is as short as possible. But the R for a given phase is dependent on the sum of the "go" intervals (including clearance intervals) for all other phases. At the same time, the "go" interval for any phase must accommodate all vehicles which arrive for that phase of movement during the cycle. The minimum length of "go" interval for any phase should allow all vehicles in the given phase to pass at minimum headways of departure. Minimum headways of departure are achieved when all vehicles for a given phase have been brought into a queue of stopped vehicles awaiting release. When each vehicle departs with a minimum headway, the maximum capacity of the signalized intersections will result.

Theoretical Cycle Computations. The number of vehicles n^1 which will arrive for any phase during a given cycle is

$$n^1 = \frac{C}{A}$$

where C = cycle length in seconds.

It will be recalled, however, that the number of vehicles which accumulate for a given interruption is $n = (R + K)/(A - D_k)$.

NOTE: $K = 4.75$ for typical passenger-car performance in previous examples, and D_k is the constant value of D, here $= 2.1$ sec.

The duration of R which will stop all arriving vehicles, therefore, becomes

$$R = n(A - D_k) - K$$

But for any given phase, n should equal n^1, under which condition the required capacity will be achieved with the shortest possible R. For any phase, then:

$$R = \frac{C}{A} (A - D_k) - K$$

The total length of a cycle must accommodate the R's of all phases. The cycle length also will be equal in duration to the sum of all G's (green intervals). Hence:

$$C = G_1 + G_2 + G_3 + \cdots , \text{ etc.}$$
$$= \sum_{x=1}^{\phi} G_x$$

where ϕ is the number of phases.

But for any phase, $G = C - R$.

Hence,

$$C = (C - R_1) + (C - R_2) + (C - R_3) + \cdots$$
$$= \phi C - (R_1 + R_2 + R_3 + \cdots)$$
$$\therefore C = \frac{R_1 + R_2 + R_3 + \cdots}{\phi - 1}$$

Hence a cycle length is the sum of the individual *red intervals* in seconds divided by the number of phases less 1, which is also the sum of all green intervals.

Substituting for R,

$$C = \frac{\left[\frac{C}{A_1}(A_1 - D_{k1}) - K_1\right] + \left[\frac{C}{A_2}(A_2 - D_{k2}) - K_2\right] + \cdots}{\phi - 1}$$

$$= \frac{\displaystyle\sum_{x=1}^{\phi} \int K_x}{\left(\displaystyle\sum_{x=1}^{\phi} \frac{A_x - D_{kx}}{A_x}\right) - (\phi - 1)}$$

For two-phase operation this simplifies to

$$C = \frac{K_1 + K_2}{\left(\dfrac{A_1 - D_{k1}}{A_1} + \dfrac{A_2 - D_{k2}}{A_2}\right) - 1}$$

If ideal pure passenger-auto performance is assumed for each phase, D_k becomes 2.1, and K becomes 4.75, hence,

$$C = \frac{9.5}{\left(\dfrac{A_1 - 2.1}{A_1} + \dfrac{A_2 - 2.1}{A_2}\right) - 1}$$

But since A (seconds per vehicle) is the reciprocal of volume in *vehicles per second*, $A = 3{,}600/V$, where V is volume in vehicles per hour. In general, then,

$$C = \frac{3{,}600 \displaystyle\sum_{x=1}^{\phi} K_x}{3{,}600 - \displaystyle\sum_{x=1}^{\phi} V_x D_{kx}}$$

and for ideal passenger-auto performance with two-phase operation,

$$C = \frac{34{,}200}{3{,}600 - (V_1 + V_2)2.1}$$

When the sum of vehicles per hour in both phases of operation produces volumes approaching $3{,}600/D_k$ (or one vehicle each 2.1 sec) over the collision point, capacity of the intersection will have been achieved, and the cycle length becomes infinitely large. Minimum cycle length is reached when volumes approach zero and C approaches the time loss due to sluggishness in starting. A given volume on one phase establishes the limit of volume in the other phase.

The cycle lengths indicated for various total volumes are shown in Fig. 21-8. When maximum limits of practical cycle lengths (about 100 sec) are reached, maximum volumes of about 1,500 passenger autos per hour are developed (theoretically) over a given conflict point. This value may be approached in practice.

From the above calculations of cycle length, four factors of concern are apparent:

1. The sluggishness factor in starting a queue of stopped vehicles
2. The headways of arrival, or volume
3. The headways of departure
4. The number of phases

As sluggishness is increased because of left turns, heavy vehicles, pedestrian interference, etc., the required cycle length increases; as headways of arrival decrease, the required cycle length increases; as headways of departure increase, required cycle length increases; as the number of phases increase, the required cycle length increases. And, in any event, a given cycle length has its equivalent capacity which cannot be exceeded.

Theoretical Cycle Length (Random-arrival Headways). Since traffic streams tend toward a random arrangement of vehicles, in any interval of

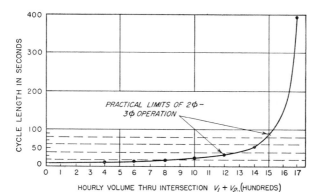

FIG. 21-8. Theoretical cycle lengths based on

$$C = \frac{K}{1 - (V_1 D_1 + V_2 D_2)}$$

where $K = 9.5$, $D_1 = D_2 = 2.1$, $V_1 + V_2 =$ vehicles per second.

time the number of vehicles arriving at a given point can vary considerably from the average used in the uniform-arrival method developed above. The extent of this variation and the probability of its occurrence can be estimated by use of the Poisson distribution function (see Chapter 7):

$$P_x = \frac{e^{-a}a^x}{x!}$$

Thus a cycle composed of intervals constructed to accommodate only the average number of vehicles may be (1) too long when fewer than average number of vehicles arrive or (2) too short when a greater than average number of vehicles arrive. In the former case, green time may be lost and resultant capacity impaired; in the latter case, vehicles will accumulate and incur additional delay. In order to achieve required capacity and avoid needless accumulations with incident delay, the cycle must necessarily be increased.

The average number of vehicles a which will arrive in a given cycle length C is dependent on volume V. The maximum number of vehicles x which might arrive in a given cycle length C is not only dependent on *a* but also on the detailed longitudinal arrangement of the vehicles in the flow. Since this vehicle arrangement can be generally assumed to follow Poisson's law of distribution, the probability P of any number x or less arriving on any cycle may be computed. This probability P may be interpreted as the percentage of all cycles in which x or less vehicles will actually arrive. If x is the capacity of a particular cycle length, the remaining percentage of cycles will receive more than their capacity and accumulation will result.

It is obvious that for any reasonable cycle length some failures must be expected. The number of such failures that are tolerable may be selected as a criterion for cycle length. Accordingly, it is necessary to adopt some value of P which will be found satisfactory in practice. If the length of a cycle is such that a given "go" interval within it has just enough capacity to pass the average number (or less) of vehicles which may arrive in cycle C, the cycle may fail to yield adequate capacity up to half the total time. This will lead to excessive accumulation of vehicles waiting for the "go" interval, and delay will be needlessly high.

For slight increase in cycle length, relatively high increases are achieved in the probability that each interval will be able to discharge the actual number of vehicles which will arrive during the cycle. The question thus asserts itself as to the probability value to be employed which will actually incur the minimum delay. Practice indicates that when P values are established so that about 95 cycles out of each hundred will not fail to discharge as many vehicles as arrive, satisfactory performance is achieved.

The determination of cycle length, assuming random-arrival headways, can best be illustrated by a practical example:

Problem: At a given intersection, which is to be placed in two-phase operation, the critical lane of approach carries 600 vph in phase 1 and 200 vph in phase 2 during peak periods. A cycle length is desired in which each "go" interval will accommodate as many vehicles as arrive in the cycle about 95 per cent of the time.

Solution: If uniform spacings in each flow were assumed, the theoretical cycle length would be computed as follows:

$$C = \frac{34,200}{3,600 - (600 + 200)2.1} = 17.8 \text{ sec}$$

Then phase 1 is expected to handle a vehicles, or $(600 \times 17.8)/3,600$, or 2.97 vehicles per cycle, and phase 2 is expected to handle a_2 vehicles, or $(200 \times 17.8)/3,600$, or 0.99 vehicle per cycle.

If 4.75 sec is allowed as the K factor to get each flow under way, and headways of departure are assumed to be 2.1 sec, phase 1 will be $(2.97 \times 2.1) + 4.75$, or 10.95 sec, and phase 2 will be $(0.99 \times 2.1) + 4.75$, or 6.85 sec.

Thus the capacity of each of these "go" intervals is exactly equal to the *average* number of vehicles which arrive; that is (in whole numbers), phase 1 will accommodate 3 vehicles and phase 2 will handle 1 vehicle per cycle. The probability that these intervals will accommodate the actual number of vehicles which might arrive if random spacings are assumed may be computed by summing the probabilities of x vehicles arriving between $x = 0$ and $x =$ capacity of the respective interval.[1]   When this

TABLE 21-1. CYCLE COMPUTATION
(Illustrative Example)

Values of C, sec	Phase 1					Phase 2				
	Go, sec	Stop, sec	Cap. veh.	Arrival veh.	Values of P	Go, sec	Stop, sec	Cap. veh.	Arrival veh.	Values of P
18.0	11.1	6.9	3	2.97	0.64	6.9	11.1	1	0.99	.74
22.2	13.2	9.0	4	3.7	0.69	9.0	13.2	2	1.2	.88
26.4	15.3	11.1	5	4.4	0.72	11.1	15.3	3	1.5	.94
28.5	17.4	11.1	6	4.8	0.80	11.1	17.4	3	1.6	.92
32.7	19.5	13.2	7	5.5	0.81	13.2	19.5	4	1.8	.96
34.8	21.6	13.2	8	5.8	0.87	13.2	21.6	4	1.9	.95
36.9	23.7	13.2	9	6.2	0.90	13.2	23.7	4	2.0	.95
39.0	25.8	13.2	10	6.5	0.93	13.2	25.8	4	2.2	.93
43.2	27.9	15.3	11	7.2	0.94	15.3	27.9	5	2.4	.96
45.3	30.0	15.3	12	7.6	0.95	15.3	30.0	5	2.5	.96

is done, P for phase 1 is 0.64 or 36 per cent failure; P for phase 2 is 0.74 or 26 per cent failure.   The cycle is obviously too short, and this percentage of failures will create staggering accumulations of vehicles.

By increasing interval lengths to the point where their capacities are at least as great as the maximum number of vehicles which might arrive during 95 per cent of the time, more practical cycle lengths will be found. In Table 21-1, theoretical cycle length has been extended from 18 sec to 45 sec to obtain the desired level of satisfactory performance.   For the ideal conditions of pure passenger-car operation with no interferences from turns or pedestrians and no losses of time due to inattention or erratic behavior, this becomes the proper cycle length for the indicated volumes.   Similar computations for other values lead to the results presented in Fig. 21-9.   For two-phase operations the correct theoretical cycle length for any pair of volumes may be found from this figure.

Practical Cycle Length.   All previous computations have been based on passenger-car operations with no time losses other than those inherent in

---

[1] Such computations are tedious and lengthy.   These summations may be found in E. D. Moline, *Poisson's Exponential Binomial Limit*, D. Van Nostrand Company, Inc., New York, 1945.

the requirement for setting a queue in motion. Studies have shown that these losses, which average nearly 5 sec for every phase, adequately account for inertia of starting and chance variations in individual performance. There are two phenomena occurring at nearly every intersection, however, which must be given special consideration: left turns and the presence of trucks and other large or sluggish vehicles in the traffic flows.

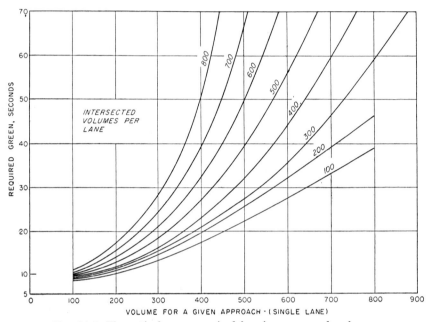

FIG. 21-9. Theoretical green required for given approach volume.

Studies have shown that because of delay enforced by opposing traffic, left turns generally require an average of 1.3 additional sec to clear the intersection. In view of the fact that this additional required time is imposed upon the entire stream, it becomes the equivalent of an additional partial vehicle in the flow requiring only 1.3 sec to pass instead of the basic 2.1 sec. Thus a left-turning vehicle may be thought of as the equivalent of $(2.1 + 1.3)/2.1 = 1.6$ straight-through vehicles.

In computing cycle lengths, then, a flow of 400 vph with 10 per cent left turns may be thought of as an equivalent volume of

$$(400 \times 0.90) + (400 \times 0.10 \times 1.6) = 424 \text{ vph}$$

For sluggish or large vehicles, it has been found that each truck or bus consumes approximately 1.5 times the amount of departure time required for a passenger vehicle. Each such vehicle is equivalent to 1.5 passenger vehicles.

For a flow of 400 vph, of which 10 per cent is trucks, the equivalent value becomes $(400 \times 0.90) + (400 \times 0.10 \times 1.5) = 420$ vph.

These equivalent volumes then may be used directly in connection with Fig. 21-9 to estimate a required cycle length adapted to more realistic conditions.

Assume that in the previously discussed example of 600 vph intersecting 200 vph there were 15 per cent left turns and 10 per cent trucks in each flow. The equivalent volumes are 684 vph and 228 vph. With these two equivalent volumes, Fig. 21-9 indicates that 38 sec of "go" interval is required for the major phase, and 17 sec for the minor phase. The total practical cycle length is, therefore, 55 sec. If such cycle length is used, it may be expected that not more than 5 times out of 100 would more vehicles arrive on either approach during one cycle than the phase for that approach could discharge. Nominal lengthening of the cycle would not greatly reduce this failure ratio, whereas only slight shortening would greatly increase it with resultant large accumulations and additional delays.

Cycle-length Limitations. All foregoing computations assume that it would be feasible to install any cycle length desirable under the stated criteria. There are, however, practical considerations which are generally accepted as limiting maximum and minimum cycle lengths which may be employed.

The needs of pedestrians have a profound effect upon minimum cycle lengths. Studies have shown that the average pedestrian walking properly in the crosswalk with the "go" interval walks approximately 3 to 4 ft/sec. In addition, pedestrians exhibit a starting time requirement of 3 to 4 sec after receiving the "go" indication. It is axiomatic that at least those pedestrians who are waiting at the curb must be allowed time to cross the roadway comfortably. Thus, roadway width and pedestrian speed become the limiting factors in minimum interval length. For ordinary road width of 30 to 40 ft, the interval must be 13 to 20 sec in length. This minimum length should be computed individually for every signal-timing problem. In general, it has been found that rarely is an interval of less than 15 sec permissible, resulting in a minimum practical cycle length of 30 sec.

Whenever any green interval is established on pedestrian clearance time, rather than on a vehicle-accommodation basis, it is necessary to make compensating increases of green interval(s) in the complementary phase(s) of a given cycle. In the use of Fig. 21-9, entry is made from the ordinate values of green time, and equivalent volumes will be read on the abscissa scale. For example, at the intersection of 200 vph with a stream of 500 vph, 13 sec is indicated for the lesser stream. If 15 sec is required for the sake of pedestrian accommodation, the equivalent vehicle flow will be 250 instead of 200, and the larger stream will require 26 sec instead of 24.

Maximum cycle lengths do not have as well-defined limits as do minimum lengths. Examination of Fig. 21-9 shows that total volumes in excess of 1,100 to 1,200 vph require cycle lengths which increase greatly for only small increases in volume. Even at a total of 1,200 vph, cycle lengths are from about 135 to 160 sec, depending upon the proportion in each flow. It is recognized that with long cycles such as these, psychological factors of annoyance and impatience play an increasingly important part in driver's attitude toward signal performance. For these reasons good practice indicates adoption of an upper limit of about 120 sec for cycle length. This admittedly increases the failure ratio of the cycles, but at those levels that seems less serious than imposing longer waits for any vehicle. It should be understood, however, that cycles set at this length, where much longer cycles are indicated, cannot be expected to give as good performance as could be desired. In such a case the intersection is being forced to handle volumes in excess of its practical signalized capacity, and improvements beyond mere signalization are indicated.

Multiple-phase Operation. All foregoing computations may be applied to finding cycle lengths for situations in which more than two phases are to be employed. The method outlined in Table 21-1 may be extended to any desired number of phases. The data of Fig. 21-9 are not applicable, but may be used for a first try to obtain a reasonable cycle length on which to base the more tedious computations previously shown in Table 21-1.

An example is shown below:

Phase	*Volume*
A	200
B	300
C	400

In Fig. 21-9:

Consider:	*Required green*
200 vs. 700	16
300 vs. 600	24
400 vs. 500	27
Total	67

Cycle	Phase A				Phase B				Phase C			
	Go	Cap.	Av.	Prob.	Go	Cap.	Av.	Prob.	Go	Cap.	Av.	Prob.
68.05	17.35	6	3.8	0.91	23.65	9	5.7	0.93	27.05	11	7.7	0.91
74.35	19.45	7	4.1	0.94	25.75	10	6.6	0.95	29.15	12	8.3	0.92
80.65	21.55	8	4.5	0.96	27.85	11	6.7	0.96	31.25	14	9.0	0.96

The required cycle found by considering each of the flows as opposed by the total flow of the other phases is not long enough to satisfy the probability requirements. Furthermore, the difference cannot be accounted for solely by consideration of the additional 5 sec lost time imposed by the third phase. A given volume, then, requires more time when split into more than one flow than it does when kept as a single flow, owing solely to the effects on the probabilities of arrivals. This is a strong argument in favor of using as few phases as possible for any given set of conditions.

21-4. Signal Coordination. Changes caused by a signal installation in the longitudinal distribution of traffic from *random* to *platoon* flow profoundly affect the flow through nearby signalized intersections. Adjacent signals, therefore, should be coordinated. When two or more signal

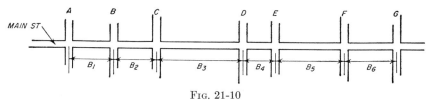

Fig. 21-10

installations on any single route have a fixed time relationship among the several intervals, a signal system is created. Coordination of signals may be achieved by the operation of two or more signalized intersections as a system in which there exists a desirable and definite time relationship among the diverse phases at all intersections, that is, when the flow of traffic on a given phase of movement at one intersection is accommodated by a "go" phase on its arrival at the next signalized intersection.

Elementary Relationships. Perhaps the simplest form of desirable coordination, and one which is readily attainable, is illustrated by the coordination of signals on a one-way street (Fig. 21-10). Signals are installed at each intersection and traffic flows one way only from A to G at a speed of S mph. Let B_1, B_2, B_3, B_n be the block lengths in feet between adjoining signalized installations. The "go" phase for Main Street at station A may be initiated at any instant, known as an instant of *time reference*. But on release of a *wave* of Main Street flow at A, there develops a logical lapse in time called an *offset* for the initiation of the Main Street "go" interval at subsequent stations B, C, D, . . . , G. This offset is such that for each subsequent station the offset in seconds will vary directly as the length of blocks B and inversely as the speed of movement S. If B is in feet and S in miles per hour, offset $L = 0.68B/S$.

If all block lengths are uniform, the offset from station to station is equal, provided speed remains constant. But regardless of block length or speed, the offset for each station should be such that the "go" interval for Main Street appears at the moment the lead vehicle of a wave or "platoon" of Main Street traffic approaches. This process for dealing with one wave or platoon of Main Street traffic may be repeated again and again, as Fig. 21-11 illustrates:

1. An infinite number of patterns of coordination may be developed for a given signal system by variation of cycle lengths C and offsets L.

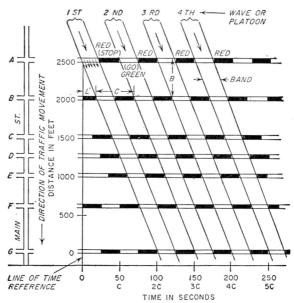

Fig. 21-11. An elementary time-space diagram for signal coordination on a one-way street.

1. The cycle length C must be equal for each intersection.

2. The offset L may vary due to block length and/or speed (throughout the entirety or any part of the system).

3. While B is fixed for any case, the speed S may be varied to suit circumstances by changing L, since $L = KB/S$ and $S = KB/L$.

4. The "band" marks the limits of vehicle passage in each wave or platoon. Its limits are fixed by the minimum "go" interval on Main Street. It is a measure of the capacity of progressive movement since the duration of each band limits this flow. This capacity equals band width, $\sec/H_a$.

2. There is a unique pattern of coordination for a given signal system which is best suited for a given speed of traffic flow from station to station.

3. For a given pattern of coordination, and in order similarly to accommodate successive platoons of traffic movement, cycle length must remain constant throughout the system.

4. In any interstation section or block B there exists a mathematical relationship between the maximum speed of progression S and the offset L, so that the average speed at which traffic may move when progressing

through the system at its maximum over-all speed is:

$$S = \frac{0.68B}{L}$$

where S = speed of progression, mph
    B = block length or interstation distance, ft
    L = offset, sec, and is less than C

There are other possible rates of traffic speed between any pair of signals which will permit traffic movement without signal interruption to progress, in which the average speed $S' = 0.68B/(L + nC)$, where n is any whole number.

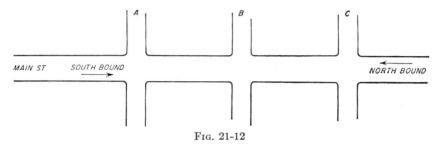

Fig. 21-12

5. For any given cycle length there is developed a *band* of progression within which traffic may move continuously on a "go" signal. This band, measured in seconds, is a measure of the progressive capacity of the system, so that $V = 3,600W/CD$, where V is volume in vehicles per hour, C is cycle length in seconds, W is band width in seconds, and D is headways of departure at signal.

6. The intersection demanding maximum time for cross flows or red intervals establishes the critical dimension of band width and system capacity.

Progression in Two-way Systems.   When traffic movement is permitted to flow in both directions, there is enforced a symmetry of pattern of the time-space diagram which is not required in one-way operation (Fig. 21-12).

Assume that a wave of traffic leaves station A, southbound toward B, at a given instant O. If a wave of traffic leaves station C, northbound toward B at the same instant, both waves will arrive at B some time later. If the speeds of both waves are equal, and if the block lengths $A–B$ and $B–C$ are equal, both waves will arrive at station B at the same instant, $L = 0.68B/S$ sec after leaving stations A and C. If the Main Street "go" interval is initiated at B, at this instant, L progressive movement will be achieved for both directions of movement. However, the arrival times of both waves at B are the same only if block lengths and speeds are equal and if starting times for both waves are simultaneous.

Case I. Equal speed and uniform block length. When the block lengths between stations are equal and speed of travel in each direction is sustained and of equal rate, the pattern of progression described above for a series of three signal stations may be extended indefinitely, as shown in Fig. 21-13. It is clear that the lag or offset at the beginning of the "go" intervals from block to block in either direction of travel must equal exactly some multiple of $C/2$ under these conditions.

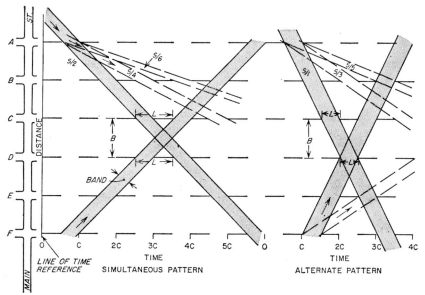

Fig. 21-13. Simultaneous and alternate patterns, equal speed—uniform block length (Case I).

When the offset from intersection to intersection is made equal to $C/2$ the pattern is called *simple alternate* and the speed of progression becomes $S_a = 0.68B/(C/2)$. This is the fastest progressive speed which will yield capacity flow in two-way systems without interruptions.

When the offset from intersection to intersection is made equal to C, or $2C/2$, the pattern is called *simultaneous* and the speed of progression becomes $S_s = 0.68B/C$.

Secondary bands of progression exist where the speeds for the simultaneous pattern are $S_s = 0.68B/[n(C/2)]$, where n is any even integer, and in the alternate pattern $S_a = 0.68B/[n'(C/2)]$, where n' is any odd integer. For example, if for a given block and cycle length

$$S = \frac{0.68B}{[n(C/2)]} = 48 \text{ mph}$$

then progression with a simultaneous pattern may occur at 24, 12, or 8 mph and with the alternate pattern at 48, 16, or 9.6 mph.

There exists, therefore, a very exact relationship between traffic speed, intersignal distances, and cycle length and the elementary pattern of coordination (Fig. 21-14).

Case II. Uniform block lengths, unequal speeds. The symmetrical pattern of the time-space diagram which yields equal speed of progression

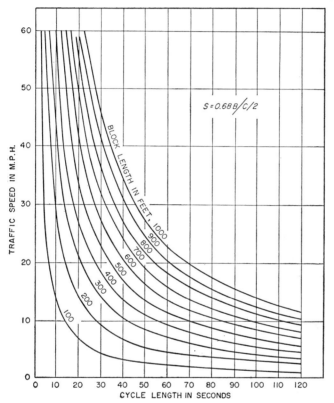

Fig. 21-14. Variation in traffic speed with respect to block cycle length.
NOTE: This chart is based on the alternate pattern of signal timing. The same curves hold for synchronous timing if the block lengths are multiplied by 2.

for both directions of travel may be made unbalanced to develop a faster or slower speed in one direction (Fig. 21-15). The difference in offset selected for one direction of movement must be balanced by the offset for countermovement, so that the primary speeds of progression become $S = 0.68B/(C/2) \pm K)$, and secondary progression speeds will also be possible so that $S' = 0.68B/[nc + (C/2 \pm K)]$.

Case III. Equal speeds, irregular block lengths. In one-way systems, variation in block lengths or intersignal distances may be readily compensated by adjustment of the offset from station to station, so that the desired band of progression is maintained at the desired rate of speed.

Under two-way conditions, however, there is enforced a symmetry in the time-space diagrams, so that the lag for flow and counterflow is identical for each interstation run or block if equal bands of progression at equal rates of speed for the two directions of movement are maintained.

It has already been shown that this interstation lag must be zero, or some value of $nC/2$ (where n is a whole number), if balanced conditions of

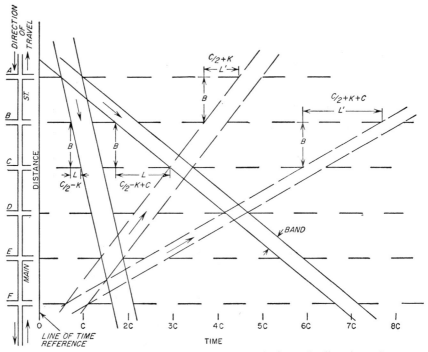

Fig. 21-15. Uniform block lengths—unequal speeds in each direction (Case II).

two-way flow are to be maintained. As Fig. 21-16 shows, when block distances are irregular and equal speeds and bands of progression are sought for each direction of movement, irregularity of block length will reduce band width or affect the speed of progression, or both. The capacity for progressive movement is quickly destroyed with irregularity of block length or intersignal distance.

Case IV. Unequal speeds, irregular block lengths. In this case, because of the fracturing of the band of progression which is inherent in irregular block lengths under two-way traffic, coordination is usually designed so that good accommodation is provided in the direction of major flow, and the counterflow (minor) is accommodated by the resultant configuration of progression for its direction of movement (Fig. 21-17). This plan is frequently found desirable to accommodate the peak movements of traffic on radial streets or whenever there occurs a major movement in one direction and a minor movement in the opposite direction. In the illustration given, the speed of favored movement is $2B/C$, while the countermovement becomes either $3B/2C$ or $6B/5C$, dependent on relative time position of the individual vehicle in the band of progression from K toward A.

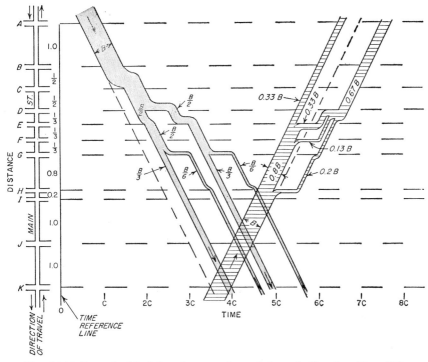

FIG. 21-16. Irregular block lengths—equal speeds in each direction (Case III).

21-5. Network Coordination. The time-space relationships in signal coordination thus far developed have been concerned only with a single route. However, two or more routes thus signalized may cross at a common intersection. At such intersection, it is evident that the cycle, phase arrangements, and apportionment of time interlock the intersecting routes, thus producing a signal network. Signal networks may be classified as *open* or *closed*.

An open network contains only one conjugate or interlocking intersection for any pair of signal systems throughout their entire lengths. A closed network encloses an area bounded by the two or more systems involved (Fig. 21-18).

Case I. The open network. In the open network, phase arrangements and their duration at the interlocking intersection establish the time reference points for each pair of signal systems. If equal length "go" (green and yellow) phases are employed for each system, the beginning of green for one of the signal systems must lag by one-half cycle the beginning of green for the other signal systems. If three signal systems are involved at the interlocking intersection, the beginning of green for systems X, Y, Z becomes 0, $\frac{1}{3}C$, and $\frac{2}{3}C$ respectively, if system X is chosen as the reference system. When unequal phase durations are employed, the offsets of the initial instants of green will be pC, where p

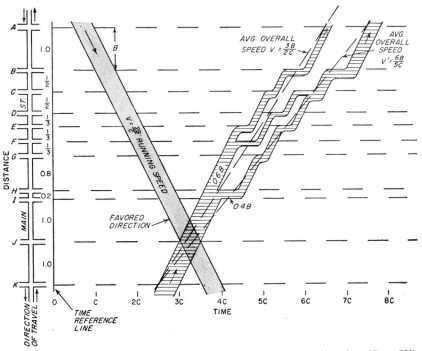

Fig. 21-17. Irregular block lengths—unequal speeds in each direction (Case IV).

is the percentage of cycle (expressed as a decimal fraction) which is employed for green and yellow in the intervening phase or phases taken from the time reference point.

For example, in Fig. 21-19, if A is taken as the time reference point and is the initial instant of green for system X–X, then the initial instant for Y–Y green is pC sec later, and for Z–Z initial green is at $(p + p')C$ sec later. These initial-green instants thus establish the offsets for each signal system with respect to each other.

Two Systems. Consider the simple closed network of only two signal systems in Fig. 21-20, assuming that conjugate intersections at A and B are operated on two phases each and that the cycle splits are pC (green and amber) for X–X at A and $p'C$ for X–X at B. Let the beginning of the "go" interval for X–X at A be taken as the time reference point.

OPEN NETWORKS

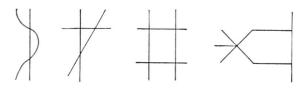

CLOSED NETWORKS
FIG. 21-18

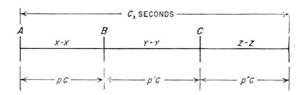

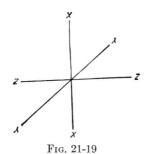

FIG. 21-19

Referring to the time relationship diagram (Fig. 21-20) from A and B, the desired system coordination pattern for X–X required an offset L_x sec at B. Because of the cycle splits at A and B, an offset L_y will result in system Y–Y. Hence the initiation of "go" in Y–Y at B is at $L_x + p'C$ sec after the time reference point. It will be seen that

$$L_y = L_x + p'C - pC$$

Under these circumstances the two systems will interlock at both conjugate intersections.

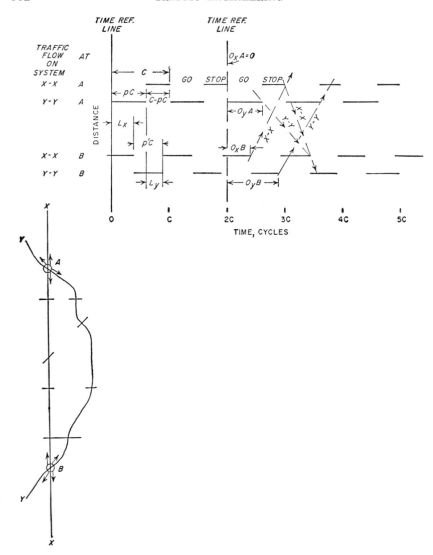

FIG. 21-20. A closed network of two systems and time relationships at conjugate intersections.

$A + B$ are conjugate signalized intersections of systems Y–Y and X–X. (Closed network.)

NOTE: *Intrasystem* offsets between conjugate intersections are L_y or L_x. *Intersystem* offsets using beginning of green for system X–Y at A as O are shown as O_yA, O_xB, O_yB.)

X–Y = indicated speed for primary progressive movement $= \dfrac{D_x}{L_x}$ or $\dfrac{D_y}{L_y}$.

In the design of a closed network involving only two-signal systems, the patterns of coordination must be such that $L_y = L_x + (p'C - pC)$, or the offset in the system is equal to the offset in the other system plus the difference in splits at the two conjugate intersections. If the splits are made equal, and if a simultaneous pattern is employed for system L_x, then L_x and L_y are $C/2$, and L_y is also $C/2$.

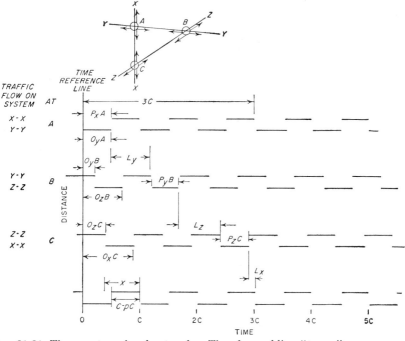

FIG. 21-21. Three-system closed network. Therefore, adding "terms"

$$P_xA + L_y + P_yB + L_z + P_2C + L_x = 3\ C$$

NOTE: The lag between conjugate intersections is

$$Y-Y = L_y$$
$$X-X = L_x$$
$$Z-Z = L_z$$

Multiple Systems. Similar reasoning may be applied to closed networks involving several systems. The sum of lags in the several systems plus the sum of the corresponding "go" intervals must always equal a complete whole number of cycle lengths. Figure 21-21 represents a typical application to a closed network consisting of three systems.

21-6. Flexibility in Signal Timing. It will be recalled that there are three cyclical variations in traffic volume (see Chapter 5). Since desirable signal-cycle division and duration are logically derived from momentary traffic demands on the approaches to the signalized intersection, signal equipment should be designed and operated to meet the fluctuations in traffic demand.   Flexibility in signal equipment to achieve this desirable performance is a primary consideration.

Historically and categorically, there have been two principal theories of signal development.   The earlier, less expensive, and more frequently employed system is based on predetermined cycle length and division and is known as the *fixed time* system.   It is readily adaptable to signal coordination in systems and networks requiring a fixed cycle length for a given pattern and speed of progression.   However, it lacks flexibility in meeting minute fluctuations in traffic demand.   Present designs of fixed-time equipment are constructed to provide change in cycle duration, cycle division, and offset, to afford up to three patterns of coordination.   These changes are effected by a clock device set on a predetermined schedule.

The more recently developed, more expensive, and less frequently employed system of signal control is known as *traffic-actuated*.   Its greatest functional advantage is its automatic response and adjustment, not only to the daily, weekly, and seasonal variation in traffic flow, but also to the momentary surges which are inherent in traffic demand.   Division and duration of cycle automatically vary within limits, as required by the momentary demands of traffic approaching the signal.   Detectors placed in the approaches, at suitable distances in advance of the signal, register the approaching vehicles.   Push buttons are provided for pedestrians.   The timing mechanism automatically apportions time to the diverse phases of movement in accordance with traffic demands.   Accordingly, delay is held to a minimum, and maximum capacity of the intersection is achieved.   The greatest disadvantage of traffic-actuated equipment is its inherent failure to provide signal coordination in system and network, because such coordination requires constant, rather than random, cycle lengths.

Both systems have been developed so that they may be modified in operation to gain advantages not inherent in the basic theory of their design.   Traffic-actuated equipment may be made to perform in semiactuated fashion to provide progression in a coordinated pattern, and fixed-time equipment may be modified to insert by push button an occasional pedestrian phase,[1] for example.

[1] For a fuller discussion of fixed-time and traffic-actuated control, see *Manual on Uniform Traffic Control Devices for Streets and Highways*, U.S. Bureau of Public Roads, August, 1948, and *Traffic Engineering Handbook*, Institute of Traffic Engineers, New Haven, Conn., 1950.   See also manufacturers' catalogues and specifications for diverse models of signal-control equipment.

CHAPTER 22

ROAD LIGHTING

From the traffic operations point of view, road illumination should provide pleasant and accurate night-seeing conditions so that traffic movements may be made easily and safely. Further considerations may affect the character of a given installation or the policies concerning street lighting in a given community; lighting is a factor in discouraging crime and in promoting business. Generally speaking, the levels of illumination acceptable for crime prevention are lower than those adequate for traffic purposes, while those employed for business attraction are higher.

22-1. Basic Factors in Seeing. Visibility is the term employed to denote the distance at which a road user with normal vision, traveling at a given speed, will detect and recognize road details and objects. There are four primary factors in visibility: (1) Time available to perceive and recognize an object, dependent on speed of travel and distance of visibility. Discounting glare-recovery time, safety requires PIEV and stopping time as a minimum. Perception requires greater time at low levels of illumination. (2) Size of the object to be seen, ranging from a small object such as a brick to the bulk of a large tractor-trailer. (3) Contrasts between object and background. (4) Level and conditions of illumination, both of which derive from (a) the level and uniformity of pavement brightness, (b) the brightness of the obstacle to be seen, and (c) the amount of glare to the observer. These in turn depend on the source of illumination (location, design, and lumen output of the lighting fixture) and on the reflection factors and characteristics of the pavement and objects thereon.

Means of Discernment. In road lighting, there are three principal means of discernment: (1) silhouette, (2) reverse silhouette, and (3) surface detail. When the obstacle to be seen has a lower brightness than its background, it is most easily recognized because of the silhouette it casts. In the lower levels of illumination achieved under most street-lighting conditions, seeing by *silhouette* is the predominant method of recognition. When the obstacle to be seen has a brightness higher than its background, but of such low level that surface detail is lost and it is recognized largely because of its general shape and size, it is said to be

355

seen by *reverse silhouette*. When the brightness of the object is such that much of the surface detail is visible, it becomes most easily recognized because of this detail and it is said to be seen by *surface detail*.

22-2. Pavement Brightness. Because of the importance of discernment by silhouette, it is clear that apparently uniform pavement brightness of adequate level is desired. From the road user's point of view, this is dependent on (1) the reflection characteristics of the pavement, (2) the spacing, mounting height, design, and photometric characteristics of the complete lighting device.

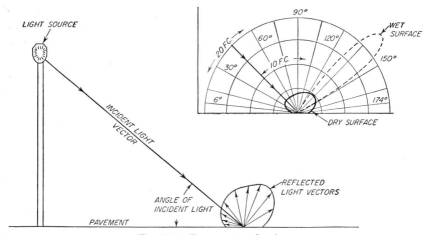

FIG. 22-1. Pavement reflection.

Pavement Reflection. The amount of light reflected to the observer from the pavement is dependent on (1) angle of incidence, (2) position of observer relating to the incident rays, and (3) reflection factor and characteristic of pavement surface.

In practice the angle of incidence on the pavement varies from 0 to $\pm 75°$ from the vertical. The reflection factor and the characteristics of reflection are dependent on the pavement surface and its conditions. Generally speaking, the reflection factors (light reflected/light incident) of pavements are very low, usually ranging from about 20 per cent for clean concrete to 3 to 10 per cent for asphalt, and may be diffuse or specular. The relationships involved in pavement brightness are illustrated in Fig. 22-1. It is evident from Fig. 22-1 that changes in the reflection characteristics of pavements should be recognized in the design of lengthy systems. The forward motion of the vehicle results in a continuous change of the angles of incidence and reflection.

22-3. Luminaire Design and Placement. The total output of a point-light source will naturally radiate uniformly from the point with equal intensity in all directions. It is evident that such light distribution would be inefficient and uneconomical when the purpose is to direct the light toward a relatively limited area, such as the pavement of a street. Furthermore, if light is allowed to distribute radially from the source with equal intensity in every direction, the amount of light falling upon

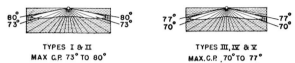

TYPES I & II
MAX. C.P. 73° TO 80°

TYPES III, IV & V
MAX. C.P. 70° TO 77°

FIG. 22-2. Recommended vertical light distributions in planes of maximum candle-power. (From *American Standard Practice for Street and Highway Lighting*, Illuminating Engineering Society, 1952.)

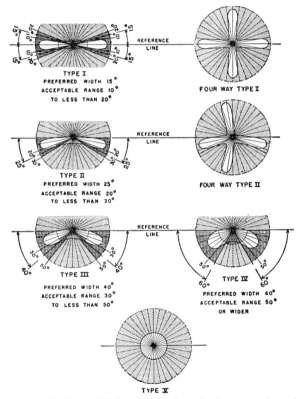

TYPE I
PREFERRED WIDTH 15°
ACCEPTABLE RANGE 10°
TO LESS THAN 20°

FOUR WAY TYPE I

TYPE II
PREFERRED WIDTH 25°
ACCEPTABLE RANGE 20°
TO LESS THAN 30°

FOUR WAY TYPE II

TYPE III
PREFERRED WIDTH 40°
ACCEPTABLE RANGE 30°
TO LESS THAN 50°

TYPE IV
PREFERRED WIDTH 60°
ACCEPTABLE RANGE 50°
OR WIDER

TYPE V

FIG. 22-3. Recommended lateral light distributions in the cone of maximum candle power. (From *American Standard Practice for Street and Highway Lighting*, Illuminating Engineering Society, 1952.)

a particular point in any plane will vary inversely with the square of the distance of that point from the light source.

The designers of luminaires limit the directions in which the light is distributed and regulate directional intensities insofar as possible to obtain sufficient distribution of light on the plane of the roadway. To accomplish these purposes, five basic light-distribution patterns have been adopted as American standards. In Fig. 22-2 the relative intensities in the vertical plane of maximum candlepower are shown by length of radial lines. An angle of about 75° from the vertical is usually adopted as the angle at which maximum candlepower will be directed. Little or no light is emitted above this angle for reasons of glare control and efficient distribution of the available light.

MOUNT-ING HEIGHT	CONVERSION FACTOR			MOUNT-ING HEIGHT	CONVERSION FACTOR			MOUNT-ING HEIGHT	CONVERSION FACTOR		
	I	II	III		I	II	III		I	II	III
15	1.78			22	0.83	1.29	1.86	29		0.74	1.07
16	1.56			23	0.76	1.18	1.70	30		0.69	1.00
17	1.39			24	0.69	1.09	1.56	31			0.94
18	1.23	1.93		25	0.64	1.00	1.44	32			0.88
19	1.11	1.73		26		0.93	1.33	33			0.83
20	1.00	1.56		27		0.86	1.24	34			0.78
21	0.91	1.42		28		0.80	1.15	35			0.74

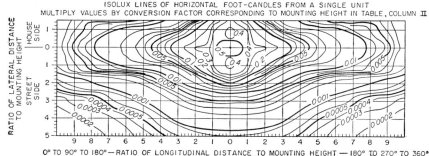

ISOLUX LINES OF HORIZONTAL FOOT-CANDLES FROM A SINGLE UNIT
MULTIPLY VALUES BY CONVERSION FACTOR CORRESPONDING TO MOUNTING HEIGHT IN TABLE, COLUMN II

0° TO 90° TO 180°—RATIO OF LONGITUDINAL DISTANCE TO MOUNTING HEIGHT — 180° TO 270° TO 360°

FIG. 22-4. Calculated street-illumination data. (From General Electric Co., Illuminating Laboratory, Schenectady, N. Y., H-135318.)

The five basic lateral distributions are shown in Fig. 22-3. In these patterns, the patterns of distribution represent the light intensity, measured in all directions radially from the center or light source and as directed over the surface of the cone of maximum candlepower, which has an apex angle of approximately 75°. It is evident that this cone does not receive equal light intensities in all directions, but rather the light is restricted to a highly directional pattern.

All lesser "cones" are similarly restricted so that there results an efficient and effective distribution of light upon the road surface.

A typical distribution pattern of light incident upon a pavement as a result of luminaire design is shown in Fig. 22-4. The integration of the various contours of light intensity will result in the total effective lumen of light on the pavement. The percentage of useful light, then, will vary in accordance with the ratio of street width to luminaire height. A typical *utilization* curve is shown in Fig. 22-5.

22-4. Glare. While uniformity of pavement brightness is a desired objective in road lighting, the level of brightness for a given traffic condition is modified by the amount of glare.   If little or no glare is present, lower levels of brightness will suffice.   Glare not only causes discomfort

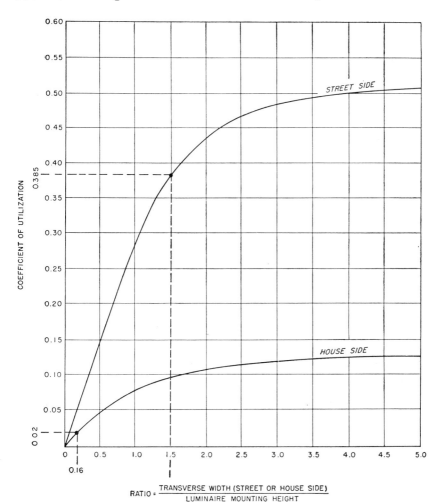

FIG. 22-5. Example of coefficient of utilization curves for a luminaire providing a Type III light distribution.   (From *American Standard Practice for Street and Highway Lighting*, Illuminating Engineering Society, 1952.)

but also reduces visibility.   The amount of glare present is dependent on (1) brightness of glare source, (2) angle of incidence with respect to the normal line of vision, (3) the general level of illumination and resultant eye adaptation, (4) area of glare source, (5) distance of glare source, and (6) distance of glare source from the observer.

The glare from luminaires may be controlled by (1) mounting height, (2) shielding the light source, and (3) reducing the brightness contrast of the light source with that of the general level of illumination. The shielding of the light source is dependent on design and manufacture of luminaire. For a given level of illumination, the brightness contrast is related to lamp size and the design of reflectors and refractors. Mounting height remains, then, as the principal corrective of glare in a given application. The relationship of blinding effect and mounting height is shown in Table 22-1.

TABLE 22-1. RELATIVE BLINDING EFFECT OF GLARE FROM STREET-LIGHTING SOURCES

Mounting height, ft	Relative blinding effect
30	1
25	1.4
22.5	1.7
20	2.1
17.5	2.7
15	3.8
12.5	6.5
10	13.1

22-5. Traffic Criteria in Roadway Lighting. In the final analysis, practical judgment prevails in the establishment of roadway-lighting criteria since no instruments will objectively measure the visibility of street users under practical roadway conditions. Accordingly, the Illuminating Engineering Society has adopted traffic volumes and pavement-reflection factors as the two principal criteria on which to base roadway-lighting intensities.[1] These values have been approved as an American standard. In these standards roadways are classified by vehicular- and pedestrian-traffic volumes as shown in Tables 22-2 and 22-3.

TABLE 22-2. CLASSIFICATION OF VEHICULAR TRAFFIC FOR
ROADWAY-LIGHTING PURPOSES

Classification	Volume of vehicular traffic (max. night hour both directions)
Very light traffic.................	Under 150
Light traffic....................	150–500
Medium traffic.................	500–1,200
Heavy traffic..................	1,200–2,400
Very heavy traffic..............	2,400–4,000
Heaviest traffic................	Over 4,000

Based on the classification of roadways in accordance with Tables 22-2 and 22-3 above, the recommended average horizontal foot-candles for roadway lighting are shown in Table 22-4.

[1] *American Standard Practice for Street and Highway Lighting*, Illuminating Engineering Society approved Feb. 27, 1953, American Standards Association for Standards and Details of Application.

22-6. Design of Systems. When the average foot-candle intensity (lumens per square foot) desired on the pavement is determined, a lighting system can readily be designed by use of a utilization-efficiency chart

TABLE 22-3. CLASSIFICATION OF PEDESTRIAN TRAFFIC

Classification	Pedestrian volumes crossing vehicular traffic lanes
None............	No pedestrians, as on express highways
Light............	As on streets in average residential districts
Medium..........	As on secondary business streets
Heavy............	As on main business streets

for a given luminaire design and output. The utilization chart derived from luminaire characteristics illustrated in Fig. 22-4 will vary for each luminaire design and is usually furnished by the manufacturer. A typical utilization curve (Fig. 22-5) shows the percentage of lumen output

TABLE 22-4. CURRENT RECOMMENDED AVERAGE HORIZONTAL FOOT-CANDLES (LUMENS/SQUARE FOOT)

Pedestrian traffic	Vehicular traffic classification			
	Very light (under 150)	Light (150–500)	Medium (500–1,200)	Heavy to heaviest (1,200 up)
Heavy.................	*	0.8	1.0	1.2
Medium...............	*	0.6	0.8	1.0
Light or none†..........	0.2	0.4	0.6	0.8

* This condition is unusual, but if it should occur, the foot-candle figures appearing in the column to the right may be used.

† Lighted highways and expressways at grade should have illumination similar to that on urban streets with comparable traffic flow, either vehicular or pedestrian. An average of 0.6 ft-c is recommended for highways with full control of access as defined by the American Association of State Highway Officials, such as elevated or depressed expressways from which pedestrians are excluded. The values given for very light vehicular traffic pertain to roadways such as those in residential areas where vehicular speeds are low, of the order of 15 to 25 mph.

NOTE: The foot-candle values are based on a pavement having a reflectance of 10 per cent. When reflectance is poor (3 per cent), the average foot-candle level should be increased 50 per cent. When reflectance is unusually high (20 per cent or more), the recommended values may be decreased by 25 per cent.

which falls on the pavement area. This percentage is the sum of coefficients for the "house" side and the "street" side. It is dependent on the transverse location of the luminaire, the street width between curbs, and mounting height. It follows then that

$$\frac{\text{Lumen output} \times \text{coefficient of utilization}}{\text{Spacing} \times \text{width of pavement}} = \text{average foot-candles}$$

The average foot-candles required are based on traffic volume, the coefficient of utilization results from luminaire design, and for a given installation the street width is fixed. In a given case, then, the design must balance lumen output and luminaire spacing.

BASED ON INITIAL LAMP LUMENS AND A MAINTENANCE FACTOR OF 80 PER CENT

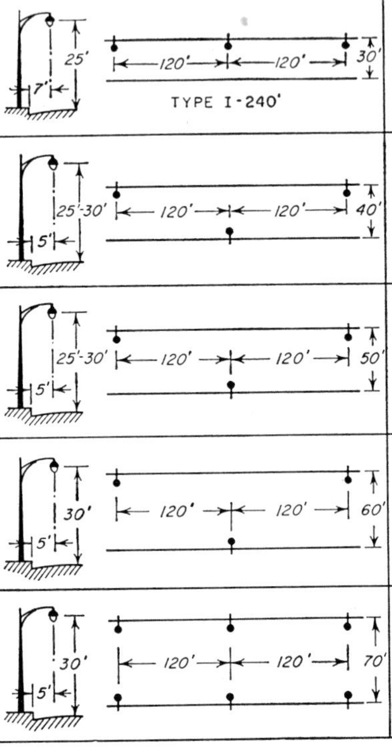

	DIST. TYPE	LAMP SIZE	SPACING	** AVG. FT-C
VERY LIGHT TRAFFIC / LIGHT PEDESTRIAN	I	6,000L FILAMENT	240' ONE SIDE	0.2
	II	2,500L FILAMENT	120' ONE SIDE	0.2
LIGHT TRAFFIC / LIGHT-TO-MEDIUM PEDESTRIAN	III	6,000L FILAMENT	120' STAGGERED	0.4
	III	10,000L FILAMENT	120' STAGGERED	0 7
MEDIUM TRAFFIC / MEDIUM PEDESTRIAN	III	15,000L FILAMENT	120' STAGGERED	0.8
		15,000 L * MERCURY		
HEAVY TRAFFIC / MEDIUM PEDESTRIAN	III	20,000L * MERCURY	120' STAGGERED	1.0
HEAVIEST TRAFFIC / HEAVY PEDESTRIAN	III	20,000L * MERCURY	120' OPPOSITE	1.8
	IV	25,000L FILAMENT	120' OPPOSITE	1.5

FIG. 22-6. Typical street-lighting layouts.
   * A-H1 and E-H1 mercury lamps are rated 15,000 and 20,000 lumens respectively, in vertical operation. Lumens are slightly lower in horizontal operation.
   ** The footcandle values are based on a pavement having a reflectance of 10 per cent. When reflectance is poor (of the order of 3 per cent), the average footcandle level should be increased 50 per cent. When reflectance is unusually high (20 per cent or more), the recommended values may be decreased by 25 per cent. (From *American Standard Practice for Street and Highway Lighting*, Illuminating Engineering Society, 1952.)

Because of loss of lumen output due to dirt on glassware and reduction of lamp efficiency, the actual coefficient of utilization is reduced by a maintenance factor usually taken as 20 per cent. Typical street-lighting layouts are as shown in Fig. 22-6.

22-7. Types of Light Sources. There are four types of light sources presently employed: (1) incandescent filament, (2) sodium vapor, (3) mercury vapor, and (4) fluorescent. The filament type predominates because it has lowest first cost, permits good light control, and is available in a wide variety of sizes. Sodium vapor has found some preference in locations where traffic hazard is high, such as bridges, intersection areas, interchanges, sharp curves, etc. The light is nearly monochromatic and of a characteristic yellow color. It has a relatively high lumen output per watt, being about $2\frac{1}{2}$ times that of incandescent. Mercury vapor also yields high lumen output per watt (about twice that of filament). It yields a bluish-white light color and is increasingly used in high-level illumination. Fluorescent light sources are relatively new and are used in limited areas. They have low glare, long life, and relatively high output per watt, but the number of lamps required is high in comparison with other types because of the low output per lamp.

22-8. Roadway Lighting and Traffic Operations. Good visibility is a prerequisite to good traffic operation. Highway lighting encourages use of full roadway width, the proper use of lanes, and the acceptance of available overtaking and passing opportunities. Studies of accident frequency show that the night ratio of accidents per vehicle-mile is three to four times that of daytime experience. Before-and-after studies of accident occurrence at night, both on urban thoroughfares and at isolated intersections, show significant accident reductions with improved lighting.

DESIGN

CHAPTER 23

FACTORS IN TRAFFIC DESIGN

Traffic design is that phase of design in which the geometry of the facility is related to traffic demands and performance. The traffic designer strives to develop, within reasonable economic limitations, the dimensional or geometric layout which will permit optimum efficiency in operations.

Traffic design includes not only the creation of new facilities but also the redesign of existing facilities. It applies to terminals as well as to streets and highways. It is not directly concerned with the structural aspects of design and construction. The materials used and procedures of their placement are of significance in traffic design only insofar as they affect the physical design of the facility. Traffic design follows over-all planning and precedes construction. The effect of the facility on land development or redevelopment, the composition and quantity of traffic to be served, and the general nature and quality of service to be provided are prescribed during the planning stage. In the traffic-design stage, the physical-site characteristics, traffic data, capacity, and, if pertinent, benefit-cost ratios are studied to determine (1) precise location, (2) the type of facility required to serve traffic needs, and (3) the geometric design of the facility. The balancing of grades, drainage computations, and preparation of precise plans and profiles are of secondary importance in traffic design.

23-1. The Benefit-cost Analysis. Planning, traffic design, and construction may each be compromised in view of considerations or limitations imposed by the others. The ideal location from a traffic-planning viewpoint may present design difficulties or extensive construction costs. One design may create more construction or planning problems than others. Decisions as to exact locations and details of geometric design must be based on a benefit-cost analysis that takes account of all the factors concerned and determines whether or not the maximum benefits provided by the facility are consistent with the cost involved.

Costs are computed by establishing the price of materials and labor required and estimating the expenditures necessary for the construction, maintenance, and operation of the road with each particular design for each possible location. It is more difficult, however, to evaluate benefits

367

in economic terms. Since such indeterminants as savings in travel time and increased safety must be dealt with, it is necessary to assign many arbitrary values.

A decision between alternative courses of action should be based not on absolute comparisons of benefits and costs but on a comparison of benefit-cost ratios. If more than one design will provide a surplus of benefits over costs, it is necessary to determine which will provide the highest ratio of benefits over costs. Some authorities consider improvements justified economically only if this ratio is 2:1 or more.

A common time period must be used in the measurement of benefits and costs. The designer could conceivably estimate the life of the structure and compare total costs with total benefits, but the comparison is usually made on the basis of a year, since road revenues are received annually, maintenance and operating costs are computed on an annual basis, and the yearly cycle is convenient for estimating traffic demands.

Costs. The total cost of a highway is the sum of the initial construction costs and the costs of maintenance and operation, minus the salvage value at the end of the life of the highway. To compute the initial construction costs on an annual basis, the expenditure is spread out over the estimated life of the highway with a yearly amortization of capital expenditure. Since a road may become useless because of structural or functional obsolescence, an estimate of the highway's service life is little more than an educated guess. Most highway designers take a period of 20 or 30 years as a basis for figuring the time of amortization. Salvage is primarily important with regard to right of way, embankments, and road substructure, which suffer little or no deterioration, but these assets are of no value if the location becomes obsolescent. The salvage value of a road depends on how well the original planning anticipated future needs in the way of location and design characteristics. Since salvage value is highly doubtful, designers generally do not take it into account in estimating costs.

The annual cost of a highway or improvement, then, may be considered equal to the annual costs of maintenance and operation plus the annual amortization of the initial capital expenditure, which equals the annual earning power of the expenditure plus the yearly retirement of the capital. This retirement may be made in several ways, depending on whether bonds are retired with a decreasing cost of interest or whether the annual amortization is made with equal yearly installments from user revenue. Whichever method of finance is used, the average annual cost of the capital construction is generally assumed to be the following:[1]

$$C_{ac} = Cr + \frac{Cr}{(1 + r) - 1}$$

[1] C. B. McCullough and J. Beakey, *The Economics of Highway Planning*, Oregon State Highway Department, Technical Bulletin 7, Salem, rev. September, 1938. p. 35.

where C_{ac} = annual cost of initial construction
C = initial capital cost of construction
r = interest rate
n = estimated service life of the development

Some authorities disagree with this analysis, arguing that a pay-as-you-go plan paid out of road-user revenues indicates that the initial capital expense should merely be divided over the estimated service life and that there is no foregone earning power of the capital, since the government would not have that capital available for alternative uses.

The annual cost of maintenance is less controversial, since it is possible to compare the maintenance-cost records of highways constructed in the same way and subject to traffic conditions similar to those anticipated on the proposed highway. Annual costs of operation include all necessary incidental services (exclusive of physical maintenance), including traffic control, policing, and investigational work. Total annual costs may be considered equal to the sum of the annual costs of capital, maintenance, and operation.

Benefits. The problem of estimating benefits is a question of (1) who will benefit, (2) in what ways they will benefit, and (3) how much they will benefit. The latter step is difficult to determine, but it is the crucial element. Since road users will gain in certain ways and others will benefit in other ways, for complete analysis the magnitude of all benefits must be estimated.

Road users, whether driving for business or pleasure, may gain through benefits directly concerning themselves or benefits concerning their vehicles. Most easily measured are benefits concerning the vehicle, including better surface, decreased distances, fewer stops, and decreased grades. It is more difficult to evaluate improvements not directly connected with the vehicle, including (1) savings in time, (2) increased comfort and convenience, and (3) increased safety. The value of savings in time depends on each individual driver, how important he considers his time, how much of a hurry he is in to get to his destination, and what use he makes of the time saved. Distinction is frequently made between commercial and private vehicles, and various arbitrary values have been suggested. In the design of the Gulf Freeway, 2 cents per minute for each passenger vehicle and 5 cents per minute for each commercial vehicle were the values used. Attempts have been made to determine how large a toll people will pay to save time, but it is difficult to isolate the variables involved.

The measurement of comfort and safety is even more difficult, and in past analyses these factors have not been accounted for except by common sense and judgment. There is some indication that safety is being brought into analysis, but accident reports are so incomplete and the various factors so difficult to evaluate that we can rarely say that a particular design will reduce accidents by a certain amount. Some authorities have placed arbitrary values on the costs of accidents which reflect the costs of property damage, medical expenses, loss of earning power, etc.[1]

In addition to the road user, the rest of the community must be considered, whether it is the city, the state, or the nation. The value of the land immediately adjacent to the improvement will probably be increased, a benefit that has long been recognized and evaluated in terms of improvement taxes. In addition, roads will make available to people in outlying areas the goods and services from the population centers. Industry will benefit by the increased mobility of labor, the easier access to raw materials, and the improved channels of distribution. The defensive position of the nation as a whole will be improved by an up-to-date communications network.

Considerable research has been conducted on economic analysis for highway improvements, but much more research is required before it will produce an infallible tool for highway planning and design. The committee on planning-and-design policies of the American Association of State Highway Officials has recommended[2] that benefit-cost analyses be used only to compare the desirability of utilizing alternate locations and alternate design elements for the same facility. Since such comparisons involve the same general area, terrain, traffic, etc., much of the error in assumed values is eliminated. Under these limitations, there is sufficient information to make benefit-cost analysis an important and useful design tool. On important projects, when other factors do not determine warrants, the benefit-cost analysis is frequently based on vehicle operational costs and time costs, without direct evaluation of the less tangible benefits, as comfort, convenience, and safety.

23-2. Road Classification for Design. Road classification indicates the standards of operation required and is a valuable aid in the administration of a road system. It is the basis for inventories of developed mileage and is used to prepare programs, to establish priorities, and to estimate financial needs.

The first highways to be constructed in the United States were those connecting major population centers. The nature and general location of these roads were established by legislative action, and they were designed with little regard for the traffic they were to serve. It soon became obvious that there was a need for a system of secondary roads between cities of lesser importance, connecting these cities with the primary system. These secondary roads, intrastate in nature, were known

---

[1] National Conference on Uniform Traffic Accident Statistics, Committee on Uses of Developed Information, *Uses of Traffic Accident Records*, Eno Foundation for Highway Traffic Control, Inc., Saugatuck, Conn., 1947, p. 39.

[2] *Road User Benefit Analyses for Highway Improvements*, part I, "Passenger Cars in Rural Areas," informational report of the Committee on Planning and Design Policies, American Association of State Highway Officials, Washington, 1952.

as state roads, while the primary system had been integrated into a national network of roadways through Federal-aid requirements. When the demand for primary and secondary highways was satisfied, the need for improved land-service roads to serve the rural population with improved means of getting to city markets became evident. These local service roads were known as farm-to-market roads, supplementary roads, and "butter-and-egg" roads.

Primary roads were given the highest standards of design, while the lowest standards were allotted to the farm-to-market system. Since cost of construction increases with higher standards of design, it was reasoned that the primary system connecting large population centers naturally would carry more traffic, and savings derived by higher design standards would be enjoyed by the larger number of road users. (High-type roadways carrying large amounts of traffic actually cost less per vehicle-mile of travel over the road than the lowest type of farm-to-market roads which carry only a few vehicles per day.) Since the same design standards cost more to construct on mountainous than on flat terrain, primary, secondary, and farm-to-market roads were subclassified with respect to terrain, and design standards were developed for each class of highway through level country, rolling and hilly country, and mountainous regions.

Functional Classifications. Increase in roadway use during the past half century has been accompanied by an increasing need for legislative controls. The legal foundation for these controls is the common English law pertaining to the rights of the public and abutting property owners to the use of a public way which says in essence that a public way is a strip of land, open, as a matter of right, to the public for purposes of travel and over which the abutting property owners have the right of light, air, and access. This connotes free use of the road by all types of vehicles and preserves the full rights of property owners along the road in their pursuits of business and pleasure.

Increased use of roads has necessitated rules of the road, limitations on vehicles, and regulations of drivers in the interests of public welfare and safety. These legal controls are established by state statutes and municipal ordinances.

Use of the stop restriction at intersections has resulted in a differentiation between through (often called preferential) and nonpreferential roads and streets. A through facility or thoroughfare is a street or highway over which traffic movement is controlled but all vehicular traffic is required to stop before entering or crossing, except where an officer or traffic signals are directing the intersecting movement. Nonpreferential, or stop streets and highways, are primarily for access to adjacent properties and are not generally used by through traffic following a continuous route.

Access control to reduce congestion and accidents caused by intersection traffic and roadside development has been recently achieved on some of the more important roadways. It may be realized in various degrees and is always accompanied by high design standards such as medial dividers, grade separation, etc. Access control requires enabling legislation because it denies abutting property owners the right to direct use of the road for which they must be compensated. Denial of the rights of light and air may also be a cause of legal considerations in developed areas where elevated structures are employed.

High-type facilities on which access is controlled are called expressways. An expressway is a divided multilane highway for through traffic with full or partial control of access and generally with grade separations at intersections. Expressways with access partially controlled may be protected from future encroachments by development easements, which do not deny access to abutting property, but control is usually obtained by "freezing" the development of roadside property to its present stage and is usually limited to a strip several hundred feet wide, adjacent to both sides of the roadway, for which the rights of development are purchased by the road authority. The property may be used by the owner or his agent within the limits of the easement agreement, and the property may be sold or inherited provided that the development restriction remains in effect. Zoning and other legal methods may also be employed to control access.

Full control of access means that the public authority has the jurisdiction to provide access connections only with public roads and only at those points which best serve the traffic for which the limited-access facility is intended. Expressways with fully controlled access are called freeways. A freeway is a divided multilane roadway for through traffic with all crossroads separated in grade and with full control of access.

The term parkway, as it is used today, stems from a legal device used before enabling legislation for limited access was available. The earliest legal definition designated a parkway as a strip of land dedicated as a public park, open to the public and permitting access to the scenic beauty and other amenities of the park area, around which abutting owners have no right of light, air, and access. The intent of this legal definition is to include the road as an incidental part of the area, which suggests a winding scenic drive, and to exclude any commercialism or other development which will mar the natural beauty and other amenities of a park area. In the absence of limited-access legislation, strips of land many miles long have been procured and designated through the resolutionary powers of governors and state park authorities to be parkways. Thus a modern parkway is an expressway or a freeway, except that use by commercial vehicles is generally prohibited.

Limited-access roads serve traffic, while roads of unlimited access serve the land. Since both traffic and land must be served, limited-access facilities are warranted only when traffic volumes are exceptionally large, and even then they should not seriously disrupt the operations of local roads. In developed areas, grade separations along limited-access facilities at local road crossings are sometimes so close together that depressed or elevated construction is employed. Local-frontage roads are frequently provided along the sides of limited-access facilities to furnish access to abutting property and to collect and distribute traffic desiring to cross the facility. In most cases, the provision of frontage roads brings about an appreciable reduction in the damages awarded to property owners for denial of access.

Design Designation. While the classification of highway systems and road functions serves administrative purposes and indicates in a general fashion the nature of design, it is not descriptive enough of the service to be performed by the road to enable the designer to meet the traffic requirements of individual routes.

In 1954, the American Association of State Highway Officials recommended use of a highway-design classification based on anticipated traffic volume, character of traffic, and design speed of the road.[1] This design classification, which is termed *design designation*, is a refinement of the classification recommended by the Association in *A Policy on Highway Classification* published in 1940. It consists of (1) the anticipated average daily traffic for the current year and the future (design) year, (2) the design-hour volume as well as the directional distribution of traffic expressed as a percentage of the design-hour volume in the predominant direction of travel, (3) the composition or character of traffic with each class expressed as a percentage of the design-hour volume, and (4) the assumed design speed of the facility.

To be of maximum value to the designer, a road classification must present a complete description of the services to be provided by each class of road. For example, the traffic designer must know where pedestrians, transit, curb parking, and road-side services must be accommodated in design. These matters cannot be expressed in a precise classification and must be prescribed during the process of planning the facility.

23-3. Traffic-design Data. The analysis of traffic volume is related directly to the number of lanes required to handle traffic demands and is used as a basis for determining the economic warrants for higher standards of design, such as better alignment and easier grades.

Future Traffic. To realize the benefit of high investment with due recognition for the life span of the facility, future traffic needs of 20 to 25 years hence must be anticipated and provided for in design. These

[1] *A Policy on the Geometric Design of Rural Highways*, chap. 2, "Design Controls and Criteria," American Association of State Highway Officials, Washington, 1954, p. 109.

needs depend on the different traffic volumes that will occur and their frequency. The design of a roadway must be based to a certain extent on peak-hour traffic, when it is carrying its maximum load. On the other hand, it is uneconomical to design for the hours of greatest traffic since

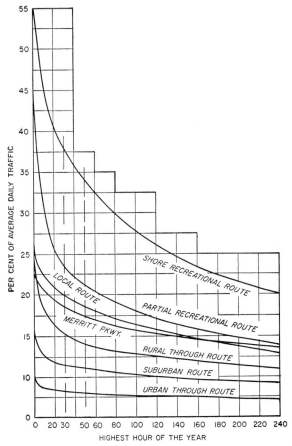

FIG. 23-1. Hourly traffic volumes as a per cent of the 1947 average daily traffic on various Connecticut highways. (From *Hourly Traffic Volumes as a Basis for Design*, Connecticut State Highway Department, Hartford, 1948, mimeographed.)

they occur such a small proportion of the time. Figure 23-1 shows several typical curves relating peak traffic volumes to frequency. The most equitable ratio between the service provided by the road and its cost will be achieved when the design volume is selected near the knee of the curve.[1] Studies made by the Highway Research Board Committee

[1] *Hourly Traffic Volumes as a Basis for Design*, Connecticut, State Highway Department, Hartford, 1948. (Mimeographed.)

on Highway Capacity[1] indicate that it is desirable to design average rural through highways for about the 30*th highest hour* for some future year, depending on the probable life of the roadway. Since the average of fifty highest hours is approximately the 30th highest hour and there are 52 weeks in a year, a rural road designed for its 30th highest hour would be capable of carrying the average week-end traffic peak during its year of maximum volume. Study of Fig. 23-1 will disclose, however, that urban facilities may be designed for the 10th highest hour without much increase in construction cost, while with routes of other types the rate of increase in cost related to degree of traffic accommodation may reach an optimum value at the 50th or 100th highest-hour point. Each route has its own volume characteristics, and design-volume criteria should not be arbitrarily applied.

Highway-capacity values are used to determine the number of traffic lanes required to accommodate the design volume at a satisfactory speed. Since speed and capacity are affected not only by volume but also by grades, curves, sight distances, and other design elements, the design volume is a factor in all phases of highway design.

For convenience, most volume data are expressed in terms of annual average daily traffic. The design-hour volume may be expressed as a per cent of annual average daily traffic for the future year for which the facility is to be designed. Since it is a two-direction volume, information on the directional distribution of traffic during the design hour is required to relate highway capacities to predicted maximum directional flows.

It should be realized that a road may be drastically overdesigned for present traffic if it is designed for traffic volume 20 to 30 years in the future. The highway engineer is obligated to serve the public as efficiently as possible in the conservation of public funds, so he may be justly criticized for the creation of a facility which is far too spacious in keeping with the short-range views of the public. For this reason, he should keep in mind the possibilities of stage highway construction and development. It is entirely possible that a highway may be designed on the basis of traffic demands for the immediate future and at the same time be part of an over-all plan for expansion to handle traffic demands efficiently throughout its useful life. A roadway may be initially constructed as a two-lane highway with additional right of way and provisions for drainage and other structures, so that two more lanes may be constructed to provide a four-lane divided roadway when increased traffic demands require it.

Class of Traffic. Trucks affect the traffic design of roads by limiting the maximum grades, requiring greater width of lane and greater vertical

[1] *Highway Capacity Manual,* Highway Research Board and U.S. Bureau of Public Roads, Department of Traffic and Operations, Committee on Highway Capacity, 1950, p. 132.

clearances, reducing capacity, and other considerations resulting from their greater size, heavier loads, and lower speeds.

For capacity considerations, trucks are rated into two classes. Small delivery trucks, pick-ups, and other types having only four wheels are considered in the same class as passenger vehicles. Larger trucks with dual rear wheels or the semi- and full tractor-trailers are given a special factor in capacity computations to account for their greater demand on road capacities and may be considered equal to two, four, or more passenger cars, depending on grade and other design elements.

To determine the speed of trucks on grades, more specific information concerning weight and power is required. Classifications as to light-, medium-, and heavy-powered trucks as well as gross weights from 10,000 to 60,000 lb are commonly employed.

Vehicle wheel-base lengths, widths, and tracking characteristics affect the design of edges of turning pavements at intersections. For the design of intersection turns, vehicles are classed as P (passenger), SU (single-unit trucks); C43 (semitrailer with 43-ft wheel base), and C50 (tractor truck or semitrailer with 50-ft wheel base).

Design designation with respect to the class of traffic to be served should include (1) the per cent of trucks with dual wheels or larger types by direction of flow during the design-hour as a part of the annual average daily traffic, (2) information on the types and weights of trucks where considerable gradient is involved, and (3) information on the wheel-base lengths and other dimensions of vehicles to be accommodated.

23-4. Speeds Used as Design Criteria. The designer must estimate probable speeds before he can think in terms of forces and time-distance relationships. The speed criteria in design may be (1) design speed, (2) running speed, (3) operating speed, or (4) relative speed.

Design Speed. The design speed of a highway directly affects curve design, sight distances, and grades. It indirectly affects the width of pavement lanes, the design of road shoulders, the control of access, and many other design elements and phases.

The American Association of State Highway officials defines the assumed design speed of a road as "a speed determined for design and correlation of the physical features of a highway that influence vehicle operations. It is the maximum safe speed that can be maintained over a specified section of highway when conditions are so favorable that the design features of the highway govern."[1] The Highway Capacity Committee of the Highway Research Board suggests a more inclusive definition: "a speed selected for purposes of design and correlation of those features of a highway such as curvature, super-elevation, sight distance, upon which the safe operation of vehicles is dependent. It is the highest

---

[1] *A Policy on the Geometric Design of Rural Highways,* chap. 2, "Design Controls and Criteria," p. 79.

continuous speed at which individual vehicles can travel with safety upon a highway when weather conditions are favorable, traffic density is low, and the design features of the highway are the governing conditions for safety."[1]

Running Speed. Running speed, the average speed of traffic over a given distance, is obtained by dividing distance by time exclusive of all time elapsed during stops. The average running speed for all traffic is the summation of distances divided by the summation of running times. Running speed is generally lower than design speed because it is influenced by design. It also varies with weather conditions and is influenced by traffic volume and driver speed desires. Ideally, to encourage uniform speeds, all highways should be designed with a constant design speed.

TABLE 23-1. ASSUMED RELATION BETWEEN DESIGN SPEED AND RUNNING SPEED

Design speed, mph	Running speed, mph
30	27
40	34
50	40
60	45
70	49

SOURCE: *A Policy on the Geometric Design of Rural Highways,* chap. 2, "Design Controls and Criteria," American Association of State Highway Officials, Washington, 1954, p. 83, Table 2-8.

Design speed is rarely reduced in the design of highway elements along the *main-stream* traffic flow. In practice, design speed may be reduced in the design of highway elements accommodating *unusual* or *minor* traffic flow in the interests of economy. For example, lower design speeds are often employed in the design of speed-change lanes, intersection curves, and uphill lanes for trucks, because drivers tend to slow when using these facilities and are more cautious—and the compromise may effect large savings in right-of-way and construction costs.

The amount of reduction in design speed to be considered where minor or unusual flows are to be accommodated depends on the judgment of the designer, but the average running speed is used as a point of reference. It is presumed that faster drivers may safely reduce speed, and design will accommodate the majority of drivers without speed reduction when the average running speed is used as a minimum design value. There is, of course, no absolute relationship between assumed design speeds and running speeds, but spot-speed studies show that design speed has a considerable influence on running speed under free-flowing traffic conditions. The American Association of State Highway Officials has established an assumed relationship between design speed and running speed for design considerations (Table 23-1) when traffic is free-flowing.

[1] *Highway Capacity Manual,* p. 17.

Operating Speed. Design speed is somewhat higher than the satisfactory operating speed assumed for the determination of design capacities. Operating speed is the highest over-all speed, exclusive of stops, at which a driver can travel on a given highway under prevailing conditions. It is the same as design speed when atmospheric conditions, road-surface conditions, etc., are ideal and when traffic volumes are low. Roads should be designed to accommodate their design volume in a manner which will be considered satisfactory by motorists. Therefore, when design volume is related to design capacity, the relationship should be made in terms of a desirable operating speed.

To avoid excessive road costs, the desirable operating speed of a facility must be anticipated for its design. The majority of drivers are satisfied with operating speeds of 35 to 50 mph in urban areas. On rural highways, where longer trips are made, operating speeds from 50 to 60 mph may be required to meet driver desires. Even better operating conditions may be expected on toll or other special facilities.

It should be remembered that for design purposes (1) the assumed design speed is the maximum critical speed used in the design of highway curves, sight distances, etc., (2) the running speed is the expected average speed of traffic on the road and may be considered along with design speed in special cases when economy dictates, and (3) operating speed is the speed at which drivers will be able to travel, if they so desire, and is usually related to the design-hour volume.

Relative Speed. Relative speed pertains to intersection design. It is the speed of convergence of vehicles in separate flows as they approach a point of potential collision. It is also the difference in speed of vehicles in the same flow. Relative speed is dependent on the absolute speeds of intersecting vehicles and the angle of their intersecting paths. The absolute speed of a vehicle is defined as its speed in relation to a fixed point, such as the point of potential collision. It is influenced by the physical design of the intersection, the road user, vehicle conditions, interference to movement caused by vehicles in the intersected flow, and traffic controls. The angle of intersecting vehicles' paths is determined by the design of the intersection.

The absolute speed taken for each of the intersecting flows in intersection design may be either the design speed or the running speed of the intersecting roads, depending upon economy, importance of intersecting flow, etc.

Determination of Design Speeds. Since higher design speeds require higher design standards and accompanying increased construction costs, the assumed design speed of a road should be limited to the highest speed that is economically feasible with due regard for the speed desires of motorists, which is indicated by their speed performance on the road. Figure 23-2 shows the distribution of speeds found where traffic is flowing freely without interference from speed limits, from other traffic, or from any physical limitations in the highway. It will be noted that the speed desires of nearly all drivers would be satisfied on a road designed for about 70 mph.

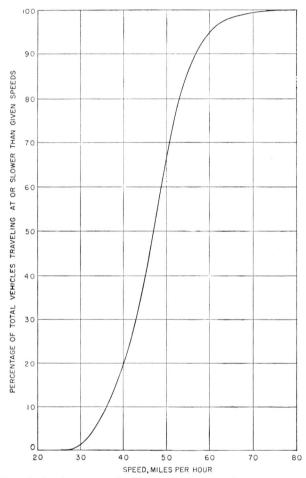

FIG. 23-2. Cumulative-frequency distribution of free speeds on level tangent sections of four-lane divided highways in Illinois. (From O. K. Norman, "Results of Highway Capacity Studies," U.S. Bureau of Public Roads, *Public Roads*, vol. 23, no. 4, June, 1942, p. 60.)

It is as necessary for the traffic designer to predict the future in speed trends as it is to predict future volume in the interest of effective planning and efficient operation of the facility throughout its useful life. For 15 years prior to World War II the average rate of increase in speed was about 1 mph per year. During the war speeds dropped and since then have climbed back to exceed prewar levels (see section 4-2).

Several traffic authorities[1] have pointed out that the greatest over-all improvement in the speed of traffic movement will be obtained through increases in speed throughout the entire speed range, perhaps through the adoption of minimum speed limits rather than through the increase of speed of the faster group of drivers. This may be interpreted to mean that more construction cost is justified for providing wider cross sections for frequent passing and flatter grades for better truck speeds, along with elimination of intersection and roadside conflicts, than for providing high-speed curves and sight distances to accommodate speeds faster than the present normal range.

The Interregional Highway Committee recommends a desirable design speed of 75 mph for passenger use and 65 mph for trucks.[2] Since the cost of a road designed for a given speed is greater in rough terrain or in developed areas, selection of design speed frequently is an economic problem.

Safety Considerations. The human element rather than design economy may determine the ceiling on design speed, since large percentages of accidents occur on straight sections of road under ideal driving conditions. Most of these accidents cannot be directly attributed to design but are caused by inattention, fatigue, and poor judgment. Accidents caused by mechanical failures are a minority.

Analysis of traffic-accident statistics has not shown conclusively that speed is related to the frequency of accidents. Apparently drivers are more attentive when traveling at higher speeds, and perhaps the fast vehicle is able to clear the path of other vehicles more quickly to avoid an accident.

Accident severity, however, is closely related to speed. Accidents resulting in fatalities, serious injuries, and excessive property damage may limit the maximum speeds permitted on public roads.

23-5. Capacity. The capacity of a road under crowded operating conditions differs radically with that offering unrestricted freedom of traffic movement. Since a high degree of accommodation is desirable, design capacity should be that which permits the best operating conditions feasible within economic limitations. Maximum possible capacities are larger than design capacities but are attained with a degree of traffic accommodation below the standard considered satisfactory by traffic designers.

Variables in Capacity. The capacity of a roadway depends on three factors: (1) the effect of design on operations (i.e., the influence of cross section, alignment, sight distances, and intersections on traffic flow), (2)

[1] H. F. Hammond, "Post War Automobile Speeds," *Proceedings of Thirtieth Annual Highway Conference,* University of Michigan, Ann Arbor, Mich., 1944, p. 124.

[2] *Interregional Highways,* National Interregional Highway Committee, Washington, 1944, p. 94.

the effect of environmental factors incidental to stream flow (i.e., weather and interruptions in flow caused by pedestrians and parking), and (3) the effect of inherent stream characteristics which decrease the fluidity of traffic movements (i.e., types of vehicles, speed ranges found within the traffic stream, the number of overtaking and passing maneuvers required to maintain a desired speed, and the interferences between vehicles in performing these maneuvers). These variables and the wide ranges in conditions presented by each highway prevent the development of design capacities which may be arbitrarily applied to any design problem.

Theoretical and Practical Design Capacities. It is convenient to consider capacities in two steps: *theoretical* design capacity, which assumes ideal design, environmental, and inherent stream conditions, and *practical* design capacity, which includes the influence of the cross section, sight distances, and the alignment of the road, as well as of the types of vehicles and environmental conditions of the facility to be designed. Theoretical design capacities will be discussed in this chapter. The modifying factors which determine practical design capacity will be discussed in the following chapters concerned with each element of highway design.

Categories of Design Capacity. Highway capacities are considered in two general categories: (1) capacities for *uninterrupted* flow and (2) capacities for *interrupted* flow. For purposes of design, uninterrupted-flow conditions apply to sections of roadways along which intersectional flows do not interfere with continuous movement. Interrupted-flow capacities apply to intersections at grades where intersecting flows interfere with each other.

The theoretical design capacity of uninterrupted-flow traffic depends on the number of traffic lanes and the operating speed of traffic. Table 23-2 shows the theoretical design capacity for two-, three-, and multilane roads.

The theoretical design capacity for interrupted-flow intersections depends on the number of intersecting traffic lanes and the relative speed of the intersecting flows. The influence of these factors on capacity will be discussed in the chapters on intersection design. The theoretical design capacity of two one-way one-lane intersecting flows ranges from about 2,400 to 1,200 passenger vehicles per hour (both flows combined), depending upon relative speed.

Operating speeds of 45 to 50 mph and theoretical design capacities of 900 passenger vehicles per hour for both lanes of a two-lane road or 1,000 passenger vehicles per hour per lane for multiple lane roads are generally employed in the design of *rural* highways; while operating speeds of 35 to 40 mph with theoretical design capacities of 1,500 passenger vehicles per hour for both lanes of a two-lane road or 1,500 passenger vehicles per hour per lane for multiple-lane roads are generally employed in the design of *urban* facilities.

23-6. Objectives of Traffic Design. The practical usefulness of any design-classification system depends upon the ability of the designer to translate the indicated traffic services into a physical design capable of providing those services with optimum efficiency. There are three basic considerations in traffic design: (1) the human element, (2) the vehicle performance, and (3) effect of design on operations.

TABLE 23-2. THEORETICAL DESIGN CAPACITY FOR UNINTERRUPTED FLOW

Number of lanes	Operating speed, mph	Theoretical design capacity, passenger vehicles per hour
2 (two-way)	35–40	1,500 (both lanes regardless of distribution by direction)
2 (two-way)	40–45	1,200 (both lanes regardless of distribution by direction)
2 (two-way)	45–50	900 (both lanes regardless of distribution by direction)
2 (two-way)	50–55	600 (both lanes regardless of distribution by direction)
3 (two-way)	35–40	2,000 (all three lanes)
3 (two-way)	45–50	1,500 (all three lanes)
4 (or more)	35–40	1,500 (per lane in direction of heavier flow)
4 (or more)	40–45	1,250 (per lane in direction of heavier flow)
4 (or more)	45–50	1,000 (per lane in direction of heavier flow)

SOURCE: *Highway Capacity Manual*, Highway Research Board and U.S. Bureau of Public Roads, Committee on Highway Capacity, 1950.

The Human Element. Because of habit, design should consistently require the same behavior of the road user at similar locations. Since decisions become more difficult to make as the complexity of the situation increases, combinations of design elements which require drivers to contend with a number of things at one time or in rapid succession tend to cause confusion and mistakes in judgment. Surprise is also a factor in many accidents, because only a part of the attention of drivers is directed toward driving. Sudden changes in path or speed are made only under conditions of emergency and are contrary to the desires of motorists. Finally, motorists' desires are criteria of traffic design (safe, comfortable, convenient, and economic travel).

Vehicle Performance. The acceleration and deceleration ability of vehicles, turning radii at various speeds, grade-climbing ability, maximum speeds, and over-all dimensions, as well as headlighting, range of driver vision, etc., may affect traffic operations through design of the road.

It would be uneconomical to design roads for individual vehicles with unusual performance characteristics. Instead, design vehicles which are nearly all inclusive in each class (passenger cars and large, medium, and small trucks) are usually assumed for purposes of design, except that extreme vehicle requirements become important when matters such as vertical clearance are involved.

Effect of Design on Operations. Traffic accidents and congestion are caused by inefficiencies in operations produced by inadequate design. It is the responsibility of the traffic designer to create route and terminal facilities which are capable of meeting the requirements of road users and their vehicles both as discrete units and en masse.

CHAPTER 24

ROAD SURFACE

24-1. Friction. The friction between motor vehicle tires and road surfaces is a determinant of safe speeds and distance requirements in stopping, starting, and turning. When a vehicle starts from a stop or accelerates to a higher speed, it is the gripping action between the tire and the road surface which sustains the accelerating force used to overcome the inertia of the vehicle. In deceleration it is the momentum of the vehicle which must be overcome, and the same considerations apply as in acceleration. When a vehicle is rounding a curve, however, the friction is the gripping action which opposes centrifugal force and adds to the effect of superelevation. Road-surface friction acts at and parallel to the road surface. In acceleration and deceleration, the force is called the *frictional force;* on curves it is also called the *cornering force.*

Road-surface friction may be expressed as a ratio of the frictional force and the weight of the vehicle acting perpendicular to the road surface and is called the coefficient of friction. For the traffic designer, the maximum coefficient of friction offered by the surface under various driving conditions is significant with regard to skid resistance. In the case of braking, the maximum value of friction between the road surface and tire is utilized only if the brakes are in good condition and are capable of developing an equal or greater decelerating force than can be sustained by the gripping action of the tire on the pavement. Thus the critical factor in minimum stopping distance may be either the friction offered by the road surface or that within the brake mechanism.

In normal driving most drivers decelerate at a much slower rate than that which would utilize the maximum friction offered by the road surface or brake mechanism. Since the coefficient of friction $f = F/W$, then $F = fW$, and $F = Ma$ or $F = Md$. The rate of acceleration or deceleration is directly proportional to the coefficient of friction developed.

Skidding. Two conditions of skidding are of interest to traffic designers: (1) when the force causing the skid is parallel to the line of travel (involved in acceleration and deceleration) and (2) when it is normal to the longitudinal axis of the vehicle (involved in turning).

Skidding caused by acceleration is not a critical factor in design because the acceleration force employed by drivers is rarely large enough

384

to cause skidding. However, forward skid resulting from deceleration, usually in the form of braking, is frequently encountered in driving and is important in traffic design. Forward skid takes place up to and including 100 per cent slippage. The maximum coefficient of friction offered by various types of hard road surfaces to forward skid usually occurs at from 6 to 16 per cent slippage, as Fig. 24-1 shows. On most surfaces the coefficient of friction increases rapidly up to the 6-to-16 per cent range

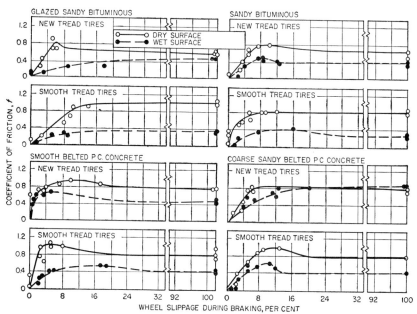

FIG. 24-1. Wheel slippage and coefficient of friction measured in towing-braking tests on various surface types (speed 20 mph). (From R. A. Moyer, "Motor Vehicle Operating Costs, Road Roughness and Slipperiness of Various Bituminous and Portland Cement Concrete Surfaces," *Proceedings of Highway Research Board*, Washington, 1942, vol. 22, p. 50.)

and gradually decreases until 100 per cent slippage, when the wheels are locked, is reached. On soft-surface roadways where tires dig in, the friction developed at 100 per cent slippage may be greater than that with less slippage.

Forward skids accompanied by 100 per cent slippage are called straight-forward sliding skids, and those made at less slippage are called impending-forward skids. While the coefficients of friction under conditions of impending skid on hard-surfaced roads are greater than those accompanying sliding skids, this increased resistance to skidding is usually not utilized by drivers because emergency stops are usually made with the wheels locked when the brakes are capable of locking them.

When rounding a curve, centrifugal force acts along a line normal to the line of travel. At high speeds, only part of the centrifugal force is balanced by the *bank* of the road, and the remainder is counteracted by friction between the tire and the road surface. To develop this friction, the plane of the wheel must be turned at an angle to the line of travel, called the slip angle. The maximum coefficient of friction (cornering ratio) on hard surfaces varies with the slip angle (see Chapter 3, Figs. 3-7 and 3-8).

When side thrust and forward deceleration or acceleration forces are combined, the cornering force resisting side skid is changed. The resultant force required to resist side skid is greater than normal when the vehicle is decelerating. However, the difference is negligible except under very extreme conditions.

Surface-friction tests run by R. A. Moyer[1] indicate that the maximum side-skid coefficients are higher than the coefficients obtained for forward skid in emergency-braking tests on the same surfaces. Since it is obvious that drivers require higher friction factors for braking than for rounding curves, it follows that the ability of the surface to resist straightforward skidding when braking with 100 per cent slippage is the most critical requirement having to do with surface friction.

Factors Affecting Friction. The maximum friction offered by a pavement surface is affected by a number of factors in addition to the type of skid. Air and pavement temperature, as well as tire pressure, affect friction to some extent. There is a small decrease in the coefficient of friction with increases in temperature, tire pressure, and load.

An appreciable reduction in the coefficient of friction occurs with increased speed and with moisture on the road surface. Most hard (but not bumpy) road surfaces offer about the same resistance to skidding when they are dry; when they are wet, however, some become dangerously slippery. The relation between speed of travel and straightforward sliding-skid coefficients of friction for two surfaces with widely different friction characteristics when wet are shown by Fig. 24-2. The effect of speed is less noticeable on dry than wet surfaces, except on extremely slippery ones, such as those covered with ice, when the coefficient of friction varies little with speed.

Contrary to general belief, smooth tires offer higher friction factors on a dry pavement than do new tires, since they present a larger area of surface in contact with the pavement than tread tires with a portion of the rubber cut away or indented to provide the tread. On wet pavements, however, tires with good tread give higher friction factors, since the indentations give the water trapped between the tire and pavement an

[1] R. A. Moyer, "Motor Vehicle Operating Costs, Road Roughness and Slipperiness of Various Bituminous and Portland Cement Concrete Surfaces," *Proceedings of Highway Research Board*, Washington, 1942, vol. 22, p. 51.

opportunity to rush out. This "squeegee" action tends to reduce the lubricating effect of the water film.

The ability of tires to resist melting from the heat generated by friction is an important factor in stopping distances. Recent studies[1] of stops from high speeds show as much as 30 per cent difference in stopping distance for tires of the same tread pattern and same rubber compound but with different hardness produced by the curing of the rubber.

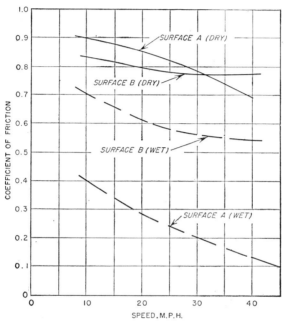

Fig. 24-2. The relation between speed of travel and straightforward-sliding-skid coefficient for two surfaces with widely different friction characteristics when wet. (From R. A. Moyer, "Motor Vehicle Operating Costs, Road Roughness and Slipperiness of Various Bituminous and Portland Cement Concrete Surfaces," *Proceedings of Highway Research Board*, Washington, 1942, vol. 22, data from Figs. 30 and 31, p. 46, for surfaces A_1 and C_6, stopping-distance tests, smooth tires.)

Foreign material on the pavement is also a major factor in surface friction. Ice or sleet offers extremely low resistances to skidding, especially if the surfaces are wet, clean, and smooth. Friction on snow depends on how hard it is packed: tires tend to bite into the softer surface. Since mud or wet leaves on a pavement may be as slippery as ice, a particularly dangerous situation exists where dirt roads intersect paved roads and mud is tracked onto the pavement. A similar condition may

[1] O. K. Norman, "Braking Distances of Vehicle from High Speed," U.S. Bureau of Public Roads, *Public Roads*, vol. 27, no. 8, June, 1953.

be found where pavement drainage is carried on the roadway and drop inlets do not take the water away fast enough to prevent a deposit of silt. *Design of Surface Friction.* From the standpoint of friction, the surface texture due to the particular proportioning of ingredients and method of placement is more important than the type of surface material

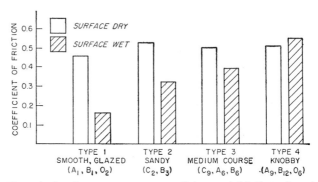

FIG. 24-3. The relation between average coefficients of friction of various surface textures under conditions of wet and dry pavements with a speed of 40 mph, smooth tires, and forward-sliding skid.

A_1 Asphaltic concrete, excess asphalt on surface
B_6 Asphaltic concrete, open type
A_9 Asphaltic concrete, open type, coarse aggregate
B_1 Bituminous macadam
B_3 Bituminous macadam with oil-mat surface
B_6 Bituminous macadam, rougher texture
B_{12} Bituminous macadam, coarse texture
C_2 Portland cement concrete, float finish
C_9 Portland cement concrete, broomed with coarse wire
O_2 Oil mat
O_6 Oil mat, not sealed

(From K. M. Klein and W. J. Brown, *Skid Resistant Characteristics of Oregon Pavement Surfaces*, Oregon State Highway Department, Technical Report 39-5, 1939, Salem.)

used. This applies mainly to wet surfaces, but it is the coefficients of friction on wet pavements which are generally the least and thus the controlling values in traffic design. Figure 24-3 shows the coefficients of friction for various surface textures on wet and dry pavements[1] with forward skid at 100 per cent slippage.

[1] These data were obtained from a study made by the Oregon State Highway Department, Technical Report 39-5, *Skid Resistant Characteristics of Oregon Pavement Surfaces*, December, 1939, using the following classifications:

Type 1: Glazed surfaces, usually occurring on asphaltic pavements which have a waterproof seal course of well-graded, small-sized aggregate. The bleeding during hot weather may result in the formation of a film of asphalt, giving a smooth, glazed surface.

Type 2: Sandy or sandpaper surfaces, usually occurring on floater belt-finished portland cement concrete, especially after considerable service, and on asphaltic

The coefficients of friction on dry pavements do not vary appreciably with the type of surface, but a smooth, sandpaper surface gives a slightly higher coefficient under dry conditions, probably because of its abrasiveness and greater area of tire contact. With a wet pavement, however, a surface with continuous grooves or an open-grained surface with subsurface drainage gives the highest friction coefficients. Roads with glazed surfaces caused by excessive asphalt offer dangerously low coefficients when wet. Relatively smooth, harsh abrasive surfaces are the most effective in the reduction of skidding when the pavement is dry, and well-drained surfaces which allow the escape of moisture trapped between the tire and surface are most effective when the pavement is wet. A surface with texture offering both of these features is highly desirable for safety in traffic operations.

24-2. Hardness and Smoothness. The hardness and smoothness of a road surface affect the cost of vehicle operation and the driver's comfort and safety.

Cost of Operation. Perhaps the most important way in which road surface affects operating cost is in fuel consumption. A soft road may so increase the tractive resistance of a vehicle that an appreciable increase in fuel consumption results from the greater power needed to negotiate a given distance of road. Fuel consumption on various road surfaces is shown in Fig. 24-4. Paved surfaces display very similar results, while earth, gravel, or crushed-stone surfaces vary widely in their effects upon fuel consumption.

Increased tire wear is another source of increased operating expense. On paved roadways, tire wear increases on the more abrasive surfaces: those surfaces which provide the greatest resistance to skidding also cause the most tire wear. However, from a practical point of view, the difference in tire wear on various paved surfaces is negligible when decreased safety is at stake. Tire wear is also affected by tire pressure, temperature, wheel alignment, and brake adjustment. It increases with

---

pavements having a fine-grained wearing surface with a minimum amount of low-penetration asphalt. Any bleeding in this type of asphaltic pavement may change it into a glazed surface.

Type 3: Medium-coarse surfaces, in which the top is composed of aggregate particles of approximately ½ to ¾ in. in size, with grooves or channels around the particles that allow water to drain away without submerging the surface. These surfaces are represented by the open-type asphaltic concretes, in which the top is not sealed and the base course is the waterproof layer, and by the bituminous macadams and oil mats. The same result is accomplished by the deep-broomed portland cement concrete pavements with continuous transverse grooves.

Type 4: Knobby surface, composed of particles of so large a size that each particle becomes the skidding surface. An extreme example is the cobblestone pavement. Asphaltic concrete, bituminous macadam, or oil-mat pavements may approach this texture if too large aggregates are used in the surface.

speed and is greatly affected by the braking and speed habits of individual drivers.

Another economic factor is the physical deterioration of a vehicle when it is driven over a rough surface. Jolting over a bumpy road will shorten

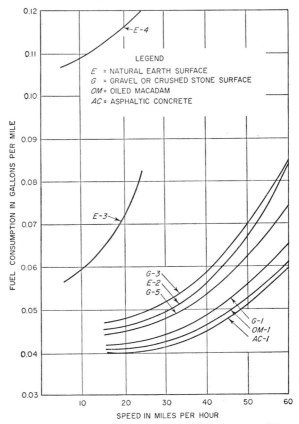

FIG. 24-4. Fuel consumption on various surfaces (1935 Ford). (From K. M. Klein and T. A. Head, *The Effect of Surface Type, Alignment and Traffic Congestion on Vehicular Fuel Consumption,* Oregon State Highway Department, Technical Bulletin 17, Salem, 1944, p. 88.)

the life of a vehicle appreciably, and the greater power needed to move a vehicle over a soft road causes faster engine deterioration.

Costs of operation, as well as costs represented by increased travel time on soft or rough roads, frequently justify change of surface.

Psychological Effects on Drivers. Rough surfaces increase fatigue, thus adding to the number of accidents. An instrument has been standardized by the U.S. Bureau of Public Roads to measure surface roughness,[1] by

[1] J. A. Buchanan and A. L. Catudal, "Standardizable Equipment for Evaluating Road Surface Roughness," *Proceedings of Highway Research Board,* Washington, 1940, vol. 20, p. 621.

vertical oscillations in inches per mile. Driving tests with this instrument on a variety of road-surface types indicate a wide range in the roughness of different roads having the same surface type. However, the smoothness of the concrete pavements shows a marked superiority over the smoothness of bituminous, brick, and gravel surfaces.[1] The tests indicate that a roughness index of 100 in./mile or less is desirable, that 160 in./mile is satisfactory up to 65 mph, and that more than 200 in./mile is very uncomfortable at speeds above 35 mph.

Roughness as a Danger Warning. Since drivers tend to seek the smoother surface when given a choice, their attention is aroused by a sudden change in vehicle operation when rougher surfaces are encountered. Roads may therefore be deliberately roughened as a means of warning traffic at critical locations. This procedure is frequently proposed and would have considerable merit if some of its serious disadvantages could be overcome. Roughness designed in the traveled portion of the roadway should be severe enough to cause only a slight vibration and rumbling to gain the driver's attention. Even small irregularities in the pavement surface, however, become a problem for snow-removal equipment and are rendered ineffective as soon as they fill up with pavement dirt.

Roughness to Designate Traffic Lanes. The State Highway Department of New Jersey has experimented with a precast ribbed divider of white concrete placed flush with the road surface at the edges of the center-pavement lane on three-lane highways. The raised surfaces were originally intended to reflect headlights back to the driver, thus indicating the lane limits, but the glare from the headlights of vehicles approaching on adjacent lanes was found to be so much greater than the light reflected from the divider that this device was ineffective from the standpoint of reflectorized light at the most critical time. However, the rumble and vibration caused by the ribs proved quite effective in advising drivers that they were straying out of the traffic lane, but in a number of cases the noise and jarring effect led drivers to believe that they had developed tire trouble.

Traffic may also be induced to follow desirable paths by varying the smoothness of the road surface. On four-lane highways where the texture of the surface of the inside or passing lane is rougher than the surface of the outside lane, passing traffic returns to the outside lane as soon as the passing maneuver is completed, in order to utilize the smoother surface. It has also been found that if the shoulder surface is as smooth as the roadway proper, drivers tend to use the shoulder as a through traffic lane. Roadway shoulders are intended for emergency and stopping lanes only, and it is difficult to construct just the right surface texture that will discourage shoulder use by through traffic and yet provide a reasonably safe condition for emergency stops.

[1] Moyer, *op. cit.*, p. 41.

Speed Control. Self-styled safety proponents frequently recommend the design of severe roughness in the road surface to control speeds. In at least one city, valleys or open drains have been placed in the paved surface at intersections, creating a severe hazard under the high-speed conditions of today. While it is desirable to control speed, to do so by placing a serious hazard in the roadway will do little toward the reduction in accidents.

24-3. Light-reflecting Characteristics. Since most accidents occur at night because of reduced visibility, the traffic designer must strive to improve nighttime visibility in every way he can. An important factor is the amount of light which is reflected by the road surface to the driver's eyes. When the driver is meeting another vehicle, the amount of light reflected by the pavement surface from the approaching headlights (the degree of pavement glare) is determined by the pavement's light-reflecting characteristics. Even where there is no glare from approaching vehicles, the distance of vision is also a function of the light-reflecting characteristics of the pavement surface.

Glare. Glare caused by the reflection of oncoming headlights is negligible on a dry pavement but is an important factor when the pavement is wet. A film of water on a smooth, impervious pavement surface creates a mirror effect called specular reflection, with an intensity essentially the same as intensity of the incident light. In essence, the same characteristics which reduce surface friction also increase the specular reflective power of the pavement. The amount of reflection is determined by the surface texture and the drainage. Figure 24-5 shows the relative amount of light reflected on wet and dry pavement surfaces of various types from headlights of oncoming vehicles.

Discernment. The discernment of objects under headlights without glare from approaching headlights is accomplished both by silhouette and by direct vision. A black object is silhouetted on the white-pavement background because the amount of light reflected back to the driver from the pavement is of greater intensity than that from the black object. Under conditions of street-light illumination, light reflected from the luminaire in the area of pavement back of the object reveals the object in silhouette. When an object is seen in silhouette and is not illuminated by headlights or street lights, it will appear black regardless of its color.

In direct vision the object is seen under the direct illumination of the headlight or under the direct illumination of the street light.

Tests indicate that the type of road surface has little effect upon night-perception distance for an obstacle projecting above the road under illumination of only the driver's headlights.[1] With no glare or light behind the obstacle, the driver almost always sees an unexpected obstacle

[1] V. J. Roper and E. A. Howard, "Seeing with Motor Car Headlamps," *Transactions of Illuminating Engineering Society*, vol. 33, New York, May, 1938.

by direct vision even when the background pavement is light and the obstacle is dark in color. Most obstacles are seen at a greater distance by silhouette vision when the driver is expecting them, particularly when the pavement surface is light in color, because most obstacles encountered on the road are of a dark color.

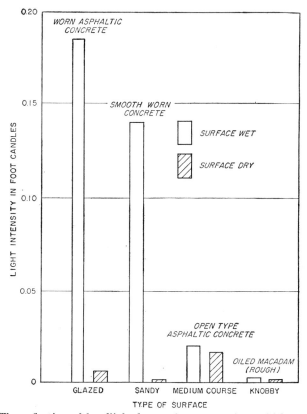

Fig. 24-5. The reflection of headlight beams from oncoming vehicles on pavement surfaces, using a distance of 130 ft. (From G. S. Paxson and J. D. Everson, *Light Reflecting Characteristics of Pavement Surfaces*, Oregon State Highway Department, Technical Bulletin 12, Salem, 1939, p. 30.)

The road itself is visible at a greater distance ahead under headlight illumination when its color is light. The higher reflection factor of a light road surface also increases the efficiency of street lighting.

Since white pavements cause eyestrain during daylight hours when the sun is near the horizon and is brilliantly reflected on the roadway surface, lampblack is often used to reduce the brightness of cement concrete pavements. However, when lampblack is used, the color of the pavement should retain much of its light appearance if night vision is to be enhanced. In the case of direct sunlight reflected on the surface of the pavement, the angle of incident light is so great that specular reflection presents no problem and the surface texture ceases to be an important factor.

From the standpoint of pavement-surface design, a light-colored surface of rough, well-drained texture is desirable for best night visibility. Such surfaces do not become specularly reflective when wet, but they do reflect more of the light from the headlights back to the eyes of the driver for silhouette seeing and provide more pavement luminosity under street-light illumination.

24-4. Contrasting Pavement Colors. Contrasting pavement colors may be utilized to indicate preferential use of traffic lanes. Cheaper road surfaces are generally black because of the use of oil-mat or tar products. Since drivers have become aware of the fact that the more important highways have light-colored surfaces, dark surfaces are often used for acceleration and deceleration lanes, other turning lanes, and shoulders. A light color on the through lane normally indicates preferential traffic movement.

It should be noted, however, that a driver tends to follow the same pavement color: having driven some distance on a light or dark surface, he expects to remain on a surface of that same color until he arrives at a major junction point. If a dark retread on a light pavement is started at an intersection with a minor road, the driver's first tendency will be to avoid the change in color by turning on to the minor road (many will make the turn regardless of directional markers). Thus surface color may be used to "pull" traffic into desired channels and, at the same time, to indicate those channels which the driver should use if he is not a part of the through-traffic stream.

24-5. Surface Slope. Cross slope on roadway surface is necessary to drain surface water from the pavement. Excessive surface moisture increases the specular reflection glare of approaching headlights and splashes dirty water on the windows and windshields of following and passing cars. Poor surface runoff may cause ice formation in freezing weather.

There are four considerations of surface slope design: (1) on tangents, (2) at intersections, (3) on curves, and (4) on surface transitions from tangents to curves. The last two will be discussed under horizontal curve design in Chapter 28. At intersections, the design of pavement-surface slope chiefly concerns the problem of warping the slopes of the approach-roadway surfaces together within the intersectional area for good drainage and smooth profiles. The treatment of the subject here is concerned only with the design of surface slopes on tangents.

A surface cross slope creates a side force caused by the weight component of the vehicle acting down the slope. When the surface slope is excessive, this force affects steering and exerts an uncomfortable side thrust on automobile occupants. In addition, the crown line where different side slopes come together may abruptly *throw* vehicles when crossed obliquely at high speeds, if the algebraic difference of the slopes at the crown line is excessive.

For two-lane pavements crowned in the center, the cross slope of each lane usually employed is from $\frac{1}{8}$ to $\frac{3}{16}$ in./ft of width.[1]   A cross slope of $\frac{3}{16}$ in./ft or flatter has no noticeable effect on steering.   However, the throw effect at the crown line of two lanes sloped in opposite directions at $\frac{3}{16}$ in./ft is excessive on high-speed roads when the planes of the slopes are brought together without rounding off the crown.   The rounding of the cross-section profile at the crown is a matter of judgment, since it reduces the slope while smoothing the surface profile.

When two or more lanes are sloped in the same direction, each successive lane should be given increased slope to accelerate runoff.   Lane slopes may be increased by $\frac{1}{16}$ in./ft, but the slope should not exceed $\frac{1}{4}$ in/ft in any case.

The direction of drainage of the pavement surface on tangents has not been standardized for divided highways.   The pavement on each side of the divider may be crowned at or near its center, or all pavement surface water may be drained to the divider or to the shoulder.   If narrow dividers are used, it is particularly desirable to carry all drainage away from the divider so passing vehicles will not splash water across the divider onto the windshields of oncoming drivers.   If wide dividers are employed, a crown line in the center of each pavement has a slight advantage because drainage is split, providing a ridge at the crown which is free of surplus water during heavy rainstorms.   The center crown also decreases the range in elevation of the pavement from the high point to the low point and blends well into superelevated sections on horizontal curves.   Drainage facilities on each side of the pavement are required when it is crowned.

[1] Most of the data used in this section were obtained from "A Preliminary Discussion on Arterial Highways in Urban Areas," tentatively approved by the Committee on Planning and Design Policies, American Association of State Highway Officials, Washington, February, 1950, part 1, p. 23. (Limited distribution.)

CHAPTER 25

CROSS SECTION

For the purposes of analysis, the cross section of a highway may be divided into four elements: (1) the traveled portion of the road, which may be further broken down into lane width and the number of lanes to be provided, (2) traffic separators or dividers, used on multilane roads to keep traffic in prescribed lanes, (3) curb design, and (4) the design of the road margin, which includes the design of shoulders, parking lanes, frontage roads, driveways, sidewalks, guardrails, slopes, and roadside beautification. In this chapter, a straight highway between intersections is assumed. Cross sections on horizontal curves and at intersections will be discussed in later chapters.

25-1. Lane Width. In the past, the width of a traffic lane has generally been determined empirically by observing traffic operations on various widths of roads. Since wider lanes cost more money, designers of early roads tended to sacrifice width for length. However, lanes which are too narrow increase the hazards and reduce lane capacities because of close clearances in the meeting and passing of vehicles, and maintenance costs tend to be high on extremely narrow roads where vehicles often drive with one set of wheels just off the edge of the pavement, causing the less permanent surface to give way and form a *drop-off*.

Present Design Practice. A lane width of 12 ft is now considered ideal for all high-speed roads and for low-speed roads with a high percentage of trucks. There are still no absolute standards to determine lane width, and there may be conditions which make the 12-ft lane impractical. An 11-ft width may be considered satisfactory on urban low-speed roads where the cost of right of way and other difficulties are major considerations. Nine- and ten-foot lane widths can accommodate substantial lane flow, but driver tension and mental discomfort render them impractical in new design.

Traffic designers have attempted to relate lane width to traffic operations through the three ways in which the driver is affected, his feeling of comfort, his safety, and his operating speed.

Effect of Lane Width on Driver Comfort. Driver comfort depends on the various components of the traffic situation. Transverse placement tests of vehicles on lanes of different widths, conducted on two-lane highways in rural areas under high-speed conditions with no parking or other hazards, are the basis for the conclusions drawn here. The only zones of influence were the stream of oncoming traffic and the edge of the pavement, which included a good shoulder and no curb or other obstacles.

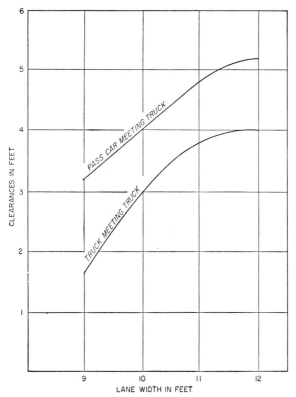

FIG. 25-1. The effect of lane widths on vehicle clearances: body clearance between meeting vehicles on two-lane highways. (From A. Taragin, "Effect of Roadway Width on Vehicle Operations," U.S. Bureau of Public Roads, *Public Roads*, vol. 24, no. 6, October, November, December, 1945, p. 143.)

As lane width was increased, drivers would move a certain distance to the right. This distance was rather constant until the wider widths were reached, when the distance moved with each increase in width fell off rapidly. As Figs. 25-1 and 25-2 seem to indicate, with the larger widths the driver was more concerned with being in a good passing position and was reluctant to swerve continually to increase his clearance with oncoming cars. At these widths the driver apparently was not particularly uncomfortable, either from the nearness of the left-hand traffic or the edge of the road to his right. The lane width at which drivers did not take a proportionate advantage of increase in width depended on the character of the traffic. From the point of view of driver comfort, 11 ft is about the ideal lane width for passenger-car traffic, and 12 ft is ideal where there is mixed traffic of trucks and passenger vehicles.

Drivers did not appreciably reduce speed when meeting and passing other vehicles on 18-ft two-lane rural highways. Apparently they were willing to take the chance that oncoming drivers would not invade their lane in spite of the nervous tension accompanying the maneuver.

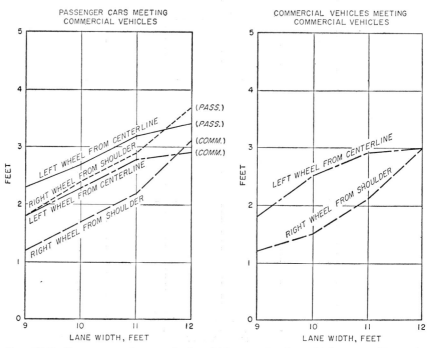

Fig. 25-2. Transverse placement of vehicles on two-lane highways. (From A. Taragin, "The Effect of Roadway Width on Vehicle Operations", U.S. Bureau of Public Roads, *Public Roads*, vol. 24, no. 6, October, November, December, 1945, p. 143.)

Effect of Lane Width on Safety. Existing data on the relation of lane widths to accidents are limited to a single study of traffic accidents related to lane widths made by the National Safety Council. This survey indicates that the rate of traffic accidents decreases with increases in pavement width, but it is a case of diminishing returns, as Fig. 25-3 shows. That is, the improvement in safety diminishes as the 24-ft-wide (two-lane) pavement is reached.

Effect of Lane Width on Capacity. Narrow-lane widths appreciably lower the capacity of a highway, especially on two-lane roads, since drivers tend to take more time to pass another vehicle on a narrow road than on a wide road and travel at greater headways between vehicles.

On multilane roads, capacity is reduced because drivers tend to straddle narrow lanes. This also holds true for two-lane roads but to a lesser extent, because of the threat of oncoming traffic.

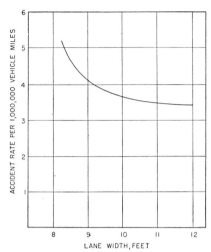

FIG. 25-3. The relation of lane width to rate of accidents on two-lane roads. (From David M. Baldwin, "The Relation of Highway Design to Traffic Accident Experience," National Safety Council mimeographed paper presented to Committee on Planning and Traffic Engineering, American Association of Highway Officials, Dec. 18, 1946.)

The results of a study of the effect of lane widths upon capacity by the Highway Capacity Committee of the Highway Research Board are presented in Table 25-1. Capacities are reduced proportionately more on two-lane than on multilane highways.

TABLE 25-1. EFFECT OF LANE WIDTH ON THEORETICAL DESIGN CAPACITY

Lane width, ft	Two-lane rural roads (percentage of 12-ft lane capacity)	Two lanes for one direction of travel on divided highways
12	100	100
11	86	97
10	77	91
9	70	81

SOURCE: *Highway Capacity Manual*, Highway Research Board and U.S. Bureau of Public Roads, Committee on Highway Capacity, 1950, p. 53.

Effect of Curbs and Lateral Obstructions. So far in the analysis of lane widths and transverse placement it has been assumed that the only zones of influence are oncoming traffic and the edge of the road. When curbs are provided at the edge of the road, however, their zones of influence affect transverse placement so that the width of the lane must be increased to allow for the new zone of influence. The extra lane widths required to offset the effect of various types of curbs on transverse placement are given in section 25-4 on curbs.

Roadside obstructions such as retaining walls, bridge trusses, guard-rails, and parked cars may have a decided effect upon transverse placement if they are close to the edge of the road, as Table 25-2 shows.

TABLE 25-2. EFFECT OF ROADSIDE OBSTRUCTIONS ON EFFECTIVE PAVEMENT WIDTHS

Clearance from pavement edge, ft	Effective width of two 12-ft traffic lanes, ft
6	24
4	23
2	21
0	18

SOURCE: *Highway Capacity Manual*, Highway Research Board and U.S. Bureau of Public Roads, Committee on Highway Capacity, 1950, p. 53.

If the operation of traffic is not to be hindered, obstructions should be placed at least 6 ft from the edge of the pavement, although obstacles 4 ft and more from the edge of the pavement have only a minor effect on operations.

TABLE 25-3. COMBINED EFFECT OF LANE WIDTH AND EDGE CLEARANCES ON THEORETICAL DESIGN CAPACITIES
Capacity expressed as a percentage of the capacity of two 12-ft lanes with no restrictive lateral clearances)

Clearance from pavement edge to obstruction	Obstruction on one side				Obstruction on both sides			
	12-ft lanes	11-ft lanes	10-ft lanes	9-ft lanes	12-ft lanes	11-ft lanes	10-ft lanes	9-ft lanes
Practical design capacities of two-lane highway								
6	100	86	77	70	100	86	77	70
4	96	83	74	68	92	79	71	65
2	91	78	70	64	81	70	63	57
0	85	73	66	60	70	60	54	49
Practical design capacities of two lanes for one direction of travel on divided highways								
6	100	97	91	81	100	97	91	81
4	99	96	90	80	98	95	89	79
2	97	94	88	79	94	91	86	76
0	90	87	82	73	81	79	74	66

NOTE: Effects of lane widths and lateral clearances on driver comfort, accident rates, etc., are not included in these relations.

SOURCE: *Highway Capacity Manual*, Highway Research Board and U.S. Bureau of Public Roads, Committee on Highway Capacity, 1950, p. 54.

Table 25-3 shows the combined effect of lane width and edge clearance on theoretical design capacities. For the capacity of interior lanes, clearance values used in this table may be assumed to be the distance from the edge of the lane to a vehicle centered in the adjacent lane.

Limitations in Lane Width. Although the principal limitation to the construction of wide roads has been the unnecessary cost, there are also operational reasons why lane widths may be too great. If too much freedom of movement is offered to drivers, they may tend to make improper maneuvers and perhaps try to squeeze in another lane. A two-lane road having a width of 26 ft is practically equivalent to a three-lane road with 9-ft lanes. The presently accepted width of 12 ft is probably very near the ideal lane width for mixed, high-speed traffic.

Lanes of 12½- and 13-ft widths outside of the zones of influence of curbs, guardrail, etc., have been used on some high-speed multilane facilities, and they seem to afford greater ease of operation. Sometimes combinations of 11- and 12-ft or even wider lanes are used for these facilities. The wider lanes may be located next to the median to facilitate passing or next to the shoulder to accommodate slower-moving trucks.

25-2. Number of Lanes. The number of lanes required in the design of a highway depends on the predicted traffic demand for the highway and the predicted traffic capacity of various roadway widths. A very powerful influence in design capacity for uninterrupted flow is the opportunity for faster vehicles to overtake and pass the slower ones. The number of opportunities to overtake and pass on two-lane roads depends to a considerable extent on the volume of oncoming traffic. On three-lane roads, the center lane, which is used exclusively for passing, offers more opportunities for passing than a two-lane road and, therefore, has a higher theoretical design capacity. Multilane roads of four or more lanes offer still greater opportunities for passing since oncoming traffic does not affect the passing maneuver.

Theoretical design capacities for two-, three- and multilane roads are shown in Table 23-2. Reduced sight distances, narrow lanes, excessive grades, sharp curves, pedestrians, trucks, parking, and similar factors also affect opportunities for passing, but these are not present under the ideal conditions for which theoretical design capacities are assumed.

Three-lane Roads. The subject of three-lane roads is very controversial. Because it provides a center lane for passing, this type of road has a greater capacity than the two-lane road, but the high accident potential of three-lane roads does not generally permit their use in design. It seems likely that the use of a common center lane for passing by drivers traveling in opposite directions would present hazards. However, investigation of the accident rates on two-, three- and four-lane roads does not conclusively prove this assumption, possibly because (1) drivers may be aware of the greater hazard and are consequently more cautious or (2) some three-lane roads in flat terrain have almost continuous passing sight distances.

At any rate, the three-lane road presents other disadvantages. It cannot be efficiently expanded to a four-lane divided roadway when a raised divider is to be employed. If it is to accommodate traffic volumes substantially greater than those accommodated by a good two-lane road, passing sight distance must be almost continuous over the length of the road. Finally, the popular concept that a three-lane road lends itself particularly well to unbalanced flow conditions (when one-third of the traffic is in one direction and two-thirds in the other) has been disproved. Under maximum traffic conditions, three-lane roads carry more traffic when the volume is equally divided in direction.

The American Association of State Highway Officials has concluded that three-lane roads are appropriate when all of the following conditions are met: (1) nearly continuous passing sight distance may be feasibly provided, (2) the design volume exceeds the capacity of a two-lane road, and (3) the expected future volumes do not exceed the three-lane-highway capacity.[1]

It is now the general consensus that two-lane highways should be converted to divided highways. Few three-lane highways are being constructed in present practice.

Roads of More Than Four Lanes. There is a belief among traffic designers that when more than two lanes for traffic in one direction are constructed adjacent to each other the volume-carrying efficiency of each lane is reduced, because traffic tends to wander about from lane to lane, thus cutting down the rate of flow. This concept has been disproved to some extent by limited observations which show no such reduction under heavy traffic conditions.

Nevertheless, it is desirable to construct traffic lanes on multilane roads in pairs when feasible, since this provides a means of channelizing traffic entering and leaving the roadway and tends to segregate fast and slow traffic. For example, a portion of state Route 100 in New Jersey has eight-lane construction. The cross-section design is a four-lane divided roadway with shoulders, bordered on each side with a separator, and another two-lane pavement with a shoulder on the outside edge. Frequent connections to local roads are provided on the outside lanes, while openings in the separator between the inside and outside pairs of lanes are fairly infrequent. This encourages the use of the outside lanes by short-distance traffic, which is inclined to be slower, and isolates the inside or fast lanes for long distance through traffic, where delays caused by traffic entering and leaving the highway will not be encountered. At major junction points interchange of traffic from slow to fast lanes is

[1] *A Policy on the Geometric Design of Rural Highways*, chap. 5, "Highway Types," American Association of State Highway Officials, Washington, 1954, p. 234.

provided by special designs which permit the gradual weaving of intersecting traffic streams. The design differs from a four-lane freeway with frontage roads on each side, in that the outside lanes are an integral part of the through roadway and the standards of design for the outside lanes are as high, or nearly so, as those for the inside lanes. In contrast, frontage roads frequently have connections with the through roadway at widely separated points, and their design provides for parking and other roadside services.

Roads of more than four lanes may be designed with more traffic lanes in one direction than in the other if design volumes show a substantial unbalanced peak traffic demand always in the same direction. More frequently, the need for reversible-direction unbalanced flow is indicated by design volumes. On urban arterials, traffic peaks are sometimes extremely directional (the inbound traffic rush to the business district in the morning and the reverse direction outbound rush in the afternoon). Roads of five or seven lanes may be used for this purpose, but movable dividers are required if opposite directional flow is to be separated at all times. Three or four lanes, respectively, may be used in the direction of peak flow.

A particularly interesting application of the movable-divider principle is the hydraulic divider used on Lake Shore Drive in Chicago,[1] an eight-lane facility with dividers placed on the center line and two lanes over on each side. Any one of the dividing lines can be raised by means of hydraulic jacks from a position flush with the road surface to a height of 8 in. above the surface. When there is balanced flow in each direction, just the center line may be raised, providing four lanes in each direction. When directional peaks occur, however, either of the two outside lanes may be raised, providing six lanes in the direction of heaviest traffic. While the theoretical advantages of this system are tempting to the traffic designer, mechanical difficulties encountered in their operation have discouraged their use in favor of nonmovable types of separators.

Nonmovable separators may be used in the design of six- and eight-lane pavements for reversible unbalanced flow. The lanes are constructed in pairs and may be operated with four or six lanes, respectively, in the direction of major flow with separation of opposing flows.

When the unbalanced-directional-flow principle is used, traffic signals and neon or other signs are used to indicate to motorists the direction of flow at different times on reversible lanes.

25-3. Traffic Separators. The principal function of traffic separators is to prevent collisions between vehicles traveling in opposite directions. They are also used to prevent U turns and dangerous maneuvers at intersections and may help to channelize traffic into efficient streams, provid-

[1] O. K. Jellinek, "Hydraulically Operated Dividers Adjust Lanes to Traffic Flow," *American City*, March, 1940, p. 95.

ing shadowing of crossing and turning traffic, segregation of slow traffic, and pedestrian protection. The methods used to separate traffic may be markings on the road, physical dividers, or area dividers.

Road Markings, Permanent and Temporary. On two-lane roads, traffic in both directions is required to use both lanes in order to overtake and pass. Separation of opposing-direction traffic is accomplished on these roads only through regulation, the keep-to-the-right rule applying, except

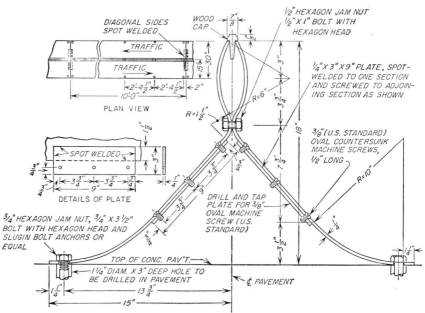

Fig. 25-4. Experimental traffic separator. (From Miller McClintock, "Mechanical Traffic Deflectors," *Proceedings of Highway Research Board,* Washington, 1938, vol. 18, part 1, p. 340.)

in passing. A painted or permanent line in the pavement is essential to mark the center of the road.

Regulation by road markings is also the method used in three-lane highways, but here the problem is more acute since drivers in both directions feel that they are entitled to use the center lane. In such cases, regulation is accomplished by the establishment of no-passing zones, usually indicated on the pavement by a painted or permanent line as in the case of two-lane highways.

In design, neither painted nor permanent lines are recommended for centerline marking on multilane highways of four or more lanes, however, because such roads should be divided with mechanical or area separators.

Mechanical Separators. On multilane highways, mechanical separators may be used to prevent centerline encroachment. Their design is still in the development stage, and only a few have been actually used. Figures 25-4 and 25-5 show two different designs that have been used.

The properly designed mechanical divider should control vehicles by deflecting them with as little damage as possible and without throwing them back into the stream of moving traffic. Assuming that the device has the necessary strength and anchoring to withstand impacts, the most important design consideration is the shape and resulting effects upon vehicles at likely angles of encroachment for different speeds. A parabolic shape permits encroachment of vehicle wheels without allowing any part

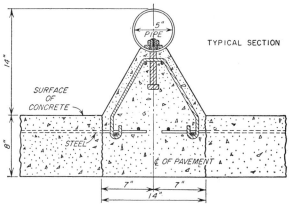

Fig. 25-5. Center divider, Long Island State Parkways, 1947. (Taken from drawing submitted by Long Island State Parkway Commission.)

of the body of the vehicle to scrape, and a knob at the top of the divider prevents climbing or tipping over at lower speeds but would not deflect the vehicle into moving traffic. These principles were incorporated in the design shown in Fig. 25-4, which was tested experimentally on Route 19 some years ago in Michigan. The operational characteristics of other designs have not yet been made available.

The main advantage of the mechanical divider is that it prevents head-on collisions with a minimum use of road width. It is particularly applicable for existing four-lane undivided roads, where a wider divider would be very costly, or for bridges and tunnels.

The disadvantages in the use of mechanical dividers arise from their appearance and from narrowness. They are unsightly and exert a zone of influence upon vehicle placement, causing a less effective lane width. Where mechanical dividers are used, an accident may block all lanes in one direction, and traffic following such accidents is not able to cross the center line to bypass crippled vehicles. Mechanical dividers do not present a wide enough haven for crossing and turning vehicles and may sometimes present undesirable barriers for pedestrians crossing the road. In view of all these objections, mechanical dividers are recommended only where wider area separation is not economically feasible.

Area Separators. The separation at the center of a multilane roadway of traffic going in opposite directions is better handled by an area divider than by a mechanical deflector at or near the edge of the traffic lane. The area separates opposing streams of traffic in time and distance, rather than by a more hazardous physical barrier, and should be as wide as is economically practical. These area dividers are also called center malls, medians, esplanades, parkway strips, and center or dividing islands.

Separation of traffic by area holds a number of advantages over that by a mechanical divider: (1) a wide separator provides a large neutral zone in which drivers may gain control of their vehicles in case of accidents, (2) area separators are much more satisfactory with regard to driver comfort and (3) they are more pleasing to the eye, particularly if planting is included, and (4) the distance and use of screening reduces headlight glare from vehicles in the opposing direction. At intersections, the wider island provides a zone of safety for pedestrians instead of an obstruction as in the case of the mechanical divider. The longer distance between opposing traffic lanes facilitates channelization at the intersection, providing more room between streams of traffic for the shadowing of crossing and left-turn traffic. Wide-area separators are therefore more appropriate on a road with many intersections than on limited-access facilities where pedestrian traffic, crossings, and left turns by vehicles at grade are prohibited.

There is no one width that can be designated as optimum for area separators. Conditions vary in each case, and the estimated benefits of a wide divider must be balanced with the increased cost of right of way. Narrow dividers substantially reduce the cost of constructing a road through highly developed areas, but the additional expenditure for wider separations may be justified through reduction of congestion and accidents. Unfortunately, few data have been collected relating efficiency of operation to increased widths of separators.

Since one of the principal functions of the area separator is to give drivers an opportunity for gaining control when they have run off the road without entering into the stream of traffic coming in the opposite direction, between intersections (where this function is the main consideration) the minimum desirable width depends upon design speed, possible angles of departure from the road, and the time required to regain control.

Headlight glare is appreciably reduced when the width of the separator is only 20 ft, but this is not an important factor in the determination of width, since planting can be used to screen glare on even the more narrow dividers.

Area separators desirably should be at least 25 ft and 45 ft wide to shadow crossing passenger cars and trucks (respectively) at intersections. Less divider width is required to shadow vehicles turning left from the divided road, and more width may be required to accommodate U turns. There are no clearly defined design standards for the construction of area separators. The area may consist of dirt or may be a paved surface. Whatever material is used, it is important that the island color contrast with that of the roadway.

Area separators consisting of dirt are usually stabilized by a good coverage of grass, with shrubbery and other low planting frequently included for the purpose of beautification and the screening of headlight glare. The shrubbery should be high enough to block headlights but not high enough to restrict sight distances at intersections. Growth should be held to a level of from 3 to $3\frac{1}{2}$ ft above the surface of the pavement. Planting should not develop into an obstruction or a hazard to traffic. When planted in a continuous line, it may collect debris and become unsightly, so that shrubbery is sometimes planted in clumps at frequent intervals along tangent sections with continuous lines around curves where headlight glare may be most objectionable. Special effects in the placement and type of shrubbery may be effectively used to designate intersections and curves for the benefit of the approaching driver.

25-4. Curbs. Since the design of curbs has generally been left to the discretion of the individual designer, there have been almost an infinite variety of curb designs. From a functional point of view, however, there should be just three classes: (1) mountable, (2) low-speed barrier, and (3) high-speed barrier. Curbs of design intermediate to Class 1 and Class 2 have been used, but their purpose is not clear to drivers who may cross over them for emergency parking. Drivers with flat tires, for instance, are reluctant to drive over the semibarrier presented by a curb of this type and as a result will change tires on the through-traffic lanes.

Mountable Curb (Class 1). The purpose of this type of curb is to encourage traffic to remain on the through-traffic lanes and yet allow drivers to enter the shoulder area with little difficulty in case of emergency. It is also useful in the control of drainage along with drop inlets and underground sewers, all of which prevent the softening or washing of dirt shoulders. The slight bump is enough to warn the driver when he is driving off the edge of the pavement, without driver discomfort or damage to the vehicle. It is also possible to construct the curb of white concrete, possibly with a ribbed surface to delineate the edge of the pavement at night. As shown in Fig. 25-6, the slope of this type of curb should be two-to-one or flatter. The width is variable.

Class 1 curbs have a very small effect on transverse placement. Studies of placement of vehicles on pavements with lip curb indicate a reduction in effective pavement width of 1 ft during daytime and no reduction at night.[1]

Accordingly, the width of the lane next to the curb need not be any wider when Class 1 curbs are used (except, of course, for the extra width created by the curb itself). The chief drawback of the Class 1 curb, as compared with no curb at all, is the increased cost of construction, including additional drainage facilities.

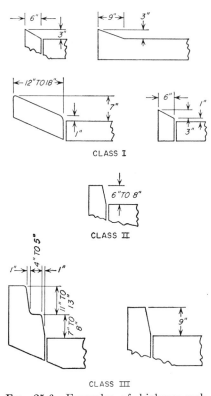

CLASS I

CLASS II

CLASS III

FIG. 25-6. Examples of highway-curb types.

Low-speed Barrier Curb (Class 2). This type of curb (Fig. 25-6) consists of an almost vertical face, 6 to 8 in. high, with a 1- in. batter to prevent scraping of the side walls of tires. It is frequently called an *urban parking curb,* but is used wherever it is desired to prevent sidewalk encroachment by slow-speed moving or parked vehicles and to provide a parking barrier where no sidewalks are involved. It may be driven over in an acute emergency, but may result in severe damage to the vehicle with possible injury to the occupants. Like other curbs, it may be reflectorized to delineate the edge of the road. It controls drainage but should not be used for this reason alone. It has an effect on transverse placement much greater than the Class 1 curb because of its greater zone of influence. It is recommended that widths of lanes adjacent to Class 2 curbs be increased 1 to 2 ft to compensate for the effect of this curb on transverse placement.

High-speed Barrier Curb (Class 3). This type of curb, from 9 to 20 in. high or higher, is intended to prevent vehicles from leaving the roadway under any circumstances. It is used at unusually dangerous locations, such as bridges or mountainous roads, where the damage inflicted by holding the vehicle on the road by physical means would be less than that suffered if the vehicle were to leave the road. The effects of Class 3 curbs on vehicle placement are similar to those caused by obstructions along the road, as shown in Table 25-2.

[1] A. Taragin, "Effect of Roadway Width on Vehicle Operation," U.S. Bureau of Public Roads, *Public Roads,* vol. 24, no. 6, October–December, 1945, p. 152.

Reflectorized Curbs. The New Jersey Highway Department has been instrumental in developing a white concrete reflecting-surface design for all types of curbs. The same principle is used in the design of reflectorized curbs, where ribs about $1\frac{1}{2}$ in. apart reflect light back to the eyes of the driver. When the curb is wet, the reflective efficiency increases.

Reflectorized curb has been given rather general use, but opinion differs as to its value under certain conditions. Where the reflectorized curb is used along the center divider and where this dividing area is narrow, the amount of light reflected by the curb may not be sufficient to overcome the glare from approaching headlights, and this is true to a certain extent of curbs along the edge of the road as well.

25-5. Road Margins. There are eight elements in the design of road margins: (1) the road shoulder, (2) parking lanes, (3) frontage roads, (4) driveways, (5) sidewalks, (6) guardrails, (7) slopes of embankment, and (8) roadside landscaping.

Shoulders. Shoulders are provided along the edge of a road because they serve as an emergency lane for vehicles forced or driven off the roadway and act as service lanes for vehicles that have broken down.

On roadways without adequate shoulders, disabled vehicles must stop on the through-traffic portions of the road, causing a high accident potential. On the Arroyo Seco Parkway in California, for example, where no shoulders were provided, during the period 1941 to 1947, 15 per cent of the total accidents on the road involved cars parked on the traveled way because of mechanical trouble.[1] Similar experience on the West Side Highway in New York City has necessitated construction of special turnouts for disabled vehicles at great expense.

In the design of shoulders, there are five factors to be considered: continuousness, width, bearing strength, smoothness of riding surface, and color. When a continuous shoulder is not possible, turnouts should be provided frequently and should be designed to enable vehicles to decelerate prior to stopping without interfering with through movement.

Most truck drivers will not park less than 1 ft from the outside edge of the shoulder, and since most large trucks are 8 ft wide, a 15-ft-wide shoulder leaves a distance of 6 ft between the parked truck and the outside edge of the pavement. This is the distance necessary to preserve the effective width of the road during periods of vehicle breakdowns (Table 25-2). In the case of passenger cars, the necessary distance is provided with a 13-ft shoulder. Unfortunately, in many cases these shoulder widths are not economically feasible, and lesser widths are used with loss in capacity and safety. (The rounding of the shoulder edge and ditch slope may require reasonable adjustment of these desired shoulder widths.)

The bearing strength of a shoulder should be sufficient to support the

[1] *Report of an Investigation of Accidents on the Arroyo Seco Parkway,* California Division of Highways, Sacramento, Apr. 23, 1948. (Mimeographed pamphlet.)

weight of a loaded truck in wet weather. If shoulders do not appear stable, drivers will prefer to take the chance of parking on the paved roadway rather than becoming stuck in the soft shoulder. The added expense of providing adequate subbase and all-weather surface is more than offset by the improved appearance, safety, and reduced maintenance.

The riding surface of shoulders should not be as smooth as that of the through-traffic lanes, or drivers will use them for normal driving. There have been a number of cases in which smooth shoulders have been used as through-traffic lanes, the shoulder being destroyed as an emergency lane and soon becoming a maintenance problem, since shoulders are not designed to carry regular traffic.

Shoulder slopes may range from $\frac{1}{4}$ to 1 in./ft, depending on the surface material used. A slope of $\frac{1}{4}$ in./ft will drain properly, but greater slopes are used on gravel and turf shoulders to offset the irregularities in the shoulder surface and settlement. The color of the shoulder should contrast with that of the road surface for better day and night delineation of the edge of the road and to indicate to the driver that the shoulder is not a through lane. Where acceleration and deceleration lanes for turning traffic at intersections or where bus-loading bays are provided, a three-color combination is desirable, one color for the through roadway, one for the acceleration or deceleration lane, and a third for the road shoulder.

Parking Lanes. On urban streets of unlimited-access type, curb parking in spite of its inefficient use of street space must usually be tolerated. However, parallel parking should be insisted upon because it is safer than angle parking and causes less congestion. The important factor in the width of parking lanes is the effect of the parked cars upon the capacity of the highway. Since the same reasoning applies here as in the determination of desirable shoulder widths, a parking lane of 13 to 15 ft is recommended. A further reason for this width is the possibility that at some future time parking may be prohibited and the lane will become a through-traffic lane. Wider parking lanes also decrease interference with through traffic when vehicles are parking and unparking.

Frontage Roads. Frontage roads are used to furnish access to properties along an expressway or freeway facility and to distribute traffic desiring to cross or enter the facility. They are local roads generally parallel to but isolated from the through facility, with connections to the through facility provided only at selected points, which preferably are grade-separated.

Where feasible, expressways or freeways may be located through the center of city blocks so the existing parallel streets may serve as frontage roads. Otherwise, frontage roads are constructed along the same general horizontal alignment of the through road and are isolated from it by a separator.

The separator between an expressway or freeway and its frontage road prevents the interference of through movement by local traffic. It should physically discourage crossings from one road to the other except at selected points where interchange is permitted.

The separator width includes the shoulder of the through road, bus-loading zones, and acceleration or deceleration lanes, as well as the width of the curbs employed. Many of the same factors which affect the width of medial separators also apply in this case. Absolute minimum width is, of course, the width of the shoulder of the through road plus the frontage-road curb. Greater widths are desirable for reduction of headlight glare, for better channelization at interchange points, etc.

Frontage roads may be designed for one-way or two-way operations. Two-way operations create more vehicle-stream conflicts at intersections and interchange points; headlight glare thrown on the through road may cause a hazardous condition; and the general operations of the frontage road are not as efficient as with one-way traffic.

The standards of cross-section design for frontage roads are lower than those for the through roadway, because low speed and congestion caused by parking and pedestrians are usually to be expected. The character of the frontage road, however, must be considered in design because it may range from a mere alleyway to a major street or relatively high-speed rural road.

Driveways   The location and design of commercial and private driveways is important in traffic design and operations, but little has been done to standardize these matters.

Driveways should be located at a distance from an intersection in keeping with the speed of through traffic, type of intersection control, and sight-distance requirements at the intersection. Between intersections they should not be located over the crest of a hill or on the inside of curves where a sight distance is restricted or where vehicles parked in the driveway will restrict sight distance.

Where more than one driveway is needed to serve a commercial establishment, such as a filling station, they should be separated by as much distance as practicable. The separating island should be raised to prevent vehicle encroachments and should extend into the property far enough to permit adequate sight distance for pedestrians on the sidewalk and for motorists leaving the property. In these cases, one-way driveways are desirable for obvious reasons.

The width of driveway should be held to a practical minimum to reduce the length of crosswalks, but curb radii should be long enough to allow vehicles to turn into and leave the property on the curb lane of the road.

On long-distance limited-access facilities, such as the Merritt Parkway in Connecticut, filling stations are sometimes permitted at infrequent intervals. The stations and pumps are duplicated on each side of the road with pedestrian subways for the convenience of personnel. Deceleration and acceleration lanes are provided with sufficient storage space off the parkway to contain all the activities of the filling station.

Most traffic authorities require builders to obtain a driveway permit before driveways may be constructed. Inspection of the construction plans and field inspections are made to check the construction from the standpoint of traffic safety. Frequent driveways on major, heavily traveled facilities are a great detriment to operations even with the best driveway designs. Limited-access, zoning, and roadside-development easements seem to be the answer. Drive-in theaters are particularly pertinent to this problem.

A study conducted in Michigan[1] found positive correlation between accidents and roadside developments such as gas stations, taverns, etc., including billboards. Roads with none of these features averaged 3.7 accidents per million vehicle-miles, while those with less than 4 per 1,000 ft averaged 9.1, and those with 4 or more per 1,000 ft averaged 13.5.

Sidewalks. There are no specific criteria which indicate when sidewalks are to be provided along a roadway. Their inclusion in cross-section design is generally based upon the importance of the road and the number of pedestrians to be served. There have been several attempts to work out formulas which determine just when sidewalks are to be used. One formula recommended by the Oregon State Highway Department[2] correlates volume of vehicular traffic, number of pedestrians, speed, and other factors in terms of pedestrian accidents eliminated by the provision of sidewalks.

Where sidewalks are constructed, a Class 2 curb should be used to prevent vehicle encroachment. In developed areas, parking is usually common, and Class 2 curbs are used for this purpose. Along high-speed roads in rural areas, appropriate barriers should be constructed between the outside edge of the shoulder and the sidewalk, unless the distance separation is sufficiently great for pedestrian safety.

From a functional point of view, sidewalks should be designed with a surface smoothness equal to or better than that of the roadway. If it is rough or does not drain properly, pedestrians will use the roadway instead. It should be at least 4 ft in width if the sidewalk is to be kept more convenient than the roadway for walking purposes.

Guardrail. Where a road is constructed on fill, a guardrail is usually used at the edge of the shoulder to prevent vehicles from running off the embankment. Most highway design standards call for guardrail when the fill exceeds 10 ft in height.

[1] J. C. McMonagle, *Traffic Accidents and Roadside Features*, Highway Research Board, Bulletin 55, Washington, 1952.

[2] J. Beakey and F. B. Crandell, *A Study of Rural Sidewalks*, Oregon State Highway Department, Technical Report 44-3, Salem, August, 1944.

Of the many different types of guardrail, the two most frequently used are (1) steel cables mounted on guardrail posts and (2) steel plates mounted on posts in such a manner that they will provide a continuous line of plates on the traffic side of the guardrail. A serious objection to the cable type of guardrail is the tendency for slack in the cable to grip a vehicle and throw it into a side spin. The drawback to the steel plates is that a sharp impact may break the guardrail. There have been occasions when the sharp end of the broken part has pierced the body of the vehicle.

In recent years the California State Highway Department has experimented with a guardrail of the parabolic deflector type. The California type is similar to one-half of the mechanical divider or deflector, shown by Fig. 25-4. Its chief advantage is that it controls vehicles at their wheels; if properly designed, it may be unsurmountable and yet be able to hold the vehicle on the road without deflecting it back into traffic or causing it excessive damage.

Slopes of Embankment. For the purposes of safety, the slope of embankment outside the shoulder should be as flat as economic considerations permit. Slopes are expressed in terms of feet of horizontal distance per foot of rise or fall and should not be steeper than 4:1 where ditches are provided for drainage outside the shoulder. Ditch backslopes of 4:1 or less are desirable, but economy often dictates steeper slopes in heavy cuts or where the cut is made in rock. Ditches are usually eliminated on urban roads, and storm sewers and drop inlets are used for drainage.

Roadside Landscaping. Though aesthetic qualities of the roadside, of traffic separators, or of highway bridges are not the traffic designer's direct responsibility, it is clear that roadside beautification can improve form and appearance of the road, safety, and driver peace of mind as well as economy in maintenance.

Appropriate plantings on islands and separators may be used to block headlight glare. Plantings on separators as well as along the road may be used to mark crossovers, curves, and intersections. Trees on the tangents of curves and at the dead end of T intersections may indicate change of alignment at a greater distance than the condition itself could be seen. It is possible that a thick vigorous growth of shrubbery may cushion the impact of vehicles with far greater safety and less damage than mechanical separators.

Roadside plantings, however, may cause hazards if they are not intelligently applied and maintained. Shrubbery and trees on the inside of curves and at intersections may block vision if they are allowed to grow tall and are not trimmed. Plantings too close to the traveled way may affect vehicle placement, and trees may become lethal obstructions no matter how small they were at the time of planting.

CHAPTER 26

SIGHT DISTANCES

Restrictions of the greatest distance a driver can see (sight distance) may stem from such causes as parking, billboards, shrubbery on the inside of curves and at intersections, or may be attributed to hill crests, curves, and intersections located too close to sight obstructions of a permanent nature, or poorly designed traffic rotaries and grade-separated junctions. Short sight distances may seriously affect operations because drivers do not see hazards of the roadway or turn-off points in time to stop or to make proper maneuvers.

The sight distance required by drivers applies to both design and regulation of traffic facilities. It depends upon a number of driver and vehicle factors, chiefly (1) the PIEV time of drivers, (2) the time required to avoid a hazard or hazardous situation, and (3) speed. In the consideration of safe sight distances for passing and at intersections, the hazard is an approaching vehicle and lends the additional factor of the speed and distance of that vehicle from the point of potential conflict.

PIEV time (see Chapter 2) represents the perception, intellection, emotion, and volition of a driver expressed in seconds. It varies from $\frac{1}{4}$ to $1\frac{1}{2}$ sec for various types of laboratory tests. In most cases these tests involved a single stimulus and response with little element of surprise and no element of indecision, and the subjects knew they were being tested, so were unusually alert. For regulatory and design purposes, longer PIEV times are required to allow for indecision and inattention which drivers exhibit under actual driving conditions. Values range from 2 to 3 sec for the average driver.

Time required for drivers to accelerate, decelerate, and turn is the second factor of importance in sight-distance requirements. Speed is the third. If the PIEV time and the time required to perform the necessary maneuver are known, the distance required may be computed from the initial speed and rate of acceleration or deceleration. For design purposes, the initial speed of the driver and, when passing sight distance is involved, of the approaching vehicle is usually taken as the assumed design speed of the road.

414

Three sight-distance situations are provided for in design: (1) the emergency stop that requires a safe-stopping sight distance (safe-stopping sight distance is sometimes also called nonpassing sight distance because it is not long enough to permit a safe passing maneuver), (2) safe-passing sight distance, when vehicles overtake and pass each other, and (3) safe sight distance for entrance into intersections.

26-1. Safe-stopping Sight Distance (Daytime). The required safe-stopping sight distance of a vehicle is equal to the sum of the PIEV distance and the distance required to bring the vehicle to a stop.

For some time it has been a design policy to provide safe-stopping sight distances throughout the length of all improved roads. On some older roads, however, the speed assumed at the time of design was too low for present speed practices, with resulting need for speed zoning or redesign to decrease hazard. This deficiency has developed in spite of the fact that present-day four-wheel brakes are much better than the two-wheel type which were in use when many of these roads were designed.

In determining braking distances, it is assumed that road-surface friction, not the friction in the brake mechanism, is the most critical.

PIEV Distance. The American Association of State Highway Officials recommends a value of 2.5 sec for the PIEV time of drivers to be assumed in the computation of safe-stopping sight distances.[1] This value is applied throughout the range of assumed design speeds.

PIEV distance, in feet, is obtained from the formula $Sp = 1.47PV$, when Sp is the PIEV distance in feet, P is the PIEV time in seconds, and V is the assumed design speed of the road in mph. Since it is believed that the PIEV times used for design purpose include a factor of safety, when this formula is used for speed zoning and other regulatory purposes, a lower PIEV value is sometimes used. So little factual data have been obtained on PIEV time under actual driving conditions that the values used for regulation and design are subject to varying interpretations.

Braking Distance. Minimum braking distance on a level road is derived from the formula $S_B = V^2/30f$, where S_B is the minimum braking distance in feet, V is the assumed speed of the road in miles per hour, and f is the coefficient of surface friction assumed for maximum braking. This formula is developed in Chapter 3.

The AASHO has recommended a maximum value of 0.36 for f when V is 30 mph, ranging to 0.29 when V is 70 mph. The Association also recommends the following design-speed equivalents for wet pavements:[2]

[1] *A Policy on the Geometric Design of Rural Highways,* chap. 3, "Elements of Design," American Association of State Highway Officials, Washington, 1954, p. 112.

[2] *Ibid.,* p. 115.

Design speed of road, mph	Assumed wet-pavement equivalent, mph
30	28
40	36
50	44
60	52
70	59

The safe-stopping sight distances derived by the use of these speed and friction values are longer than those based on actual design speeds and the friction offered by dry pavements.

The friction values assumed by the Association for wet pavements are based on the results of stopping-distance studies made in the field for a wide variety of tires, surface types, and surface conditions. They were selected to conform with the minimum friction offered by smooth tires on the vast majority of clean, wet-surface types in the case of straightforward skid with 100 per cent slippage when stopping from various speeds of travel. They are higher than, and do not provide for, the low friction offered by foreign materials on pavements, such as ice, mud, etc., nor for the low friction offered, when wet, by improperly designed "bleeding" surfaces.

A recent study[1] has indicated that although kinetic energy is a function of the square of the speed, braking distance, with increased speed, increases at a rate somewhat greater than the square of the speed. Some engineers believe that safe-stopping sight distances based on AASHO assumptions should be increased 20 to 30 per cent.[2]

It is clear that the safe-stopping sight distances employed in highway design are subject to different assumptions. While there have been recent changes by the AASHO in the values assumed for speed, PIEV time, and friction, the derived safe-stopping sight distances recommended by the Association remain the same as they have been for many years.[3] Accident experience over the years has not indicated that these values are inadequate.

Minimum Requirements. Adding the PIEV and braking distances together, the minimum safe-stopping sight distance required by drivers becomes $SSSD = 1.47PV + V^2/30f$. The safe-stopping sight distances shown for various assumed design speeds on the left side of Table 26-1 are minimum recommended distances, and it is the desire of the AASHO

---

[1] O. K. Norman, "Braking Distances of Vehicles from High Speeds," U.S. Bureau of Public Roads, *Public Roads,* vol. 27, no. 8, June, 1953, p. 159.

[2] *Highway Practice in the United States of America,* U.S. Bureau of Public Roads, 1949, p. 96.

[3] The AASHO recommended policy discussed herein is according to current revised policy. A previous policy published in 1940 employed different assumed values but resulted in the same safe-stopping sight distances as currently recommended.

to suggest longer distances whenever it is economically feasible to provide them. The right-hand column of Table 26-1 shows comparable minimum distances recommended by the National Interregional Highway Committee. The increase is approximately 15 per cent, as compared with the suggested 20 to 30 per cent mentioned above.

TABLE 26-1. RECOMMENDED MINIMUM SAFE-STOPPING SIGHT DISTANCES (DAYTIME)

Assumed design speed, mph	Level road distances, ft	
	AASHO	NIHC
30	200	
40	275	
50	350	400
60	475	525
70	600	700

SOURCE: *Highway Practice in the United States of America*, U.S. Bureau of Public Roads, 1949, p. 96, and *Interregional Highways*, U.S. Bureau of Public Roads, Interregional Highway Committee, Appendix V, p. 151, Table 2.

The Effect of Grade. The component of weight acting on a vehicle along the plane of a grade is equal to approximately 20 lb/ton/per cent grade (Chapter 3). Thus when the effect of grade G is included, the safe-stopping sight distance formula becomes $SSSD = 1.47PV + V^2/30(f + g)$, where g is per cent grade $\times 20/2,000 = G/100$.

Gradient has a significant effect on safe-stopping sight distances. For example, assuming a friction factor of 0.30, the braking distance on a 6 per cent upgrade from a vehicle speed of 50 mph is 232 ft as compared with 348 ft on the same downgrade from the same speed.

26-2. Safe-passing Sight Distance. Safe-passing sight distance applies only to two- and three-lane roads where common lanes are used by traffic moving in opposite directions. Where safe *daytime*-passing sight distances are available, *night* sight distances are greater, because the presence of approaching vehicles is indicated by the glow of their headlights some time in advance of their actual appearance.

While it has been standard practice to provide safe-stopping sight distances throughout the length of all improved roads, a similar policy for safe-passing sight distances, which are generally from three to five times longer, would present a severe economic problem. The practical impossibility of providing continuous safe-passing sight distances on rolling or hilly terrain or through developed areas presents the decision of how often passing sight distances should be provided.

Safe-passing sight distances involve many variables: the speeds of all vehicles involved in the maneuver, the number of vehicles passed or passing in one maneuver, clearances, driver reaction and skill, rates of acceleration, etc.

No significant factual studies have been made of passing requirements on three-lane roads, of passing maneuvers involving more than one passed or passing car, or truck-passing practices.

Theoretical Studies. The most critical passing situation occurs when the passing driver must slow down to the speed of the passed car before starting his maneuver and when he is required to hurry back into the

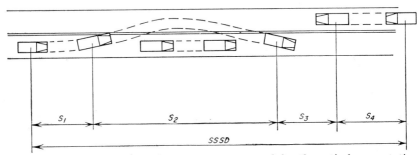

Fig. 26-1. Overtaking-and-passing maneuver assumed for theoretical computations on two-lane roads during daylight conditions.

right lane by an oncoming car as he finishes the maneuver. This situation is assumed for the theoretical consideration of passing sight distances (Fig. 26-1).

The distance required for a passing maneuver includes (1) the distance traveled during the passing driver's PIEV time, (2) the distance traversed while the passing driver is in the left lane, (3) clearance distance between the passing vehicle and an oncoming vehicle at the end of the passing maneuver, and (4) the distance traversed by the oncoming vehicle during the passing maneuver.

PIEV Distance. PIEV time, in the case of passing maneuvers, is assumed to be the time elapsed between the beginning of the perception interval and the encroachment of the passing vehicle on the left lane. The distance traversed during this PIEV time is $S_1 = t_1 [v_1 + (at_1/2)]$, where S_1 is the PIEV distance in feet, t_1 is the PIEV time in seconds, v_1 is the speed of the passing vehicle at the beginning of the maneuver in feet per second, and a is the rate of acceleration, ft/sec².

Distance Traversed in Left Lane. The distance traversed while the passing vehicle is in the left lane is $S_2 = v_2t_2$, when S_2 is the distance traversed in the left lane in feet, v_2 is the average speed of passing vehicle while in the left lane in feet per second, and t_2 is the time the passing vehicle occupies the left lane in seconds.

Clearance Distance. Clearance distances between the oncoming vehicle and the passing vehicle at the end of the maneuver are

$$S_3 = t_3 \left(\frac{v_3}{2} + \frac{v_4}{2} \right)$$

when S_3 is the clearance distance in feet, t_3 is the clearance time in seconds, v_3 is the speed of the passing vehicle at the end of the maneuver in feet per second, and v_4 is the speed of the oncoming vehicle in feet per second.

Distance Traversed by Oncoming Vehicle. The sight distance should be long enough to reveal oncoming vehicles close enough or traveling fast enough to interfere with the pass. This distance is $S_4 = v_4 t_4$, where S_4 is the distance traversed by the oncoming vehicle in feet, v_4 is the speed of the oncoming vehicle in feet per second, and t_4 is the time passing vehicle must have either to complete the pass safely or decelerate and return to the right lane without passing, in seconds.

Minimum Design Values. The total sight distance required for overtaking and passing maneuvers is $S_1 + S_2 + S_3 + S_4$. It is evident that values for various speeds, time intervals, and acceleration must be known before the formulas may be solved to obtain theoretical safe-passing sight distances.

The AASHO has adopted values to be used in the theoretical computation of minimum safe-passing sight distances.[1] These values were obtained from extensive field studies of actual overtaking and passing maneuvers.[2] Based on these empirical values, adjusted for consistencies, the elements shown by Table 26-2 were established. In the application

TABLE 26-2. ELEMENTS OF SAFE-PASSING SIGHT DISTANCES FOR
TWO-LANE HIGHWAYS

Speed group, mph	30–40	40–50	50–60
Average passing speed, mph	34.9	43.8	52.6
a_1, ft/sec²	1.40	1.43	1.47
t_1, sec	3.6	4.0	4.3
t_2, sec	9.3	10.0	10.7
S_3, ft	100	180	250
S_4 (⅔ of S_2), ft	315	425	550

of these elements for the computation of minimum theoretical safe-passing sight distances, the Association assumed the following conditions:[3]

[1] *A Policy on the Geometric Design of Rural Highways*, p. 119.

[2] C. W. Prisk, "Passing Practices on Rural Highways," *Proceedings of Highway Research Board*, Washington, 1941, vol. 21, p. 366.

[3] *A Policy on the Geometric Design of Rural Highways*, p. 118

1. The passed vehicle travels at a uniform speed during the maneuver.

2. The passing vehicle trails the overtaken vehicle and starts the passing maneuver from the same speed as the passed vehicle.

3. The passing vehicle accelerates during the maneuver, and its average speed during occupancy of the left lane is 10 mph higher than that of the passed vehicle.

4. The oncoming vehicle travels at the same speed as the passing vehicle, and the critical time element t_3 is two-thirds of the time that the passing vehicle is in the left lane.[1]

Minimum safe-passing sight distances recommended by the AASHO are shown by Table 26-3. No data are available related to passing requirements on three-lane roads. The AASHO assumes that the minimum passing sight distances desirable for three-lane-highway design are based on the same elements used for the theoretical computations of passing distances for two-lane roads, except that the distance S_4 is omitted. Table 26-3 shows the minimum safe-passing sight distances recommended by the Association for both two-lane and three-lane highways.[2] The differences between the theoretical values for two-lane roads shown in the table and the values obtained by actual studies of overtaking and passing on two-lane rural highways are small.

TABLE 26-3. MINIMUM PASSING SIGHT DISTANCES FOR THE DESIGN OF
TWO- AND THREE-LANE HIGHWAYS
(Recommended by AASHO)

Design speed, mph	Assumed passing speed, mph	Minimum passing sight distance, ft	
		Two-lane roads	Three-lane roads
30	30	800	
40	40	1,300	
50	48	1,700	1,200
60	55	2,000	1,400
70	60	2,300	1,600

In Table 26-3, the assumed passing speed considered equivalent to the assumed design speed is arbitrary but based on observations[3] that the majority of passing maneuvers are made at speeds which are below that assumed for design.

Table 26-4 shows empirical values of time and distance (in left lane) of passing maneuver for different combinations of speeds. It is interesting to note that the time and distance values become longer with increased

---

[1] It is assumed by AASHO that the passing driver could return safely to the right lane without completing the passing maneuver during the first one-third of the time he occupies the left lane.

[2] *A Policy on the Geometric Design of Rural Highways,* pp. 121, 122.

[3] *Highway Capacity Manual,* Highway Research Board and U.S. Bureau of Public Roads, Committee on Highway Capacity, 1950, p. 58.

speed of the passed vehicle, but that the time decreases while the distance increases as the speed of the passing vehicle becomes higher. It is significant that the appropriate distances obtained empirically (Table 26-4) are very similar to those obtained for S_2 when the assumed values shown in Table 26-2 are employed.

TABLE 26-4. RESULTS OF OVERTAKING AND PASSING STUDY, AVERAGE LEFT-LANE PASSING DISTANCES AND TIMES

(All types of passes—two-lane highways)

Speed of passing vehicle, mph	Speed of passed vehicle, mph									
	0–19		20–29		30–39		40–49		50–59	
	Dist., ft	Time, sec	Dist., ft	Time, sec	Dist., ft	Time, sec	Dist., ft	Time, sec	Dist., ft	Time, sec
50–59	...	...	660	8.2	730	9.0	820	10.1	900	11.1
40–49	560	8.5	550	8.3	620	9.4	850	12.4		
30–39	410	8.0	450	8.8	600	10.6				
20–29	330	9.0	380	10.3						

Adopted from C. W. Prisk, "Passing Practices on Rural Highways," *Proceedings of Highway Research Board*, Washington, 1941, vol. 21, p. 372, Fig. 3.

Frequency of Provision. The frequency of safe-passing sight distances on two- and three-lane highways has a definite effect on the capacity and safety in operations. Table 26-5 indicates the effect of sight-distance restrictions on the capacities of two-lane roads.

TABLE 26-5. EFFECT OF PASSING SIGHT-DISTANCE RESTRICTION ON THEORETICAL DESIGN CAPACITIES OF TWO-LANE ROADS

Sight distance restricted to less than 1,500 ft; percentage of total length of road	Practical design capacity in passenger cars per hour	
	Operating speed 45–50 mph	Operating speed 50–55 mph
0	900	600
20	860	560
40	800	500
60	720	420
80	620	300
100	500	160

SOURCE: *Highway Capacity Manual*, Highway Research Board and U.S. Bureau of Public Roads, Committee on Highway Capacity, 1950, p. 58, Table 11. (NOTE: "Practical capacity" in the *Highway Capacity Manual* is the same as "practical design capacity" used here.)

When the location of a road presents frequent sight restrictions, provision of the safe-passing sight distances required for high speeds is often not economically practical. In some cases, construction costs would be little greater and traffic operations would be greatly improved by the construction of a four-lane road—on which operations are not affected

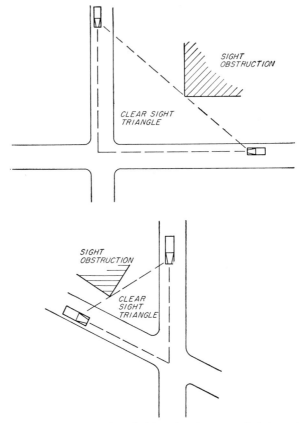

FIG. 26-2. Typical examples of sight triangles at grade intersections.

by safe-passing sight distances—rather than the construction of a two-or three-lane road where provision of adequate passing sight distances, at considerable expense, is required.

26-3. Intersection Sight Distances. The most critical sight requirement at intersections usually occurs at their corners. Thus it is important that drivers on all approaches of intersection roadways be provided a clear view across the corners created by adjacent approaches, beginning at a sufficient distance from the point of intersection to avoid a collision. This area of unobstructed sight is called the *sight triangle*. Typical sight triangles are shown in Fig. 26-2.

No distinction is made between daytime and nighttime conditions of vision at intersections. It seems reasonable to assume that headlights of cars approaching the intersection will be seen across the corner area, and the headlight beams would usually indicate their presence before they actually come into view.

The AASHO has established three possible conditions upon which the design of sight distance at intersections may be based:[1] (1) enabling vehicles to change speed, (2) enabling vehicles to stop, and (3) enabling stopped vehicles to cross a preference highway.

The first condition arbitrarily assumes that 2 sec are required for driver PIEV time and an additional second is required to make the change in speed. To satisfy this condition, the two sides of the sight triangle which fall along the intersection approaches become the distance from the conflict point the intersecting vehicles would travel in 3 sec when traveling at the assumed design speed of the respective roads. This sight distance is the absolute minimum and is not recommended except on lightly traveled roads, where additional sight distance would be too costly in terms of the importance of the road.

Condition 2 assumes that the driver on either road must see the other vehicle in time to stop before reaching the point of collision. In this case, the two sides of the sight triangle become the same safe-stopping sight distances shown in Table 26-1 taken for the assumed speed of the intersecting roads. This sight distance is considerably greater than that provided by condition 1 and therefore provides greater safety because the drivers may not only change speed but may also stop if necessary.

Enabling Vehicles to Stop or Proceed. Condition 2 is intended by the AASHO to apply to intersections without positive traffic control and assumes that one or both vehicles will stop if necessary. It seems desirable, when economically practical, to consider this condition also in terms of the *relative speed* of the intersecting vehicles. At nearly every uncontrolled intersection, one of the roads is a preference (or through) road, and drivers on the other (or minor) road have the responsibility of entering the road safely.

Drivers crossing or entering a major traffic stream at an uncontrolled intersection must decide to stop or proceed at a distance from the point of potential collision at least as great as the safe-stopping distance. When the decision is made to cross or enter, the crossing or entering driver must also be able to see to the side a vehicle in the major stream flow which will arrive at the collision point when he does, plus an additional distance equivalent of 3 sec (see Chapter 8, sections 8-3 and 8-4).

Figure 26-3 is a diagram of the sight triangle involved, and the critical conditions for minimum sight distances are (1) the speed of both vehicles

[1] *A Policy on the Geometric Design of Rural Highways*, chap. 8, "At Grade Intersections," p. 313.

is the assumed speed of the respective roads; (2) the decision to stop or proceed is based on the potential time advantage computed without reduction in speed; and (3) the responsibility for safe entrance rests with the driver on the minor road.

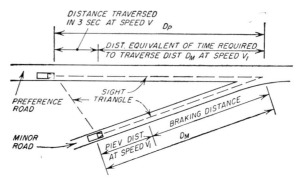

FIG. 26-3. Suggested sight triangle distances for crossing or entering a preference road. (V = assumed design speed of preference road, mph; V_1 = assumed design speed of minor road, mph.)

The dimension D_M in Fig. 26-3 is the safe-stopping sight distance listed in Table 26-1. D_p then becomes

$$D_p = \frac{D_M V}{V_1} + 4.41V$$

where D_p is length of minimum sight triangle along preferential road in feet, D_M is length of minimum sight triangle along minor road in feet, V is assumed design speed, preferential road, in miles per hour, and V_1 is assumed design speed, minor road, in miles per hour.

It will be noted that the above formula does not consider the speed differential which may occur between vehicles at the point of potential conflict. Merging vehicles should maneuver at about the same speed if conflict is to be avoided. For this reason, V_1 and V should be about the same (within approximately 10 mph of each other) if the formula is to be employed for merging traffic. It is reasonable to assume that both vehicles can accelerate or decelerate sufficiently to overcome a 10-mph speed differential within the distances provided by the formula. When V_1 and V differ widely, greater sight distances may be required for speed adjustments.

Table 26-6 shows the minimum dimensions of the sight triangle for various assumed speeds when the differentials in speed are 10 mph. These dimensions should allow the entering or crossing driver to proceed safely into the intersection without stopping when no interfering vehicles are visible within the distance D_p. The assumptions made in the development of the formula for D_p and, accordingly, the values shown in

Table 26-6 are subject to considerable controversy. Thus, this analysis is not employed as a criterion in current design. It does, however, indicate need for further research.

TABLE 26-6. MINIMUM DIMENSIONS OF SIGHT TRIANGLE FOR CROSSING OR ENTERING MAJOR TRAFFIC STREAMS AT UNCONTROLLED INTERSECTIONS

Design speed, mph, minor road V_1	D_M, ft	Design speeds, mph, preferential road V	D_p
30	200	30	332
		40	443
40	275	40	452
		50	565
50	350	50	570
		60	685
60	475	60	739
		70	863
70	600	70	908

NOTE: This table is based on the previously discussed formula and assumes that the 3-sec gap acceptance criterion will apply when the speed differential is 10 mph or less.

Enabling Stopped Vehicles to Cross a Major Highway. This is condition 3 suggested by the AASHO for intersections where traffic on the minor road is controlled with a "stop" sign. It also applies at an uncontrolled intersection when drivers crossing the major flow reject gaps in that flow and must stop.

Intersections should not be located near hill crests or on horizontal curves where sight distance is restricted. A critical situation exists when a vehicle stopped on the crossroad must accelerate across the preference road. Sight distance for this situation involves the PIEV time of the crossing driver, acceleration time, length of crossing, and length of vehicle, on one hand, and the speed of an approaching vehicle on the preference road, on the other.

Thus the minimum sight distance along the preference road is equal to the product of the assumed design speed of the preference road and the time required for the crossing vehicle to clear the intersection. By formula this becomes

$$S = 1.47V(P + T)$$

where S = required sight distance along preference road, ft
V = assumed design speed of preference road, mph
P = PIEV time of the crossing driver, sec
T = time elapsed while vehicle is moving to clear intersection, sec

The AASHO recommends a PIEV time of 2 sec for this situation.[1] *Distance Traversed While Clearing the Intersection.* The distance a crossing vehicle requires to clear the preference road depends upon (1) distance stop line is from edge of pavement, (2) width of pavement of preference road, and (3) length of vehicle.

Where medial dividers are capable of shadowing crossing vehicles in the middle of the preference road, the crossing is made in two steps and the distance factor must be adjusted accordingly. Also, the zone of danger for the crossing vehicle may be reduced when the sight-distance restriction is in one direction only, as in the case of a hill crest.

Time for Vehicle to Clear Intersection. The acceleration of vehicles crossing a through road from a stopped position has been observed at three intersections in New Haven and Hartford, Conn., and New York, with surprising consistency in results[2] (Table 26-7). These intersections were traffic-signal controlled and accordingly were indicative of normal-acceleration practices. The normal-acceleration characteristics of buses resemble very closely those values in the table shown for passenger cars.

TABLE 26-7. NORMAL AVERAGE ACCELERATION RATES FROM A STOP AT
INTERSECTION, EXPRESSED IN DISTANCE AND TIME
(Does not include PIEV time)

Passenger vehicle and buses		Trucks	
Distance, ft	Time, sec	Distance, ft	Time, sec
30	3.5	30	3.9
50	4.7	50	5.4
70	5.7	70	6.5
90	6.8	90	7.6
110	7.8	110	8.8
130	8.8	130	9.7

Minimum Requirements. The time shown in Table 26-7 may be substituted as the value of T in the formula $S = 1.47V(P + T)$ provided that the distance entered from Table 26-7 includes the width of pavement, distance from stop line to pavement, and length of vehicle. For example, a 40-ft undivided roadway with stop line 20 ft from the pavement edge calls for an acceleration distance of 80 ft for passenger cars and 100 ft for trucks. The value of $P + T$ in the formula then becomes $2 + 6.2 = 8.2$ sec and $2 + 8.2 = 10.2$ sec, respectively. If crossing trucks contribute to the hazard and the design speed of the preference road is 70 mph, this

[1] *Ibid.*, p. 316.
[2] B. D. Greenshields, D. Schapiro, and E. L. Ericksen, *Traffic Performance at Urban Street Intersections*, Yale University, Bureau of Highway Traffic, Technical Report 1, New Haven, Conn., 1947, p. 19.

example indicates a minimum sight distance along the preferential road of $(1.47)(70)(10.2)$ or 1,050 ft.

Under no circumstances should a value of $P + T$ less than 6 sec be used. Empirical studies of minimum gaps required by drivers to enter or cross a moving traffic stream from a stopped position have shown that the average driver requires a 6-sec gap between vehicles in the moving stream.[1] This is sufficient evidence to establish the absolute minimum of 6 sec for very narrow roadways.

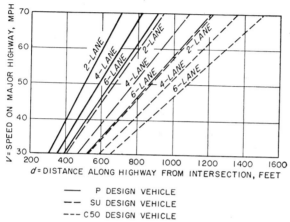

— P DESIGN VEHICLE
— — SU DESIGN VEHICLE
— — — C 50 DESIGN VEHICLE

Fig. 26-4. Sight distance at intersections, Case III: required sight distance along major highway. (From *A Policy on the Geometric Design of Rural Highways*, American Association of State Highway Officials, Washington, 1954, chap. 8, Fig. 8-5.)

Figure 26-4 shows the sight distances along the major road required for condition 3, when values for P and T recommended by the AASHO are used.

Enabling Vehicles to Enter a Preference Highway from a Stop. Vehicles turning from the minor road become a part of the preferential flow, and their entry continues to be a source of hazard and interference until they have reached the speed of the through traffic on the preference road. Therefore, the minimum sight distance required for entering a preferential flow from a stopped position involves:

1. The PIEV time of the entering driver
2. The time required to accelerate to the speed of the preferential flow from a stopped position
3. The distance traveled by the vehicle in the preferential flow
4. A desirable headway clearance at the end of the maneuver

The time-space relationships of these factors for a merging vehicle are shown by Fig. 26-5.

By formula,

$$S = 1.47V(t + P) + 1.47VC - D_1$$

[1] *Ibid.*, p. 71.

where S = sight distance required along preference road, ft

D_1 = distance traveled during acceleration from stopped position to speed of preferential flow, ft

t = time of acceleration from stop to speed of preferential flow, sec

P = PIEV time for stopped driver, sec

V = assumed speed of preference road, mph

C = clearance-time headway at end of maneuver, sec

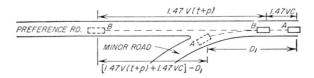

A = VEHICLE STARTING FROM STOP ON MINOR ROAD
B = VEHICLE ON PREFERENCE ROAD TRAVELING AT V MPH

Fig. 26-5. Suggested sight distances for merging from a stopped position with a preference road.

PIEV Time. The PIEV time of drivers crossing a preference road from a stop—2 sec—is also recommended for entering a preference road under the same speed conditions.

Time and Distance Required to Accelerate to Speed of Preference Road. Table 26-8 shows normal and full acceleration rates for passenger vehicles

TABLE 26-8. SPEEDS REACHED, DISTANCES, AND ACCELERATION TIMES OF VEHICLES ACCELERATING FROM A STOP

Speed reached, mph	Time required, sec		Distance traveled, ft	
	Normal accel.	Full accel.	Normal accel.	Full accel.
20	9	6	140	89
30	15	9	376	222
40	23	14	778	448
50	32	19	1,393	812

SOURCE: D. W. Loutzenheizer, "Speed Change Rates of Passenger Vehicles," *Proceedings of Highway Research Board*, Washington, 1938, vol. 18, part 1, p. 90.

starting from a stopped position in terms of the speed reached, distance traveled, and time required. The data for this table were obtained by field observations. The observations included passenger vehicles traveling along a straight path. No conclusive information is available on the acceleration characteristics of turning vehicles and trucks.

By entering in Table 26-8 the assumed speed of the preference road, values for t and D_1 may be found for substitution in the formula.

Clearance Headway at End of Maneuver. Figure 7-8 indicates that a headway of 2 sec is a reasonable average minimum headway to assume for clearance without appreciable interference with preferential flow.

Minimum Requirements. From the previous formula and with the assumed and measured values, sight-distance requirements for vehicles entering a preference road from a stop may be computed.

Table 26-9 shows the sight-distance requirements for various assumed speeds of the preference road when normal acceleration and full acceleration are used. Sight-distance requirements are reduced considerably at higher assumed speeds when full-acceleration values rather than normal-acceleration values are used.

TABLE 26-9. DESIRABLE SIGHT DISTANCES FOR STOPPED VEHICLES ENTERING A PREFERENCE ROAD

Assumed speed of preference road, mph	Suggested sight distance	
	Normal accel.	Full accel.
20	242	205
30	462	351
40	810	610
50	1,253	879

The sight distances indicated in Table 26-9 are not employed as a practical sight-distance control in current design practice. In the first place, intersection approaches for merging with high-speed traffic should not be designed for stop control. Secondly, when vehicles merge from a stop there is a tendency for drivers in the intersected flow to reduce speed or change lane, on multilane facilities, to make way for the merge. The table does provide a theoretical illustration of the sight distances involved when stopped vehicles are required to merge with through nonstop traffic.

26-4. Night Sight Distances. Night vision on the highway is limited by headlight ranges, and matters such as the reflection factor of objects; their brightness, brightness contrast, and visual size become far more critical at night than during the daytime.

In the absence of a better understanding of this subject, any discussion concerning it can be no more than a description of its most important elements: (1) obstacle factors, (2) driver factors, (3) headlight factors, and (4) highway factors.

Critical Night-visibility Conditions. Safe-passing sight distances and sight distances at intersections are not critical at night because the presence of approaching vehicles is usually announced in advance by their headlights. On the other hand, visibility distances available for safe stopping during nighttime conditions are much less than during daylight.

Night visibility is the least when drivers are subjected to glare from approaching headlights or other extraneous light sources. Atmospheric conditions may also greatly reduce visibility at night. However, these conditions should probably not be taken into consideration for design purposes because they occur too seldom to make it economically feasible to design for them. Then, too, it may be assumed in design that such abnormal conditions are compensated for by drivers through reduced speed and increased alertness.

Obviously, illumination of an obstacle from street lights or the head-lights of other vehicles will increase the visibility distance, provided glare is not present. Reflectorization of an obstacle with materials of adequate reflective qualities will also warn drivers in advance of the distance at which the obstacle actually becomes visible. It is when these aids are not present that the need for adequate stopping sight distance is greatest.

Obstacle Factors. The reflection factor of a "white" diffusing surface is better than 90 per cent. A pedestrian in dark clothing has a reflection factor of about 2 per cent. Most objects encountered in night driving have a reflection factor of about 7 per cent.[1]

Brightness contrast (the difference between an object and its background) is also a factor in night visibility. When the only illumination is the driver's own headlights and the reflection factor of the obstacle is greater than that of the pavement, the obstacle appears brighter than the pavement and is seen by direct vision.

Driver Factors. Not much is known about the time required for a driver to perceive and react to a hazard at night. Eyestrain and glare, along with the physical and mental fatigue experienced at the end of the day, probably are responsible for a longer reaction time at night than during the day. Visual stimuli under low levels of illumination are much weaker than during daytime and therefore do not compel as much attention.

Quantitative measurements of driver-perception distances at night have been made in an exploratory investigation by engineers at Nela Park in Cleveland. These studies[2] show that drivers see the expected obstacle, at night, at about twice the distance that they see an unexpected one. If expected, the obstacle is often seen by silhouette (it appears darker than the background pavement). More brightness and brightness contrast are required to gain the driver's attention if the obstacle is unexpected. Most unexpected obstacles are perceived only after the vehicle is close enough to illuminate the hazard sufficiently for direct vision. The difference between the perception distance of the expected and of the unexpected obstacle takes on more meaning when it is realized that the

[1] Val J. Roper and E. A. Howard, "Seeing with Motor Car Headlamps," *Transactions of the Illuminating Engineering Society*, vol. 33, New York, May, 1938.

[2] *Ibid.*

level of illumination upon the obstacle varies inversely as the square of the distance from the light source.

Headlight Factors. The hazard caused by headlight glare has necessitated limitations on beam candlepower and the pattern of light emitted by headlights.

The distribution of light within the headlight pattern is controlled by both the Uniform Vehicle Code and specifications set forth by the Society

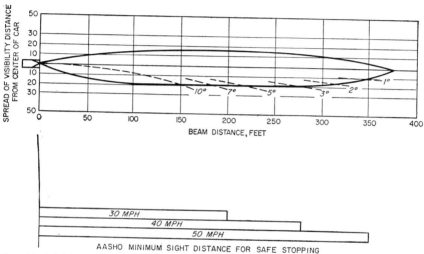

FIG. 26-6. AASHO minimum sight distance for safe stopping. (Dotted lines are horizontal curves with *P.C.* at 0 distance; headlight distribution based on 1949 SAE specifications.)

of Automotive Engineers. Figure 3-13 shows the candlepower distribution of the driving (upper) beam of two 20,000-candlepower head lamps in excellent condition.

By applying the inverse square rule to this distribution, foot-candle contours may be located along the horizontal and vertical planes of the combined headlight beam. This was done to obtain the visibility distances shown in Figs. 26-6 and 26-7. An intensity of 40,000 candlepower was used, since it is probably more representative of the headlighting of average vehicles in the traffic stream.

The mounting height of headlights is about $2\frac{1}{2}$ ft above the surface of the road. This low position, doubtless, was chosen to aid in the reduction of headlight glare. At this height, headlights are about 2 ft below the level of the driver's eyes. As the top of a hill is approached, the distance of surface illumination within the headlight range is limited to the length of a line from the headlight to a point of tangency with the vertical curve of the road. Applying the same principle, a driver, from his higher vantage point, is in a position to see the road surface for a greater distance if it is illuminated by his headlights.

Higher mounting of head lamps near the driver's eye would also increase the reflective brilliancy of retrodirective reflectorizing materials used on signs and obstacles.

Highway Factors. Figure 26-6 shows the beam distance at which a driver should see an unexpected obstacle with 7 per cent reflection factor

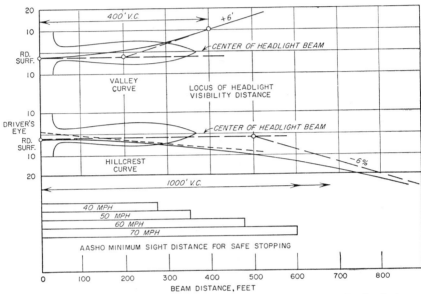

Fɪɢ. 26-7. Visibility distance of unexpected object with 7 per cent reflection factor on vertical curves with 40,000 beam candlepower headlights, upper beam. (Headlight distribution based on 1949 SAE specifications.)

and with 40,000 headlight-beam candlepower on the straightaway and on various horizontal curves. Fig. 26-7 shows the same information for a valley and hill-crest vertical curve. On both figures, the minimum safe daytime stopping distances recommended by the AASHO are given in graphic form for comparison with nighttime visibility distances. For these figures it was assumed that the "hot spot" of the headlights is directed straight ahead. Vehicle loads will change the vertical angle of headlight beams, and erratic steering as well as bumps in the road surface will cause the beam to shift around on the road ahead. It was also assumed that a reflection factor of 7 per cent applies to the average obstacle and that it will be seen, when it is not expected, at one-half the distance of the 0.075-ft-c contour.[1]

[1] These assumptions are based upon the limited information available on the subject. It is believed that they give a reasonably accurate measurement of night-driving visibility conditions.

The figures clearly show that more illumination than that provided by modern headlights is required on straightaways for safe-stopping sight distances on roads designed for speeds above 50 mph. They also show that wider distribution of the headlight beam on the horizontal and vertical planes, along with increased candlepower, is essential for improved night sight distance on vertical and horizontal curves.

Effect on Highway Design. The major difference between the adequacy of stopping sight distances provided by the road during daytime as compared with nighttime is the amount and area of illumination. Accordingly, traffic designers may be justified in assuming the night sight-distance improvement should be achieved through better headlighting and street lighting or the use of reflective devices rather than through design.

Figures 26-6 and 26-7 indicate that, in view of present headlighting, any gain through design would be necessarily small. This relatively small gain would be achieved at disproportionate expense caused by the need for almost straight and level alignment.

Traffic islands and other traffic channelizing controls have demonstrated their usefulness in the reduction of accidents and congestion. Poor nighttime visibility discourages their application on high-speed roads where they are needed the most. Figures 26-6 and 26-7 show why a raised island or any other design element which is an obstruction becomes particularly lethal at night when located on a horizontal or vertical curve.

CHAPTER 27

VERTICAL ALIGNMENT

Vertical road geometry is the profile of a road along its longitudinal axis. *Vertical alignment* is the elevation of the road surface, usually applied to the center of the road for two- and three-lane pavements and to the center of each pavement on divided roads. This elevation is called *profile grade*, and all other elevations of the cross section are referred to it. Vertical alignment consists of grades and vertical curves. It influences vehicle operations through its effect on vehicle speed, acceleration, deceleration, turning, stopping, and driver's sight distance and comfort.

27-1. Grade. Grade is the number of feet of vertical rise and fall in each 100 ft of horizontal distance. Since road plans usually read from left to right, grades shown on the profile are taken as plus or minus, depending on whether the road rises or falls in the right-hand direction on the plans. If the profile grade rises at a rate of 5 ft vertically for each 100 ft of horizontal distance, it is a plus 5 per cent grade.

Grade Ability of Trucks. Grades over 10 per cent are rarely employed in highway design except on unusual mountainous roads or special ramps. Since passenger vehicles are equipped with sufficient power to ascend grades up to 10 per cent without appreciable reduction in speed, they do not present a problem in gradient design.

Medium and heavy trucks, however, show a substantial reduction in speed on grades of 3 per cent and over, where their power is severely taxed in overcoming the grade resistance of their gross weights. Slow trucks on grades may seriously limit the speeds of passenger cars and faster trucks, especially on two- or three-lane roads. Because grades are generally accompanied by limited sight distances at hill crests, unsafe passing practices on two- or three-lane roads may result when drivers of faster vehicles become impatient while following slow trucks.

Most truck drivers enter ascending grades at the highest speed considered safe under prevailing conditions in order to utilize as much momentum as possible in overcoming the resistance of the grade. However, on long grades the momentum is eventually lost, and a vehicle operating at full throttle reaches a constant speed called the vehicle's *sustained speed.*

Sustained Speeds on Long Grades. Sustained speed occurs when the maximum horsepower developed by the engine is equal to the horsepower required to overcome resistance to motion on the grade [the sum of the grade, rolling, and air resistance (see Chapter 3)]. If extra power is available, the truck will be accelerated. If insufficient power is available, the truck will stall.

Table 27-1 shows the average grade abilities of 30 individual trucks obtained by actual grade tests made with new 1938, 1939, and 1940 model

TABLE 27-1. SUSTAINED TRUCK SPEEDS ON GRADES

Truck type	Gross weight, lb	Sustained speed classified by gradient, mph					
		2%	3%	4%	5%	6%	7%
Light	10,000	44	41	36	32	29	27
	20,000	30	27	21	17	16	14
	30,000	23	17	15	12	10	7
	40,000	17	14	10	7	5	4
Medium	10,000	Over 45	44	41	37	33	30
	20,000	34	29	24	22	19	16
	30,000	25	22	17	14	13	11
	40,000	21	16	13	11	9	8
Heavy	15,000	45	40	37	33	29	26
	20,000	40	35	31	26	22	20
	30,000	30	26	21	18	15	13
	40,000	24	20	17	13	11	9

SOURCE: C. C. Saal, "Hill Climbing Ability of Motor Trucks," U.S. Bureau of Public Roads, *Public Roads*, vol. 23, no. 3, May, 1942, p. 39.

trucks with gross weights of 5 to 20 tons on grades ranging from 2 to 7 per cent. Figure 27-1 shows the influence of length and steepness of grade on the speed of a truck with 30,000 lb gross load.

Implications in Design. The problem of reduced truck speeds on grades may be approached by (1) changing the characteristics of the trucks themselves or (2) improving the design of the road. Road tests have proved that "for motor trucks even to approach reasonable speeds on grades, engine power must be more than doubled or gross weights must be reduced excessively."[1]

The designer may improve truck speeds (1) by limiting the steepness and/or length of grades and (2) by providing extra lanes on steep grades for trucks.

[1] C. C. Saal, "Hill Climbing Ability of Motor Trucks," U.S. Bureau of Public Roads, *Public Roads*, vol. 23, no. 3, May, 1942, p. 33.

Most trucks, legally loaded and in good condition, are able to maintain a sustained speed better than 20 mph on grades under 3 per cent. Accordingly, grades should be held to 3 per cent or less whenever possible.[1] Steeper grades may be used when their length is short, such as ramps at grade separations. Typical grade standards for high-type roads, recommended for the National Interregional Highway System in rural areas, are shown in Table 27-2. The effect of topography on the economics of grade design is apparent.

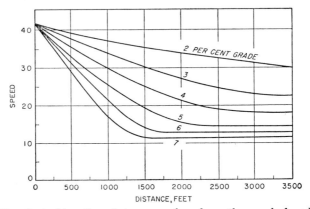

FIG. 27-1. The effect of length and steepness of grade on the speed of medium trucks with a gross load of 30,000 lb. (From A. Taragin, "Effect of Length of Grade on Speed of Motor Vehicles, *Proceedings of Highway Research Board*, Washington, 1945, vol. 25, p. 344, Fig. 3.)

While design is affected to a considerable extent by topography, the designer has some opportunity to adjust the grade (within economic limitations) to obtain the best truck operations. It has been suggested that long grades should be broken with the steepest grade at the top of ascent, sustained grades should be broken by short intervals of lighter grade, and a series of short steep grades is better than one long sustained grade. These opinions are based on observations that sustained grades cause reduced truck speeds in ascent and the overheating of brakes in descent. However, the present consensus is that gradient should be designed to fit the natural terrain and that the total effect of gradient on the operation of trucks is essentially the same for equal amounts of rise and fall regardless of possible combinations of broken gradient.

Extra truck lanes (or truck-climbing lanes) have been provided on existing two- or three-lane highways on particularly troublesome grades in a few states with good results.[2] It must be acknowledged that the

[1] *Ibid.*

[2] W. H. Miller, "Operating Characteristics and Suggested Design Considerations of Extra Lanes for Trucks on Ascending Grades," thesis submitted as part of course in Traffic Engineering, Yale University, Bureau of Highway Traffic, New Haven, Conn., 1949.

extra road width may encourage the passing of high-speed passenger cars and slow-speed trucks at a location where sight distances are usually limited, and that the end of the truck lane presents the same bottleneck problem in design as any transition where the road narrows.

TABLE 27-2. RECOMMENDED STANDARDS FOR MAXIMUM GRADIENTS
(Interregional Highway System—rural)

ADT	No. of lanes	Relatively level, %	Rolling, %	Mountainous, %	Short grades, %
1,000–2,000	2	3	4	6	7
2,000–5,000	4	3	6	6	7
Over 5,000	4 and 6	3	4	5	7

SOURCE: *Interregional Highways*, U.S. Bureau of Public Roads, Interregional Highway Committee, 1944, Appendix V, p. 154.

Carefully designed truck lanes may overcome these difficulties and have the effect of increasing the amount of passing sight distance along a section of road and may appreciably add to the smoothness of operations and increase capacity. The American Association of State Highway Officials recommends[1] truck lanes on highways handling substantial truck traffic when critical lengths of grades are exceeded and the volume exceeds the practical capacity because of the effect of gradient on truck operations. Critical lengths of upgrades are assumed to be those which will cause a 15-mph reduction in the speed of trucks below the average running speed on the approach to the grade. Critical lengths of grades where trucks are concerned are shown for various per cent grades in Table 27-3.

TABLE 27-3. CRITICAL LENGTHS OF UPGRADES FOR TRUCK OPERATIONS

Upgrade, per cent..............	3	4	5	6	7	8
Critical length of upgrade, ft....	1,600	1,100	800	650	550	500

SOURCE: *A Policy on the Geometric Design of Rural Highways*, chap. 3, "Elements of Design," American Association of State Highway Officials, Washington, 1954, Fig. 3-19.

Grade Reduction. The reduction of excessive grades requires costly construction, and its application in a particular case should be determined by a benefit-cost analysis. There are four methods of reducing grades: '(1) tunnels which keep grades to 3 per cent or less, (2) excavation, (3) filling on both sides of the hill, and (4) a compromise of methods 2 and 3, which uses material removed from the top of the hill for the fill.

[1] *A Policy on the Geometric Design of Rural Highways*, Chap. 5, "Highway Types," American Association of State Highway Officials, Washington, 1954, p. 227.

Very little improvement in truck speeds may be expected when grade reductions are small. Table 27-1 shows that when a long grade of 7 per cent is reduced to 6 per cent, the improvement in sustained speed for medium trucks is only 2 to 3 mph. An average speed of about 30 mph is possible only when the grade is reduced to 3 per cent.

Acceleration. When grade resistance is added to tractive resistance, longer distances are required for acceleration. In the case of trucks, grade resistance may utilize all of the residual power at relatively low speeds and make acceleration impossible. Therefore allowances for the decreased acceleration ability of trucks on ascending grades must be made in the design of acceleration lanes, passing sight distances, and other features of the road.

Loss of the acceleration ability of passenger vehicles on gradients usually employed becomes serious only when very high speeds or high differences in speed are involved.

Acceleration resulting from the grade effect on downgrades sometimes becomes a serious problem in truck operations. Long, steep downgrades sometimes found in mountainous areas offer serious hazards when descending trucks slip out of gear and their brake mechanisms burn out.

Static Sliding. A vehicle will stand on a grade without sliding only when the frictional force (fW) is equal to, or greater than, the weight force acting down the grade.

The coefficient of friction of pavement surfaces covered with ice varies considerably but has been assumed to be 0.05 lbs/lb for design purposes. Grades in excess of 5 per cent are therefore not recommended in areas subject to ice and snow, but may be used if they are "sanded" whenever an icy condition develops.

Curve Compensation. Curve compensation is the "breaking" of long grades to create a lesser grade throughout the length of horizontal curves. Several arbitrary formulas[1] have been devised as the basis for determining the absolute amount of grade compensation required to improve truck operations on ascending grades by reducing grade resistance (1) to compensate for increased resistance to motion caused by turning and (2) to prevent loss of momentum caused by reduced speed required to traverse a horizontal curve. It is obvious that curve compensation for these purposes is not important within the range of truck speeds on ascending grades, except when very steep grades and exceedingly sharp turns are employed on low-standard roads in mountainous areas.

Curve compensation does seem important, however, from the point of view that truck drivers are prone to descend grades at high speeds in order to gain momentum for the next ascending grade. Sometimes these downgrade speeds approach or exceed the critical safe speeds of horizontal

---

[1] For an example see Arthur G. Bruce and John Clarkeson, *Highway Design and Construction*, 3d ed., International Textbook Company, Scranton, Pa., 1950, p. 133.

curves. A small amount of curve superelevation becomes ineffective when vehicles round curves at high speeds on down grades. The "attitude" of the vehicle, resulting from slip, causes the vehicle wheels on the outside of the curve to assume a position slightly downgrade with respect to the inside wheels. Also, the force required to resist side skid is greater than normal when vehicles decelerate on horizontal curves. Finally, the stopping distances are greater and rates of deceleration of vehicles are lower than normal on downgrades.

27-2. Vertical Curves. The transition from one grade to another is accomplished gradually by means of a vertical curve called a *crest*, if the point of intersection of the two grades is above the road surface, and a *sag*, if it is below.

The grade angle of a vertical curve is fixed since it is the angle between intersecting grades, but the locus of the vertical curve is determined by its length. The length of vertical curve for a given combination of grades and assumed design speed is computed by the traffic designer. Minimum lengths are based on sight distance, vertical acceleration, and other factors.

Theory of Types. Several types of curves could be used in vertical curve design: (1) circular arcs, (2) parabolas, and (3) spirals. The change in slope of a circular arc is not constant but is essentially constant for the grade angles and lengths of curves employed in roadway vertical-curve design. The rate of change of the slope of a parabola is a constant.

The cubic parabola is a representative form of spiral. Its simplest equation is $y = ax^3$, and the rate of change of its slope is $d^2y/dx^2 = 6\,ax$, which is a variable. The radius of a spiral curve is infinity at the beginning of the curve and gradually approaches its minimum finite value at the end of the curve. When used as a vertical curve, two spirals must be placed end to end with their point of infinite radius at the grade tangent points on each side of the vertical curve.

The choice of type of curve to be employed for vertical curves must be based upon (1) the economy of earthwork, which may be related to the length of curve, (2) consistency in length of sight distance along length of curve, (3) riding comfort, and (4) simplicity of calculations.

Economy of Earthwork. On crest curves, where minimum safe-stopping sight distances usually determine the minimum length of vertical curves, the type of vertical curve that gives the longest sight distance with the least roadway excavation is the most economical. The circular curve is superior in this respect. A slightly longer sight distance may be obtained with a circular curve than with a parabola of the same length. Appreciably longer spiral curves are required to obtain the same sight distance over the crest. Figure 27-2 shows a graphic comparison of a circular arc, spiral, and parabolic vertical curve at greatly exaggerated scales.

Consistency in Sight-distance Length. The sight distance along a circular arc is constant throughout its length because its radius is constant. Within practical limitations, the parabola also offers a constant length of sight distance at the small gradient angles used in roadway design. The length of sight distance along a spiral vertical curve varies throughout its length because the radius of a spiral curve is constantly changing. This is a disadvantage, particularly on crest curves, because drivers usually fail to make allowances for variable sight distances as they travel over vertical curves of considerable length.

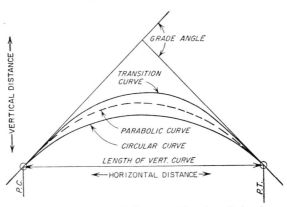

Fig. 27-2. Graphic comparison of parabolic, transitional, and circular vertical curves. (Scale: schematic only.)

Riding Comfort. The riding qualities of crest vertical curves differ from those of sag curves because the centrifugal force generated by the vehicle moving along the curve is neutralized by gravity in the first case and adds to gravity in the second.

The circular arc is at a disadvantage with respect to riding qualities because the sudden change from a straight grade to a curved path (at constant radius) may result in excessive impact and discomfort on sag curves of short lengths and at high speeds. The parabola has only slightly better transitional characteristics. The spiral, however, is fully transitional because it affords a gradual change in vertical acceleration. Thus the spiral is capable of introducing the centrifugal force caused by the vertical curve gradually, with very little impact.

Simplicity of Calculations. The parabola is the simplest curve to lay out because offsets from the tangents to the curve vary as the square of the distance along the tangent from the point of tangency. For this reason, the parabolic curve has been *almost universally* adopted for the design of crest vertical curves. Some authorities point out, "within the range of highway vertical curves, the curve although technically 'parabolic' is practically indistinguishable from a plain circular arc and should be treated as such."[1]

Design Practices. Since the additional impact experienced at the beginning of sag curves may be effectively "feathered" by a spiral-type curve and since there is no sight-distance limitation in sags (during daytime) the length of curve may be decreased for equal or better economy of earthwork. On this basis, spiral curves for sags are recommended in England[2] and other countries.

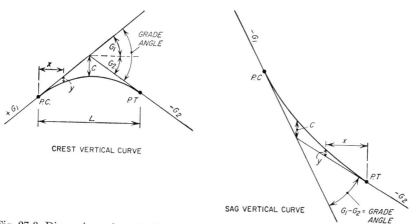

Fig. 27-3. Dimensions of parabolic vertical curves.

G = grades, ft/100 ft
L = length of curve, stations
P.C. = beginning of curve
P.T. = end of curve
C = center correction, ft
x = tangent distance, ft
y = tangent offset, ft

It is believed in America that the impact at the beginning of parabolic sag curves is negligible at highest speeds with the grade angles and minimum curve lengths employed in highway design. American practice, therefore, is to utilize the parabolic curve in the design of both crest and sag vertical curves.

In conformance with American practice, the following discussion of vertical curves will be limited to the design of parabolic vertical curves.

Fundamental Equations. Figure 27-3 shows the dimensions of several representative parabolic vertical curves.

The length of the curve is *always* taken horizontally to conform with

[1] F. G. Royal-Dawson, *Vertical Curves for Roads*, E. and F. M. Spon, Ltd., London, 1946, p. 15. This reference contains comparisons of circular, parabolic, and spiral vertical curves.

[2] *Ibid.*, p. 91.

the highway practice of measuring all longitudinal distances in a horizontal plane. For convenience in plotting, offsets from the tangent to the curve are *always* taken in the vertical plane. While this procedure varies from those required to obtain a true parabolic curve when gradients are not equal, the distortion is very small and may be considered negligible.

The center correction of any vertical curve for highway work is

$$C = \frac{LA}{8}$$

where C is center correction in feet (always taken vertically), L = length of curve in stations (always taken horizontally), and A is algebraic difference in grades (grade angle). Tangent offsets are obtained from the formula $Y = (2x/L)^2 \times C$, where Y is any tangent offset in feet (always taken vertically), L is length of vertical curve in stations (always taken horizontally), x is shortest horizontal distance from tangent offset point to $P.C.$ or $P.T.$ in stations, and C is the center correction in feet.

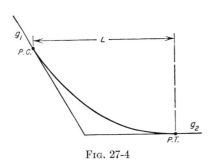

FIG. 27-4

Additional curve data and the derivation of equations may be obtained from any good highway-surveying handbook.

Minimum Lengths. The minimum length of vertical curves may be determined by vertical acceleration, daytime sight distances, nighttime sight distances, and riding comfort.

Vertical Acceleration. Vertical acceleration is the rate of increased rise or fall. As Fig. 27-4 shows, the change in vertical velocity is $1.47Vg_1 - 1.47Vg_2$, where V is assumed design speed in miles per hour and g_1 and g_2 are per cent grade ÷ 100. Also, the time required to traverse vertical curve = $L/1.47V$, where L is length of curve (approximately) in feet. Then vertical acceleration (ft/sec^2) = $(1.47V)^2(g_1 - g_2)/L$.

Any value used to define the limit of vertical acceleration must be broad, since it deals with driver comfort. One authority[1] has suggested that 0.5 ft/sec^2 should be a maximum vertical acceleration for high-type roads and 1.0 ft/sec^2 a maximum for other roads, allowing a maximum of 1.5 ft/sec^2 under exceptional conditions.

Vertical acceleration is not critical on crest curves where adequate stopping sight distances are allowed nor on sag curves of minimum

---

[1] David G. Price, *The Mathematical Design of Vertical Curves for Highways*, The Institution of Civil Engineers, London, 1942, p. 38.

allowable lengths. Therefore, the rate of vertical acceleration is important only for special or substandard designs.

Daytime Sight Distances. Daytime safe-stopping sight distance is not a factor in the design of sag vertical curves because drivers can see from one hill crest to the next, but it is an important factor in the design of crest vertical curves.

The height of the driver's eye above the road surface and the height of the hazardous object must be assumed before daytime-stopping sight distances may be related to the geometry of crest vertical curves.

The height of the driver's eye is about 4.5 ft for the average passenger car, somewhat higher for trucks, and somewhat lower for some of the "bantam"-type cars. Current design practice has adopted the average height of 4.5 ft.[1]

For daytime safe-stopping sight distances, the height of a hazardous object on the road is assumed to be at least 4 in.[2]

Distances of clear sight, at the appropriate assumed design speed, must therefore be provided over the crest from a vertical height of 4.5 ft on one end and vertical height of 4 in. on the other.

When grades are short and sight distances extend over several vertical curves, adequate curve lengths to provide a given sight distance are selected by trial and error. Grades are plotted to a suitable scale on a profile sheet, and a straight edge is held between the two points plotted at the respective height of eye and object. This line should clear the surface of the road throughout the section in question when crest curves of adequate length are applied to the breaks in grade.

Figure 27-5 is a chart for the calculation of minimum length of crest vertical curves when the daytime safe-stopping sight distance and grade angle are known. This chart may be conveniently used *when only one vertical curve* is involved. Assume that the grades are a plus 1 per cent and minus 1 per cent and that the assumed design speed is 60 mph. From Table 26-1 the minimum safe-stopping sight distance (daytime) is 475 ft. Reading from left to right, the curve length required (225 ft) is found at the bottom edge of the chart. Vertical curve lengths are usually rounded off to even 25-ft distance intervals.

Since sight distances required for safe passing are much longer than those for safe stopping, provision of safe passing distances over crests is limited by economics except where small grade angles are involved.

Passing on a crest drivers first see the tops of approaching vehicles. Most passenger vehicles have an over-all height of about $5\frac{1}{2}$ ft, while the height of trucks ranges up to 12 ft. It is arbitrarily assumed that vision requirements are satisfied when both drivers see each other at the

---

[1] *A Policy on the Geometric Design of Rural Highways,* chap. 3, "Elements of Design," p. 125.

[2] *Ibid.*

eye height of passenger vehicles. Thus the height of eye and of object are both assumed to be at a vertical height 4.5 ft above the road surface for safe-passing sight distances.[1] While this height of object is used in current design, it is believed to be more applicable to safety in passing than to efficiency of traffic flow. Recent studies show that drivers seldom pass in areas of restricted sight distance until they can see the pavement surface ahead for a sufficient distance to permit safe passing.[2]

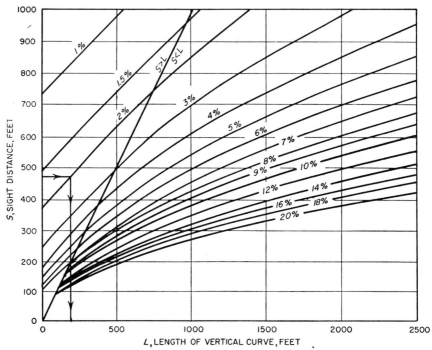

Fig. 27-5. Chart relating algebraic difference of grades—length of vertical curve and safe-stopping sight distance. (From *A Policy on Geometric Highway Design*, "A Policy on Sight Distance for Highways," American Association of State Highway Officials, Washington, 1954, p. 18, Fig. 3.)

Minimum safe-passing sight distances are shown in Table 26-3. These distances are related to crest-curve design by the same procedure previously described for daytime safe-stopping sight distances. The only difference is that the vertical height of object is 4.5 ft instead of 4 in. Where only one vertical curve is involved, the chart shown by Fig. 27-6 may be used. It will be noted that a curve length of 2,225 ft would be required to provide minimum safe-passing sight distance when applied to the previous example.

[1] *Ibid*, p. 126.

[2] *Highway Practice in the United States of America*, U.S. Bureau of Public Roads, Washington, 1949, p. 97.

There is good reason to combine vertical with horizontal curves on two- or three-lane roads when no hazard is created. The combination reduces the number of locations with restricted-passing sight distance and therefore increases the practical capacity of the road. Sharp horizontal curvature should not be employed near a pronounced vertical

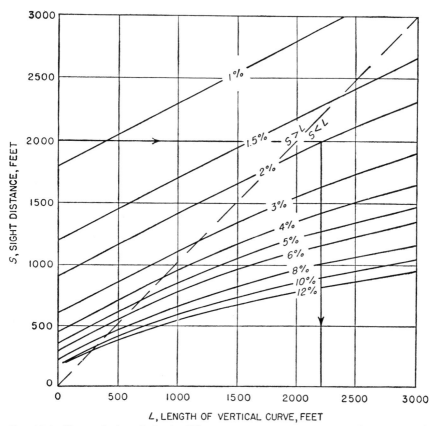

FIG. 27-6. Chart relating algebraic difference of grades—length of vertical curve and safe-passing sight distances. (From *A Policy on Geometric Highway Design*, "A Policy on Sight Distance for Highways," American Association of State Highway Officials, Washington, 1954, p. 19, Fig. 4.)

curve where sight-distance restrictions on the vertical curves will conceal the change in horizontal alignment.

Night Sight Distances. For night driving the vehicle headlight range on straight and level roads is insufficient to meet safe-stopping distance requirements at speeds in excess of 50 mph. In general, a small improvement in night visibility would be obtained at disproportionate expense if road alignment were designed for maximum utilization of the visibility made available by current headlighting practices. Under these circumstances, nighttime sight distance requirements are generally ignored in the design of vertical curves, but some engineers have devised formulas to obtain minimum vertical curve lengths on the basis of sight distances at night. Certain broad assumptions are made in the application of these formulas to obtain specific design values: (1) driver PIEV time is 1 sec longer at night than during daytime, (2) the forward range of headlight provides unlimited visibility distance, (3) the beam intensity within which an object will be seen diverges upward at an angle of 1° from the longitudinal axis of the car, (4) the beam intensity downward is limited only by the curvature of the road, and (5) the height of object is taken at 1.5 ft for crest curves and at the surface of the road for sag curves. Figure 27-7 shows charts based on these assumptions, from which the minimum lengths of crest and sag vertical curves may be obtained.

Night sight distance has been used as the criterion for the length of vertical curves in at least one known case.[1] The charts shown by Fig. 27-7 were developed for this case. Under the assumptions made, minimum vertical curve lengths for crest curves are approximately the same as those obtained when daytime safe-stopping sight distance criteria are applied. However, in the case of sag curves, minimum lengths are appreciably longer than those usually employed in present design practice.

Riding Comfort. In the absence of accurate criteria for the design of sag-curve lengths, minimum lengths of sag vertical curves have been established arbitrarily in American design practice, dependent in each case upon the judgment of the designer. The adopted lengths are those believed to provide a smooth comfortable ride without undesirable impact at the beginning and end of the curve.[2] If the lengths of sag vertical curves are determined from Fig. 27-7 to provide adequate night-stopping sight distance, the comfort criterion is more than satisfied.

Effect of Vertical Alignment on Capacity. The effect of grades on speed, acceleration, and deceleration performance of vehicles, as well as on sight distance, influences the capacity and safety of highways. Except for sight distance considerations, vertical alignment reduces the practical design capacity of a road through its effect on the operations of trucks. The effect of dual-tired commercial vehicles on capacity of multilane roads may be equivalent to that of two- to eight-passenger vehicles, depending on the length and per cent of grade.[3] Table 27-4 shows the effect of commercial vehicles on the practical design capacity of multi-

[1] Charles M. Noble, "Pertinent Elements of Express Highways," paper presented before students of Yale University Bureau of Highway Traffic, New Haven, Conn., May 3, 1949. (Mimeographed.)

[2] Donald Thompson, "Sight Distance on Sag Vertical Curves," *Civil Engineering*, vol. 14, no. 1, January, 1944, p. 22.

[3] *Highway Capacity Manual*, Highway Research Board and U.S. Bureau of Public Roads, Committee on Highway Capacity, 1950.

lane roads. On two-lane roads, the effect of dual-tired commercial vehicles on capacity may be equivalent to 2.5 to 10 passenger vehicles. Long grades in mountainous terrain may increase the effect of trucks on capacity greatly in excess of the values shown in the table.

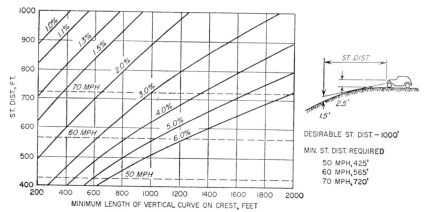

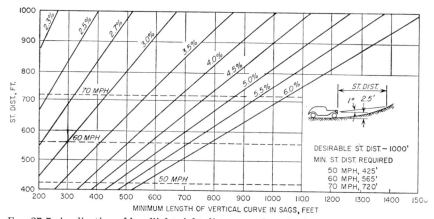

Fig. 27-7. Application of headlight sight distance to length of vertical curves. (From Charles M. Noble, "Pertinent Elements of Express Highways," paper presented before students of the Yale University Bureau of Highway Traffic, New Haven, Conn., May 3, 1949, p. 26a–b, Figs. 11 and 12.)

Effect on Safety. Though it is believed that accidents increase with increased gradient, one of the most comprehensive studies[1] reports that there does not seem to be any correlation between accidents and gradient but at the same time it questions the accuracy of the basic data. However, many locations are accident-prone because they occur near or on vertical curves or sustained steep grades.

[1] M. S. Raff, "The Interstate Highway Accident Study," U.S. Bureau of Public Roads, *Public Roads,* vol. 27, no. 8, June, 1953, p. 170.

Effect on Speed. Studies[1] of speeds on crest vertical curves with restricted stopping sight distances show that drivers reduce speed to some extent, but the reduction is far less than that required for safe operations. No direct relation seemed to exist between speeds at crest

TABLE 27-4. EFFECT OF TRUCKS ON PRACTICAL DESIGN CAPACITIES
OF MULTILANE FACILITIES

Commercial vehicles (dual-tired trucks), %	Practical design capacity expressed as per cent of passenger-car theoretical design capacity on level terrain	
	Level terrain, %	Rolling terrain, %
None	100	100
10	91	77
20	83	63

SOURCE: *Highway Capacity Manual,* Highway Research Board and U.S. Bureau of Public Roads, Committee on Highway Capacity, 1950, p. 56, Table 9.

vertical curves and minimum stopping sight distances. Apparently drivers do not realize that they are exceeding the theoretical safe speed of the vertical curve, or else the incidence of critical situations at vertical curves is so rare that drivers are not concerned.

[1] B. A. Lefeve, "Speed Characteristics on Vertical Curves," *Proceedings of Highway Research Board,* Washington, 1953, vol. 32, p. 395.

CHAPTER 28

HORIZONTAL ALIGNMENT

Horizontal alignment is the longitudinal dimension of a road with respect to the horizontal plane and consists of tangents and curves. A highway tangent is a straight section of road. A horizontal curve is a curved section of road placed between two tangents to accomplish a gradual change in direction.

Horizontal alignment is usually applied to the center of the road and as such is the *center line*. All other horizontal dimensions on the plan view of the road are referred to it. For most designs, *profile grade* in the vertical plane is at the same point on the cross section as the center line in the horizontal plane. However, distances are measured along the center line and *not* along the profile grade.

A critical factor in the operations of vehicles on horizontal curves is the centrifugal force developed. The maximum side force and the rate at which this force is developed determine the paths of vehicles and the comfort of their occupants as a change in direction is made. Comfort and vehicle paths become important if drivers are to remain in their respective lanes on horizontal curves without sudden reductions in speed. Another determinant of curve design is sight distance on the inside of curves. Either sight distance or centrifugal force may be the determinant of horizontal-curve design.

28-1. Forces on Horizontal Curves. A turning force is required to change the direction of the vehicle. If a vehicle turns along the path of a horizontal curve at constant speed, the magnitude of the velocity remains the same, but the direction changes, and a force must be applied to cause the change in direction. This force is equal to the centrifugal force acting on the vehicle as it follows a curved path. Centrifugal force acts through the center of gravity of the vehicle, normal to its path. Neglecting superelevation, it is the force with which the vehicle tends to leave the curve along a line tangent to it. The force opposing centrifugal force results from superelevation and friction developed between the tire and road surface.

Centrifugal Force. The centrifugal force acting on a vehicle following a curve of constant radius at a constant speed is obtained by the formula $F_c = 0.067WV^2/R$, when F_c is centrifugal force in pounds, W is weight of

449

vehicle in pounds, V is design speed in miles per hour, and R is radius of curve in feet (see Chapter 3).

Superelevation Force. The force acting upon a vehicle on a grade is approximately 20 lb/ton of vehicle weight for each per cent grade (Chapter 3). This same force acts upon a vehicle on a superelevated curve regardless of the fact that the gradient is at right angles to the path of the vehicle. The effect on superelevation, therefore, is to exert a down force, $F_e = (20/2,000)GW$, where F_e is force resulting from superelevation in pounds, G is per cent grade in feet of vertical rise per 100 ft of horizontal distance, and W is weight of vehicle in pounds.

Superelevation is expressed as the ratio of feet of vertical rise per foot of horizontal distance. Thus the formula becomes $F_e = We$, where e is superelevation in feet per foot.

If the angle of superelevation is neglected, $F_f = Wf$, when $F_f = $ frictional force developed in pounds, $f = $ coefficient of friction in pounds per pound, and $W = $ weight of vehicle in pounds.

Equilibrium Formula. Superelevation force and frictional force oppose centrifugal force, and the equilibrium formula is $e + f = 0.067V^2/R$, where V is speed in miles per hour and R is radius of curve in feet.[1]

Limitations in Curvature. Superelevation and friction factors have practical limitations. These limitations control the minimum radius allowable in the design of horizontal curves for a road of given assumed design speed.

Maximum Superelevation. Theoretically, superelevation could be designed to counteract the maximum obtainable centrifugal force, but superelevation on public highways must be designed for a wide range in speed and climatic conditions.

In Northern climates, the maximum practical amount of superelevation is limited by the tendency of slow-moving vehicles to slip to the inside of curves on ice. It is assumed, on superelevated curves, that even at cautious speeds some centrifugal force is developed to aid friction. A maximum superelevation of 0.08 ft/foot is recommended[2] for Northern climates. Some design authorities use a maximum superelevation of 0.10 ft/foot in Southern climates. The choice is a matter of opinion rather than fact, and values employed range from 0.06 to 0.12 (or 0.14 on ramps).

In climates where ice is not a problem, maximum allowable values of superelevation are determined by low-speed vehicles. When super-

---

[1] This derivation assumes that the angle of superelevation is the sine of that angle and that the vehicle-weight component normal to the superelevated surface equals the vehicle weight.

[2] *Interregional Highways*, U.S. Bureau of Public Roads, Interregional Highway Committee, Washington, 1944, Appendix V, "Basic Standards of Road and Structural Design, Superelevation of Curves," p. 151.

elevation is excessive, there is a tendency for the load on trucks to shift toward the inside of the curve at low speeds of travel, and even the drivers of passenger vehicles have an off-balance sensation which may lead to improper steering. The maximum superelevation recommended for Southern climates is, therefore, 0.12 ft/ft.[1] Higher values of superelevation, up to 0.16, have been used on intersection interchange ramps.[2]

Maximum Friction. The maximum friction factor used in the design of horizontal curves is based upon drivers' comfort and stability. The relationship between the amount of frictional force developed and the comfort of drivers has been established through many field tests. One of these studies was conducted by the U.S. Bureau of Public Roads.[3] Nine hundred road tests indicated that the side-friction factor of 0.16—up to and including speeds of 60 mph—was utilized, with a decrease in the factor of 0.01 for each 5-mile increment in speed above 60. Studies made on the Pennsylvania Turnpike by highly skilled drivers indicated that unstable steering conditions which might have been disastrous for less experienced drivers were reached when the side-friction factor exceeded 0.10 at 70 mph.[4]

From these and other studies, the AASHO has concluded that maximum values of side friction to be used in the design of nonintersectional horizontal curves should vary directly with the design speed from 0.16 at 30 mph to 0.12 at 70 mph.[5]

It should be stressed that side-friction factors based upon comfort are considerably less than the maximum side friction that may be developed between normal surfaces and tires. The use of the lower values, in design, is intended to discourage reductions of speed on horizontal curves and to provide a factor of safety for those motorists who travel at speeds in excess of the assumed design speed. It should also be made clear that higher friction factors are assumed for curves at intersections. In terms of the all-over trip, turns at intersections are few and far between, and drivers are willing to endure greater discomfort while making turns at intersections than when following the alignment of the through roadway.

Minimum Allowable Radius. The minimum allowable radius to be employed in the design of horizontal curves for roads having various assumed design speeds are shown in Table 28-1.

Naturally, in the design of horizontal alignment, the minimum allow-

[1] *Ibid.*

[2] *Highway Practice in the United States of America*, U.S. Bureau of Public Roads, 1949, p. 95.

[3] Joseph Barnett, "Safe Side Friction Factors and Superelevation Design," *Proceedings of Highway Research Board*, Washington, 1936, vol. 16, p. 69.

[4] K. A. Stonex and C. M. Noble, "Curve Design and Tests on the Pennsylvania Turnpike," *Proceedings of Highway Research Board*, Washington, 1940, vol. 20, p. 429.

[5] *A Policy on the Geometric Design of Rural Highways*, chap. 3, "Elements of Design," American Association of State Highway Officials, Washington, 1954, p. 133, Table 3–6.

able radius should be applied only at critical points where the topography, or cost of right of way, will not permit easier curves.

TABLE 28-1. MINIMUM ALLOWABLE RADIUS FOR VARIOUS ASSUMED DESIGN SPEEDS

Assumed design speed, mph	Maximum side-friction factor, lb/lb	Maximum super-elevation, ft/ft	Minimum allow-able radius, ft
40	0.15	0.08	464
40	0.15	0.12	395
50	0.14	0.08	758
50	0.14	0.12	641
60	0.13	0.08	1,143
60	0.13	0.12	960
70	0.12	0.08	1,633
70	0.12	0.12	1,361

28-2. The Natural Transition. The path described by drivers when traveling on horizontal curves at speeds high enough to develop "unbalanced" side thrust, even though the speed is constant, is not generally a true circular arc. Since the degree of curvature is constant, in entering a circular arc from a tangent the full effect of unbalanced side thrust would be developed abruptly, theoretically, and the vehicle occupants would be subjected to proportionate shock. To follow a circular arc from a tangent, the steering angle and slip angles of the vehicle (Chapter 3) must be developed instantaneously, or the vehicle will veer from a circular path.

Drivers prefer gradual steering and gradual development of side forces, and as the result, describe a path which is a transitional curve, rather than a circular arc. This transitional path is followed regardless of the curve alignment of the road. Thus for maximum safety and comfort of drivers on high-speed roads the alignment of the road should also be transitional. The term *transitional* is a broad expression applied to a curve whose radius gradually decreases from infinity, starting at the beginning of the curve.

Vehicle paths on horizontal curves are highly variable among drivers, and the selection of an "ideal" curve to be adopted for design purposes must be a compromise.

Forms of Transitional Curves. Since engineers are not in agreement concerning the ideal design of transitional horizontal curves, a number of curve *forms* and *types* are employed. The form of transitional curves often is determined by the simplicity of the theory or ease in computations and application.

One school of thought favors the use of a curve whose curvature increases uniformly[1] from the beginning of the curve to the point of

[1] Or, expressed in terms of radius, a curve whose radius decreases uniformly from the beginning of the curve to the point of minimum radius. (The degree of curve at the point of beginning of the curve is zero, and the radius is infinite.)

maximum curvature. Thus, when the curve is driven at a constant speed, the rate of gain in centrifugal force is constant, and the driver sharpens his curve by steering into the curve at a constant rate. The spiral has these properties and is used, almost exclusively, in American design practices.

In England, some authorities recommend the lemniscate.[1] The radius of the lemniscate gradually reduces from the beginning of the curve but reduces less rapidly than in the case of a spiral. When steering along a lemniscate transition, the rate of turning into the curve is eased off slightly as the point of minimum radius is approached. Obviously, the lemniscate results in a slightly longer curve than the spiral when a given minimum radius is to be attained.

Another modification, the cubic parabola, is quite similar to both the spiral and lemniscate for *short* transitions.

Many of the transition curves recommended in various surveying handbooks are spirals with slight difference adopted for increased ease in computations or application. The Talbot spiral[2] uses the chord definition for degree of curve. Hickerson[3] and Barnett[4] use the arc definition. The American Railway Engineering Association[5] uses the chord definition, and the spiral is divided into 10 equal chords. Searles' spiral[6] is a series of compounded circular curves. All of these are curves of the same form, and their differences are practically indistinguishable.

Since vehicles are not confined to rails and their paths on curves differ with drivers, it may be concluded that any transition form similar to the spiral or lemniscate has suitable properties for the design of horizontal curves for highways.

Types of Transitional Curves. There are two principal types of transitional highway curves. These are (1) curves which are transitional throughout and (2) transitional curves on both ends of a central circular curve. Figure 28-1 shows a typical example of each type. The maximum allowable curvature is determined by the minimum allowable radius at the *S.C.S.* or *S.C.* for the respective curves. The minimum allowable radius is determined from the equilibrium formula and should not exceed the values given in Table 28-1. For a given minimum radius, it is sometimes desirable to increase the length of the spiral by increasing the spiral

[1] F. G. Royal-Dawson, *Road Curves for Safe Modern Traffic and How to Set Them Out,* chap. 3, "Physical Characteristics of Lemniscate and Spiral," E. and F. N. Spon, Ltd., London, 1946, p. 14.

[2] Arthur N. Talbot, *The Railway Transition Spiral,* 6th ed. McGraw-Hill Book Company, Inc., New York, 1927.

[3] T. F. Hickerson, *Highway Surveying and Planning,* McGraw-Hill Book Company, Inc., New York, 1936.

[4] J. Barnett, *Transition Curves for Highways,* U.S. Bureau of Public Roads, 1938.

[5] H. C. Ives, *Highway Curves,* John Wiley & Sons, Inc., New York, 1941.

[6] W. H. Searles and H. C. Ives, *Field Engineering,* John Wiley & Sons, Inc., New York, 1936.

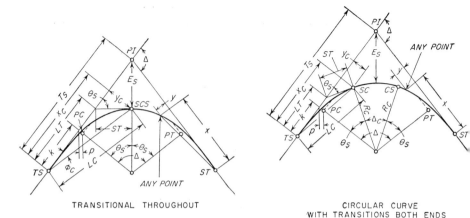

TRANSITIONAL THROUGHOUT

CIRCULAR CURVE
WITH TRANSITIONS BOTH ENDS

FIG. 28-1. Typical examples of curves which are transitional throughout and circular with transitions on both ends.

P.I.	Point of intersection of the main tangents
T.S.	Tangent spiral, common point of tangent and spiral of near transition
S.C.	Spiral curve, common point of spiral and circular curve of near transition
C.S.	Curve spiral, common point of circular curve and spiral of far transition
S.T.	Spiral tangent, common point of both spirals or mid-point of a curve transitional throughout
R_c	Radius of the circular curve
L_s	Length of spiral between T.S. and S.C.
L	Length between T.S. and any other point on spiral
L_1	Length between any two points on spiral
T_s	Tangent distance P.I. to T.S. or S.T., or tangent distance of the complete curve
E_s	External distance P.I. to center of circular curve portion, or to S.C.S. of a curve transitional throughout
L.T.	Long tangent distance of spiral only
S.T.	Short tangent distance of spiral only
L.C.	Straight-line chord distance T.S. to S.C.
p	Offset distance from the tangent of P.C. of circular curve produced
k	Distance from T.S. to point on tangent opposite the P.C. of the circular curve produced
Δ	Intersection angle between tangents of entire curve
Δ_c	Intersection angle between tangents at the S.C. and at the C.S. or the central angle of the circular-curve portion of the curve
θ_s	Intersection angle between the tangent of the complete curve and the tangent at the S.C., the spiral angle
θ	Intersection angle between the tangent of the complete curve and the tangent at any other point on the spiral, the spiral angle of any other point
D_c	Degree of the circular curve same as degree of curvature of spiral at the S.C. (arc definition)
D	Degree of curvature of spiral at any other point on spiral (arc definition)
ϕ_c	Deflection angle from tangent at T.S. to S.C.
ϕ	Deflection angle from tangent at any point on spiral to any other point on spiral
x_c, y_c	Coordinates of S.C. from the T.S.
x, y	Coordinates of any other point on spiral from the T.S.

(From Joseph Barnett, *Transition Curves for Highways*, U.S. Bureau of Public Roads, 1938, p. 15.)

angle θ_s. The maximum spiral length for a given radius occurs when $\theta_s = \Delta/2$, and the curve becomes transitional throughout.

Generally, horizontal curves which are transitional throughout are not favored in American highway-design practice. Their disadvantages are that (1) the curve often has the appearance of a sharp bend at its mid-point and (2) at no point along the curve are steering conditions constant. Based on a limited number of observations, there is evidence that drivers do not follow a completely transitional path. Studies made in England, although not conclusive, show that drivers tend to use about one-half to two-thirds of a curve as transition, and the all-transition path is comparatively rare.[1]

28-3. Design of Spiral Transitions. The need for horizontal-curve transitions is increased with increased assumed design speed and/or increased curvature. Transitional curves are recommended for the National Interregional Highway System on all curves in both urban and rural sections where the degree of curvature is sharper than 2°.[2] Curves of lesser degree resemble nontransitional circular curves so closely that transitioning is not practical.

A number of excellent surveying handbooks are available for the detailed study of spiral-curve dimensions and geometry. These data will not be repeated here, but the factors which determine the design of a spiral transition capable of the best vehicle operations will be discussed. These factors are (1) the length of the transition, (2) superelevation runoff, and (3) widening runoff.

Length of Spiral Transitions. The radius of a spiral is inversely proportional to the distance from its beginning. Therefore the magnitude of the change in centrifugal force acting on a vehicle traveling at constant speed depends upon the length of the spiral and the minimum radius employed. The minimum allowable radius for a horizontal curve (Table 28-1) is obtained from the equilibrium formula developed earlier in this chapter. The minimum radius employed in the design of a spiral transition should be equal to the radius of the circular curve to which it connects.

The length of spiral transition should meet the requirements of drivers with respect to (1) the rate at which they prefer to turn the steering wheel and (2) comfortable rate of increased "unbalanced" force. A number of different authorities have suggested various methods for the determination of spiral lengths to meet these requirements.[3]

[1] John Joseph Leeming, "The General Principles of Highway Transition Curve Design," *Proceedings of American Society of Civil Engineers,* vol. 73, no. 8, New York, October, 1947.

[2] *Interregional Highways,* U.S. Bureau of Public Roads, Interregional Highway Committee, 1944, Appendix V, pp. 151, 167.

[3] Leeming, *op. cit.,* p. 1257.

In American practice, the rate of change of radial acceleration is assumed as the critical factor. This factor is proportional to centrifugal force and to the rate of change of curvature.

A vehicle rounding a circular curve at constant speed has an acceleration toward the center of the curve equal to V^2/R. On a spiral transition, this radial acceleration is zero at the beginning of the curve, where the radius is infinity, and is V^2/R at the end of the spiral, where the radius is at its minimum value R. (The end point of the spiral is at the beginning of the central circular curve or at the mid-point of the double spiral, depending upon whether the curve has a central circular curve or is spiraled throughout.) Since the radial acceleration changes from zero to V^2/R at a uniform rate throughout the length of the spiral, the average rate of change in radial acceleration is V^2/R divided by the time required to travel over the spiral. The rate of change in radial acceleration then becomes

$$a' = \frac{3.16V^3}{RL_s} \quad \text{or} \quad L_s = \frac{3.16V^3}{Ra'}$$

where a' = rate of change in radial acceleration, ft/sec³

     V = assumed design speed, mph

     R = minimum radius of the spiral transition, ft

     L_s = length of the spiral transition, ft

A safe and comfortable rate of change in radial acceleration can be found only by field tests. One study of drivers on speedways disclosed that values of a' ranging from about 6 to 7 ft/sec³ were used.[1] In England, a value of 1 ft/sec³ is assumed. This value is based on railroad operations.[2] American practice assumes that the value employed by drivers increases on the sharp turns requiring decreased speed. Accordingly, a value of 2 ft/sec³ is recommended for through roadways,[3] and values ranging from 2.5 to 4.0 ft/sec³ for corresponding speeds of 50 to 20 mph are recommended for turns at intersections.[4]

For through highways, where a value of 2 ft/sec³ is assumed, the formula for spiral length becomes $L_s = 1.6V^3/R$, where L_s is length of spiral in feet, V is assumed design speed in miles per hour, and R is radius of the central circular curve in feet.

[1] R. A. Moyer, *Skidding Characteristics of Automobile Tires on Roadway Surfaces and Their Relation to Highway Safety*, Iowa Engineering Experiment Station, Bulletin 120, Ames, 1934.

[2] W. H. Shortt, "A Practical Method for the Improvement of Existing Railway Curves," *Proceedings of Institution of Civil Engineers*, Westminster, Great Britain, 1909, vol. 176, p. 97.

[3] *Highway Practice in the United States of America*, p. 96.

[4] *A Policy on the Geometric Design of Rural Highways*, chap. 7, "Intersection Design Elements," p. 265, Table 7-4.

AASHO recommends the following minimum degree of curve for various design speeds at which transitions are recommended:

Design speed	Minimum degree of curve
30	3.5
40	2.25
50	1.75
60	1.25
70	1.25

SOURCE: *A Policy on the Geometric Design of Rural Highways*, chap. 3, "Elements of Design," American Association of State Highway Officials, Washington, 1954, p. 147, Table 3-13.

Superelevation Runoff. A great advantage of the spiral transition curve is its suitability to the application of superelevation. Since the radius of a spiral transition is inversely proportional to the distance from the beginning of the curve, the increase in centrifugal force developed by

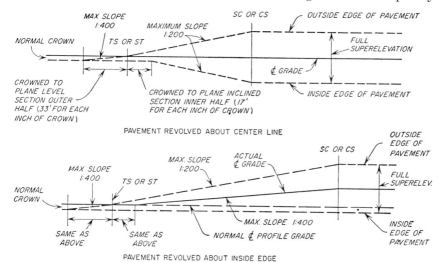

FIG. 28-2. Suggested methods of attaining superelevation along spiral transitions. (From Joseph Barnett, *Transition Curves for Highways*, U.S. Bureau of Public Roads, 1938, p. 10.)

a vehicle traveling along the transition at constant speed is directly proportional to the distance from the beginning of the curve. Thus, attainment of full superelevation slope at a uniform rate throughout the length of the transition provides an increasing resistance to side skid as centrifugal force increases.

Figure 28-2 shows recommended methods of attaining superelevation along spiral transitions when (1) the pavement is revolved around the center line and (2) it is revolved around the inside edge.

The length of spiral required for good appearance and operations is sometimes determined from the rate of superelevation runoff. For example, in the New Jersey Highway Department, a maximum 2 per cent rotational change in slope per second is recommended. Thus if the maximum superelevation attained is 0.08 ft of rise per foot of horizontal distance and the design speed is 60 mph, the recommended length of transition would be $8 \times 8\frac{8}{2} = 352$ ft.

Regardless of the formula used, the length of the transition should be sufficient to avoid the appearance that results from too rapid change in

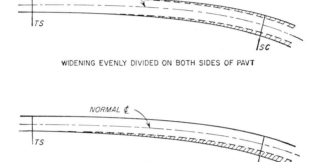

WIDENING EVENLY DIVIDED ON BOTH SIDES OF PAVT

WIDENING ENTIRELY ON INSIDE OF PAVT

Fig. 28-3. Suggested methods of attaining widening along spiral transitions (widening proportional to distance from $T.S.$).

superelevation slope. The slope of the edge of the pavement with respect to the profile of the center line should not exceed 1:150 for an assumed design speed of 30 mph; 1:175 for 40 mph, and 1:200 for higher speeds.[1]

Widening Runoff. When a horizontal curve is widened, gradual attainment of this widening should be accomplished throughout the length of the spiral transition. Widening attainment on the transition is usually accomplished on the basis of a straight-line relationship between the increasing width and distance from the beginning of the transition.[2] Since the pavement edge without widening is a spiral, after widening is added on this basis, the alignment of the pavement edge retains a pleasing appearance, except at the end of the transition where slight eye adjustments may be needed. Figure 28-3 shows suggested methods of attaining widening on spiral transitions when (1) widening is placed entirely on the inside of the curve and (2) widening is evenly divided on both sides of the pavement.

[1] Barnett, *op. cit.*, p. 11.
[2] *Ibid.*, p. 42.

Central Circular Curve. The degree of curvature D_c of the central circular curve must satisfy the conditions of the equilibrium formula. Thus its maximum value is fixed. In addition, the minimum length of spiral L_s is fixed by the desirable rate of change of radial acceleration or maximum slope of superelevation runoff. Since the external angle Δ of the road tangents is generally given, the following equations of the functions of a spiral may be used to compute the length L_c of the central circular curve. (See Fig. 28-1 for definition of symbols.)

$$\theta = \frac{L_s D_c}{200}$$
$$\Delta_c = \Delta - 2\theta_s$$
and
$$L_c = 100 \frac{\Delta_c}{D_c}$$

Superelevation along the central circular curve should be the maximum value attained and should be held constant with all superelevation runoff placed along the transition. If there is any superelevation on the road tangent, where there is no curvature, and if the superelevation along the circular curve, where the radius is constant, is not also constant, the transverse equilibrium of the vehicle is theoretically disturbed.

The minimum horizontal curvature for which superelevation is required varies in practice. Many design standards establish the minimum curve requiring superelevation at 1° or 0°45′ without regard for assumed design speed. The AASHO recognizes a 0.02 ft/ft superelevation value as the lowest of any significance and recommends that superelevation be applied only when the curve-speed equation calls for superelevation greater than this amount.[1] On flatter curves, the normal crown provides some superelevation for half of the pavement. Normal crown on the other half of the pavement causes reverse superelevation. This adverse crown may be removed by rotating the normal-crown section of the entire pavement or by simply raising the edge of the pavement to remove the crown on that side.

Widening should also be held constant at the maximum value attained, throughout the length of the central circular curve. There is a certain amount of controversy over the justification of pavement widening on transitioned horizontal curves when the modern lane width of 11 or 12 ft is used. When the transition is designed properly, motorists should have little difficulty remaining within their respective lanes when traveling at, or below, the speed for which the curve was designed, without widening. On the other hand, the clearances allowed traffic will be less on curves

[1] *A Policy on the Geometric Design of Rural Highways*, chap. 3, "Elements of Design," p. 137.

than on tangents, and if motorists are to enjoy the same degree of comfort and safety on both curves and tangents, widening on curves is indicated. There are two factors which limit clearances on horizontal curves and which determine the amount of widening necessary: (1) the off-tracking (Chapter 3) of rear wheels of vehicles at low speeds and (2) the slippage of the rear wheels toward the outside of the curve at high speeds. Since highway curves are traveled by both fast and slow drivers, the width required for both inside tracking and outside slipping should be added together to give the total amount of widening.

The additional width necessary to compensate for off-tracking may be found, for single-unit vehicles, from the formula $T = R - \sqrt{R^2 - L^2}$, where T is off-tracking in feet, R is turning radius in feet, and L is wheel base in feet.

On the basis of high-speed tests on curves, it has been suggested for outside slippage that at least 1 ft should be added for each 0.10 friction coefficient developed.[1]

Combining the indicated widths for off-tracking and slippage, the following formula is developed: $W = n(R - \sqrt{R^2 - L^2} + 10f)$, where W is pavement widening in feet, n is number of lanes, R is radius of curve in feet, L is wheel base in feet and f is "unbalanced" friction developed at assumed design speed in pounds per pound. This formula is suggested by theory, but has yet to be adopted in design practice.

A widely used formula for curve widening is:[2]

$$W = n(R - \sqrt{R^2 - L^2}) + \frac{V}{\sqrt{R}}$$

In this formula, $V/\sqrt{R}$ is an arbitrary figure intended to allow for outside slippage. It is not multiplied by the number of lanes. It is suggested, when this formula is used, that it should be considered applicable to 11-ft lanes and that the widening computed by the formula should be increased or decreased accordingly for narrower or wider lane pavements.

The American Association of State Highway Officials currently recommends widening on curves of two-lane highways based on inside tracking and vehicle overhang plus assumed extra clearances.[3]

28-4. Design of Circular Curves. Where transition curves are not needed, circular curves are employed. The maximum degree of curvature, maximum allowable "unbalanced" side force, maximum allowable superelevation, and required widening are the same as previously described for central circular curves with transitions.

[1] M. L. Fox, "Relations between Curvature and Speed," *Proceedings of Highway Research Board*, Washington, 1937, vol. 17, p. 202.

[2] Barnett, *op. cit.*, p. 46.

[3] *A Policy on the Geometric Design of Rural Highways*, p. 156.

Excellent handbooks are available on the elements and functions of circular curves having various radii and external angles.

Circular curves without transitions should not be used except on secondary roads where traffic volumes are small and lanes are not rigidly defined, or on high-type roads where the curvature is less than that requiring transitions.

However, the use of transitional curves has not been universally adopted by all highway engineers. Some of them assume that with 11- or 12-ft lanes, possibly with curve widening, sufficient leeway is allowed drivers for them to describe a transitional path of their own choosing and still remain within the vehicle lane. In most of these cases, superelevation runoff is applied for a transitional vehicle path by the placement of one-half to one-third of the superelevation runoff inside of the circular curve at each end.

One method of transitioning horizontal curves on two-lane roads is to spiral the inside and outside edges of the pavement without changing the center line from a circular curve. The principal disadvantage of this method is the "off-center" position of the construction center line, which may be more visible to drivers than a poorly maintained painted center line, and complications in staking out the curve.

Design of Curves with Curvature Less Than Maximum. The forces which are critical in horizontal curve design are developed when the maximum allowable curvature is used. So far, the analysis has been based upon these maximum forces. Such design is only necessary at problem locations where the topography or right of way will not permit flatter curves. Obviously, a large percentage of horizontal curves may be designed with curvature which is less than the maximum allowable. In all cases, the selection of degree of curvature is a compromise between driver comfort, safety, and cost of construction.

When the curvature to be employed on a horizontal curve is less than the maximum allowable, there is a leeway in how the force opposing centrifugal force is to be distributed between superelevation and "unbalanced" side thrust.

The crucial factor is the driver's speed around the curve. Drivers seem to make a succession of judgments, starting with how the curve appears as they approach it. They will judge the safe speed to some extent by the appearance of the superelevation. Then, as they begin to negotiate the curve, the side thrust will determine how secure they feel. Of course, drivers will not be affected by the degree of the curve or its radius per se, but they will be cognizant of the angle of deflection, that is, the angle at which the roadway for a given distance ahead on the curve is seen. Both the length of the curve and the sight distance around it will have a bearing upon what speed the driver will consider to be safe. Therefore, assuming adequate sight distance, drivers become experienced regarding the safe speeds of curves through a combination of the appearance of the curve and the "unbalanced" forces developed on the curve. For a given assumed design speed of the road, the unbalanced forces should be proportional to the degree of curvature if this relationship is to be established.

One method of accomplishing this is the so-called *three-quarters method*. With this method, superelevation is applied on all curves to counteract centrifugal force for a speed of three-quarters of the assumed design speed up to that point where the maximum allowable superelevation is reached.[1] This method is illustrated by Fig. 28-4. From the figure, the

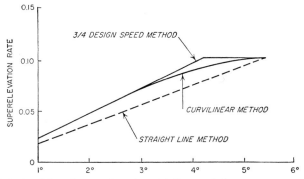

FIG. 28-4. Relationships of superelevation, friction, and degree on curves of less than maximum curvature. (Assumed design speed = 60 mph, maximum allowable superelevation = 0.10, maximum allowable friction = 0.13.)

magnitude of unbalanced side thrust increases rapidly with increased curvature after the point where maximum allowable superelevation is reached. To avoid this, another method has been proposed advocating a straight-line increase in superelevation up to the maximum allowable curvature permitted for the assumed design speed[2] (Fig. 28-4).

In theory, the *straight-line method* should assist the driver to judge the safe speeds of curves more accurately and should encourage uniform speed. However, it is known that drivers do not adopt uniform speeds, but rather they tend to drive faster on flat curves and on tangents. This tendency for motorists to *overdrive* the flatter curves has suggested that an improvement over the straight-line method would be to place a slightly higher proportion of superelevation on the flatter curves than would be indicated by the straight-line method. This refinement is recommended by the AASHO[3] and consists of a curvilinear relationship between superelevation and friction for increasing degrees of curvature. This *curvilinear method* is also shown on Fig. 28-4.

[1] Barnett, *op. cit.*, p. 69.
[2] Stonex and Noble, *op. cit.*, p. 429.
[3] *A Policy on the Geometric Design of Rural Highways*, p. 151.

28-5. Pavement-width Transitions. The design of horizontal alignment includes the design of curves and tapers to be employed when the width of the roadway is changed. Such transitions in pavement width occur when extra lanes are added or where channelizing islands are introduced into the pavement area.

It is evident that the rate of change in transverse placement adopted by drivers will vary with the time available to make the transition. For sudden, forced, or emergency situations the rate of change will be much higher than it would be if drivers were free to follow a normal transitional path.

The same forces which are critical in horizontal curve design are at work when drivers shift transversely to enter another lane or to avoid an obstruction.

A properly designed pavement-width transition should produce smooth operations, should be utilized throughout its length by drivers, and should *induce* traffic rather than *force* it into proper channels. It is evident that the lengths of such transitions depend on (1) speed, (2) amount of displacement, and (3) rate of change in placement considered to be desirable for design.

Recent studies of rates of change in placement adopted by drivers, when overtaking and passing a slower vehicle on a four-lane divided parkway on which traffic was running freely, indicate that the time rate of change in placement remains constant under these conditions for speeds ranging from 45 to 65 mph.[1] The average of values observed for all vehicles between these speeds ranged between 0.50 to 0.53 sec/ft of transverse shift. Another study[2] of similar nature found that the average driver employed 0.42 sec/ft of transverse shift. In making this change in placement, drivers were forced into the maneuver only to the extent that they were seeking the proper lane from which to make an exit from a multilane facility. In this same study, an average rate of 0.30 sec/ft was reported for vehicles which were forced to change lane by a parked car ahead.

These observations indicate that the time rate of change in placement ranges from 0.5 sec/ft of displacement for *free* maneuvers to 0.3 sec/ft of displacement for *forced* maneuvers. The American Association of Highway Officials suggests a value of ⅓ sec/ft of displacement for minimum design at speed-change lanes, ramp-nose offsets, and the *funneling* of ramps.[3] These matters are discussed in Chapter 30. For widening

[1] F. W. Hurd, "Open Discussion," Institute of Traffic Engineers, *Traffic Engineering*, New Haven, Conn., April, 1953.

[2] F. Wynn, S. Gourlay, and R. Strickland, *Studies of Weaving and Merging Traffic, A Symposium*, Yale University Bureau of Highway Traffic, Technical Report 4, New Haven, Conn., 1948.

[3] *A Policy on the Geometric Design of Rural Highways*, chap. 7, "Intersection Design Elements," p. 280.

of through-stream traffic channels and the introduction of islands affecting through traffic, more gradual transitions based on a rate of 0.5 sec/ft of displacement are recommended.

28-6. Sight Distance. There are times when a permanent sight obstruction on the inside of horizontal curves may reduce safe-stopping sight distances and determine the maximum degree of curvature that may safely be applied. Safe-passing sight distances around curves are also important on two- and three-lane roads if high degrees of capacity and safety are to be provided.

The same PIEV distances and braking distances employed in computing sight distances for vertical curves are used in the design of horizontal curves (Chapter 26).

The degree of curve that a given required sight distance demands depends upon the location of the sight obstruction with respect to the center of the inside lane of the road on the curve. Since this distance is determined by the particular curve that is chosen, cut-and-try methods must be used. The formula employed in relating sight distance to degree of curvature is $m = (5,729.58/D)$ versine $(SD/200)$, where m is offset to sight obstruction measured from center of the inside lane in feet, D is curvature at center line of inside lane in degrees, and S is sight distance in center of lane along curve in feet. Figure 28-5 shows a chart solution of this formula through the range of safe-stopping sight distances for assumed design speeds from 30 to 70 mph.

It will be noted that the heights of object and eye are not factors in the formula, as in the case of sight distances at vertical curves. The formula, of course, applies only to the sight line on the inside of the curve. The elevation of clear sight along this line, such as when an embankment is cut back, is dependent upon the height of eye and object, which are, theoretically, the same as those discussed in Chapter 27 for vertical-curve design.

28-7. Effect of Horizontal Alignment on Operations. Up to this point, the theory of horizontal-curve design has been considered only in terms of dynamic forces. Manifestation of the effect of horizontal alignment upon traffic operations becomes apparent in accidents and congestion.

Effect of Capacity. The effect of passing sight distances on the capacities of two-lane roads was shown in Table 26-5. Although horizontal curves are designed with adequate radii for safe-stopping sight distances, safe-passing sight distances are often restricted. Capacity is also lowered on curves because of the hesitancy of drivers to overtake and pass with the relatively unstable steering conditions encountered. Even when the passing sight distances are technically sufficient, the unbalanced side thrust on curves tends to deter drivers from passing. This tendency accounts for the reduced capacity of two-lane ramps having small turning radii. While no significant data are available, it has been noticed that freedom of passing is appreciably reduced on curves having a radius of 500 ft or less. As far as general alignment is concerned, both these factors—fewer passing opportunities and driver reluctance to pass on curves—may lower the over-all capacity of a highway having a large number of horizontal curves.

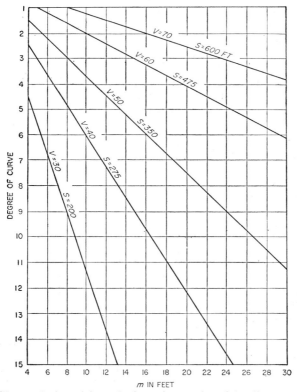

FIG. 28-5. Chart solution of formula for safe-stopping sight distance on horizontal curves. (From chart prepared by Department of Design, U.S. Bureau of Public Roads, January, 1946. Based on definitions and values of "A Policy on Sight Distances for Highways," in *A Policy on Geometric Design of Rural Highways*, American Association of State Highway Officials, Washington, 1940.)

Effect of Accidents. There is not a great deal of information available on the relationship between accidents and horizontal curves. The effects of curvature on accidents on two-lane rural highways are presented in Table 28-2. It will be noted that the table shows a continuous increase in the accident rate with increased curvature up to 6°, then a reduction in the rate up to curves at 10°, with a sharp increase on roads with light traffic for curves above 10°. The results suggest that the hazard caused by curvature becomes apparent when curves exceed 5 to 6°, and that extremely sharp curves are frequently misjudged.

The effect of both curvature and frequency of curves upon accident experience is shown by Table 28-3, which also applies to two-lane rural roads. Analysis of the data shown in Table 28-3 has indicated that there is little statistical significance in the relation of curve frequency to accident rates. However, the analysis revealed a direct relation between curvature and accident rates. The number of accidents per million vehicle-miles was found to increase about 0.15 for each additional degree of curve.[1]

TABLE 28-2. ACCIDENT RATES ON TWO-LANE RURAL-HIGHWAY HORIZONTAL CURVES, BY VOLUME, BY DEGREE OF CURVE
(Based on data from eight states)

| Degree of curve | Under 5,000 vehicles per day | Accident rate per vehicle-mile |
		5,000–10,000 vehicles per day
Less than 2	2.4	1.9
2–2.9	3.3	2.5
3–3.9	3.5	3.5
4–4.9	3.7	3.7
5–5.9	4.3	3.3
6–6.9	3.9	2.8
7–9.9	3.1	2.5
10–13.9	3.7	2.6
14–19.9	6.3	*
20 and over	7.6	*

* Sample too small for reliability.
SOURCE: David Baldwin, *The Relation of Highway Design to Traffic Accident Experience*, AASHO Convention Group Meeting, Los Angeles, 1946, American Association of State Highway Officials, Washington, p. 107, Table III.

There are numerous examples available showing that curves at the end of long straight sections are more accident-prone than others. This is generally explained on the basis that drivers become "velocitated" on the long straight sections and enter the sudden curves at higher speeds than they would normally use. It would seem desirable in such cases, therefore, if a sequence of curves were provided with those of the greater curvature preceded by those with the lesser. In other words, where it is necessary to utilize a sharp curve, the curves on either side could be arranged to achieve a gradual reduction in speed down to that speed at which the sharp turn could be safely negotiated. This principle was

[1] Morton S. Raff, "The Interstate Highway Accident Study," U.S. Bureau of Public Roads, *Public Roads*, vol. 27, no. 8, June, 1953.

applied in the design of the Pennsylvania Turnpike in the following way: "As far as practical the sequence was maintained in such a manner that the speed corresponding to a friction value of $F = 0.10$ did not vary more than five to eight miles per hour on adjacent curves."[1] No accident statistics have been collected to demonstrate the effectiveness of the measure, however.

TABLE 28-3. ACCIDENT RATES ON TWO-LANE RURAL HIGHWAYS BY DEGREE
OF CURVE AND FREQUENCY OF CURVES
(Based on data from fifteen states)

Number of curves per mile	Rate per million vehicle-miles on curves—			
	0–2.9°	3–5.9°	6–9.9°	10° or more
0–0.9	1.4	2.7	2.0	4.3
1.0–2.9	1.4	2.1	2.9	2.6
3.0–4.9	1.9	2.5	2.9	3.4
5.0–6.9	3.1	2.9	2.6	3.9

SOURCE: Morton S. Raff, "The Interstate Highway Accident Study," U.S. Bureau of Public Roads, *Public Roads*, vol. 27, no. 8, June, 1953, p. 181, Table 22.

Reverse horizontal curves should be avoided. Where it is necessary to provide curves in opposite directions, it has been arbitrarily recommended that they be placed at least 1,000 ft apart. If spiral transitions are used, there need not be as long a distance separating the points of tangency as would be required for curves without spiral transitions. In these latter cases, adequate distance should be provided between the curves to allow the normal attainment of superelevation for both curves at the assumed design speed of the road.

Broken-back circular curves should also be avoided. Such curves consist of several curves in the same direction joined by a short tangent. If the difference in radii for successive curves is large, an appreciable difference in side force and steering occurs abruptly at each break, resulting in erratic traffic operations. When such curves must be employed, better traffic operations result when the circular curves are replaced by a transition curve which introduces the change in curvature gradually.

Compound circular curves of properly chosen radii may closely approximate a transition easement and may be used instead of a spiral. Compound circular curves are also sometimes utilized in the design of curved ramps at grade separations, where it is desirable to effect a reduction or

[1] Charles M. Noble, "Design Features and Traffic Control for the Pennsylvania Turnpike," *Proceedings of Institute of Traffic Engineers*, New Haven, Conn., 1939, p. 41.

increase in speed. It is of interest that field studies have shown that drivers consistently overdrive many of the designs of this type at the present time. Design engineers are still seeking the proper combination of compounded curves for ramp design that will cause the desirable reduction in speed without creating hazardous and uncomfortable riding conditions.

CHAPTER 29

THE DESIGN OF MANEUVER AREAS

Both vehicular and pedestrian activities at intersections produce increased potential hazards and delays when compared with the hazards and delays encountered by drivers between intersections. Intersection design is the most complex problem in the traffic design of roadways.

29-1. Manuever Areas. Each intersection consists of one or more maneuver areas, the simplest elements of intersection design. They are the "building blocks" of intersection design.

Alternate types and arrangements of maneuver areas may be employed to create intersections with varying qualities of operations. In general, those intersections which offer the higher qualities of operations are more costly to construct.

Maneuver areas include not only the area of potential collision wherein the actual maneuver is performed, but also its channels of approach or departure wherein design or regulation may influence the conditions of the maneuver.

There are two kinds of maneuver areas, elemental and multiple. Elemental maneuver areas occur when *only two* one-lane one-way flows cross, merge, or diverge. Multiple maneuver areas accommodate *more than two* one-lane one-way flows (Fig. 29-1).

Relative Speed. The quality of operations of maneuver areas is dependent upon relative speed expressing the movement of intersecting vehicles as related to each other. Relative speed is expressed as a speed vector (Fig. 29-2). Table 29-1 shows comparative values of relative speed for various intersection angles and different speed ratios.

It was shown in Chapter 8 that the size of gap in traffic required by crossing and merging drivers becomes less with reduced relative speed. Observed values for average gaps accepted which were reported in that chapter ranged from 8 sec for maneuvers performed at high relative speed to 2 sec for low-relative-speed maneuvers. It was also made clear in Chapter 8 that relative speed is an important factor in driver delays and the capacities of maneuver areas.

It is axiomatic that collisions between vehicles intersecting at small angles and at about the same speed will generate much less impact than when vehicles collide at high relative speed. Also, potential collisions

469

take form more slowly when intersecting maneuvers are performed at low relative speed.

Maneuver areas, when isolated from the operational influence of other maneuver areas, are inherently safe and cause the least driver delay when they operate at low relative speed.

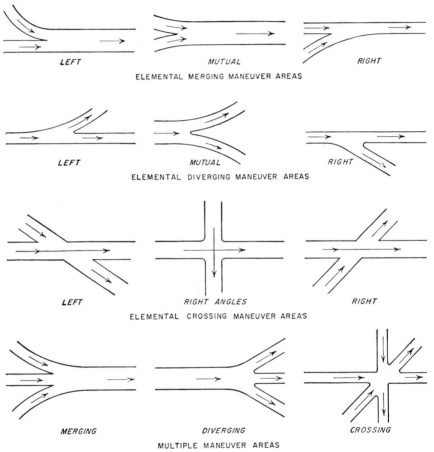

LEFT   *MUTUAL*   *RIGHT*

ELEMENTAL MERGING MANEUVER AREAS

LEFT   *MUTUAL*   *RIGHT*

ELEMENTAL DIVERGING MANEUVER AREAS

LEFT   *RIGHT ANGLES*   *RIGHT*

ELEMENTAL CROSSING MANEUVER AREAS

MERGING   *DIVERGING*   *CROSSING*

MULTIPLE MANEUVER AREAS

Fig. 29-1. Examples of elemental and multiple maneuver areas.

Multiple Maneuver Areas. Multiple maneuver areas are to be avoided in design. The use of a common collision area by more than two flows confuses drivers and produces problems in capacity and safety. Drivers may exercise better judgment when they are required to contend with only one intersecting conflict at a time.

A possible exception to this rule in design occurs in the case of multiple diverging. The diverging maneuver is relatively simple, and multiple maneuvers of this form may usually be performed in safety without delay.

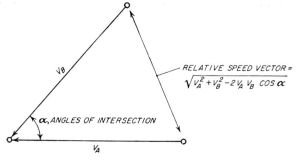

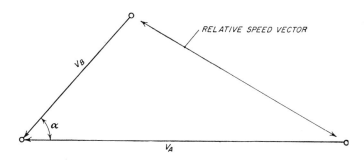

FIG. 29-2. Illustration of relative-speed vector.

It is obvious that the function of multiple-maneuver areas may be performed in design through the provision of two *elemental* maneuver areas located so as to be adequately separated from each other in distance and time.

Elemental Diverging Maneuver Areas. Compared with the other two types of elemental maneuvers, the diverging maneuver causes the least hazard and delay. The divergence of two flows stems from a common

TABLE 29-1. RELATIVE SPEED AS PER CENT OF LOWER SPEED

Intersection angle, deg	Ratio of higher speed to lower speed			
	1:1	2:1	3:1	4:1
0	0	100	200	300
30	52	124	219	316
60	100	173	265	361
90	141	224	316	412
120	173	265	361	458
150	193	291	388	489
180	200	300	400	500

flow; thus, to avoid conflicts, diverging maneuver areas should always be designed for low-relative-speed operations.

Divergence inherently takes place at a small angle of maneuver. Therefore, design for low relative speed is produced when road cross section, alignment, and sight distance along the channels of *departure*

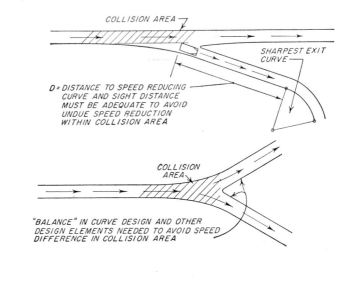

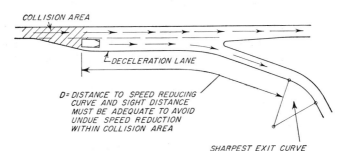

Fig. 29-3. Elemental diverging maneuver areas—schematic illustrations of design to encourage low relative speed.

from the collision area do not cause speed reductions which will be reflected back into the common flow.

Deceleration lanes are sometimes used to provide sufficient distance from design elements influencing speed when this distance cannot be provided within the normal roadway alignment. Figure 29-3 shows typical examples of diverging maneuver areas designed for low relative speed.

Elemental Merging Maneuver Areas. Mergence also inherently occurs at a small angle of maneuver. Therefore, merging maneuver area design for low relative speed is produced when cross section, alignment, and sight distance along the channels of *approach* to the collision area do not cause speed differences within the merged, or common, flow.

In the case of merging, an additional factor influencing speed, called maneuvering time, includes the time required by drivers to select a gap in the moving stream to be intersected and to use that gap without speed interference.

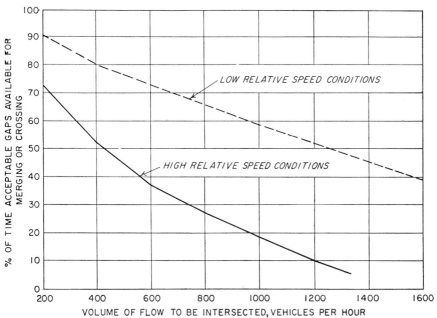

FIG. 29-4. Availability of acceptable gaps for merging and crossing at elemental maneuver areas. Applies to single lanes operating under free-flow conditions. (SOURCE: F. Wynn, S. Gourlay, and R. Strickland, *Studies of Weaving and Merging Traffic, A Symposium,* Yale University, Bureau of Highway Traffic, Technical Report 4, New Haven, Conn., 1948, p. 4, Fig. 1.)

During the maneuvering time, drivers must anticipate the time of their arrival at the collision area so as to reach the point of intersection simultaneously with an acceptable gap in the intersected flow. At the same time, they must pace the speed of the intersected flow so as to enter the gap without speed interference.

As volumes increase, fewer acceptable gaps occur in the intersecting flows. Insufficient maneuvering time causes slowdowns and stops as drivers fail to adjust their speed and time of arrival at the maneuver point to conform with acceptable gaps. Maneuver areas then become congested and operate at high relative speed until a reduction in volume permits traffic again to become fluid.

Figure 29-4 shows the availability of acceptable gaps for merging and crossing in terms of relative speed and various volumes of traffic in the intersected flow. The per cent of time shown during which intersecting maneuvers are possible takes into consideration the fact that longer gaps are required with increased relative speed. Maneuvering time may be provided through intersection sight distance, or design affording flexibility in the place of maneuver, or both. Designs with fixed and flexible place of maneuver are illustrated by Fig. 29-5.

Acceleration lanes are sometimes used to provide adequate distance between speed-reducing design elements along the channels of approach

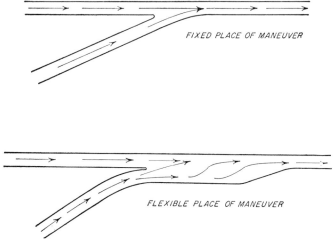

FIXED PLACE OF MANEUVER

FLEXIBLE PLACE OF MANEUVER

FIG. 29-5. Two methods of providing maneuvering time through the design of elemental maneuvering areas. (Schematic only.)

to the collision area. Such lanes also provide additional maneuvering time. Figure 29-6 illustrates typical examples of merging maneuver areas designed for low relative speed.

Elemental Crossing Maneuver Areas. Crossing maneuvers may be performed at any angle. They cause the most hazard and delay but are the easiest to control because crossing flows are inherently independent of each other. Crossing maneuver areas may, therefore, be properly designed for either low- or high-relative-speed operations.

Crossing maneuver areas designed for high relative speed should provide for crossing at approximate right angles. Drivers are more accustomed to right-angle crossings than to skewed crossings and have formed better driving habits for that angle of crossing. Vision in both directions is more convenient at right angles, and the head-on type of collision is prevented.

Crossing designed for high relative speed should be controlled by traffic control devices to promote safety. Better observance of such devices is obtained when the regulated channels of approach are designed so as to reduce speed by either *bending* or *funneling*, or both. By bending, the designer introduces curvature into the intersection-approach alignment to cause reduction in speed. The funneling of traffic channels so as to produce the psychological impression of restricted freedom of movement may be achieved by gradual narrowing of the approach cross

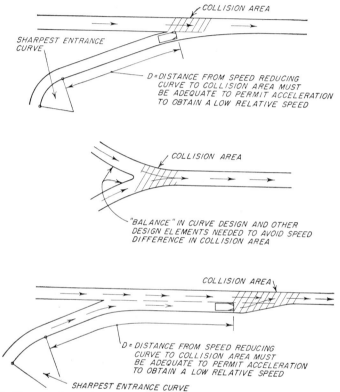

Fig. 29-6. Elemental merging maneuver areas—schematic illustrations of design to encourage low relative speed.

section. Both funneling and bending must be employed carefully to avoid increased hazard. In the application of these principles, design should always favor the heaviest and fastest flows.

The same techniques employed in the design of merging maneuver areas may also be employed in the design of elemental crossing maneuver areas intended to operate at low relative speed. When crossing maneuver areas are to operate at low relative speed, the principle of *weaving* is usually employed in their design (section 29-2).

Elemental crossing maneuver areas may be grade-separated, in which case there is no conflict between crossing flows.

Figure 29-7 illustrates high and low relative speed as well as grade-separated elemental crossing maneuver areas.

Modifying Factors. Traffic control regulations and devices are used to regiment the traffic flow and to increase safety at intersections. When elemental maneuver areas operate at low relative speed they do not require delay-producing regimentation caused by traffic signals and "stop" signs. Maneuver-area design for low relative speed is desirable

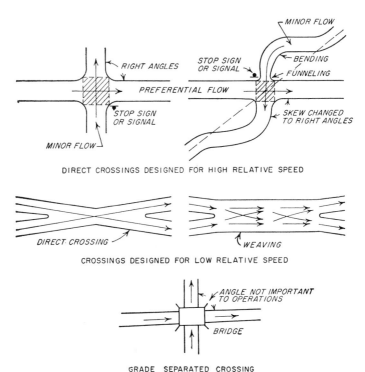

DIRECT CROSSINGS DESIGNED FOR HIGH RELATIVE SPEED

CROSSINGS DESIGNED FOR LOW RELATIVE SPEED

GRADE SEPARATED CROSSING

Fig. 29-7. Elemental crossing maneuver areas designed for low relative speed, high relative speed, and separation. (Schematic only.)

but more costly to construct than design for high relative speed, since better alignment and longer sight distances are required.

Where volumes are light or where traffic controls, particularly traffic signals, are to be employed, relative speed becomes less important in the design of maneuver areas. Parking, pedestrians, and other conditions which interfere with the flow destroy low relative speed. Low-relative-speed design of maneuver areas is not advisable at intersections with heavy pedestrian traffic unless the pedestrian-crossing conflict may be eliminated by pedestrian bridges or tunnels.

Theoretical Design Capacities. The capacities of elemental maneuver areas depend on relative speed. There are three cases in the consideration of the design capacities of elemental maneuver areas where intersecting maneuvers take place at grade. These are (1) where all intersecting vehicles must stop before performing the maneuver, (2) where vehicles in one of the intersecting flows must stop and the other flow proceeds relatively unhampered, and (3) where vehicles in both flows proceed unhampered or are equally hampered but do not stop.

In case 1, the theoretical design capacity of an elemental maneuver area is 1,200 passenger vehicles per hour for the combined flows. In this case, the right of way is more or less regularly alternated, simulating operations where traffic signals or "stop" signs operated against both flows are employed. This is the most stable operating condition, but produces the most delay.

Capacities under case 2 are dependent on the volumes of the intersecting flows. The unhampered flow may proceed at such close headways that few, if any, vehicles in the stopped flow may have opportunity to enter the intersection. This condition provides capacities approaching those of a single-lane nonintersectional flow—up to about 2,000 passenger vehicles per hour, depending on absolute speed. At volumes near this size, stopped vehicles do not have opportunity to enter, which obviously should not be considered in design because of excessive delays incurred by the stopped flow. The magnitude of delay incurred at high-relative-speed maneuver areas is discussed in Chapter 8. Actually, drivers on the minor road will tolerate only so much delay and will then crowd into the collision area producing case 1 operating conditions. Again, reference should be made to the solid-line curve of Fig. 29-4, which shows the availability of gaps for intersecting maneuvers under high-relative-speed conditions. The amount of delay which will be tolerated by drivers at intersections is not known but it is an important factor in the application of case 2 conditions to traffic regulation and design problems.

In case 3, the elemental maneuver area operates at low relative speed. In this case, intersecting drivers utilize smaller gaps in traffic, and minimum delay is incurred. Under ideal conditions, the capacity of elemental maneuver areas may approach that of a single-lane nonintersectional flow. However, when a single intersecting driver fails to utilize a gap in the flow to be crossed or entered, he must stop, and causes following vehicles to stop. The maneuver area then operates under case 1 or case 2 conditions. The ability of an elemental maneuver area to operate at low relative speed (under case 3 conditions) depends not only on the angle of the intersecting maneuver and difference in speed of intersecting vehicles but also on the maneuvering time made available through design. Sufficient knowledge of the relationship of maneuvering time and capacity is not available to permit, with any assurance, the design of maneuver areas which will operate at low relative speed without failures.

Continuity of movement, delay, and stability of operations, therefore, become the most important considerations in the design of elemental maneuver areas. It is apparent that design for case 3 operations provides the highest degree of continuity and the lowest amount of delay.

For case 3, the theoretical design capacity of elemental merging (and crossing) maneuver areas are assumed by the AASHO to be:[1]

1,000 passenger vehicles per hour for combined flows on high-speed rural roads
1,200 passenger vehicles per hour for combined flows at speeds not less than 35 mph
1,500 passenger vehicles per hour for combined flows at speeds less than 35 mph (urban)

The capacity of merging maneuver areas may be governed by the capacity of the channel of combined flow ahead of the intersection. Obviously, the volume of combined through and merging traffic should

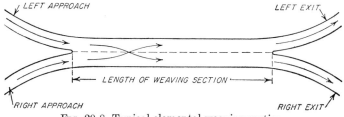

FIG. 29-8. Typical elemental weaving section.

not exceed the capacity of road section ahead. It is believed that merging drivers operating at low relative speed will force their way into through traffic by adopting headways which are shorter than they will tolerate after they leave the intersection. This has not been checked in the field.

The capacities of *grade-separated* crossings are the sum of the free-flow capacities of each of the crossing channels.

Capacities of elemental *diverging* maneuver areas are seldom a problem in design. Assuming conditions of smooth diverging flow, the capacity of the channel of combined flow determines the capacity of maneuver areas of this type.

29-2. Weaving. Weaving is the act performed by a vehicle when it moves obliquely across the path of other vehicles moving in the same direction. The same principles of design for merging and diverging areas apply in design for the weaving maneuver.

The combined merging and diverging collision area of a crossing designed for weaving maneuvers is called a weaving section. Figure 29-8 shows a typical elemental weaving section.

The quality of operations of weaving sections depends on relative speed. Weaving sections must operate at low relative speed to incur

---

[1] *A Policy on the Geometric Design of Rural Highways,* chap. 2, "Design Controls and Criteria," American Association of State Highway Officials, Washington, 1954, p. 100.

minimum delay with a high degree of safety. The length of the weaving section contributes to the maneuvering time available to weaving drivers and, where the section is of sufficient length, the same gap may be utilized by more than one weaving vehicle.

Increased lengths of weaving section increase the ability of such sections to accommodate weaving vehicles in greater volumes and at higher speeds. Table 29-2 shows the relationship of average running speed and number of weaving passenger vehicles for elemental weaving sections of various lengths. These values are employed in practice as the theoretical design capacities of weaving sections.

TABLE 29-2. LENGTH OF WEAVING SECTIONS RELATED TO VOLUME OF WEAVING VEHICLES AND RUNNING SPEED

Length of weaving section, ft	No. of weaving passenger vehicles per hour at—	
	25 mph	35 mph
100	700	350
200	1,000	600
300	1,200	750
400	1,400	900
500	1,600*	1,050
600	1,700*	1,200
800	2,000*	1,400

* Since these values were obtained from studies made at compounded maneuver areas, it is not known how much of the increased volume was produced by compounded weaving.

SOURCE: *Highway Research Board Capacity Manual*, Highway Research Board and U.S. Bureau of Public Roads, Committee on Highway Capacity, 1950, p. 115, Fig. 43. (Adjusted to running speeds.)

29-3. Compounded Maneuver Areas. Elemental maneuver areas are compounded when they are arranged in parallel to accommodate multilane flows. Figure 29-9 shows comparative examples of compounded elemental merging and diverging maneuver areas. Compounded merging and diverging maneuver areas create supplemental crossing conflicts. These crossing conflicts cause driver confusion and destroy low relative speed. The volumes of traffic accommodated by *compounded* merging and diverging maneuver areas during the time that they operate at low relative speed are little greater than those of *elemental* merging and diverging maneuver areas, and they cause greater hazard and delay.

Figure 29-9 also shows recommended design where multilane roadways converge or separate. By providing extra lanes, as illustrated, two-lane capacities for the merging and diverging channels are achieved with elemental merging and diverging. Such *balance* in cross section is an important consideration in intersection design.

Figure 29-10 shows an example of a compounded weaving section. The same multiple conflicts are produced as with compounded merging and diverging areas. It is evident that merging, diverging, and weaving maneuvers are elemental in character and should not be compounded in design when it is assumed that these types of maneuvers should be performed under conditions of low relative speed.

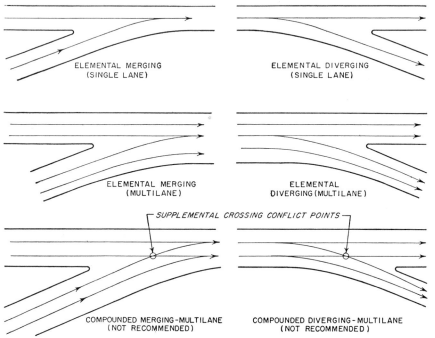

ELEMENTAL MERGING
(SINGLE LANE)

ELEMENTAL DIVERGING
(SINGLE LANE)

ELEMENTAL MERGING
(MULTILANE)

ELEMENTAL
DIVERGING (MULTILANE)

SUPPLEMENTAL CROSSING CONFLICT POINTS

COMPOUNDED MERGING-MULTILANE
(NOT RECOMMENDED)

COMPOUNDED DIVERGING-MULTILANE
(NOT RECOMMENDED)

FIG. 29-9. Typical elemental and compound merging and diverging maneuver areas. (Schematic only.)

Maneuver areas which are to operate at *high relative speed* may be safely compounded with increased capacity, particularly when *positive* traffic control is employed. *High-relative-speed operation is inherently unsafe and nearly always requires traffic control.* Positive traffic control, such as traffic signal (or officer) control, eliminates conflicts by supervising the time conflicting flows may use the collision area. Positive time controls are generally employed at high-relative-speed-operated intersections when traffic approaches capacity volumes. Traffic signal (or officer) control may be employed to eliminate the hazard and confusion caused by supplemental conflicts at compounded maneuver areas.

Figure 29-11 illustrates elemental and compounded crossing maneuver areas at grade. It is apparent that positive time control affords the same *lane-flow* efficiency to both types of intersections. The theoretical design capacity of each approach of a compounded maneuver area operated with positive time control is therefore about 1,200 passenger vehicles *per traffic lane* per hour of *green-signal indication.*

29-4. The Separation of Maneuver Areas. It is fundamental that drivers should be required to contend with only one conflicting flow at a

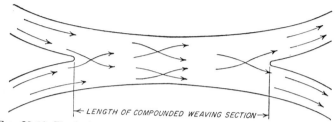

FIG. 29-10. Typical compounded weaving section. (Schematic only.)

time if they are to exercise their best judgment. Overlapping maneuver areas operate with reduced capacity of each area because only one vehicle may occupy the common collision area at a time.

Intersection hazards and delays are similarly increased when maneuver areas are too close together. There must be sufficient separation between

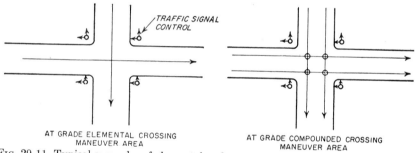

AT GRADE ELEMENTAL CROSSING
MANEUVER AREA

AT GRADE COMPOUNDED CROSSING
MANEUVER AREA

FIG. 29-11. Typical examples of elemental and compounded crossing maneuver areas. (Schematic only.)

successive maneuver areas for drivers to cope with the changing traffic situation so as to adjust speed and path to the conditions of each conflict.

Through design, maneuver areas may be separated in *space* and *time.* Obviously, the space, or distance, element is related to time through the medium of speed.

Separation in Space. Maneuver areas may be separated in space by the dispersion of intersection movements. Figure 29-12 shows typical examples of dispersion for crossing, right-turn, and left-turn movements. Through the dispersion of intersection movements, a complex intersection of a single area may be expanded into a number of simple intersections which, in combination, perform the same total intersection function, with less hazard and delay.

Time Separation. Time separation of successive maneuver areas in terms of design consists of the provision of zones of refuge located outside of the path of other flows.   These zones provide a haven where motorists or pedestrians may wait between successive maneuvers.   Such waiting

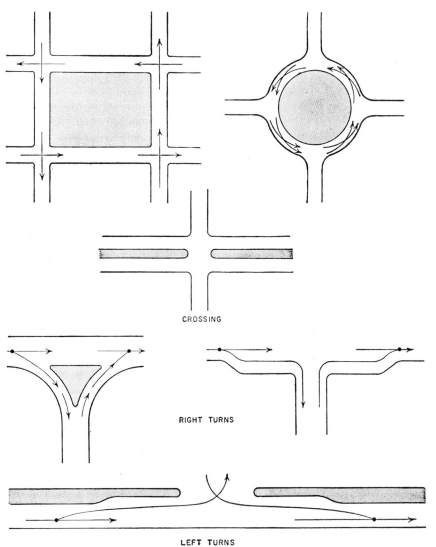

CROSSING

RIGHT TURNS

LEFT TURNS

FIG. 29-12. Typical examples of separation of maneuver areas.   (Schematic only.)

zones may be provided for turning or crossing vehicles as well as for pedestrians.   Figure 29-13 shows typical examples of zones of refuge.

Amount of Separation. The time or distance separation required between successive maneuver areas varies greatly with different condi-

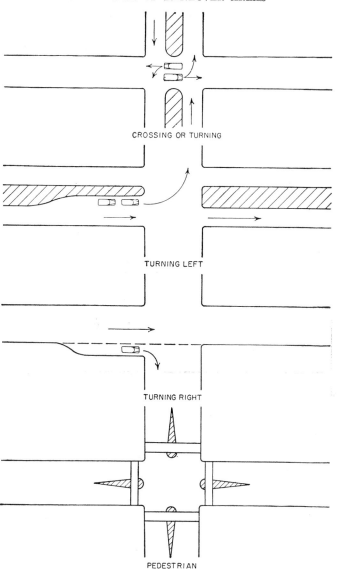

CROSSING OR TURNING

TURNING LEFT

TURNING RIGHT

PEDESTRIAN

FIG. 29-13. Typical examples of zones of refuge. (Schematic only.)

tions. Driver PIEV times vary with the complexity of the situation and nature of the response. The time required to change speed and path depends on absolute values and requirements. The distance separation required to prevent *backups* from one maneuver area into another depends on the amount of delay incurred, the volume of flow, types of vehicles, and other factors.

Each situation presented by design should be analyzed by the designer in terms of the particular time-and-distance separation required by specific traffic conditions.

29-5. Crossing and Turning Geometry. Vehicle flows may cross by (1) direct crossing at grade, (2) weaving, and (3) grade separation (Fig. 29-7). Alternate choice in intersection design is made possible when one of these types of crossing maneuver may be substituted for another. Further choice in intersection design is made possible by the

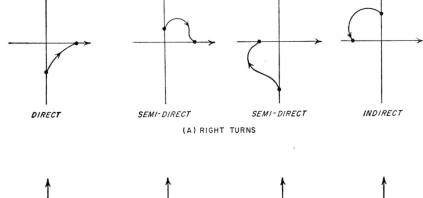

DIRECT	SEMI-DIRECT	SEMI-DIRECT	INDIRECT

(A) RIGHT TURNS

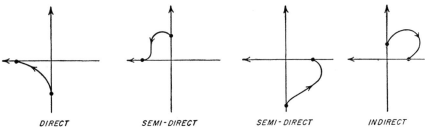

DIRECT	SEMI-DIRECT	SEMI-DIRECT	INDIRECT

(B) LEFT TURNS

FIG. 29-14. Arrangements of maneuver areas to accommodate right and left turns. (Schematic only.)

diverse ways in which turning movements may be accomplished. Figure 29-14 shows the alternate geometry of right- and left-turn movements. These turning movements are classified as direct, semidirect, and indirect in terms of the paths followed by drivers.

The direct turn to the right or left consists of a simple diverging and merging maneuver without crossing conflicts. It provides the shortest travel distance and is best understood by drivers because it follows the desired path of travel.

Semidirect and indirect turns offer longer travel distances but may be employed when physical conditions prevent the use of direct turns, or when it is desirable to consolidate crossing conflicts so they may be better or more economically controlled.

29-6. Arrangement of Maneuver Areas. The crossing conflict created either by through or turning movements is the most critical determinant

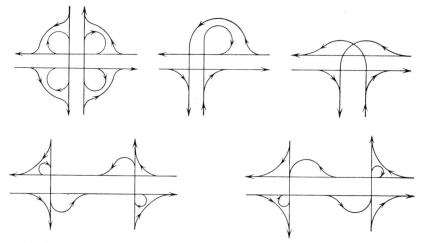

A — DIRECT CROSSING, INDIRECT OR SEMI-DIRECT LEFT TURN AND DIRECT RIGHT TURN (SCHEMATIC ONLY)

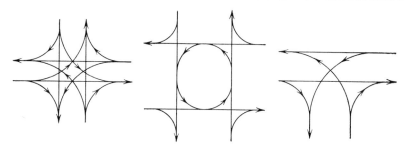

B — DIRECT CROSSING, DIRECT LEFT TURN, DIRECT RIGHT TURN

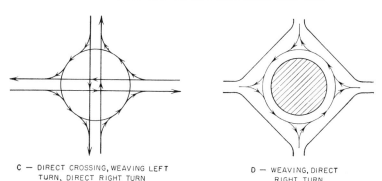

C — DIRECT CROSSING, WEAVING LEFT
TURN, DIRECT RIGHT TURN

D — WEAVING, DIRECT
RIGHT TURN

FIG. 29-15. Common arrangements of maneuver areas in intersection design.

of intersection design. The selection and arrangement of *crossing maneuver areas* to accommodate the heaviest flows produce the nucleus of the geometry of a particular intersection, and the arrangements of maneuver areas for other movements are designed to conform.

The right-turn movement presents the *least* problem in the integration of intersection movements. The *direct* right turn does not cross other intersection movements and is nearly always utilized in intersection design where site limitations do not prevent its use.

The left-turn movement presents the *greatest* problem in the integration of intersection movements. The *direct* left turn crosses other left-turn and crossing movements. These secondary crossing maneuver areas cause increased accidents and congestion. Their influence on intersection operations may be modified by employing semidirect or indirect left-turn design.

Figure 29-15 diagrams the most common arrangements of maneuver areas in intersection design, classified according to the type of design provided for crossing and turning movements. It is evident that any of the crossing maneuver areas shown here could be grade-separated.

There are many combinations of maneuver-area arrangements in intersection design. Traffic factors or physical conditions at particular intersections may warrant geometric patterns, which are not shown in Fig. 29-15.

CHAPTER 30

GEOMETRIC ELEMENTS OF INTERSECTION DESIGN

Many of the same factors pertaining to the elements and nature of highway design discussed in previous chapters also apply at intersections.

30-1. Road Surface. The friction, smoothness, and light-reflecting characteristics of properly designed pavement surfaces are adequate for both routes and intersections but may require corrective measures at intersections before a route becomes hazardous at other locations. Roughened surface as a warning device, pavement-color contrast to show preference and minor-road vehicle paths, and lighter pavement-surface color to increase the efficiency of night lighting obviously pertain to intersection design. These design elements were discussed in Chapter 24.

30-2. Vertical Alignment. Grades and vertical curves should be avoided as much as possible in intersection design. The effect of gradient on speed and braking as well as the reduction of sight distance at vertical curves should be taken into consideration as factors modifying design when gradient at intersections is unavoidable. Their influences are essentially the same for intersections as for routes (see Chapter 27).

30-3. Sight Distance. Sight distance at intersections was discussed in Chapter 26. In the cases of crossing and merging, the drivers on the minor road must see traffic approaching on the preference road. For this purpose, a 4.5-ft height of eye and 4.5-ft height of object are considered satisfactory for sight-distance requirements.

The same minimum *safe-stopping sight distances* shown in Table 26-1 for through-highway conditions applies to turning roadways at intersections. Values for the lower speeds encountered at intersections, assuming a 2.5-sec PIEV time and f varying from 0.42 to 0.36 for design speeds of 15 to 30 mph, respectively, are shown in Table 30-1.

30-4. Cross Section. Most of the factors discussed in Chapter 25 pertaining to lane widths, clearances, curbs, and shoulders also apply at intersections. Cross section at intersections may be varied to control speed, to channelize traffic, and to decrease hazard.

Funneling. Funneling may be employed to channelize merging traffic as well as to control speed. Funneling for channelization is employed in the design of low-relative-speed elemental maneuver areas to control the angle of intersection and to prevent entrance of more than one-lane flow

487

from maneuver-area approaches. It may also be employed at high-relative-speed maneuver areas to cause speed reduction in support of intersection regulatory controls.

TABLE 30-1. MINIMUM STOPPING SIGHT DISTANCE FOR TURNING ROADWAYS

Assumed design speed, mph	Level road distance, ft
15	80
20	120
25	160
30	200

SOURCE: *A Policy on the Geometric Design of Rural Highways*, chap. 7, "Intersection Design Elements," American Association of State Highway Officials, Washington, 1954, p. 295.

Figure 30-1 illustrates both applications of funneling. The width of pavement at the small end of the funnel should be selected to reduce speed in one case and to channelize vehicles without speed reduction in the other case. The transition distance should be determined by the

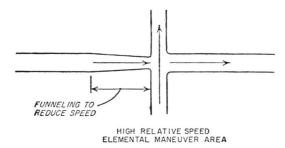

FUNNELING TO
REDUCE SPEED

HIGH RELATIVE SPEED
ELEMENTAL MANEUVER AREA

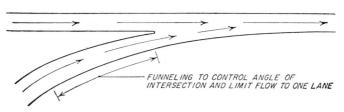

FUNNELING TO CONTROL ANGLE OF
INTERSECTION AND LIMIT FLOW TO ONE *LANE*

LOW RELATIVE SPEED ELEMENTAL
MANEUVER AREA

FIG. 30-1. Applications of funneling to control speed and to channelize traffic.

time required by drivers to accomplish the necessary change in lateral placement. A transition distance of about $\frac{1}{3}$ sec/ft of lateral shift is recommended.[1]

Acceleration Lanes. An acceleration lane is defined as extra pavement, of constant or variable width, placed parallel, or nearly so, to a merging

[1] See Chapter 28, Pavement-width Transitions.

maneuver area to encourage merging at low relative speed.   Figure 30-2 illustrates two acceleration-lane design forms used in design practice. The major difference in opinion concerning acceleration-lane design stems from lack of information on driver performance.   Field observations have indicated that drivers desire to follow the direct path inherent in Form B design shown in Fig. 30-2, as compared with the reverse curve path required at the end of the Form A acceleration lane.   Some designers contend that Form A design is superior because it affords drivers the

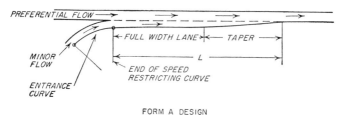

FORM A DESIGN

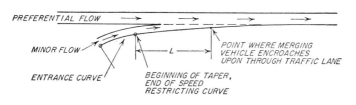

FORM B DESIGN

FIG. 30-2. Forms of acceleration lanes.

opportunity to perform their merging maneuver as soon as possible. Other designers contend that the Form A design encourages unsafe passing practices and that drivers in the preferential flow tend to "make way" for drivers entering their flow from the acceleration lane when the place of mergence is fixed as provided by the Form B design.

The length of acceleration lanes should be determined by two factors: (1) time required for drivers to accelerate to the speed of the preferential flow from the speed of entry into the acceleration lane and (2) maneuvering time required as a supplement to the sight distance which is provided in advance of the accleration lane.

Table 30-2 shows the lengths of acceleration lanes recommended by the AASHO.   These lengths apply to level roads and are based on average (over-all) rates of normal acceleration which have been observed for passenger vehicles.   The lengths apply to the limits L indicated on Fig. 30-2.   Taper distances shown in the table apply only to Form A acceleration lanes and are based upon a lateral transition time of about $\frac{1}{3}$ sec/ft of displacement.

Note that the AASHO recommends for high-volume roadways a length greater than that based upon acceleration requirements alone. This added length is for increased maneuvering time, although precise maneuvering-time requirements are not known.

The selection of assumed design speed for the design of acceleration lanes, provisions for slower trucks, and the influence of traffic volume are matters of opinion to be deferred to the judgment of designers in individual cases. However, the AASHO suggests the use of average

TABLE 30-2. DESIGN LENGTHS OF ACCELERATION LANES

Acceleration lanes	Design speed of highway, mph	Length of taper, ft	Total length of acceleration lane, including taper, in feet, for following design speed of entrance curve, mph			
			20	30	40	50
High-volume highways	40	175	400	250		
	50	200	650	500	250	
	60	225	950	800	550	250
	70	250	1,200	1,000	800	500
Other roadways	40	175	250	*		
	50	200	450	300	*	
	60	225	700	500	250	*
	70	250	900	750	500	350

* Less than length of taper.

SOURCE: *A Policy on the Geometric Design of Rural Highways*, chap. 7, "Intersection Design Elements," American Association of State Highway Officials, Washington, 1954, p. 288, Table 7-10.

running speeds (see Table 23-1) as the criteria for minimum acceleration-lane lengths, and the values shown in Table 30-2 were obtained on this basis, assuming that longer lanes may not be economically feasible.

Deceleration Lanes. Deceleration lanes are defined as extra pavement, of constant or variable width, placed parallel, or nearly so, to a diverging maneuver area to encourage diverging at low relative speed. Figure 30-3 shows three forms of deceleration lanes. Forms A and B are the same except that the lane is started abruptly in Form B to attract the attention of drivers. Some designers contend that drivers are not immediately aware of the existence of a deceleration lane when its width is adopted gradually, as in the case of Form A.

Form C provides a more direct path for turning drivers but requires more length, particularly if deceleration is not to take place within the area of combined flow. General observations have shown that drivers increasingly utilize the entire length of Form A deceleration lanes as traffic volumes and speed pressures increase. Additional research is required to prove the superiority of each design, but it seems reasonable to assume that (1) the target-value advantages of Form B could be obtained by contrasting pavement colors or other means; (2) Form A design is more economical when large speed differentials are to be overcome; and (3) Form C design is more convenient for drivers when small speed differentials are to be eliminated.

The American Association of State Highway Officials recommends design of the taper used at the beginning of Form A based on about ⅓ sec/ft of transverse displacement.

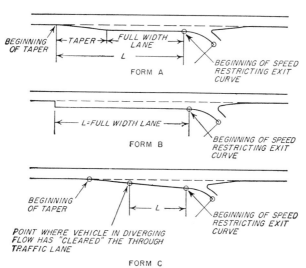

FIG. 30-3. Forms of deceleration lanes.

The lengths of deceleration lanes are based on the difference in tne speed of traffic of the combined flow (in advance of the collision area) and the speed at which drivers negotiate the critical diverging channel curve, as well as the deceleration practices of drivers. Not much is known about the deceleration practices of drivers at intersections. Certain rationale has been established assuming that drivers decelerate in gear without braking for 3 sec, then decelerate at a comfortable or leisurely rate, or combination of both, and that this performance takes place entirely within the deceleration lane, inclusive or exclusive of the taper distance, or that it starts in advance of the taper.

Table 30-3 shows the minimum lengths of deceleration lanes recommended by the AASHO. Lengths L shown in the table are based on the average running speed at the beginning and end of the deceleration lane limits, as indicated on Fig. 30-3.

These deceleration lane lengths are based on the assumed performance of passenger vehicles only. Extra allowance must be made for grades and for trucks with different deceleration characteristics.

Breakdown Lanes. On tangents or on curves, vehicle breakdown lanes should be provided parallel to the lane of travel where *single lane* channels at intersections are of appreciable length. Specific warrants for such added pavement widths have not been developed, but volume of flow and lengths of channels are the two more important considerations.

The extra cross-section width needed for breakdowns may, of course, be provided in the material of a stabilized shoulder.

30-5. Design of Merging and Diverging Pavements. In the design of merging pavements, the two merging channels should be aligned to create

TABLE 30-3. DESIGN LENGTHS OF DECELERATION LANES

Design speed of highway	Length of taper, ft	Total length of deceleration lane, including taper, in feet, for following design speed of exit curve, mph			
		20	30	40	50
40	175	250	175		
50	200	350	250	200	
60	225	400	350	250	*
70	250	450	400	350	250

* Less than taper.

SOURCE: *A Policy on the Geometric Design of Rural Highways,* chap. 7, "Intersection Design Elements," American Association of State Highway Officials, Washington, 1954, p. 288, Table 7-10.

the smallest practicable oblique angle of maneuver. Therefore, the island end separating the channels should be as narrow as possible, and the entrance channel for merging traffic may be funneled to channelize the path of merging vehicles.

The design of merging and diverging pavements is illustrated by Fig. 30-4. It should be noted that the end of the island separating the channels of divergence is shaped to discourage erratic steering, but at the same time it is offset from the preferential flow channel and is tapered to allow drivers who have unwittingly followed the right edge of the pavement an opportunity to turn back into the desired channel along a path of gradual lateral shift. A transition length of ⅓ sec/ft of transverse displacement is recommended in the design of such offset tapers.

The offsets at the end of the island formed by the channels of divergence widen the diverging channels at their entrance. Such widening is designed to make the entrance appear open to drivers so as to avoid reduction in speed. The extra pavement width to be employed is a matter of design judgment.

In the design of merging and diverging areas for uniform speeds, compounded curves or transitional pavement-edge alignment is desired to avoid sudden changes in speed and to provide better condition for gradual superelevation runoff.   Figure 30-4 also illustrates horizontal curvature at merging and diverging maneuver areas.   The curve radii and arrangement as well as length of curves and transitions vary with the design speeds assumed within the collision area and on the channels of approach and departure.

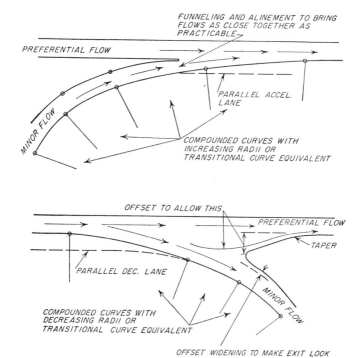

Fig. 30-4. Design of merging and diverging pavements.

30-6. Horizontal Alignment.   Intersection curves may be designed to control speed; the principle of bending to produce high relative speed in support of regulatory controls was mentioned in the previous chapter.

The speed design of intersection curves may be classed into three basic categories, as follows: (1) design of minimum-speed curves, (2) design of curves to produce a given speed, and (3) design of compounded curves or transitions to produce a given speed increase or reduction.

Minimum-speed Curves.   Curves of minimum radius may be employed in the design of urban intersections where speeds are very low or in the design of an intersection where turns are to be made under high-relative-speed conditions.   Principal design factors for minimum intersection turns are (1) the minimum turning radii of design vehicles, (2) the steering practices of drivers as they perform minimum radius turns, and (3) the wheel paths of sharply turning vehicles. Minimum turning radii for various types of vehicles are discussed in Chapter 3.

Drivers steer into low-speed minimum-radius turns along various transitional paths. Some drivers prefer to *swing wide* at a sharp turn while others tend to *cut in*.

TABLE 30-4. MINIMUM EDGE OF PAVEMENT DESIGNS FOR TURNS

Design vehicle	Angle of turn, deg	Simple circular curve	Three-centered compound curve
P	30	60	
SU		100	
C43		150	
C50		200	
P	60	40	
SU		60	
C43		100	
C50			200–75–200
P	90		100–20–100
SU			120–40–120
C40			120–40–120
C50			150–50–150
P	120		100–20–100
SU			100–30–100
C40			100–30–100
C50			120–35–120
P	150		75–18– 75
SU			100–30–100
C40			100–30–100
C50			120–30–120

Taken from *A Policy on the Geometric Design of Rural Highways*, chap. 7, "Intersection Design Elements," American Association of State Highway Officials, Washington, 1954, p. 258, Table 7-1.

It has been assumed that drivers may turn at very low speeds more conveniently when roadway design permits them to turn in such a manner that the front wheels of their vehicles follow a circular curve which is tangent to the line of approach and departure. At least one field experiment has been conducted to determine the *swept path* of vehicles turning in this manner.[1]

The off-tracking of the rear wheels of design vehicles on such turns, as discussed in Chapter 3, very nearly conforms to a symmetrical arrange-

[1] J. C. Young, *Truck Turns*, California Highways and Public Works, March–April, 1950.

ment of three-centered compounded curves of properly selected radii which are used for convenience in construction of the inside edge of minimum-turn pavements. The radii of three-centered curve design vary with different angles of turn and different types of design vehicles. Also, three-centered design may not be warranted for small angles of turn or when design vehicles are passenger cars or small trucks. Table 30-4 shows suggested minimum edge of pavement designs for various *angles of turn* and design vehicles.

The Speed Design of Intersection Curves. Friction values assumed for intersection curve design are greater than those allowed in the design of

TABLE 30-5. MINIMUM RADII FOR INTERSECTION CURVES

Turning speed of curve, mph	15	20	25	30	35	40
Assumed side-friction factor (f)	0.32	0.27	0.23	0.20	0.18	0.16
Assumed superelevation (e)	0	0.02	0.04	0.06	0.08	0.09
Minimum safe radius, ft						
Calculated	47	92	154	231	314	426
Rounded	50	90	150	230	310	430

NOTE: for continuation of table see Table 28-1, Minimum Allowable Radius for Various Assumed Design Speeds.

SOURCE: *A Policy on the Geometric Design of Rural Highways,* chap. 7, "Intersection Design Elements," American Association of State Highway Officials, Washington, 1954, p. 263, Table 7-3.

nonintersectional curves (see Chapter 28). Field observations of speeds on intersection curves have indicated that drivers tend to develop higher side-friction values when operating on curves of short radius. Table 30-5 shows minimum recommended radii for intersection curves as well as the side-friction factors and superelevation values assumed in intersection-curve design.

The superelevation values shown in Table 30-5 are those generally attainable in practice for curves of the indicated turning speeds because of limitations in superelevation runoff distance and the limited amount of pavement warping desirable along the edge of the through pavement. In specific cases, as much superelevation as practicable should be employed, preferably approaching that for open-road conditions, as discussed in Chapter 28. Also, where possible at intersection curves, superelevation should be proportionately decreased in the design of curves with curvature less than maximum.

The transitioning of nonintersection curves, discussed in Chapter 28, applies also to intersection circular curves. On curves of smaller radii, higher rates of change in radial acceleration are assumed for transitions at intersection curves. These higher rates range from 4 ft/sec³ for a circular curve of 90-ft radius to 3.0 ft/sec³ when the curve radius is 430 ft. These curve radii are satisfactory for turning speeds of 20 and

40 mph, respectively. The length of transitions may be computed from the formula for transition lengths given in Chapter 28.

Variable-speed Intersection Curves. Frequently the speed requirements at the ends of intersection curves are not the same. It therefore may become desirable to design intersection curves for variable speed to control operating conditions and to conserve on intersection right of way.

Speed change on intersection curves may be induced when a series of compounded curves with decreasing, or increasing, radii are employed in design. Of course, transitional curves may be used instead of compounded curves to accomplish the same changes in alignment. One of the applications of variable speed curves was illustrated in Fig. 30-4. Another application is in the design of interchange ramps connecting roads of widely different assumed design speeds.

The design of horizontal curvature to encourage change in speed is still experimental. However, for safe and comfortable operations on variable radii curves (1) steering requirements should be uniform, (2) required changes in speed should take place gradually without fluctuation, and (3) acceleration and deceleration requirements should not exceed the rates found in normal driving practice.

Intersection-curve Superelevation. Values for superelevation and superelevation runoff suggested for nonintersection curves also are desirable for intersection curves. Frequently, lower values of superelevation, as shown in Table 30-5, and higher superelevation runoff rates are required at intersections where the distances between horizontal curves are limited and the edge elevations of other pavements must be met. In such cases, superelevation design is subject to the judgment of the designer. Aside from possible drainage problems, a comfortable riding profile over the crown formed by adjacent pavements, having opposed-direction slopes, and pleasing appearance of the slope of the edge of the pavement are the two principal design criteria.

Intersection-curve Pavement Width. The lane width of pavement on intersection curves depends on the types of vehicles to be accommodated and the radius of the curve. For minimum-speed turns, the lane width includes (1) the vehicle tracking width, (2) extra width for front- and rear-body overhang of the vehicle, and (3) clearance on both sides of the vehicle from the edges of the lane.

Inside tracking was discussed in Chapter 28. Tracking widths include the normal vehicle width plus the off-tracking width. Values employed in design range from slightly more than the width of design vehicles on curves of 500-ft radius to 19 ft for C50 semitrailer combinations when turning on a 50-ft radius (outer front wheel).

Front- and rear-body overhang widths range between 0.2 and 2 ft depending on the type of vehicle and radius of curve. The assumed

TABLE 30-6. DESIGN WIDTHS OF PAVEMENTS FOR TURNING ROADWAYS AT INTERSECTIONS

	Pavement width in feet for —								
	Case I			Case II			Case III		
R = radius on inner edge of pavement, ft	One-lane one-way operations—no provision for passing			One-lane one-way operation—with provision for passing a stalled vehicle			Two-lane operation—either one-way or two-way		
	Design traffic condition								
	A	B	C	A	B	C	A	B	C
50	16	17	20	21	24	27	30	33	37
75	15	16	18	20	22	25	28	31	34
100	14	16	17	19	21	24	27	30	33
150	13	15	16	18	20	23	26	29	31
200	13	15	16	18	20	22	26	28	29
300	12	15	15	17	19	21	25	27	28
400	12	14	15	17	19	21	25	27	28
500	12	14	15	17	19	21	25	27	27
Tangent	12	14	14	16	18	20	22	24	24

Width modification regarding edge-of-pavement treatment

No stabilized shoulder	None	None	None
Mountable curb Barrier curb	None	None	None
One side	Add 1 ft	None	Add 1 ft
Two sides	Add 2 ft	Add 1 ft	Add 2 ft
Stabilized shoulder, one or both sides	None	Deduct shoulder width; minimum pavement width as under Case I	Deduct 2 ft where shoulder is 4 ft or wider

SOURCE: *A Policy on the Geometric Design of Rural Highways,* chap. 7, "Intersection Design Elements," American Association of State Highway Officials, Washington, 1954, p. 273, Table 7-7.

clearance distance from the vehicle overhang to the edge of the lane used in design practice ranges between 1 and 3 ft.

Table 30-6 shows pavement widths for turning roadways of various radii based on the above factors and recommended by the AASHO. Design traffic conditions A, B, and C shown in the table refer respectively to (1) occasional large trucks in the traffic flow, (2) moderate volume of trucks (5 to 10 per cent), and (3) more and larger trucks. The differences in widths for these three conditions are accounted for in terms of reduced clearances with lower percentages of trucks in the traffic flow. The widths shown in Table 30-5 may be increased to overcome the influence of curbs or may be decreased when stabilized shoulders are available for stalled vehicles. Suggested modifications for these reasons are shown at the bottom of Table 30-6.

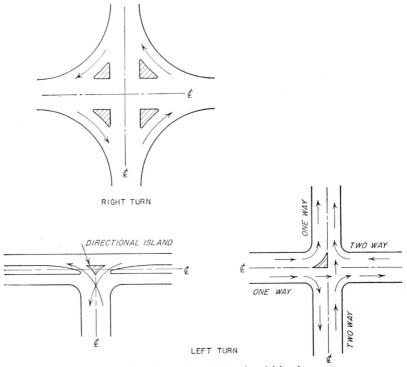

Fig. 30-5. Examples of directional islands.

30-7. Pedestrian Ways. Since pedestrians are regulated to cross the road at intersections, they are an important factor in intersection design. Without grade separation, maneuver-area design to produce low relative speed is not practicable at intersections where pedestrians are frequent.

Pedestrian crosswalks at grade should be at least 4 ft wide and are usually painted on the pavement surface. They are most frequently located as an extension of street sidewalks. It is important when practicable to locate crosswalks where the crossing is most convenient for pedestrians.

There are no established warrants for pedestrian grade separations. Experience has shown that pedestrians will not use the ramps or steps of grade separations unless their crossing on the street surface is made inconvenient by fences or heavy traffic volumes.   Generally, pedestrians prefer bridges to tunnels, which are furthermore difficult to keep clean and policed.

30-8. Intersection Islands.   Traffic islands are frequently employed at intersections.   Such islands are delimited areas between traffic lanes for vehicle and pedestrian control and refuge.   Intersectional islands may take on any shape, depending on their function and the geometry of

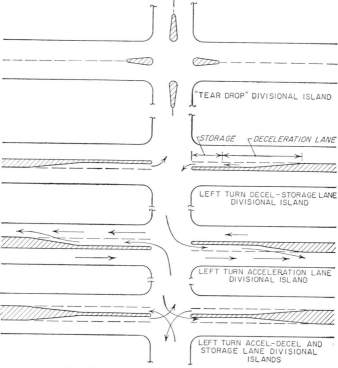

FIG. 30-6. Examples of divisional islands.

particular intersections.   They may be of the nature of (1) physical barriers, (2) deterring semibarriers, or (3) advisory markings.

There are three types of intersectional islands, (1) directional, to control turning traffic, (2) divisional, to separate traffic flows, and (3) refuge, for the protection of pedestrians.   Most islands of intersections perform more than one of these functions.

Directional Islands.   Directional islands are generally triangular in shape.   Figure 30-5 shows typical examples of directional islands for right- and left-turn movements.   Right-turn islands have particular application where (1) large intersectional areas are created by skewness of intersection legs or by long turning radii and (2) heavy pedestrian and turning vehicular flows cross.   Triangular left-turn directional islands may be employed at intersections with three legs or at the beginning or end of one-way roads, as illustrated by Fig. 30-5.

Divisional Islands. Divisional islands are placed between through traffic lanes to separate through movements and to control left turns. They may be *tear drop* in shape or may be the ends of traffic separators, such as at openings in medial dividers. Figure 30-6 illustrates typical divisional islands.

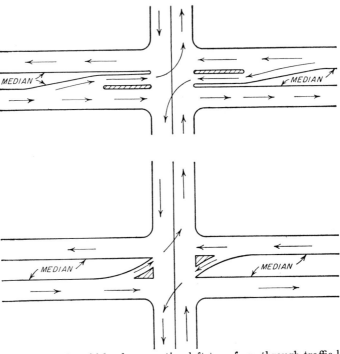

Fig. 30-7. Examples of islands separating left-turn from through-traffic lanes.

Median divisional islands may be designed with left-turn deceleration and storage lanes as well as left-turn acceleration lanes (Fig. 30-6). The design of acceleration and deceleration lanes in median islands is the same as discussed in section 30-4. There is some difference in opinion concerning the need for left-turn acceleration lanes because this movement is generally controlled by stop signs or signals at grade intersections.

The length of the storage-lane section may be estimated from traffic volume data for turning vehicles, assuming periods of accumulation for left turns and the average length of turning vehicles.

Left-turn lanes should be separated from the through-roadway lanes to discourage their use by through traffic. Figure 30-7 illustrates two types of island installations to accomplish this purpose.

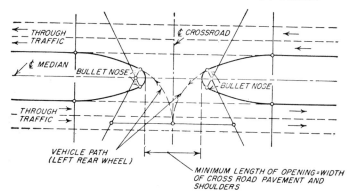

MINIMUM LENGTH OF OPENING=WIDTH OF CROSS ROAD PAVEMENT AND SHOULDERS

NOTE—MEDIAN AND SHAPE OF ITS ENDS MAY BE MODIFIED TO PROVIDE LEFT TURN LANES AND ACCELERATION LANES AS ILLUSTRATED BY FIGS. 30-6 AND 30-7

FIG. 30-8. Typical example of median-opening design.

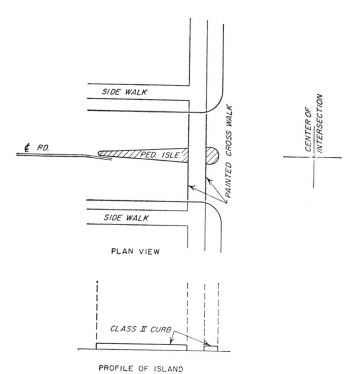

FIG. 30-9. Typical pedestrian-refuge island.

Median Openings. The design of median openings is governed by (1) width of median, (2) turning paths of vehicles, and (3) width of cross-road. The width of traffic separators was discussed in Chapter 25. Recommended design for the turning paths of vehicles is shown by Table 30-4. Excessive length of median opening is to be avoided in order to eliminate large paved areas which encourage improper maneuvers and endanger pedestrians. Figure 30-8 shows an example of median-opening

Fig. 30-10. Streetcar loading zone with approach "safety" prow and bumper.

design wherein the ends of the median islands are *bullet-nose* rather than semicircular in shape to conform to the paths of turning vehicles and to reduce the length of median opening. Minimum lengths of median openings are generally taken as the width of the crossroad pavement and shoulders.

Refuge Islands. Both divisional and directional intersection islands may serve as pedestrian refuge islands in urban areas. Barrier islands are essential when pedestrian safety is involved. Figure 30-9 shows suggested design of pedestrian refuge islands. The barrier curb is eliminated in all or part of the crosswalk area for baby buggies, etc. However, the barrier is resumed on the intersection side of the crosswalk to prevent vehicle encroachments and better to control left turns.

Where pedestrian hazards are great, massive bumpers are sometimes placed at the approach end of pedestrian islands to prevent vehicle encroachments at any cost. Such drastic treatment usually requires a *safety prow* intended to reduce the severity of vehicle collisions with the bumper. These prows have been found to increase safety to some extent but are not effective when high speeds are involved. Figure 30-10 is a photograph of a safety prow and bumper at a streetcar safety zone. Normally refuge islands are delimited by Class 2 curb (see Chapter 25).

Channelization. Channelization is the application of islands at grade intersections to direct traffic along definite paths in order to simplify operations. The purposes of channelization are:

Separation of maneuver areas. To present drivers with only one decision at a time and reduce the influence on operations caused by the overlapping of maneuver areas

Control of maneuver angle. Small angles for merging, diverging, crossing, and weaving at low relative speed; approximate right angles for crossing at high relative speed.

Control of speed. Bending or funneling to support stop regulations or to remove differentials in speed for merging, diverging, crossing, or weaving (Figs. 29-7 and 30-1)

Protection of pedestrians. To provide a haven or refuge between traffic flows

Protection and storage of turning and crossing vehicles. To shadow slow or stopped vehicles from other traffic flows

Elimination of excessive intersectional areas. Caused by multi-approaches and skewed intersection legs wherein drivers may perform unsuspected and improper maneuvers

Blockage of prohibited movements. To support regulations by making improper movements or encroachments impossible or inconvenient

Segregation of nonhomogeneous flows. To provide separate channels for turning and through, fast and slow, local and through, and opposite-direction traffic

Location of traffic control devices. To provide space and protection for control devices when ideal location is within the intersectional area

Channelizing islands should be placed and designed so the proper traffic channels seem natural and convenient to drivers and pedestrians. There should be no choice of path for the same intersectional movement. To this end, the number of islands should be held to the practical minimum to avoid confusion. The number of islands viewed by the motorist as he approaches an intersection is more important than the total number shown by the intersection plan.

Channelizing islands placed on intersection legs should extend away from the intersection a sufficient distance to induce approaching vehicles into their proper channels. The introduction of islands on horizontal and vertical curves is to be avoided.

Traffic islands require adequate approach-end treatment to prewarn drivers and to avoid need for sudden changes in direction or speed. Pavement markings, signs, illumination, and other devices employed for the approach-end treatment of intersection islands are discussed in previous chapters. Changes in transverse placement required by channelization should be gradually transitioned in the design of islands and pavement edges at intersections. The time rate of change in placement of $\frac{1}{3}$ to $\frac{1}{2}$ sec/ft of displacement (Chapter 28) is the minimum recommended value.

Deterring islands may be employed where collisions with the island offer less hazard than would be created by lack of positive channelization. Islands consisting of pavement markings alone are not very effective when encroachments increase the convenience of motorists.

Channelization has its greatest application in urban and suburban areas where speeds are lower and better illumination is available. This does not imply, however, that channelization in rural areas is not successful treatment when it is carefully designed for high-speed conditions.

Many factors affecting channelization design cannot be anticipated prior to installation. It is recommended that temporary channelization, such as stanchions, wooden platforms etc., which may be moved about, be used for a trial period before permanent islands are installed.

CHAPTER 31

INTERSECTION DESIGN

Intersections fall into two general design categories: (1) intersections at grade and (2) grade-separated intersections.

Intersections at grade are subclassified as (1) nonchannelized, (2) channelized, and (3) rotary. Grade-separated intersections are subclassified according to their geometric pattern, such as cloverleaf, diamond, trumpet, etc.

31-1. Number of Intersection Movements. The number of conflicts between intersection movements increases with increase in number of movements at an intersection. Each conflict between intersection movements is a merging, diverging, or crossing maneuver, which are sources of potential delay and hazard that may become accumulative when maneuver areas overlap or are close together.

An intersection leg (the part of a route that radiates from an intersection) may accommodate one-way or two-way traffic. The leg for traffic entering the intersection is called an intersection *approach;* for leaving traffic, an intersection *exit.*

Table 31-1 shows the relationship between the number of two-way intersection legs and the number of conflicts between intersection movements by types of maneuvers.

TABLE 31-1. RELATIONSHIP OF NUMBER OF CONFLICTS BETWEEN INTERSECTION MOVEMENTS TO NUMBER OF TWO-WAY INTERSECTION LEGS BY TYPES OF MANEUVERS

No. of two-way legs	No. of conflicts between intersection movements by types of maneuver			
	Crossing	Merging	Diverging	Total
3	3	3	3	9
4	16	8	8	32
5	49	15	15	79
6	124	24	24	172

NOTE: The number of additional pedestrian-vehicular crossing maneuver areas may be computed from the formula $2(L^2 - L)$, where L is the number of two-way intersection legs.

One-way traffic flow on intersection legs reduces the number of intersection movements and conflicts between movements. Table 31-2 shows the relationship of the number of conflicts between intersection movements for an intersection with four legs when one or more of the legs are one-way.

TABLE 31-2. NUMBER OF CONFLICTS BETWEEN MOVEMENTS AT INTERSECTIONS WITH FOUR LEGS AND WITH SOME OF THE LEGS ONE-WAY, BY TYPES OF MANEUVER

No. of one-way legs	No. of conflicts between intersecting movements by type of maneuver			
	Crossing	Merging	Diverging	Total
One leg				
Approach............................	8	6	5	19
Exit................................	8	5	6	19
Two adjacent legs				
Both approach......................	3	4	2	9
Both exit..........................	3	2	4	9
One approach—other exit............	5	4	4	13
One approach—other exit............	3	4	4	11
Two opposite legs				
Both approach......................	2	4	2	8
Both exit..........................	2	2	4	8
One approach—other exit............	5	4	4	13
All legs				
Two adjacent legs same direction.......	1	2	2	5
Adjacent legs opposite direction........	0	2	2	4

TABLE 31-3. NUMBER OF CONFLICTS BETWEEN MOVEMENTS AT AN INTERSECTION WITH FOUR TWO-WAY LEGS AND CERTAIN TURNING PROHIBITIONS

Turn condition	No. of conflicts between movements by maneuver types			
	Crossing	Merging	Diverging	Total
No left turns..........................	4	4	4	12
No right turns........................	16	4	4	24
No turns.............................	4	0	0	4
One left turn prohibited................	12	7	7	26
Two left turns prohibited..............	8, 9	6	6	20, 21
Three left turns prohibited.............	6	5	5	16
One right turn prohibited...............	16	7	7	30
Two right turns prohibited.............	16	6	6	28
Three right turns prohibited...........	16	5	5	26

The number of conflicts between movements at an intersection may also be reduced through the prohibition of turns. Table 31-3 shows the number of conflicts between movements at an intersection of four two-way legs when certain turns are prohibited.

No intersection should be planned for more than four two-way intersection legs. This general policy may be modified by regulations which reduce the number of intersection movements, as indicated by Tables 31-2 and 31-3.

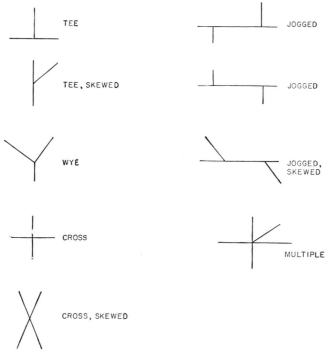

Fig. 31-1. Basic intersection forms.

31-2. Intersection Forms. Intersections occur in a number of basic forms. These forms apply particularly to nonchannelized and channelized intersections at grade. They are illustrated by Fig. 31-1.

Not only should intersections be planned or designed to have not more than four two-way intersection legs, but also it was shown in Chapter 29 that the angle of crossing maneuvers at grade should be approximately a right angle for intersections intended to operate at high relative speed. These two rules exclude all forms shown by Fig. 31-1 except those with three or four legs intersecting at right angles. However, the Y or skewed T forms may be employed in design for one-way intersection legs when the intersection is intended to operate at low relative speed, as illustrated in Chapter 29.

Jogged intersections are acceptable in design when the distance between offset legs is adequate for weaving and storage of left-turn vehicles. When this distance is adequate, design becomes that for two intersections of three legs.

It may be concluded, therefore, that *new* design is concerned with the cross and T intersection forms only when traffic is two-way and with the Y or skewed T intersection forms only when traffic is one-way.

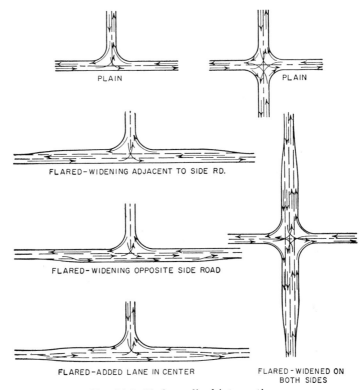

PLAIN   PLAIN

FLARED-WIDENING ADJACENT TO SIDE RD.

FLARED-WIDENING OPPOSITE SIDE ROAD

FLARED-ADDED LANE IN CENTER   FLARED-WIDENED ON BOTH SIDES

FIG. 31-2. Unchannelized intersections.

31-3. Unchannelized Intersections. This class of intersection at grade is either plain or flared. Figure 31-2 illustrates plain and flared intersections. The plain intersection is the simplest to design but is the most complex in operations. No additional pavement width is provided for turning or through traffic at intersections of this class. It is the lowest class of intersection design.

Flared intersections are widened intersections in terms of the normal pavement widths of the intersecting roads. The widening is usually in units of lane widths and may be placed on the right side, left side, or in the center of the intersection pavement area. Flared intersections provide extra lanes for through and turning traffic movements. Figure 31-2 shows four types of flared intersections.

31-4. Channelized Intersections. Channelization is achieved by the introduction of islands into intersectional areas (section 30-8). Figure 31-3 shows examples of partial and complete channelization for new

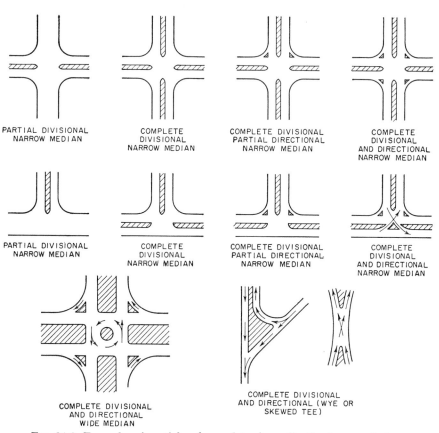

| PARTIAL DIVISIONAL NARROW MEDIAN | COMPLETE DIVISIONAL NARROW MEDIAN | COMPLETE DIVISIONAL PARTIAL DIRECTIONAL NARROW MEDIAN | COMPLETE DIVISIONAL AND DIRECTIONAL NARROW MEDIAN |

| PARTIAL DIVISIONAL NARROW MEDIAN | COMPLETE DIVISIONAL NARROW MEDIAN | COMPLETE DIVISIONAL PARTIAL DIRECTIONAL NARROW MEDIAN | COMPLETE DIVISIONAL AND DIRECTIONAL NARROW MEDIAN |

COMPLETE DIVISIONAL AND DIRECTIONAL WIDE MEDIAN

COMPLETE DIVISIONAL AND DIRECTIONAL (WYE OR SKEWED TEE)

FIG. 31-3. Examples of partial and complete channelization for new design.

design. The number of islands employed may range from one to complete channelization, and the design of channelization for wide medial dividers may call for additional "special islands." The examples may be modified to include provisions for acceleration, deceleration, and storage lanes as required by traffic conditions.

Channelization of Existing Intersections. All the basic forms of intersections shown by Fig. 31-1 may be improved by redesign. Island sizes, shapes, and locations vary widely at specific intersections, but typical channelization arrangements for multiple-approach, skewed, and jogged intersection forms are shown by Fig. 31-4. It should be noted that channelization treatment of the same intersection form varies with different traffic conditions.

It will also be noted from Fig. 31-4 that traffic channels should be bent (and funneled) to change the angle and speed of intersecting flows to produce either high- or low-relative-speed intersecting maneuvers.

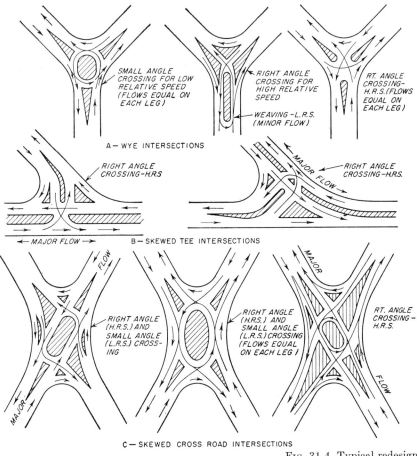

FIG. 31-4. Typical redesign

Major traffic flows should always be bent the least and minor flows the most to accomplish this. Chapter 29 shows that maneuvers at low relative speed (1) must be elemental and (2) require no regulatory traffic controls. Thus, both capacity and delay are important considerations in intersection redesign.

Intersection redesign may include improvements obtained through the actual relocation of intersection legs so as to change one form of grade intersection to another. Figure 31-5 shows examples of this kind of improvement before and after the change. It is evident that improvements of this nature may be called channelization.

Channelization-design Warrants. Even the simplest intersections provide sufficient area for drivers to follow hazardous and unexpected paths. The moderate increases in area and cost generally required to provide channelization at intersections are nearly always justified by the improved traffic operations they afford.

There are no commonly accepted warrants for channelization. Each case depends on the judgment of individual designers. Generally, need for channelization depends on the extent that islands may accomplish the purposes listed in section 30-8.

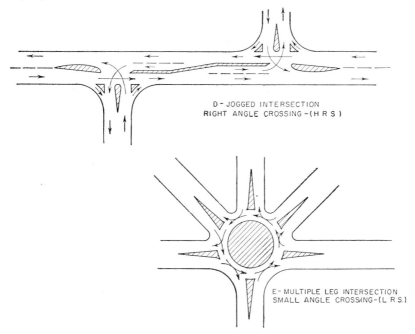

D - JOGGED INTERSECTION
RIGHT ANGLE CROSSING –(H R S)

E - MULTIPLE LEG INTERSECTION
SMALL ANGLE CROSSING–(L R S.)

*NOTE – ACCEL. AND DECEL. LANES SHOULD BE EMPLOYED
TO PRODUCE LOW RELATIVE SPEED WHERE WARRANTED*

channelization. (Schematic only.)

31-5. Capacity of Signalized Intersections. Most channelized and nonchannelized grade intersections require traffic signal control when their volumes approach their capacities. Grade intersections usually consist of a number of elemental or compounded maneuver areas located close together and operating in conjunction with each other.

The theoretical design capacities of elemental and compounded maneuver areas were discussed in Chapter 29. However, the effect of interrelated operations of maneuver areas on capacities cannot be theoretically determined.   Thus, the capacities of signalized grade intersections have been obtained through empirical methods.

The Highway Capacity Committee of the Highway Research Board has analyzed a large number of signalized grade intersections in terms of

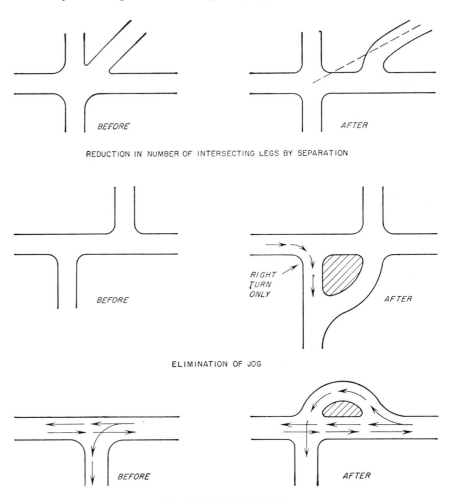

REDUCTION IN NUMBER OF INTERSECTING LEGS BY SEPARATION

ELIMINATION OF JOG

SUBSTITUTION OF ALTERNATE TURNING GEOMETRY

Fig. 31-5. Examples of channelization by relocation of intersection legs.

their capacities.   The possible capacity of such intersections under ideal operating conditions was found to be about 1,500 passenger vehicles per hour of green-signal indication per 12-ft traffic lane, or 1,250 per 10-ft lane.[1]   This value is modified in the manual to include the effect of

---

[1] *Highway Capacity Manual,* Highway Research Board and U.S. Bureau of Public Roads, Committee on Highway Capacity, 1950, part V, p. 71.

parking, turning movements, truck and bus stops, turning lanes, type of area, etc.

Practical capacity of a signalized intersection approach is defined in the *Highway Capacity Manual* as the maximum traffic volume that can enter the intersection from that approach during one hour with most of

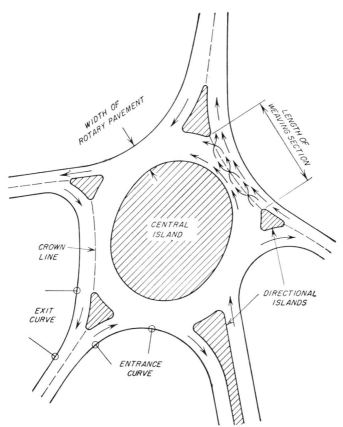

WIDTH OF ROTARY PAVEMENT

LENGTH OF WEAVING SECTION

CENTRAL ISLAND

CROWN LINE

DIRECTIONAL ISLANDS

EXIT CURVE

ENTRANCE CURVE

FIG. 31-6. Vehicle paths and design terms used in rotary design.

the drivers being able to enter without waiting for more than one complete signal cycle. Values for practical capacities under a wide variety of design and operational conditions are available in the manual. These values should be employed in design. However, in application, the designer must make allowances for the fact that they are based on average rather than on unusual conditions. For example, the practical capacity values shown in the manual should be tempered with good judgment at intersections where there are unusually heavy pedestrian volumes or at those intersections where operations are unusually complex. Practical design capacity per lane at signalized intersections is generally in the range of 250 to 600 vph.[1]

31-6. Rotary Intersections. A rotary intersection is a confluence of three or more intersection legs at which crossing and left-turn traffic movements weave on a one-way roadway in a counterclockwise direction around a central island.

Figure 31-6 shows the average path of vehicles through a traffic rotary. Right-turn traffic remains in the right lane without weaving. Left-turn and through traffic weave within the section of the rotary located next to their point of entry and just prior to their point of exit. It is not contended that all drivers follow these paths, but they are assumed for purposes of design.

The distance between adjacent entrances and exits must be sufficient for weaving at low relative speed. Where this distance is so small that vehicles cross at an oblique angle without weaving, the intersection is not classed as a rotary but rather as a channelized intersection.

Advantages and Disadvantages of Rotaries. The rotary intersection has inherent advantages and severe disadvantages. Some years ago, rotary design was considered to be a stage in the normal growth of intersection development. Experience has since shown that rotary design should be employed only at specific locations where certain critical requirements can be met.

Advantages:

1. When properly warranted and designed, rotaries operate at low relative speed with all the inherent advantages of continuous movement and safety.

2. Left-turn movements may be performed easily in complete conformance with the speed and path of through movements because both movements merge and diverge with a one-way flow.

3. Rotaries inherently separate conflict points and therefore may be designed to accommodate traffic efficiently at multi-approach intersections.

4. Rotaries offer low-relative-speed operations at a lower construction cost than grade-separated design providing for the same movements with continuous flow.

Disadvantages:

1. Rotaries are no more efficient than channelized intersections under conditions of stop-and-go operations.

2. Parking, bus loading, and other impediments to continuous traffic flow, as well as underdesign, destroy low-relative-speed operations at rotaries.

3. Continuous-flow rotary operation does not permit traffic control for pedestrian safety.

4. The capacity of a rotary, excluding right turns, is limited to that of an elemental weaving section (Table 29-2).

5. To obtain the necessary weaving-section lengths, rotaries become excessively large when weaving volumes in critical sections approach about 1,500 vph. Large rotaries increase out-of-way travel and intersection costs.

[1] *A Policy on the Geometric Design of Rural Highways,* chap. 2, "Design Controls and Criteria," American Association of State Highway Officials, Washington, 1954, p. 97.

6. To avoid excessive area requirements, reduce out-of-way travel, and accommodate the volume of weaving vehicles, rotaries usually require a substantial reduction in speed when employed at the intersection of high-speed roads.

7. Steps must be taken to reduce hazard and driver confusion at rotaries. A large, relatively flat area providing good sight distance, along with numerous traffic warning and directional signs, is required. Highway lighting is most desirable at rotaries.

Assumed Design Speed. Efficient operations at rotaries depend on relative speed. Small angles of maneuver and small differences in absolute speeds of weaving flows are essential. All the design factors and elements applicable to weaving at low relative speed discussed in Chapters 29 and 30 apply in the design of rotaries.

In the interests of economy, it is seldom feasible to design rotaries so that they may accommodate weaving at the assumed design speeds of the intersecting roadways. The AASHO has suggested compromise values for the assumed design speed of rotaries in order to reduce costs as well as the amount of out-of-way travel. These values are shown by Table 31-4.

TABLE 31-4. SUGGESTED DESIGN SPEEDS FOR ROTARY INTERSECTIONS

Design speed of highway, mph	Av. running speed on highway, mph	Design speed of rotary, mph	
		Minimum	Desirable
30	27	20	30
40	34	30	35
50 or more	40–50	35	40

SOURCE: *A Policy on the Geometric Design of Rural Highways,* chap. 8, "At Grade Intersections," American Association of State Highway Officials, Washington, 1954, p. 356, Table 8-2.

Central Island. The design of the central island of a rotary is subject to three limiting controls: (1) length of weaving section, (2) speed design of curved roadway, and (3) transverse displacement.

Lengths of weaving sections depend on the number of weaving vehicles and the rotary-design speed. Values suggested for these lengths are shown in Table 29-2. Weaving sections are arbitrarily measured between the nearest extremities of the islands at approaches and exits of the rotary (see Fig. 31-6).

In addition, the speed design of the curved roadway around the central island must satisfy the rotary-design speed requirements. This phase of intersection design, which was discussed in Chapter 30, also applies to rotaries.

Low weaving volumes and speeds or other factors may result in lengths of weaving sections too short to provide the distances needed for gradual transverse displacement. Section length based on a time interval of

4 sec at the rotary assumed design speed is considered to be the minimum length necessary for this purpose.

Both the length of weaving sections required and the minimum allowable radius criteria should be satisfied in molding the shape and size of the central island through cut-and-try methods. A smoothly curved layout, which may be circular, elliptical, or of other curved shapes, may be developed. Circular central islands are desirable when all other requirements are fulfilled because a constant radius exerts uniform speed control.

Rotary Pavement. The minimum width of roadway around the central island should be two lanes—that of an elemental weaving section (Fig. 29-8). An additional lane may be provided for right turns. The *Highway Capacity Manual* recommends extra lanes when outer flows exceed 600 passenger cars per hour.[1] The maximum recommended width of rotary pavement is four lanes, because drivers tend to wander about when greater widths are provided.

The width of the rotary pavement is measured at its point of minimum dimension. This point normally occurs midway between entrances and exits (Fig. 31-6). The outside curb lines should not parallel the curb line of the central island forming a reverse curve for right-turn traffic; instead, the portion of the outside curb between the entrance and exit radii should be tangent to them. In following a direct path, traffic will not describe a reverse curve except in cases where the distance between entrance and exit curves is so great that designs permitting direct paths are not feasible.

The number of lanes should generally be uniform around the rotary, since traffic will use one part of the rotary when traveling in one direction and the remaining part when traveling in the return direction.

The width of traffic lanes normally used for rotary pavements is 12 or 13 ft. Allowances should be made for extra clearance distance from curbs or other roadside obstructions affecting transverse placement (see Chapter 25).

Entrances and Exits. The design of rotary entrances and exits is critical in creating low-relative-speed operations within the rotary itself. Differentials in speed of entering and rotary traffic must be eliminated in the design of rotary entrances. Thus, the curved roadway of the entrance should be designed to produce a speed of entrance equal to the assumed design speed of the rotary. In addition, special care should be taken to provide adequate signing, markings, and lighting in advance of the entrance to permit safe reduction in speed.

The radius of exit curves should also be designed to produce a speed of exit equal to the assumed design speed of the rotary, or a little higher.

[1] *Highway Capacity Manual*, p. 115.

It has been observed that drivers tend to speed up as they leave a rotary, and this aids its capacity.

The channels of entrance and exit at rotaries are delineated by directional islands. Figure 31-7 illustrates typical islands of this type. Good channelization principles should be applied in the design of directional

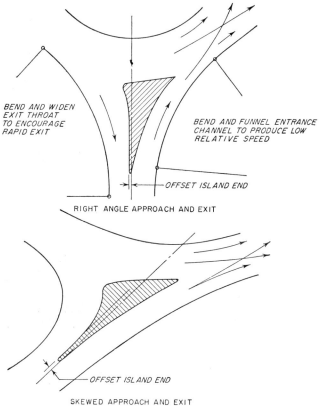

BEND AND WIDEN
EXIT THROAT
TO ENCOURAGE
RAPID EXIT

BEND AND FUNNEL ENTRANCE
CHANNEL TO PRODUCE LOW
RELATIVE SPEED

OFFSET ISLAND END

RIGHT ANGLE APPROACH AND EXIT

OFFSET ISLAND END

SKEWED APPROACH AND EXIT

Fig. 31-7. Directional islands at rotaries.

islands to control vehicle speeds rigidly and to produce small maneuver angles.

Superelevation. The design of superelevation for rotary pavements and for the entrance and exit channels of rotaries is very complex. The rotary pavement is sloped downward on each side of a crown line extending between the tips of directional islands (Fig. 31-6). The location of this crown line between islands is extremely important because superelevation in the wrong direction for some rotary movements may be produced. Normally, the crown line should pass through a point in the middle of the rotary pavement midway between directional islands. To aid in the design of superelevation, the paths of vehicles through the rotary should be analyzed in terms of crown-line location. The algebraic difference in cross slopes at the crown line is another important consideration. If the change in slope is too abrupt and is not properly rounded, objectionable vehicle body roll is induced. Severe body roll on rotaries has been known to cause accidents, including the upsetting of trucks which were loaded so as to have a high center of gravity.

Superelevation of entrance and exit channels is the same as for other intersection curves (Chapter 30) and must be warped to meet the rotary pavement edge. It is considered good practice to apply this superelevation so as to elevate the directional island as much as practicable. This provides for better visibility of the island at the intersection approach and facilitates the warping of pavement edges.

There is considerable difference of opinion concerning the magnitude and placement of superelevation on rotaries. Some designers use only a minimum drainage slope of the rotary pavement to avoid excessive body roll and to simplify the warping of pavement edges. Others advocate the use of higher, carefully selected values of superelevation to increase the safe speed of operations.[1] In either case, much is left to the judgment of individual designers.

Other Factors. The large area needed for rotaries is ideal for landscape development. While care must be taken not to reduce sight distance by plantings, landscaping may be advantageously employed to warn and assist traffic. Tall plantings in the central island opposite rotary entrances will call attention to the turning roadway in advance of the rotary, and lower plantings will screen headlight glare. Lines of plantings or utility poles which cross the central island so as to create the illusion that the rotary does not exist should be avoided.

Adequate traffic signs and markings for effective guidance and warning of motorists under both day and night conditions are especially important at rotaries. "Stop" signs and traffic signals violate the concept of rotary operations and are employed only as a palliative measure when excessive volumes or other impediments to uninterrupted flow cause rotary operation to break down to that of a channelized intersection. Lane markings should not be employed on the rotary pavement because the merging and diverging maneuvers within the weaving section are not subject to lane control.

Capacities of Rotaries. The theoretical design capacities of rotaries depend primarily on the width of rotary pavement, speed of operation, and length of weaving sections. Table 29-2 relates the number of weav-

[1] *A Policy on the Geometric Design of Rural Highways*, chap. 8, "At Grade Intersections," p. 361.

ing passenger vehicles and their running speed to lengths of weaving section.

It is clear from the discussion in Chapter 29 that weaving is elemental in character and that no appreciable increase in weaving capacity will be gained through the compounding of weaving maneuvers. However, within any section of the rotary, right-turn vehicles and those making left turns which do not wish to weave in that section monopolize the gaps which would otherwise be available for weaving vehicles, unless extra lanes are provided for those movements. Therefore, these nonweaving vehicles must be added to the volume of Table 29-2 to determine the capacity of a weaving section when extra turning lanes are *not* provided.

Capacity values obtained from Table 29-2 provide for free-flowing vehicles traveling at the running speed specified with only occasional slowdowns. There are two other conditions of operations at rotaries which occur as traffic exceeds these volumes. These are (1) when speed is greatly reduced but continuous movement is maintained with very close spacing and momentary stops and (2) when stoppage and backups occur producing stop-and-go operations. Within the ranges of volumes producing condition 1, any irregularity in flow could quickly plunge the intersection into stop and go operations.

When condition 2 occurs, the capacity of the rotary weaving section becomes that of an oblique crossing operating at high relative speed (Chapter 29).

Factors to represent the influences of trucks and other interferences with rotary operations are not available. Therefore, in the case of rotaries, adjustments to change theoretical to practical design capacity must be made at the discretion of the designer.

31-7. Grade Separations. Grade-separated intersections cause less hazard and delay than grade intersections. A highway grade separation is a bridge used to eliminate crossing conflicts at intersections by vertical separation of roadways in space. Route transfer at grade separations is accommodated by interchange facilities consisting of ramps. Interchange ramps are classified as direct, semidirect, or indirect (Fig. 29-14).

Grade-separation design is the highest form of intersection treatment. Its ultimate objective is to eliminate all grade-crossing conflicts and to accommodate other intersecting maneuvers by merging, diverging, and weaving at low relative speed. Crossing conflicts caused by through-traffic movements may be eliminated simply by a single bridge. The elimination of crossing conflicts caused by turning movements requires additional bridges as directness of turning path is improved. In some cases grade-separated intersection costs may be reduced by permitting crossing maneuvers at grade for minor traffic movements. In other cases, minor traffic movements may be prohibited to avoid complex design.

Warrants for Grade Separations. Since grade separation is the highest form of intersection design, it is warranted whenever the road user benefits—as when compared with lower forms of design, it will justify the additional costs of grade separation. Other factors to be considered are:

A freeway development. When freeway alignment passes through a network of existing roads, it is often impossible to close these roads to preserve the freeway principle. In such cases, the movement of traffic on the local system may circulate unhampered by the freeway if all crossings of local roads with the freeway are grade-separated.

Site topography. At some sites where there is a natural difference in the elevation of intersecting roads, grade separation may be accomplished at a very favorable cost.

Unusual hazard. Grade separation may be the most economical treatment of an existing hazardous location where major redesign or relocation of intersecting roads would be required with lesser forms of intersection design.

Elimination of congestion. Inability to provide the required capacity with an intersection at grade necessitates grade separation. Exact traffic volume criteria for grade-separation warrants have not been developed. It is clear that these criteria should consider future traffic as well as current traffic in order to provide and plan for the most economic and satisfactory design treatment.

Types of Grade Separations. A comprehensive classification plan for grade-separated intersection design which includes all possible geometric patterns has not yet been developed. However, the basic types which are employed most frequently in design are shown by Fig. 31-8.

Direct Connection. This type provides direct paths for left turns. Generally, ramp grades and ramp alignment at direct-connection grade separations may provide for higher ramp speeds. Left-turn traffic leaves each roadway before other left turns enter. Principal disadvantages are cost of construction and area required. This form may be warranted on high-speed roads where left-turn traffic volumes are heavy. Some designers do not favor entrance of merging traffic in the left lane, which is a characteristic of most direct-connection designs.

Cloverleaf. Cloverleaf design requires only one bridge. In this respect it is the cheapest form providing for elimination of all crossing maneuvers at grade. Principal disadvantages are circuity of travel, large areas for loops, left-turn movements which must weave, sight distances to exits on other side of bridge, and confusion caused by turning right to go left.

A partial cloverleaf may have ramps in one, two, or three quadrants. The nonexistence of ramps for turning movement in any quadrant creates crossing conflicts at grade. When ramps are provided in more than one quadrant, traffic movements may be regulated so crossing conflicts occur on either road—usually the secondary road.

Cloverleaf grade-separation design is adequate when the volume of left-turn traffic does not warrant direct-connection design. Partial cloverleaf development may be employed when crossing conflicts on the secondary road will not produce objectionable amounts of hazard and delay.

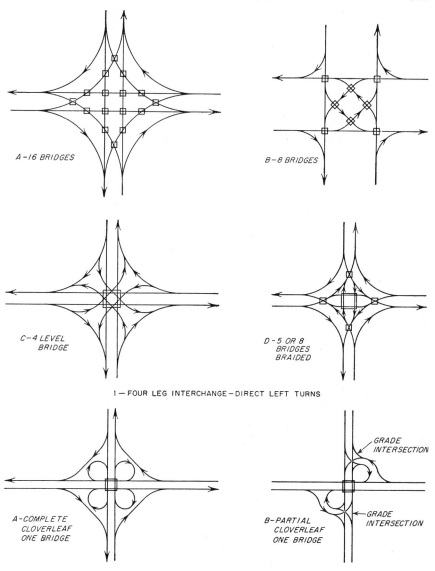

A—16 BRIDGES

B—8 BRIDGES

C—4 LEVEL
BRIDGE

D—5 OR 8
BRIDGES
BRAIDED

1—FOUR LEG INTERCHANGE—DIRECT LEFT TURNS

A—COMPLETE
CLOVERLEAF
ONE BRIDGE

B—PARTIAL
CLOVERLEAF
ONE BRIDGE

GRADE
INTERSECTION

GRADE
INTERSECTION

2—FOUR LEG INTERCHANGE—INDIRECT LEFT TURNS

FIG. 31-8. Basic types of grade separations.   (Diagrammatic.)

Bridged Rotary.   Bridged rotaries operate the same as other rotaries except that the major through highway is separated from the other roadways.   Generally, in spite of grade separation, the volume of weaving vehicles requires weaving sections (Table 29-2) of sufficient length to demand an intersectional area at least as large as that of a cloverleaf. Bridged rotaries require two bridges as compared with one bridge required

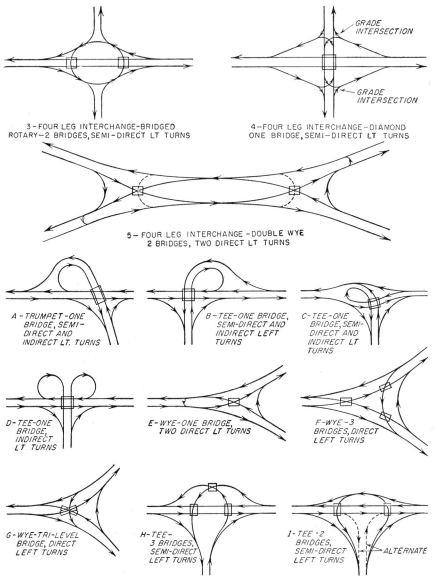

3 – FOUR LEG INTERCHANGE-BRIDGED
ROTARY–2 BRIDGES, SEMI–DIRECT LT TURNS

4—FOUR LEG INTERCHANGE–DIAMOND
ONE BRIDGE, SEMI–DIRECT LT TURNS

5 – FOUR LEG INTERCHANGE –DOUBLE WYE
2 BRIDGES, TWO DIRECT LT TURNS

A – TRUMPET –ONE BRIDGE, SEMI– DIRECT AND INDIRECT LT. TURNS

B – TEE – ONE BRIDGE, SEMI–DIRECT AND INDIRECT LEFT TURNS

C – TEE – ONE BRIDGE, SEMI– DIRECT AND INDIRECT LT TURNS

D – TEE– ONE BRIDGE, INDIRECT LT TURNS

E – WYE–ONE BRIDGE, TWO DIRECT LT TURNS

F – WYE – 3 BRIDGES, DIRECT LEFT TURNS

G – WYE–TRI–LEVEL BRIDGE, DIRECT LEFT TURNS

H – TEE – 3 BRIDGES, SEMI–DIRECT LEFT TURNS

I – TEE – 2 BRIDGES, SEMI–DIRECT LEFT TURNS — ALTERNATE

6 – THREE LEG INTERCHANGE (TRUMPET, TEE AND WYE)

Fig. 31-8. (*Continued.*)

for cloverleaf design. For these reasons bridged rotaries are seldom employed in new design.

Existing rotaries have been grade-separated with considerable success. In these cases, the elimination of the major through-traffic component of weaving volumes has reduced traffic demand to the practical design capacities of overburdened rotaries. However, adoption of this plan for stage construction is not always practical because it usually results in overdesign of the final intersection.

Diamond or Parallel Ramp. This intersection type has four one-way ramps which are essentially parallel to the major artery. The minor-road terminals are intersections at grade where "stop" signs or traffic

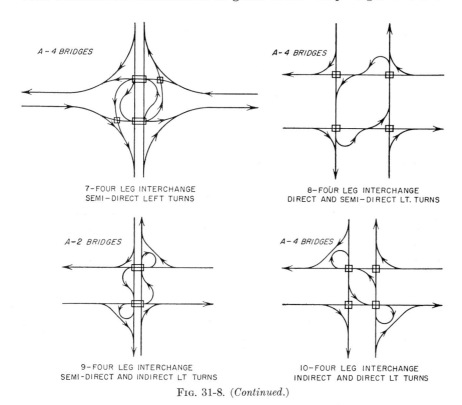

7–FOUR LEG INTERCHANGE
SEMI–DIRECT LEFT TURNS

8–FOUR LEG INTERCHANGE
DIRECT AND SEMI–DIRECT LT. TURNS

9–FOUR LEG INTERCHANGE
SEMI–DIRECT AND INDIRECT LT. TURNS

10–FOUR LEG INTERCHANGE
INDIRECT AND DIRECT LT TURNS

Fig. 31-8. (*Continued.*)

signals are usually employed. Limitation in application of this design depends on the operations of these terminals. This type of grade-separated intersection requires the least right of way. It is suitable for locations where the volume of left-turn traffic is relatively small.

Double Y. This grade-separation type combines grade separation and weaving. The design of the weaving section is similar to that for rotary intersections. The double Y requires an elongated area in the central portion of the intersection. It has special application where roads intersect on each side of a restricted area. Trilevel bridges may be used to eliminate crossing conflicts caused by left turns.

Trumpet, T and Y. These two basic types of grade-separated intersections represent the application of direct, semidirect, or indirect ramps to intersections with three legs. They have the same general advantages and disadvantages of the other similar types previously discussed.

Other Forms. The combinations and modifications of the basic types of grade-separated intersections are numberless and depend upon the judgment of individual designers. Drawings 7, 8, 9, and 10 of Fig. 31-8 illustrate this point. The combinations include the design of grade separations for intersections with more than four legs and at other locations where limitations in available area occur. The choice of grade-separation type and design depends on many factors, including traffic, topography, and cost.

Grade-separation Bridges. A highway grade-separation bridge should be designed as an integral part of the highway it serves. Years ago it was considered good economy to build bridges on only straight alignment and at right angles with the crossroad, to simplify construction, and to require minimum bridge-span lengths. It is now conceded that modern bridges should be constructed with curved alignment, variable super-elevation, or variable cross section, where necessary to conform with the design of intersecting highways.

The optimum design of grade-separation structures, from the point of view of traffic design, is reached when driver behavior is not influenced by the structure. Side clearances to bridge rails, abutments, and piers must be sufficient so as not to affect speed or placement (see Chapter 25). To this end, the AASHO has established desirable and minimum side clearances and shoulder widths at structures.[1]

The minimum vertical clearances for highway grade separations recommended by various design departments vary from 14 to 15 ft over traffic lanes. This dimension may be verified in Table 13-5 concerning legal limitations in vehicle heights. Sufficient leeway in vertical clearance must be allowed for irregularities in measurement of height of vehicle loads and for pavement resurfacing. Minimum vertical clearance at railroad grade separations is about 23 ft from top of rail.

The difference in elevation of grade-separated highways is the sum of the vertical clearance and depth of structure. Structure depth varies with type of structure and the length of span. For the preliminary design of highway grade separations, the depth of structure may be taken as one-fifteenth of the span length.[2] The total difference in elevation rarely exceeds 20 ft.

Special consideration should be given to the desirability of placing the major roadway *over* or *under* the crossroad at grade separations. The fact that greater width is generally more economical than longer span length in bridge construction is important; drainage is also a factor.

[1] *Ibid.*, chap. 9, "Grade Separations and Interchange," p. 378.

[2] *Planning Manual of Instructions*, part 7, "Design," California State Department of Public Works, Division of Highways, Sacramento, May, 1952.

However, in conformance with topographic and other conditions, it is most important to favor the major highway in the design of alignment, sight distance, and other roadway or intersection elements.

Vertical sight distance at sag vertical curves located at underpasses may be restricted by the overhead structure. Except for substandard

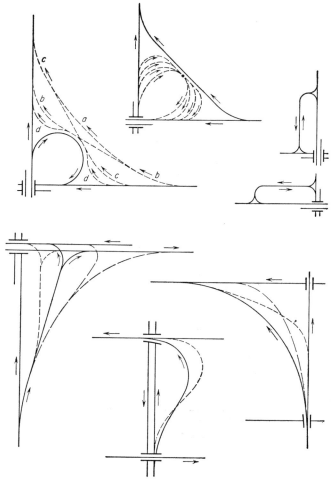

FIG. 31-9. Common shapes for grade separation ramps. (From *A Policy on the Geometric Design of Rural Highways*, American Association of State Highway Officials, Washington, 1954, p. 568, Figs. 9–11.)

vertical curve design (see Chapter 27), the sight distance available under the bridge is usually sufficient to meet safe-stopping sight distance requirements. However, where bridge clearances are provided only for passenger cars or where passing sight distance is desired, sight distance under the bridge may be critical.

Interchange Ramps. Interchange ramps may have many shapes depending on the type of grade-separated intersection and the designer's interpretation of requirements and limitations at specific locations. Figure 31-9 illustrates some of the different ramp shapes which may be employed in design.

All of the elements of maneuver-area design in Chapters 29 and 30 may be applied in the design of interchange ramps. Safe-stopping sight distances should be provided throughout their length. Grades on ramps should be as flat as possible, and the factors limiting gradient on through roadways (see Chapter 27) also apply on intersection ramps. However, ramp grades are usually short in length and do not appreciably affect truck operations. One agency specifies that grades on interchange ramps should be controlled by provision of safe-stopping sight distance rather than maximum allowable gradient.[1] However, even short grades of 5 per cent or steeper require sanding under icy-pavement conditions.

The assumed design speed of interchange ramps should equal that of the intersecting roads if speed reduction for turning traffic is to be avoided. High-speed interchange ramps require large radii, and it is frequently necessary to compromise ramp speed and area requirements in the interests of economy. When this is done, the average running speed of the through road is generally assumed to be the minimum design speed for ramp entrances and exits.

The AASHO has established guide values for ramp design speeds in terms of road design speeds, as shown in Table 31-5. The desirable ramp

TABLE 31-5. GUIDE VALUES FOR INTERCHANGE RAMP ASSUMED DESIGN SPEEDS

Highway design speed, mph..................	30	40	50	60	70
Highway running speed, mph................	27	34	40	45	49
Ramp design speed, mph					
Desirable.............................	25	35	40	45	50
Minimum.............................	15	20	25	30	30

SOURCE: *A Policy on the Geometric Design of Rural Highways,* chap. 9, "Grade Separation and Interchanges," American Association of State Highway Officials, Washington, 1954, p. 393, Table 9-2.

speed values shown in the table approximate the running speed of the highway, while the minimum values are based on speeds which will be generally accepted by motorists. Acceleration and deceleration lanes are required at ramp terminals to accommodate merging and diverging at low relative speed when the minimum values shown in the table are employed.

The minimum radii to be employed in interchange ramp design are

---

[1] *Ibid.*

determined by speed-design or variable speed-design methods (see Chapter 30). Superelevation design on interchange ramps is the same as for other intersection curves, except that where curves in the same direction are separated by a relatively short tangent section the pavement should not be crowned, but rather a minimum drainage slope (about 0.02 ft/ft) in the direction of curve superelevation is maintained between the adjacent extremities of superelevation runoff.

Ramps are usually designed for one-lane operations because their horizontal curves and grades discourage overtaking and passing and because elemental merging and diverging maneuvers at ramp terminals are desired. Recommended ramp widths are given in Table 30-6.

Ramp Terminals. The design of merging and diverging pavements discussed in Chapter 30 applies at ramp terminals. Where ramp terminals must operate under high-relative-speed conditions, such as at connections with city streets, appropriate traffic design and control should be employed. All types of ramp terminals are subject to channelization.

The sight-distance requirements at intersections for merging and crossing vehicles discussed in Chapter 26 have application at ramp terminals. In addition to these requirements, sufficient distance must be provided between grade-separation structures and ramp terminals, where traffic leaves the roadway on the far side of the structure, for drivers to see the turnoff and maneuver into it. Exclusive of deceleration-lane requirements, it is assumed that this distance should not be less than that traveled in $3\frac{1}{2}$ sec at the assumed design speed of the roadway.

Capacities of Grade-separated Intersections. The capacities of grade-separation structures and through roadways at grade separations can be made to approach or equal the open-road capacities of the intersecting roads.

The capacity of interchange ramps depends on (1) the capacity of the ramp between its terminals or (2) its ramp-terminal capacities. Where ramps between their terminals have long radii, with good grades and adequate sight distance permitting speeds of 30 mph or more during peak traffic hours, their theoretical design capacities may approach those of the open road, provided that capacity is not reduced at ramp terminals. The long radii required for this condition of operations may generally be provided, in a practical sense, only when direct ramps are employed. The design limitations encountered with indirect and most semidirect ramps restrict the theoretical design capacity of such ramps to about 1,200 passenger cars per lane per hour.[1] Where sharp curvature, or other conditions, on the ramp restrict operations to one lane, the capacity is reduced to that of a single lane regardless of the number of lanes provided by design. The data required to translate theoretical into practical

[1] *Highway Capacity Manual,* part VII, "Ramps and Their Terminals," p. 117.

design capacities for interchange ramps are not available, but it is reasonable to apply the same factors used for free-flow conditions with commonsense modifications.

The capacities of ramp terminals are the same as those for merging, diverging, and crossing maneuver areas (see Chapter 29). Merging and diverging maneuver areas at ramp terminals should be designed for low-relative-speed operation wherever possible, and compounded maneuver areas (Fig. 29-9) should therefore be avoided.

It is apparent that the capacities of ramp terminals on multilane facilities are greatly affected by lane distribution. Lane distribution on multilane facilities within the influence of intersections varies with traffic volumes, type of design, and traffic control, as well as light conditions and other factors. It is characteristically different at merging and diverging maneuver areas. Figure 31-10 shows one example of lane distributions at interchange ramp terminals.

31-8. Accidents at Intersections. Intersections account for a high percentage of the total traffic accidents. Insufficient data are available to compare accidents for the various classes and types of intersections. General experience has justified the ranking of intersection classes according to their accident potentials. Nonchannelized grade intersections are considered to be the most subject to accidents. Channelized and rotary intersections fall in the intermediate rank. Grade-separated intersections, while not accident-free, are the safest.

A limited amount of information of a comprehensive nature on accident tendencies at nonrotary-type grade intersections is available.[1] These data indicate that accident rates increase at grade intersections as the number of intersection legs increases and the percentage of cross traffic increases. Statistically significant comparisons of accident rates for various angles of intersection or types of traffic control at grade intersection have not been made.

It is known that accident experience at rotaries consists mostly of failure to traverse entrance curves into the rotary safely and of relatively minor sideswipe collisions between weaving vehicles. Most grade-separation accidents occur on sharp curvature or at ramp terminals.

31-9. Procedures in Intersection Design. It is clear that there may be a number of alternate solutions to intersection-design problems. Intersection design is a cut-and-try process. Knowledge of the theory and practice of intersection design aids in the selection of designs to be analyzed, but a benefit-cost analysis of each trial solution should be used for the yardstick in selecting the optimum design.

Traffic data for intersection design consist of design-hour volumes and vehicle classifications for each intersection movement, as well as assumed

---

[1] Morton S. Raff, "The Interstate Highway Accident Study," U.S. Bureau of Public Roads, *Public Roads*, vol. 27, no. 8, June, 1953, p. 170.

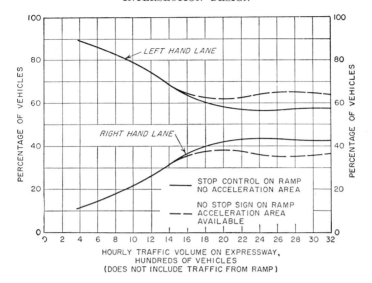

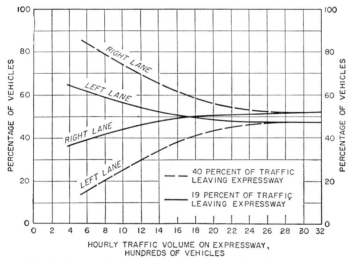

FIG. 31-10. Typical example of lane distribution at interchange ramp terminals. (From *Highway Capacity Manual*, part VII, "Ramps and their Terminals," Highway Research Board, and U.S. Bureau of Public Roads, Committee on Highway Capacity, Washington, 1950, pp. 124 and 127, Figs. 47 and 49.)

design speeds and other road-classification information (see Chapter 23). Physical data concerning the site topography, property lines, general bearing of intersecting roads, etc., in appropriate scale-drawing form, provide a study base for all likely designs.

During the preliminary stages of intersection-design development, drawings may consist of simple sketches with only enough detail shown to permit cost estimates and evaluation of traffic operations. Benefit-

cost analyses were discussed in Chapter 23.   Considerations in addition
to those normally employed in such analyses include:

1. Appropriateness of design related to topography, to operations of intersecting
roads, and to conformity with design of adjacent intersections
2. Maintenance of traffic during construction
3. Stage development
4. Influence on adjacent property

The AASHO has published detailed instruction on intersection-design
procedure.[1]

[1] *A Policy on the Geometric Design of Rural Highways*, Appendix A, "Intersection
Design Procedure," p. 607.

CHAPTER 32

THE TRAFFIC DESIGN OF PARKING TERMINALS

Accommodating traffic at its terminals is a critical factor in highway transportation, since parking at the curb monopolizes street space which is increasingly needed for moving traffic. Modern highway terminals provide for off-street parking in parking lots and garages and are usually designed for passenger vehicles only.

Each floor of a multistory garage is similar to a parking lot in operations, except for the influence of floor-to-floor travel.

32-1. Highway-terminal Operations. The traffic operations of a parking lot or garage consist of the following steps: (1) entrance, (2) acceptance, (3) storage, (4) delivery, and (5) exit. Step 1 includes the intersectional movement which takes place as vehicles leave the street traffic stream and enter the inbound reservoir space of the terminal. Step 2 takes place in the inbound reservoir space where the vehicle is usually checked in and may change drivers. Storage involves the travel of the vehicle from the inbound reservoir space to a parking stall and the parking of the vehicle in the stall. Delivery is the unparking of the vehicle and travel to an outbound reservoir space where the vehicle is checked out and may change drivers. The exit, step 5, consists of departure from the outbound reservoir space and includes the intersectional movement as the vehicle again enters the street traffic stream.

Two methods of operations may be employed in both parking lots and garages: (1) attendant parking and (2) patron parking. In attendant parking, the patron operates his vehicle only through the entrance and exit steps; employees of the terminal accept the vehicle, store, and deliver it. In a patron parking terminal, the patron handles his car during all five operational steps with direction and assistance by terminal employees as required. The method of operation employed has an important influence on the design of parking lots and garages.

32-2. The Geometry of Stalls and Aisles. Designing parking stalls and aisles involves (1) dimensions and turning radii of vehicles, (2) clearance between vehicles parked in their stalls, (3) angle and direction of parking, (4) clearances between parked and parking vehicles, (5) minimum aisle widths for one- or two-way traffic movements.

The Design Vehicle.   While the dimensions of vehicles range between those of undersize to those of oversize types, it is necessary to develop a standard size which will reflect the space demands of a majority of parkers.   This standard, or design, vehicle should not have the dimensions of an average car but rather dimensions equal to or greater than the largest standard, or common, models of cars.

Current trends in vehicle design are toward longer and wider vehicles with longer minimum turning radii.   The changes which may be made in the dimensions of future vehicles depend largely on public demand and therefore are difficult to predict.

Table 32-1 shows the dimensions of the design vehicle which are critical in the computations of its parking-space requirements.

TABLE 32-1. CRITICAL DIMENSIONS OF DESIGN PASSENGER VEHICLE FOR
PARKING TERMINALS

Dimension	Symbol
Over-all length	L
Over-all width	W
Wheel base	B
Front overhang (front axle to bumper)	O_f
Rear overhang (rear axle to bumper)	O_r
Side overhang (center of rear tire to fender)	O_s
Rear tread (center to center of tires)	t_r
Front tread (center to center of tires)	t_f
Minimum turning radius	
Inside rear wheel	r
Inside front wheel	r'
Outside point, front bumper	R
Outside point, rear bumper	R'
Bumper depth from maximum turning point, front	b_f
Bumper depth from maximum turning point, rear	b_r

NOTE: The dimensions listed in Table 32-1 are illustrated on Fig. 32-1.

Clearances between Parked Vehicles.   The width of stall affects (1) the width of aisle for maneuvering into and out of the stall and (2) clearance between parked vehicles sufficient for drivers to get into and out of the car.

More clearance between cars is required for patron parking than for attendant parking because patrons (1) frequently carry bundles, (2) are usually less agile than attendants, (3) may carelessly strike side of adjacent car with door, (4) do not like to squeeze through narrow openings or against dirty car exteriors, and (5) frequently do not park their vehicle in the center of the stall.

Minimum stall widths of 8 ft for attendant parking and 8.5 ft for patron parking are recommended.[1]   At commercial developments, such

[1] Edmund R. Ricker, *The Traffic Design of Parking Garages*, Eno Foundation for Highway Traffic Control, Inc., Saugatuck, Conn., 1948.

as shopping centers, stall widths of 9 ft are sometimes used as an added convenience to patron parkers and an inducement to shoppers.

Angle and Direction of Parking. Cars may be parked with front or rear end toward the aisle. When the angle of parking is 45°, an overlapping or *herringbone* parking arrangement may be used. Figure 32-2 illustrated the various angles and directions of parking frequently employed in terminal design.

Regardless of parking angle or direction, the stall depth *measured perpendicular to the aisle* is given by the formula $L' \sin \theta + S \cos \theta$, where θ is angle of parking in degrees, L' is length of stall in feet, and S is width of stall in feet.

The dimension of stall width *measured parallel to the aisle* is given by the formula $S \div \sin \theta$, where θ is angle of parking in degrees and S is stall width in feet.

Both formulas apply also to the same dimensions (perpendicular and parallel to the aisle) of the rectangle covered by a car when the vehicle width and length (with due consideration for the corner rounding of bumpers) are employed.

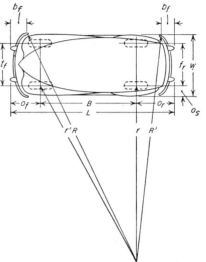

FIG. 32-1. Dimensions of design vehicle for parking-terminal design. (From Edmund R. Ricker, *The Traffic Design of Parking Garages*, Eno Foundation for Highway Traffic Control, Inc., Saugatuck, Conn., 1948.)

Clearance between Parked and Parking Vehicles. Aisle widths depend on (1) the stall width, (2) the angle and direction of parking, and (3) the clearances between parking and parked cars.

For satisfactory quality of terminal operations, it is believed that parking and unparking maneuvers should be accommodated in a single movement, whether backward or forward. Adequate clearances should be provided between a parking (or unparking) vehicle and those parked in the adjacent stalls, as well as those across the aisle opposite the stall, if the maneuver is to be made easily.

It is evident that unskilled drivers require larger clearances than those who are more experienced. Therefore larger clearances should be allowed when patron parking is employed than when attendant parking is employed. In either case, the amount of time required to park and unpark a vehicle may be expected to increase with decreased clearances.

Aisle Widths Based upon Clearances. The geometry of parking in a single drive-in or back-in movement at angles greater or less than the critical parking angle is shown by Fig. 32-3. The critical parking angle is defined as that angle at which the aisle width—required for parking or unparking—to provide a given clearance to the car in the stall on the *left* is equal to the aisle width required to provide the same clearance to the car in the stall on the *right*.

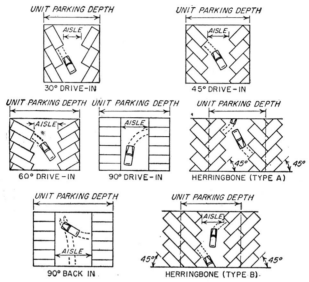

FIG. 32-2. Various parking angles and directions commonly employed in parking terminals.

The critical parking angle for a given design vehicle, stall width, and clearance may be computed.[1] For practicable use the critical parking angle may be taken as 45°.

When the stall width, parking angle, minimum clearances, and design-vehicle dimensions are known, the minimum aisle width required to accommodate parking maneuvers in a single backing or forward movement may be computed from the following formulas:[2]

1. For drive-in stalls at angles of parking less than the critical parking angle,

$$\text{Aisle width} = R' + c - \sin\theta[b_r + \sqrt{(r - O_s)^2 - (r - O_s - i + c)^2}] - \cos\theta(r + t_r + O_s - S)$$

[1] *Ibid.*

[2] To simplify the formulas certain assumptions are made:

1. The minimum turning radius of right and left turns and backward and forward movements are the same.

2. The movements into and out of a stall are identical.

3. The car is driven along a straight path until the clearance path is reached, and the wheels are turned the maximum amount for the turning portion of the movement.

4. Slippage of tires on the pavement is negligible.

(From *ibid.*)

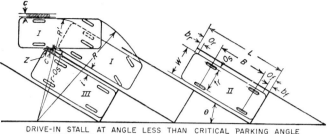

DRIVE-IN STALL AT ANGLE LESS THAN CRITICAL PARKING ANGLE
MOVEMENT IS LIMITED BY CAR IN STALL TO RIGHT

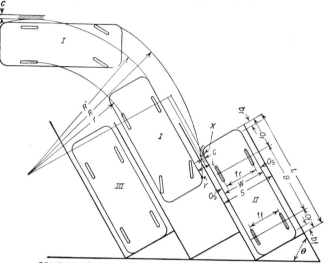

DRIVE-IN STALL AT ANGLE GREATER THAN CRITICAL PARKING ANGLE
MOVEMENT IS LIMITED BY CAR IN STALL TO LEFT

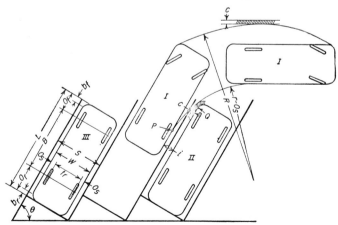

BACK-IN STALL (NO CRITICAL ANGLE)

FIG. 32-3. Geometry of drive-in and back-in parking movements with respect to critical parking angle. (From Edmund R. Ricker, *The Traffic Design of Parking Garages*, Eno Foundation for Highway Traffic Control, Inc., Saugatuck, Conn., 1948, p. 88.)

where θ is parking angle in degrees, c is clearance between cars, as one moves into or out of stall, i is intercar distance $= S - w$, S is stall width, and other symbols are as defined in Table 32-1.

2. For drive-in stalls at angles of parking greater than the critical parking angle,

$$\text{Aisle width} = R' + c - \sin \theta[b_r - \sqrt{R^2 - (r + t_r + O_s + i - c)^2}] - \cos \theta(r + t_r + O_s + S)$$

3. For back-in stalls,

$$\text{Aisle width} = R + c - \sin \theta[b_f + \sqrt{(r - O_s)^2 - (r - O_s - i + c)^2}] - \cos \theta(r + t_r + O_s - S)$$

Minimum Aisle Width. The aisles of parking facilities must not only be wide enough for parking and unparking, but they must also function efficiently as the traffic circulation system of the terminal. When the parking angle is shallow, aisle widths based on given clearances while parking or unparking may be inadequate from the point of view of traffic circulation. A distance of at least 3 ft to parked and other moving cars should be provided for vehicles as they move along the aisles of parking facilities. Minimum suggested aisle widths for parking facilities therefore become 12.5 ft for aisles designed for one-way movement of traffic and 22.0 ft for two-way traffic movement.[1]

The design of aisles at their intersections is determined by the turning radii of vehicles.

32-3. Parking Areas. Minimum suggested stall and aisle dimensions for parking-terminal design are shown in Table 32-2. Smaller stall and aisle dimensions than those suggested in the table for various angles and directions of parking would seriously hamper traffic operations, while larger dimensions would increase the convenience of drivers.

Unit Parking Depth. The unit parking depth shown in Table 32-2 is a useful unit of measure in fitting the various patterns of parking to the dimensions and shapes of land parcels. It is the width of aisle plus the depth of parking stall on each side measured perpendicular to the aisle (Fig. 32-2). It should be noted that the herringbone pattern (both Types A and B in Fig. 32-2) reduces the unit parking depth as compared with the nonoverlapping 45° angle parking pattern.

Area per Car. The area per car shown in Table 32-2 is the area of the stall plus one-half of the area of the aisle adjacent to the stall. This area does not include that consumed by aisles which are parallel to the sides of stalls nor does it include the triangular areas which occur at the ends of stall rows when the angle of parking is not parallel to the side of the land parcel.

[1] The over-all width of a majority of 1953 cars is 6.5 ft or less.

The area-per-car values in Table 32-2 show that 90° back-in parking requires the least area per car; 45° herringbone parking requires the least area per car for the drive-in types of parking.

Convenience and Ease of Parking. It is generally conceded that drive-in parking at an angle, such as 30, 45, and 60°, provides more convenient parking than 90° parking angles. Some experts also believe that drivers are better able to see empty spaces along a parking row when cars are

TABLE 32-2. MINIMUM DIMENSIONS SUGGESTED FOR THE DESIGN OF
PARKING FACILITIES

Width of stall, ft	Angle of parking, deg	Direction of parking	Width of aisle, ft	Depth of stall perpendicular to aisle, ft	Width of stall parallel to aisle, ft	Unit parking depth, ft	Area per car, sq ft
8.0	30	Drive-in	12.5	15.1	16.0	42.7	341.6
8.0	45	Drive-in	12.5	18.0	11.3	48.5	274.4
8.0	45*	Drive-in	12.5	15.7	11.3	43.9	249.0
8.0	60	Drive-in	21.0	19.7	9.2	60.4	278.4
8.0	60	Back-in	19.0	19.7	9.2	58.4	270.0
8.0	90	Drive-in	25.0	19.0	8.0	63.0	252.0
8.0	90	Back-in	22.0	19.0	8.0	60.0	240.0
8.5	30	Drive-in	12.5	15.1	17.0	42.7	363.0
8.5	45	Drive-in	12.5	18.0	12.0	48.5	292.0
8.5	45*	Drive-in	12.5	15.7	12.0	43.9	264.0
8.5	60	Drive-in	20.0	19.7	9.8	59.4	292.0
8.5	60	Back-in	18.0	19.7	9.8	57.4	282.0
8.5	90	Drive-in	24.0	19.0	8.5	62.0	263.5
8.5	90	Back-in	22 0	19.0	8.5	60.0	255.0

* Herringbone (both Types A and B, Fig. 32-2).

parked at angles less than 90°. Without question, back-in parking requires more driving skill than drive-in parking at any angle. For this reason some designers are reluctant to use 90° back-in parking except for facilities where attendant parking is to be employed. However, 90° back-in parking has generally been accepted by patron parkers who have been introduced to it. Back-in parking at angles less than 90° is seldom used for obvious reasons. Studies of patrons' attitudes and the time required for parking and unparking maneuvers are needed to develop a policy on the convenience and ease of parking related to its angle and direction.

One- and Two-way Aisles. Drive-in parking angles at less than 90° require continuous aisles because unparking vehicles will head in their original direction. The best traffic circulation plan for parking facilities with drive-in parking angles less than 90° is a continuous system of alternating-direction one-way aisles. One-way aisles are desirable because they require less area per car, and head-on as well as crossing vehicular conflicts may be eliminated.

When 90° parking is used, vehicles may be unparked to the right or left and may therefore use the aisle in either direction. The principal advantage of two-way-aisle operations is reduction in travel distance. In some instances, where aisle space is restricted, dead-end aisles are warranted. In these cases, 90° parking is always employed.

General Traffic Circulation. The system of traffic circulation produced by the arrangement of parking aisles and stalls should be designed to *reduce travel distances* as much as possible and should *minimize the number of turns.* Turns cause increased hazard because they have restricted sight distance, and many inexperienced drivers are not able to judge accurately the inside-tracking requirements of vehicles on minimum radius turns.

The aisle system should also be designed to disperse activity within the parking area. Centers of movement through which all, or a large part, of the terminal traffic must pass to enter and leave the parking area become congested. This congestion is increased when vehicles perform parking or unparking maneuvers at stalls adjacent to the center of movement. Activities within the parking area may be dispersed by strategic location of the entrances and exits for the facility and by careful planning of the layout of aisles and stalls.

Continuous-aisle arrangements become important in patron parking facilities because drivers may progress from one row of stalls to another until a vacant space is found. Elaborate communication systems have been set up at some patron parking facilities to direct parkers to vacant stalls so as to avoid congestion and unnecessary travel in parking aisles.

Figure 32-4 compares a good and a poor aisle layout utilizing the same area. The example illustrates how travel distances may be reduced, turns minimized, and activity dispersed through careful design of aisle layout.

32-4. Reservoir Space. Reservoir space is the area adjacent to the entrance and exit of parking facilities which is used for the acceptance and delivery of vehicles. Lack of sufficient reservoir space causes street congestion and the turning away of potential parkers even when the facility has unused parking capacity.

Where a parking fee is charged, vehicles are usually checked in while within the inbound reservoir space. In attendant parking facilities, inbound cars must be stored in the reservoir space until attendants are able to move them to their parking stalls. Therefore, the size of reservoir space is more critical in the operations of attendant parking facilities.

At parking lots, reservoir space is usually provided in the entrance aisle of the lot. If the lot is large, the entrance aisle may be made somewhat wider than otherwise required so as to increase its reservoir capacity. At parking garages, the reservoir space must serve several floors of parking and, when attendant parking is employed, often occupies most of the main-floor area.

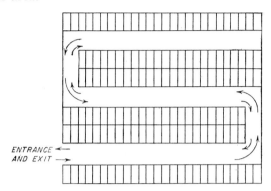

POOR AISLE LAYOUT WITH LONG TRAVEL DISTANCES-MANY TURNS AND CENTERS OF MOVEMENT

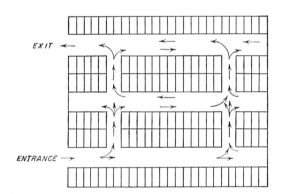

BETTER AISLE LAYOUT MINIMIZING TRAVEL DISTANCES AND TURNS-WITH ACTIVITY DISPERSED

FIG. 32-4. Examples of good and poor aisle layout.

To determine the required size of reservoir space, several traffic operational values must be known or assumed: (1) the average rate of arrival of vehicles to be parked during the peak parking period, (2) the average time required for an attendant to dispose of cars waiting in the reservoir and return for the next car, and (3) the number of attendants to be employed for storage operations.

The rate of storage may be computed from the formula $R = 60N/T$, where R is rate of storage in cars per hour, N is number of attendants, and T is average time required to park each car and return to the reservoir space in minutes. When the rate of storage equals the average rate of arrival of vehicles to be parked, efficient operations occur so long as the arrival times of vehicles are equally spaced. But this is rarely the case, and reservoir space is actually needed to accommodate cars that arrive in groups exceeding the average rate of arrival and to smooth out the effects of nonaverage storage times.

Studies of vehicle arrivals at garages during peak parking hours have indicated that the probability with which a given number of vehicles will arrive in excess of the average rate may be approximated by application of Poisson's theory of probability. This theory proposes that the

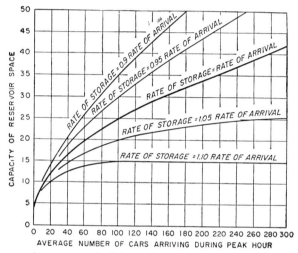

FIG. 32-5. Reservoir space required for various vehicle-arrival rates with overloading less than 1 per cent of the time. (From Edmund R. Ricker, *The Traffic Design of Parking Garages*, Eno Foundation for Highway Traffic Control, Inc., Saugatuck, Conn., 1948, p. 66, Fig. 18.)

number of arrivals at one time which will not be exceeded 99 per cent of the time is[1]

$$A = 2.4 \sqrt{R_1}$$

where A = number of vehicles which will arrive in excess of the average rate

R_1 = average arrival rate, vph

The value of A in the above formula is the required reservoir capacity when the rate of storage equals the average rate of arrivals. Figure 32-5 shows the relationship of needed reservoir capacity to average number of cars arriving during peak hours. The figure also shows the reservoir-capacity requirements when the rate of storage is greater or less than the average rate of arrivals.

[1] The selection of 1 per cent probability of failure is arbitrary, but it is taken on the basis that the overloading of reservoir space only 1 per cent of the time during peak parking periods is desirable.

The parking-facility design which offers the most efficient traffic movement at stalls and on the aisles and ramps of attendant parking facilities requires the fewest attendants and the least reservoir space. The optimum relationship between the size of reservoir space and number of attendants depends mostly on economic factors.

Reservoir space for outgoing vehicles does not pose a large problem, even in attendant parking facilities. Little delay is experienced in the outbound reservoir space except when patrons are not on hand to receive

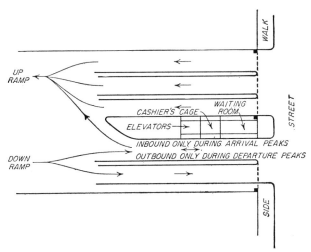

Fig. 32-6. Illustration of reversible-lane-reservoir usage.

their vehicles promptly or when they take extra time for loading or other purposes. Outbound reservoir space must be at least two lanes wide—so moving vehicles may pass stopped vehicles—and in large garages more than two lanes may be needed.

Periods of peak parking seldom coincide with those of peak departures. For this reason reservoir space may be arranged to operate with reversible lanes to be used for inbound movements at some times and for outbound at other times. Figure 32-6 shows an example of reservoir-space design which utilizes the reversible-lane principle.

This example also illustrates desirable main-floor layout because access between reservoir space and ramps is direct and pedestrian islands along with strategic location of the cashier's booth increase the safety and utility of reservoir-space operations.

32-5. Ingress and Egress. The design of entrances and exits for parking facilities embodies many of the principles of intersection design. Vehicles entering and leaving the facility should diverge from and merge with the street traffic stream at low relative speed. A one-way-street system is often employed to facilitate these diverging and merging maneuvers. Where street speeds are high, acceleration and deceleration lanes may be utilized to accommodate diverging and merging at low relative speed.

Terminal exits and entrances should be located as far as practicable from street intersections. When this distance is not adequate, a line of left-turn vehicles waiting to enter the terminal may extend into the street intersection, impairing its operations; or terminal exits may be blocked by street traffic delayed on the intersection approach. Terminal entrances and exits placed within intersectional areas so as to create multileg intersections should be avoided, for reasons given in Chapter 29.

At parking facilities with coinciding heavy flows of arriving and departing vehicles, the arrangement of entrances and exits should avoid the crossing of these flows on the street as well as in the terminal. Where such crossing movements are unavoidable, entrances and exits should be adequately separated so that the crossing flows may weave.

Pedestrian safety is often a problem at terminal entrances and exits. Good terminal operations may demand several entrances and exits which usually are multilane in width. These numerous and wide vehicular sidewalk crossings increase pedestrian hazards. Pedestrian safety may therefore be an important factor in the orientation and design of terminal entrances and exits, and the shadowing of pedestrians by islands located in the reservoir space and, sometimes, the control of pedestrians by traffic signals are possible solutions to this problem.

Parking facilities may cause a greatly increased number of left turns at adjacent street intersections. It may be necessary to prohibit some or all of these turns to reduce street congestion. The prohibition of left turns into and out of the parking facility may also be necessary for obvious reasons.

Many of the principles of terminal ingress and egress design are in conflict with each other. Some may be impractical to achieve under given conditions. In each particular case, the effect of terminal traffic on adjacent streets and intersections must be thoroughly studied to arrive at the best design.

32-6. Vehicle Interfloor-travel Facilities. Multilevel parking garages require means of interfloor travel by pedestrians and vehicles.

Interfloor travel facilities for *vehicles* are of two types: (1) elevators and (2) ramps.

Elevators. In completely mechanical garages, the elevator may be designed to move both vertically and diagonally. Power dollies move vehicles onto and off the elevators to and from the reservoir and parking spaces or the vehicles may be driven on or off. Such operations may be push-button controlled from the elevator or from a control panel located in the cashier's booth.

The elevators in completely mechanical garages usually open directly into parking stalls on both sides.   These stalls are seldom more than two deep to permit efficient mechanical operations.   When both stalls are occupied and a rear vehicle is to be unparked, the front vehicle must be moved to another floor.   Figure 32-7 is a photograph of a completely mechanical garage.   Such garages are highly specialized in design and operations and are patented.

FIG. 32-7. A completely mechanical garage.   (Courtesy of John Fahey, The Pigeon Hole Parking Co., Spokane, Wash.)

The ordinary elevator-type garages have patron- or attendant-operated elevators used as the means of vertical travel from floor to floor and with all horizontal movement on each floor performed by drivers.

Ramps.   Garage ramps are sloping surfaces between floors.   A ramp system includes all ramps as well as the floor aisles connecting ramp ends which are used by vehicles going up or down from floor to floor.   Regardless of the ramp system used, vehicles moving from floor to floor follow a rotary path.

Garage ramps and ramp systems have a number of classifications:

One-way or two-way.   Two-way ramps are sloping surfaces used by vehicles traveling in both directions.   One-way ramps are single ramps used exclusively for one-way movement.   Two-way *undivided* ramps are unsafe because drivers tend to cut corners and swing wide at turns, with possible head-on collisions.   They should not be used except, perhaps, in small garages where ramp volumes are small.

Parallel or Opposed, Adjacent or Separated. Parallel ramps are two one-way ramps placed between garage floors on the same sloping plane. Parallel ramps may be *adjacent*, whereupon they become a two-way divided ramp, or they may be *separated* with floor area between them. Opposed ramps are two one-way ramps pitched in opposite directions between garage floors. Opposed ramps may also be *adjacent* or *separated*.

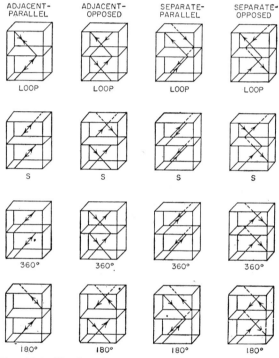

FIG. 32-8. Classification plan for straight ramp systems.

Straight or Curved. Ramps are further classified by their alignment. When ramp alignment is straight, all the turning movements on a ramp system take place on garage floors. This turning path may be semi-circular (180°), circular (360°), a *loop*, or an S curve.

There is a consistency in the design of ramp systems for garages. To save on space and construction costs, the ramp system established between any two floors where adjacent ramps are used, and between three consecutive floors where separated ramps are used, is generally adopted for the entire structure regardless of the number of floors. Figure 32-8 is a classification plan for straight-ramp garage systems. It is evident that the best system with respect to distance of travel from floor to floor is the 180°-turn group. However, this group presents increased construction problems and may reduce parking accommodation because two different stacks of ramps are required. The 360°-turn group requires more floor-to-floor travel distance, but only one stack of ramps is needed for both the adjacent-parallel and the adjacent-opposed classes. The loop and S-curve classes are too inefficient in operations to be considered in design, except when unusual physical limitations are encountered.

Curved ramps are subclassified as *continuous* or *noncontinuous*. Continuous circular ramps are commonly referred to as spirals. They are complete ramp units with entrances and exits at the various floor levels. The up and down ramps may be parallel or opposed. The entrances and exits onto and off a continuous ramp system may be located on the same side or on opposite sides of the cylindrically shaped ramp column. Figure 32-9 shows the most commonly used continuous circular ramp systems.

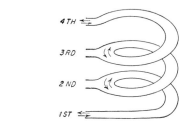

The noncontinuous curved ramps generally employed in garage design are semicircular. The same classification plan for straight ramp systems may be used for noncontinuous, semicircular ramp systems. The straight ramps classed as 180° and 360° in Figure 32-8 are most

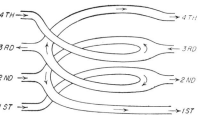

Fig. 32-9. Continuous circular ramp systems in common use.

fitted for noncontinuous semicircular ramp application in garages.

Curved ramps require more space than straight ramps, but they inherently offer better traffic operations. The radius of curved ramps is usually uniform. Uniform curvature along the ramp provides gradual turning as compared with the sharp turns which must usually be made at the ends of straight ramps. In addition, superelevation at the ends of straight ramps requires undesirable warping of floor areas. Thus space-saving advantages of straight ramps may often be offset by the operational advantages of curved ramps.

Clearway and Adjacent Parking. All ramp systems may be either clearway or adjacent-parking types depending on the amount of interference between vehicles traveling from floor to floor and those which are parking or unparking. A clearway system provides a path for the interfloor travel of vehicles which is completely separated from parking aisles. Adjacent-parking ramp systems are those in which the path for interfloor travel passes along aisles used for parking and unparking maneuvers. The amount of interference to ramp movement in *adjacent-parking* garages is determined for the most part by the number of stalls adjacent to the ramp system.

Adjacent-parking ramp systems require less area per car parked because aisles may serve the double purpose of stall access and interfloor travel. The clearway type operates more safely and with less delay, so it is to be preferred for all patron parking garages and for large attendant parking garages.

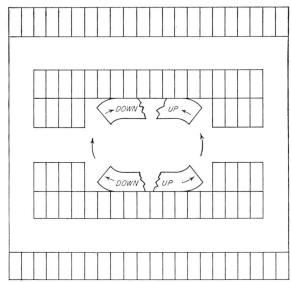

FIG. 32-10. Parking floor of a curved ramp (separate-opposed) garage with clearway operations.

Continuous circular ramp systems are inherently clearway in operations. Straight and curved noncontinuous ramp systems may become clearways when parking stalls are located only along aisles which are not a part of the ramp system. A curved noncontinuous ramp system with clearway operations is shown by Fig. 32-10.

Concentric or Tandem. Ramp systems are concentric when the paths of vehicles moving up or down revolve around the same center. They are tandem when up movements revolve around a different center than down movements. Straight ramps classed as the 360° type may be operated in tandem, or tandem operations may be achieved through two completely separate arrangements of up and down ramps. In every case only one up-and-down ramp is placed between floors. Figure 32-11 shows an adjacent-parallel (360°) ramp system with tandem operations. The figure also shows a tandem system with completely separate ramp arrangements for up and down movements.

Tandem ramp systems are desirable because up and down interfloor movements operate independently. They require larger floor areas than are available for most garages and, when designed as clearways, require more area per car than other ramp-system types.

Single- and Double-ramp Systems. It is obvious that a garage may be provided with more than one system of ramps. In other words, more than two one-way ramps may be used to connect adjacent garage floors. Most garages employ single ramp systems because the capacities of properly designed one-way ramps may equal or exceed 700 vehicles per hour. Double ramp systems may be used in garages of unusually large area to reduce travel distances and to disperse activity on parking floors.

TANDEM OPERATIONS WITH TWO (ADJACENT-PARALLEL - 360°) RAMPS

PARKING FLOOR VIEW

TANDEM OPERATIONS WITH FOUR RAMPS

PARKING FLOOR VIEW

FIG. 32-11. Examples of tandem ramp operations.

Staggered-floor Garages. The staggered floor, or half-story, ramp principle has been used extensively in garage design. The garage building is built in two or more adjacent sections; the floor levels of the sections are staggered one-half story. The intermediate half story is usually smaller than the other section and may be limited to the area of the ramp system. The ramps are much shorter than those required to connect floors with full-story height and therefore may be considerably steeper.

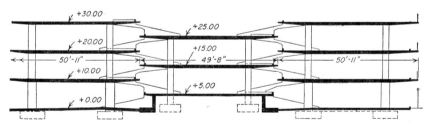

FIG. 32-12. Illustration of staggered-floor garage design. (From R. L. Weed and G. K. Newberg, "Parking Garages," *Architectural Forum*, September, 1954.)

Cars are usually parked on the intermediate level and may be overlapped to some extent by cantilevering the half-story sections. Figure 32-12 illustrates the staggered-floor ramp system. Most existing staggered-floor garages consist of only two sections rather than the three sections shown in the figure. Any of the types of straight ramp systems shown by Fig. 32-8 may be adapted to staggered-floor design.

Garages with the staggered-floor system usually require less space per car than those with other ramp systems. To achieve this increased parking capacity the ramps must be of the adjacent-parking type. Delays caused by parking off of ramps and by the increased number of turns to travel from one full story to another offset much of the advantage of increased capacity.

Sloping-floor Garages. Sloping-floor garages are designed particularly for patron parking. In sloping-floor design the garage is designed as a continuous rectangular spiral ramp. The ramp has two-way travel with parking (usually 90°) on both sides. This adjacent-parking feature causes delay, but a sloping-floor garage has much flatter grades than other types of garage ramps. Travel distances may become excessive in sloping-floor garages.

The operations of a sloping-floor garage are similar to that of a parking lot. Figure 32-13 is a photograph of the exterior and a section of ramp of a sloping floor garage.

Other Ramp Systems. Numerous variations in ramp systems may be required to meet existing physical conditions. Some garages may be located on sloping ground where direct street access may be obtained on separate levels at each end or opposite sides of the garage structure. Such location, with ramp orientation to meet natural differences in elevation, is desirable because the ramp system is simplified.

Underground garages may require unique ramp and street-access treatments to be fitted to the special case. At some garage locations, space for parking may be increased by cantilevering upper-story ramps over adjacent property.

32-7. Pedestrian Interfloor travel Facilities. Facilities to provide for the interfloor movement of pedestrians in garages must accommodate attendants or patrons, depending on the method of parking. In the case of attendant parking, an inbound and an outbound vehicle may be handled in one round trip between the reservoir space and parking areas. Generally, inbound and outbound peaks occur at different times, and attendants must travel from or to the parking areas as pedestrians. With patron parking, half of the round trip is always made as a pedestrian.

A principle design consideration for interfloor pedestrian travel facilities is to provide sufficient operational capacity. In attendant parking garages this required capacity may be computed by the formula

$$C = \frac{N}{T}$$

where C = required capacity of interfloor pedestrian travel facility, attendants per min
N = number of attendants engaged in storing cars
T = average time required to handle each car, min

In patron parking garages, this capacity is estimated from the peak rate of demand with allowances made for extra passengers.

The pedestrian interfloor travel facility should be oriented, located, and designed to conserve on the energy of patrons or attendants and

FIG. 32-13. Exterior and ramp view of sloping-floor garage.

reduce walking distances and time as much as possible. The reservoir area on the main floor of the garage and the centroid of the parking areas on other floors establish the most convenient and desirable location, although a compromise location is usually required.

Interfloor pedestrian travel facilities for patrons must be more convenient and attractive than those used by attendants. In large terminals, escalators may be used, but most garages employ passenger elevators.   Passenger elevators or man lifts are usually employed in attendant parking garages.   It is obvious that the speed of the elevator and the time required to load and unload at various floors have considerable influence on the rate of attendant operations.

Stairways may be required as an auxiliary in garages where pedestrians and vehicles are served by elevators in case of emergencies.   Stairways or pedestrian ramps are sometimes used as the only means of pedestrian interfloor travel in garages of two or three floors.   Patrons and attendants alike may tend to avoid stairways by using vehicle ramps.   Fire poles are sometimes used in attendant parking garages.

32-8. Miscellaneous Design Elements.   *Surface.*   Parking lots should be surfaced and well drained.

Illumination.   Parking areas and walkways leading to them should be well lighted at night to reduce pilferage and increase the personal safety of patrons.

Landscaping.   Attractive landscaping and screening around parking areas preserve the values of adjacent property.   Some city ordinances require fences of specified type around parking lots.

Pedestrian Islands.   Raised pedestrian islands between rows of parked cars are desirable for the safety of patrons.   Widths vary, but islands at least 7 ft wide are needed to allow for the bumper overhang of vehicles and for two people to walk abreast.   Unless these islands lead directly to the pedestrian entrances and exits of the lot, many patrons will not use them.   The space involved in providing these islands is not included in the minimum dimensions shown by Table 32-2.

Bumpers.   Some form of bumper is required at each stall as a reference point and to prevent overrunning the stall.   Since the bumper overhang of vehicles varies with different manufacturers and models, encroachment onto pedestrian islands or onto areas where parking meters may be installed is better controlled with post-type bumpers.   The herringbone parking types require rather intricate bumper arrangements, which is the major disadvantage of this system.   Herringbone Type A parking arrangements may produce especially damaging collisions unless adequate bumpers are installed.

Stall Markings.   Parking-lot stalls should be marked on the surface of the parking area with paint or by other means.   This is particularly important with patron parking to avoid inefficient space use caused by straddling of stalls.   Even where attendant parking is used, the marking of spaces leads to more orderly parking.

Signs.   Traffic signs and aisle markings should be installed to indicate one-way aisles, entrances and exits, parking-time limits, fees, and other regulations.

Garage Cashier's Booth.   A small cashier's booth or cage should be located conveniently for attendants and patrons adjacent to the reservoir space. To reduce the space requirements of this booth its activities should consist only of the handling of parking checks and collection of fees. Other bookkeeping and personnel activities should be conducted in a manager's office located in less premium space.

Waiting and Rest Rooms. Attendant parking garages should provide an attractive waiting room adjacent to the outbound reservoir space where drivers will receive their cars conveniently. Public rest rooms are usually located on the second floor or basement of the waiting room. Employees' dressing rooms and lockers may be located in "wasted space" adjacent to the ramp system on a top parking floor.

Architecture. Most modern garages are either of fabricated steel (Fig. 32-7) or reinforced concrete construction with concrete curtain walls (Fig. 32-13). This open-type architecture is of pleasing appearance, reduces construction costs, and eliminates the need for heat and ventilation.

Ventilation. Underground or other enclosed garages must be well ventilated to exhaust deadly carbon monoxide fumes. Automatic warning and exhaust or intake systems are sometimes used.

Heating. Waiting rooms, rest rooms, and offices should be heated. Outside ramps may be heated to eliminate ice and snow in Northern climates. This may be done electrically or by forcing hot water or oil through pipes imbedded in the ramp.

Communications. An adequate method of communications between garage offices and parking floors is desirable in case of "lost" cars and emergencies. In elevator garages, electrically operated panel boards located in the elevators and cashier's booth may be used to direct parking and unparking operations.

Lighting. Garages should be well lighted, particularly in the vicinity of ramps and reservoirs.

Building Codes. Local building codes generally specify minimum garage-construction requirements with respect to fire protection, safety, health, appearance, floor loadings, etc.

32-9. Parking-terminal Design Factors and Procedures. *Parking Lots versus Garages.* The choice between a parking lot or garage usually depends on the value of the land to be occupied by the facility. High-cost land frequently justifies a multistory garage because the land cost per vehicle is less than that for a parking lot. Conversely, expensive garage construction is seldom warranted on low-cost land because *garage capacities* may be obtained on a larger land area at lower capital cost. The availability of land, cost of maintenance and of operations, and demand for parking may modify this general rule.

Attendant versus Patron Parking. It has been shown that attendant parking requires more reservoir space than patron parking, but aisles and stalls may be less spacious, steeper ramps are permissible, and other minimum dimensions may be employed successfully because attendants are more skilled in handling cars in restricted areas than are ordinary drivers.

Further maximum use of space may be gained with attendant parking because cars are not locked and may, therefore, be parked behind those adjacent to aisles. This conserves space because one aisle serves more than two rows of stalls. With attendant parking, also, the aisles may be loaded with parked vehicles during periods of peak-parking accumulation. (Both of these practices may be prohibitive when they destroy freedom of movement and require extra attendants to provide satisfactory service.)

For the above reasons, the capacity of an attendant parking facility may be 30 to 50 per cent higher than that of a patron parking facility of the same size.

The choice between attendant and patron parking usually depends on the extent to which the cost of attendants' salaries is justified by increased parking revenues produced by the greater parking capacity. Attendant parking will most likely be found desirable where space is at a premium and parking demand is heavy.

Ramp Design. No one ramp system is best. The utility of a ramp system depends on how it is fitted to the size and shape of the available land parcel. Generally, those ramp systems which provide the greatest freedom of movement require the most space and highest construction cost but produce lower operating costs.

Elevators versus Ramps. The place for elevator garages in terminal design has not been established. An important difference between the operations of elevator and ramp garages is in their rate of handling vehicles. Ramps may operate satisfactorily at continuous 5-sec headways, while elevators are able to handle only one vehicle at a time.

The storage rate of elevator garages varies with the number of elevators and height of structure, but an average rate of 1 to 2 min per car per elevator is rarely exceeded. Elevator garages are also subject to mechanical failure.

The principal advantages of elevator operations are (1) savings in space otherwise required for ramp systems, (2) elimination of long travel distances on numerous ramps in tall buildings, and (3) minimum number of garage employees required. It is believed that elevators may be used most advantageously in garages on small or narrow parcels of land or where the garage structure will exceed six or seven stories.

Design Procedures. In the traffic design of parking facilities, the size and shape of the available land parcel, its differences in elevation, and location of its road access must be known. The desired parking capacity, rate of parking turnover, the origin and destination of street and terminal traffic and the influence of terminal traffic on the operations of adjacent streets and street intersections should be anticipated.

The choice between parking lot or garage construction and between attendant and patron parking methods may be based on local customs and preferences or on economic analyses. When attendant parking is considered, the rate of vehicle storage must be estimated for facilities of alternate design in order to design the reservoir space properly. The desired capacity of interfloor travel facilities for both vehicles and pedestrians may be determined from storage and turnover rates.

For garages, *90° parking is generally employed*, and the aisles, interfloor travel facilities, reservoir space, street access, and other elements are designed and located as an integrated unit for optimum operations.

The angle and arrangement of stalls in parking lots are often governed by compromises between maximum capacity and greatest accessibility. The area-per-car values shown in Table 32-2 indicate the efficiency with which available area is utilized by parking only when the unit parking depths shown in the table are multiples of the dimensions of the area. For example, 90° back-in parking, which requires the least area per car in Table 32-2, may not result in the most efficient use of available area of irregular shape or with no dimension equal to a multiple of 60 ft. Often it is necessary to use trial and error in the design of parking lots for the maximum utilization of available areas. Different angles and direction of parking should be fitted to the size and shape of the lot—sometimes more than one parking pattern is required to achieve the most efficient use of space.

32-10. Design Data. Recommended minimum dimensions for parking stalls and aisles are given in Table 32-2. Other suggested maximum and minimum design values particularly applicable in the design of parking garages are:

Lane widths
    Entrances and exits................ 12 ft
    Reservoir space.................... 12 ft
    Straight ramps.................... Minimum 10 ft between crubs, flared at ramp ends
    Curved ramps..................... 14 ft between curbs
    Turning radii, floors or ramps......... 60 ft diameter to outside curb, 32 ft diameter to inside curb
Ramp grades
    Full story......................... 15 per cent maximum
    Half story........................ 15 per cent maximum
Story height....................... 7 ft 6 in. minimum clear-ceiling height

Wide shallow beams (in slab- and beam-type construction) or flat-slab floor construction is preferred to reduce floor-to-floor heights.

32-11. Other Types of Terminals. Parking lots and garages for passenger vehicles are the most prevalent types of highway terminals, but traffic engineers are sometimes concerned with the traffic design of truck-loading bays, truck terminals, and bus terminals.

TABLE 32-3. APRON SPACE REQUIRED FOR 90° PARKING OF TRUCKS OF
VARIOUS LENGTHS

Over-all length of tractor trailer, ft	Width of stall, ft	Apron space required, ft
35	10	46
	12	43
	14	39
40	10	48
	12	44
	14	42
45	10	57
	12	49
	14	48

SOURCE: Fruehauf Trailer Company, "Modern Docks for Modern Transport," *Architectural Record*, October, 1947.

Truck-loading Bays. Little information is available on the minimum area required to maneuver trucks to a loading dock or platform. Table 32-3 shows the apron space required for tractor-trailers of various lengths to park at a 90° angle with a loading dock. Apron space is defined as

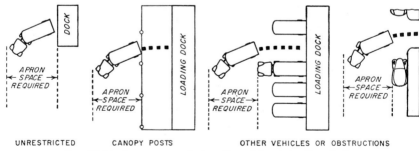

UNRESTRICTED   CANOPY POSTS   OTHER VEHICLES OR OBSTRUCTIONS
FIG. 32-14. Apron-space designation at truck-loading docks.

the unobstructed aisle width adjacent to the dock. Figure 32-14 illustrates different applications of the apron-space dimensions given in Table 32-3, depending on conditions of parking.

The floor heights of heavy trucks vary from 45 to 60 in. above the pavement. Dock heights from 44 to 50 in. are employed because it is believed more desirable to have dock levels below rather than above the bed level of most trucks.

Truck Terminals. Truck terminals are used to facilitate the handling of freight in urban areas. The function of such terminals is to receive long-distance truck shipments and distribute their loads to smaller trucks having regular areas of delivery. This reduces the number of heavy trucks on city streets and the mileage of truck deliveries.

Truck terminals usually consist of one or more piers of loading and unloading docks with good accessibility. Some efficient type of conveyor operated on each pier is used for the distribution of freight between docks. Much office space is required to process bills of lading, etc.

Truck terminals must be located carefully to minimize the distance of travel in congested areas and on streets which are not designed for truck use. Ramps are to be avoided in truck operations, so most truck terminals are single-story structures except when truck elevators are provided.

Bus Terminals. Bus terminals may serve both long-distance and local buses. For local buses, which operate on close headways, one or more raised pedestrian platforms of sufficient length to permit rapid parallel parking and unparking of buses is provided. The width of platform depends on the size of crowds waiting for buses and on whether both of its sides are bus-loading zones. The length of platform depends on the number of buses loading and unloading simultaneously and on bus-patron walking distances to and from loading points.

Long-distance buses are parked for longer periods and are usually accommodated in regular parking stalls located in an area completely separated from the local-bus operation. One-way circulation of buses in and out of the terminal is necessary, and pedestrian movements within the terminal, or at its entrances and exits, should be separated from vehicular movements as much as possible.

ADMINISTRATION AND PLANNING

CHAPTER 33

TRAFFIC ENGINEERING FUNCTIONS
AND ORGANIZATIONS

Traffic functions occupy a major position in state and city governments. This condition was largely developed over the last quarter century as the problems of traffic accidents and congestion became a principal concern of the public and of public officials. Many of the administrative techniques and specialized functions, including traffic engineering, have resulted from sheer necessity—from laborious trial-and-error methods. As new traffic activities were established, they were fitted conveniently into some already existing arm of governmental organization, where enlargement of staff and responsibilities made it possible to meet the immediate problem. However, in traffic, as in other areas of public activity, it is recognized that the best techniques and plans will not effect satisfactory solutions without developing an adequate structure as a basis of operations in the existing government.

These methods of developing traffic organizations have inevitably produced a wide diversification of responsibility and activities in traffic matters. While the grouping of traffic activities is desirable in organization and administration, it must be recognized that the traffic function cuts across the established duties of many agencies. A strict objective of traffic administration is to coordinate the many and varied activities having both a direct and indirect influence on traffic planning, construction, and operation, and at the same time require all appropriate departments and agencies to assume effective responsibility for traffic functions which fall within their framework. In some instances, drastic reorganizations are required, while in others the changes can be gradually and simply fitted into existing routines.

Traffic engineering is a relatively new branch of engineering. By title it first appeared about 1924 when two cities and a state highway department[1] created the position of Traffic Engineering. An outgrowth of the great acceptance of automotive transportation, traffic engineering resulted from demands of the public for expert handling of traffic problems. Traffic engineering has often been referred to as the third logical stage in highway development. First, there was the job of construction

[1] Pittsburgh, Pa.; Seattle, Wash.; and the state of Ohio.

559

which became apparent when automobiles were accepted generally. Second, there was the need for physical maintenance of the public way to protect investments. As in other forms of transportation, there is a need for a third area of activity—operations. Most of the work of the traffic engineer deals with operations, even though in many instances it is concerned with planning and design.

The growth of traffic engineering has been phenomenal. From a beginning in 1924, traffic engineering organizations are now found in practically all the state highway departments; most cities of 50,000 to 100,000 population have traffic engineers, and some smaller cities have traffic engineering service. Rather than create separate traffic engineering departments, a few cities have employed part-time traffic engineering assistance, utilizing consultants or sharing with other cities the time of a traffic engineer.

Titles given the individuals and the divisions doing the traffic engineering work vary. Although *traffic engineering* is becoming the most common title, it should be noted that many fine traffic engineering jobs are being performed under other titles. After all, it is the activity that is of principal concern, not formal titles. With public officials, bankers, business leaders, and educational institutions increasingly using the term traffic engineering to include traffic studies and the functions of highway operations, it is evident that in a short time there will be no misunderstanding as to the intent and activity of traffic engineering.

Numerous colleges and universities are providing training in traffic engineering. Courses are offered as electives in undergraduate engineering curricula. Increasingly, graduate work is being offered in traffic engineering, with several institutions, such as Yale University, giving a full academic year of graduate study. Demands for trained traffic engineers are great, so that there is every reason to assume that colleges will extend their engineering courses to meet these needs.

The Institute of Traffic Engineers is an engineering society with more than 900 active members. It is a well-known and recognized professional society, patterned after the founder engineering societies in organizational structure, membership requirements, policies, and publications.

33-1. Traffic Engineering Functions. The functions and duties of traffic engineering were initially centered around control devices and fact-finding surveys. Today they have grown to include many other activities. These vary according to the basic organization, legislative authority, and resources of the individual traffic engineer. Obviously, the functions also vary between city, state, and county departments. In cases where traffic engineering is established outside these areas of government, it might be found that the functions are more varied, even though more limited in scope.

A review of traffic engineering departments will show that they deal with many functions.   Great care is required to assure a proper evolution and assignment of functions to a particular traffic engineering bureau or department.   Typical functional charts are shown in Figs. 33-1A and 33-1B.

Collection and Analysis of Factual Data.   One of the principal objectives of the traffic engineer is to replace opinions with facts in various traffic situations.   It is likely, therefore, that the work of every traffic engineer will entail a substantial effort in the collection and analysis of traffic facts. These studies may be aimed at the development of treatments or solutions for a particular problem, or they may be of a sustained, continuing character intended primarily to determine trends and to make basic comparisons.

The simpler types of surveys include those having to do with traffic volumes (vehicular and pedestrian), speeds, and accidents.

One of the principal survey tools of the traffic engineer is the origin-and-destination survey.   Because this study provides so much basic information, it is extensively used by the traffic engineer in the solution of immediate problems, as well as in the planning and design of long-range traffic improvements.

Physical and economic data are acquired by the traffic engineer to measure existing conditions and to provide a basis for estimating future characteristics and needs.   Trends in traffic growth, sufficiency of existing facilities, and the level of traffic performance are determined and evaluated for all planning and development work.

In some surveys and studies, the interest of the traffic engineer overlaps those of the highway-planning-survey officials in state highway organizations and of city planners.   In such cases, care should be exercised to avoid needless duplication and overlapping of functions.   The common interests should permit utilizing the special talents and resources of the different groups.

The increasing emphasis on parking and terminal matters in recent years has brought the traffic engineer very close to parking activity. Many of the surveys and studies undertaken by traffic engineers are in the field of parking.   The survey techniques of traffic engineering permit an objective determination of the parking deficiencies and needs and an accurate analysis of the trends and likely needs of the future.   Such studies become especially important when used as the basis for financing large investments in off-street facilities, particularly with regard to revenue-bond issues.

The traffic engineer should exercise care not to become so involved in the collection of traffic facts and figures that he completely overlooks the applications.   Very little good is apt to come from the accumulation of large masses of statistical data.   The collection of the data should be in accord with needs and specific problems.   Unless collections are carefully

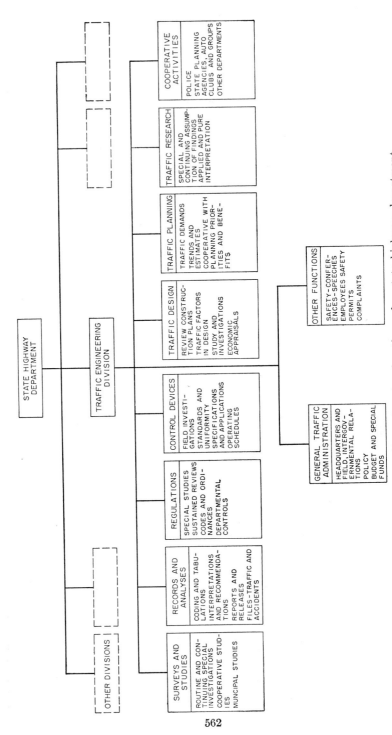

FIG. 33-1A. Typical traffic engineering functions of a state highway department.

562

planned and well utilized, they become a great expense and a grossly inefficient operation. In other words, the analysis and interpretation of the data should always be key factors in plans for the collection.

The traffic administrator is especially concerned with the effectiveness of various traffic treatments. For this purpose, the traffic engineer should be alert for opportunities to measure factually, by before-and-after studies or otherwise, the extent to which a particular treatment has succeeded. While simple, this is an effective administrative tool available to the traffic engineer through the various surveys and studies with which he is daily concerned and from which he can obtain many benefits.

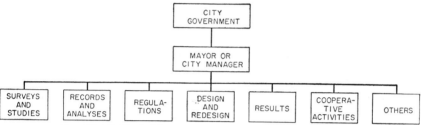

FIG. 33-1B. Desirable traffic engineering functions in a city.

Traffic Regulation. The traffic engineer is the logical person to supervise the application of many traffic regulations, such as one-way streets, curb-parking controls, pedestrian controls, turning restrictons, through streets, and the regulation of mass transportation. He can determine through surveys and studies the type of regulation which appears to be most desirable or best suited to a given set of conditions. He can then measure the effectiveness. He is usually better equipped than any other public official to determine, on the basis of sound studies, traffic demands and characteristics and to evaluate them in terms of the various regulations which might be applied.

In dealing with traffic regulations, the traffic engineer must consider the close relationship of the controls to enforcement and must work closely with the police in planning and applying most regulations.

Control Devices. There is general agreement that the traffic engineer should have primary supervision over the application of signs, signals, markings, and traffic control islands. Many public officials also agree that the traffic engineer should have close supervision and control over street and highway lighting because of the close relationship of lighting to safety.

Analysis of traffic engineering activities in cities reveals that a more effective and efficient supervision can be achieved if the traffic engineering department is given both the line and staff function. However, in most states the assignment of line functions in traffic control devices does not appear essential, and in some cases it becomes highly inefficient, and therefore uneconomical, to place the line responsibilities in the traffic engineering division. In states it would appear that the traffic engineer need only have supervisory or staff control so that he can regulate the general standards, both as to the design and application of the devices. This will enable the traffic engineer to direct research and to develop over-all uniformity in the application of devices. It places on him the responsibility for adequate warrants and standards. At the same time the actual erection and maintenance of the devices might be far more effectively achieved through some other agency, such as the maintenance department.

Traffic Design. The traffic engineer is primarily concerned with the operational qualities of roadway design. In the establishment of a traffic engineering bureau the proper steps should be taken to give the traffic engineer an opportunity to review plans and standards for new highway facilities so that he can suggest and recommend adequate features for safe and efficient traffic usage.

Experiences have shown that there is much that the traffic engineer can contribute from his studies and knowledge of traffic characteristics and future traffic needs that will be helpful in the work of the design departments, and also in administrative decisions relative to the reconstruction and new construction of public ways.

Traffic Planning. Most phases of traffic planning are of concern to the traffic engineer, and he should therefore be one of the principals in these activites. Types of routes, expressways, highways, and terminals which will be needed are closely related to the other activities of the traffic engineer. He should know the proper location and the probable effect of the developments on traffic needs and on competing facilities. He must deal with land uses, population distributions, and general trends in his areas of jurisdiction so as to be an effective part of the traffic planning function.

Cooperative Activities. Because traffic engineering is so broad in its interests and applications, the traffic engineer must participate in many cooperative activities. The needs of traffic engineering must be evaluated in terms of the needs of other traffic agencies. One cannot keep abreast of rapidly changing conditions and developments without maintaining a close cooperation at both local, state, and national levels. The multitude of agencies and organizations devoting time to traffic safety and highway-development matters require most traffic engineers to give a substantial part of their time to over-all traffic work and to the coordination of related activities.

Administrative Functions. In addition to basic functions of business management and personnel relations the traffic engineer must undertake in the operation of his own department, there are numerous ways in which he can assist in key administrative decisions. Few public agencies have a better opportunity to engage in public-relations activities than the traffic engineer. Through objective approaches and careful study of traffic problems, the traffic engineer can do much to effect the proper understanding between the public and official agencies in many traffic situations. He can point out the magnitude and effectiveness of the activities and programs which have been developed to reduce accidents and to provide better traffic conditions. He has at his disposal information and techniques which are of much interest to the public and which can be disseminated through the many media at his disposal. Obviously, the traffic engineer should not become so concerned with matters of public relations that he fails to produce an adequate technical and objective result. On the other hand, the many opportunities which are available to the traffic engineer in the field of public relations should not be overlooked or minimized.

Many fields of highway research involve the work of the traffic engineer. He has opportunity to undertake research in very simple ways, such as through the use of basic before-and-after studies, and he has equal opportunity to engage in basic and involved research problems. It is well known that the science of traffic engineering and traffic control has not advanced to a point where all the answers are apparent and that there is a very fruitful field for further research activities. The problems of traffic research are complicated by the human factors which are always present in traffic conditions. The traffic engineer is more advantageously situated, because of his other areas of activity, to deal with many of these human problems and measurements than other public officials.

The traffic engineer has an excellent opportunity to work with other departments of the government in which he is situated. There is not an activity of state highway departments, for example, that is not concerned with traffic. Numerous agencies of city and county government, other than the traffic engineer, are concerned with traffic conditions and problems. The traffic engineer can extend his relations and increase the effectiveness of his work by maintaining close working relationships with all agencies concerned with and interested in traffic, regardless of the area of interest or the depth of interest.

Other Functions. Traffic engineers and traffic engineering bureaus engage in other activities than those already enumerated. Almost all have a direct interest in traffic laws and ordinances. Because all of the work of the traffic engineer is dependent on the legislative codes as a basis of operation and for applications, he must concern himself with these codes and with the needs for new laws and revisions or amendments to existing laws. The traffic engineer is frequently called upon to express opinions and to render assistance in the development of codes which extend far beyond the immediate area of his work. This is because the traffic engineer is generally looked upon by public officials and by legislative leaders as one competent to deal with *all areas* of highway transportation.

Some traffic engineering departments are given the responsibility of issuing various types of permits; for example, controlling the movement of overloaded and oversized vehicles on the public ways, road openings, and access controls. On the other hand, the highway-planning survey unit of some states is better equipped, because of the availability of physical roadway and traffic data, to issue the permits. The extent to which this function is assigned the traffic engineer should be dependent upon the extent of the information and records available in his office.

The handling of traffic complaints is becoming a common traffic engineering function. Mayors, city managers, and highway directors increasingly recognize the importance of traffic surveys and studies in the effective handling of complaints.

In some instances, the work of the traffic engineer has been sufficiently broad to include the responsibilities for departmental safety programs, but this is not generally considered desirable. The activity can usually be better and more effectively controlled by the personnel department, so that the proper actions can be taken for unsafe and inefficient personnel practices.

33-2. Municipal Organizations. Traffic engineering functions in city governments can be performed within existing departments, or a new department may be established.

Police. When problems of automotive traffic first developed the police assumed many traffic functions which are now considered part of the traffic engineering program. In cities without traffic engineering units, most traffic work is still performed by the police. Since police departments do not often have technically trained men, many progressive police administrators have sought traffic engineering assistance. In general the traffic engineer has now moved into engineering divisions or into separate departments of city government. In a few cities, however, he remains as an integral part of the police function.

Effective results can be achieved with the traffic engineer reporting to the chief of police or to other enforcement officials. In general, however, his interests greatly exceed the traffic duties and responsibilities of the police, and he is much more able to undertake the broad functions enumerated above when given a position in the city's government outside of the police department.

It is generally agreed that the city traffic engineer can be more effective and render a better service, even to the police, when removed from the police department. In some cases the police do not wish to give such functions as erection and maintenance of traffic control devices to a new traffic engineering bureau. Most enforcement officials today, however, are willing to turn them over to a traffic engineer, wherever he may be placed in the city's government.

Public Safety.   There are advantages in making the traffic engineer a division head under the director of public safety, on a par with the chief of police (Fig. 33-2).   In such a position he can work closely with the police but is not subordinated to the police.   However, this does not

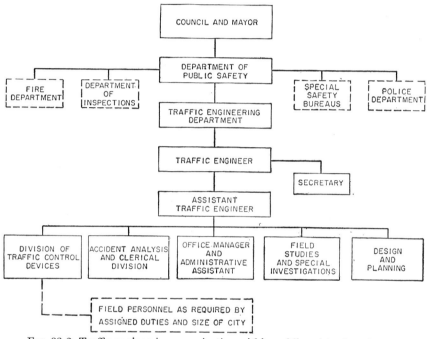

FIG. 33-2. Traffic engineering organization within public safety department.

generally remove the restrictive approach which is inherent in public safety and police work, and it does not greatly enlarge the scope of activity with which the traffic engineer may deal.   He is likely to be too limited to work on traffic control devices and basic traffic regulations and may give all of his time to the restrictive aspects of traffic rather than to the broad planning and construction aspects as well.

City Engineer or Public Works.   In many cities the traffic engineer is in the department of engineering or public works.   This is both logical and desirable, since he is dealing with other engineers and many of his interests normally fall within the scope of duties and responsibilities of the city engineer or director of public works.   However, the traffic engineer should not limit his activities to matters of roadway construction and maintenance and the other traffic functions which are likely to be found in the city engineering department.

If the traffic engineer reports to a director of public works he can be given a position on the same level as the city engineer and thereby enjoy a greater flexibility to deal with both the conventional city engineering functions, which involve traffic, and with other matters that are of principal concern only to him.

In most cities, it is satisfactory to have traffic engineering either under the city engineer or under the director of public works. However, as the

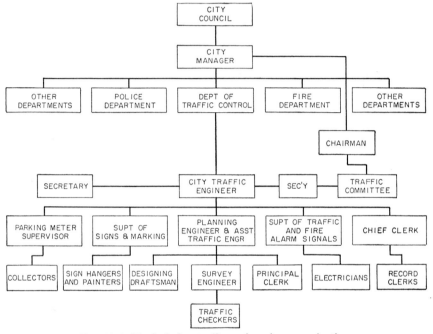

FIG. 33-3. Typical city traffic engineering organization.

function grows in scope and in personnel requirements, it is usually desirable to remove it from these departments and make it a separate department of municipal government (see Fig. 33-3 for such a plan in Dallas, Tex.).

Separate Bureau. More than half of the city traffic engineering departments now established are operating as separate divisions of municipal government. The traffic engineer, or head of the department, reports directly to the city manager or to the executive head of the city government. Under this system the traffic engineer has greatest freedom and flexibility to work with other departments and bureaus. He has the necessary prestige and standing in the over-all organization structure of the city to carry out properly assigned duties and responsibilities.

If there is no traffic engineering bureau, many bureaus and agencies may be performing traffic functions. The traffic engineer functioning

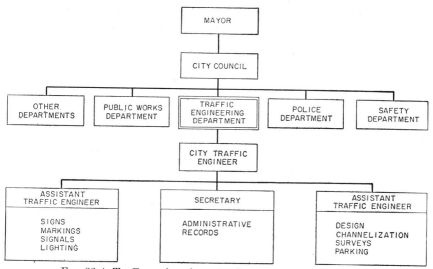

FIG. 33-4. Traffic engineering organization chart for small city.

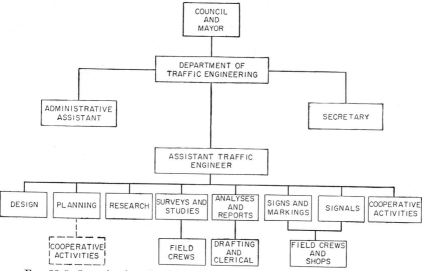

FIG. 33-5. Organization plan for separate department of traffic engineering.

in a separate department can coordinate and develop the greatest efficiency in traffic activities.

A simple plan for a small city is indicated in Fig. 33-4. A recommended plan of city traffic engineering organization is shown in Fig. 33-5. A major reorganization plan in Los Angeles, Calif., is reproduced in the plan illustrated in Fig. 33-6.

Other Locations. In a few cities traffic engineering has been made a part of the bureau of city planning. Sometimes the traffic engineer directs both the traffic engineering and planning functions of the city.

In other instances the traffic engineer is a bureau head under the director of planning. While planning is one of the principal concerns of the traffic engineer, this places him where his activities may be so limited

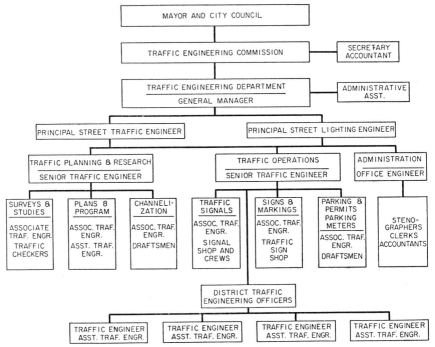

FIG. 33-6. Proposed reorganization plan for Los Angeles Traffic Engineering Department.

as to make him ineffective in many other areas. Since traffic is but one of many planning considerations, the traffic engineer can usually contribute more substantially to the over-all planning functions if he is in a separate department.

In large cities with large planning divisions, it may be desirable to have a traffic engineer work full time in the planning department. This, however, should not supplant the over-all traffic engineering function in the city, but merely supplement it.

In a few cities, special traffic committees or commissions have been created and the traffic engineer has been assigned to work directly under them. In general it is not a preferred plan, however, as carrying on day-to-day activities becomes cumbersome when reporting to committees and commissions.   Many large cities that have tried this system have abandoned it in favor of separate departments of traffic engineering.

In some cases the city traffic engineer has served under the public utilities commission or the department primarily concerned with mass transportation.   This, again, however, is too narrow an area for effective over-all traffic engineering.   It is frequently desirable to have a qualified traffic engineer work with the transit authority or similar agencies, but such traffic engineering should not replace a traffic engineering bureau for the city.

Combined Traffic and Parking Divisions.   Now that parking and terminal matters are of such paramount concern, many traffic engineers are being given full responsibilities for the development and administration of off-street parking programs.   These duties include the supervision and direction of controls exercised by the city over private parking facilities.

Adding parking to the other responsibilities of the traffic engineer has resulted in the creation of divisions of traffic and parking.   The traffic engineer becomes the principal agent of the city in dealing with off-street parking matters.   Such a system coordinates all of the basic traffic activities with parking and terminal activities, thereby giving the traffic engineer prime responsibility for both the movement and storage of vehicles in the city.   It is a sound pattern and one which promises to become increasingly popular and effective (Fig. 33-7).

33-3. Coordination of City Traffic Activities.   Many types of city traffic committees and commissions have been conceived for the purpose of coordinating traffic and safety functions, including traffic engineering. Basically, these committees or commissions are of three types: (1) they consist only of officials of city government; (2) they contain only appointed citizens and business leaders; or (3) they are a combination of officials and others.

With the first type, membership may be limited to heads of city departments having primary interest in traffic, i.e., the chief of police, the traffic judge, the city solicitor or prosecutor, the superintendent of schools, and the head of public transportation.   Ex-officio members may include the mayor, the city manager, the principal fiscal officer, and the fire commissioner.

The second type of coordinating body, which does not contain public officials, can logically have representation from the parents-teachers associations, the chamber of commerce, the merchants' association, the private transit company, civic clubs, women's clubs, truckers' associations, the local safety council, the press, and citizens.   Where the two areas of representation are included, care should be exercised to keep the membership small enough so that it can be an effective working group. It is usually desirable to include representatives from the judiciary, trucking companies, taxicab companies, insurance companies, and the press.

The most common types of committees or commissions have advisory powers only, as the Model Traffic Ordinances recommend. They serve principally to coordinate general traffic activities, to assist in safety education, to publish and distribute reports, to receive public complaints,

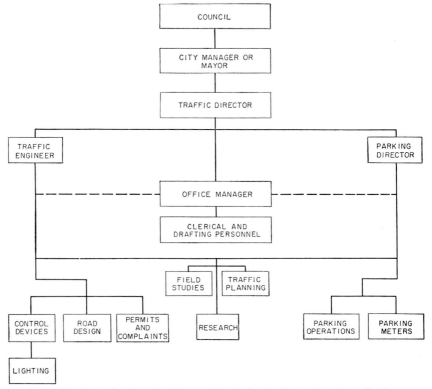

FIG. 33-7. Organization plan for combined city traffic and parking division.

and to *recommend* improvements. There are a few instances in which the traffic commissions or committees have been given powers and duties (by city ordinance or charter), such as improvement of traffic regulations, the review of traffic ordinances, and the regulation of traffic activities of public carriers.

From the traffic engineer's point of view, the advantages of a traffic commission or committee may be summarized as follows:

Effective coordination and integration of traffic activities can be achieved; better understanding of activities and needs can be a major objective; public information can be effectively procured and disseminated; needed changes in most traffic areas can be promulgated and sponsored; group opinions and thinking can be effected; some jealousies and inertia can be overcome within the over-all governmental structure; an outlet can be provided for traffic and safety propaganda; and most traffic problems can be studied and presented as community problems rather than as individual problems.

The traffic engineer is likely to find many objections to most traffic commissions and committees, particularly those which attempt to supervise his work directly. These bodies may become a "body of experts" and as such attempt to dictate basic traffic policies and objectives. They fail to distinguish between initial and long-range needs and views in local traffic situations. They may not recognize traffic engineering as a technical approach to the over-all problem. Biases and selfish views are apt to enter into the deliberations and decisions. Sometimes a strong personality will take over and so control the actions and decisions that an unbalanced program develops. Often the commissions are not inclined to accept compromises, even though the traffic engineer recognizes a compromise as a timely solution to a difficult problem. Certain traffic commissions and committees soon become indifferent, and interest reaches such a low ebb that there is no challenge to technical persons to do a vigorous job. Sometimes the traffic commissions inject a considerable amount of red tape and criticism, which tends to slow down actions. If the committee or commission is large, its activities may become awkward and unwieldy. There have been some glaring examples in which members of traffic commissions took peculiar views and ideas, asked impossible actions of officials, and then were critical of results. This situation develops very bad morale and tends to retard and upset even the best basic type of traffic operation.

Most practicing traffic engineers feel that the need for a committee or commission to coordinate a traffic engineering program is negated after the earliest beginning of a traffic engineering organization. In formative stages, the committee or commission can often act as a buffer between the public and the traffic engineer. It can become the "support group" for funds and powers which are needed to do an effective traffic engineering job. But once the desired plan of organization has been achieved, reasonable functions have been assigned, powers have been delegated, and financial support has been obtained, there is little that the committee or commission can add to the work of the traffic engineering unit. Strong executive leadership is the best support for the traffic engineering function.

33-4. Place of Traffic Engineering in Highway Departments. It is generally agreed that traffic engineering at the state level should normally be a part of the state highway department. However, there are a variety of schemes or plans whereby traffic engineering may be established within the structure of the state highway organization.

Separate Divisions. In approximately one-half the states, a bureau or division of traffic engineering has been established in the highway department.   The head of this division, usually the state traffic engineer, reports directly to the chief engineer (Fig. 33-8).   At headquarters he maintains a position comparable to that normally held by the construction engineer, the maintenance engineer, the engineer of design, and other principal department heads.   This gives him the ease of working with

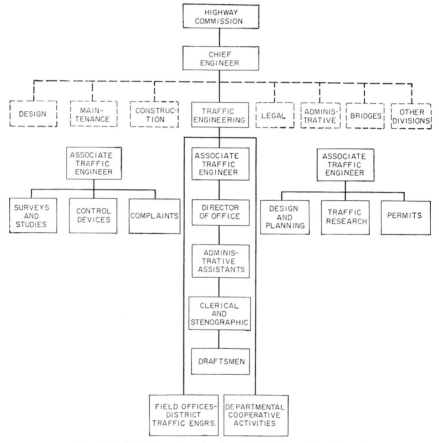

Fig. 33-8. Traffic engineering in state highway department.

all other divisions on an equal plane and makes possible the fast growth and expansion usually needed to keep pace with the growing traffic problems and needs.   Red tape is kept at a minimum, and all of the work of the traffic engineer can be expedited by direct contacts with chief administrative officers of the highway department.   A typical plan of organization for a state highway department with traffic engineering as a separate division is shown in Fig. 33-8.

A special study, which was undertaken by a Highway Research Board Committee to reorganize a State Highway Department, proposed "Traffic" as a key unit of the department (Fig. 33-9).

Because of the tendency to establish many departments and divisions in the rapidly growing state highway departments, thereby placing a great demand on the time and abilities of the chief engineer, there is a trend toward consolidating divisions for easier administrative direction.

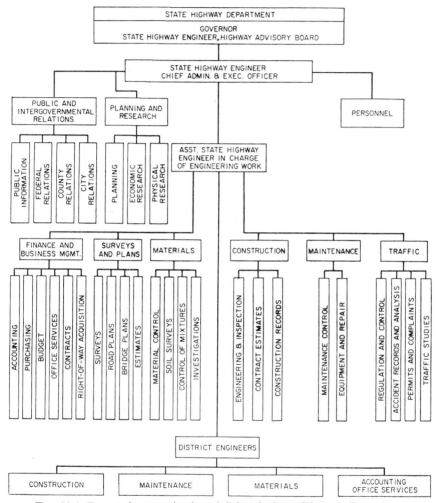

FIG. 33-9. Proposed reorganization of Colorado State Highway Department.

Under such a plan, traffic engineering becomes a part of a division of planning (Fig. 33-10). At headquarters level, the traffic engineering function in such consolidation might be included with such other activities as design, highway-planning surveys, and budgeting.

Part of Existing Division. In some states, the traffic engineering function is made a part of an existing division of the state highway organization, most frequently the maintenance division, because this division has responsibility for traffic control devices.   Also, the maintenance department is most likely to receive and act on traffic complaints, and in some cases is responsible for the collection and analysis of accident records.   The only objection to this system is that, as the traffic engineer is subordinated to an organizational level below that of other department heads, he may be denied sufficient flexibility and breadth of action to do an effective over-all traffic engineering job.

In a few states, the traffic engineer has been placed under the design department, but since this is only one of the many activities of the state highway organization with which the traffic engineer is concerned, it is

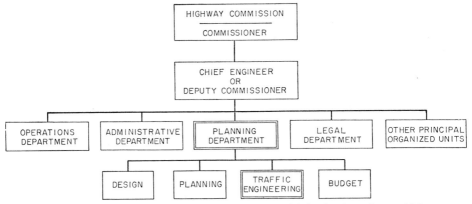

Fig. 33-10. Traffic engineering organization varies with basic plan of state highway department.

not the proper place for the total traffic engineering function.   It is only one of the many activities of the state highway organization with which the traffic engineer is concerned.

One or two states have assigned the traffic engineer to an administrative division of the highway organization, such as the office engineer, where he has leeway and authority to work throughout the entire highway organization.   This, however, does not give him the prestige and responsibility he has as a department head.

A few highway administrators have expressed belief that trained and experienced traffic engineers should be assigned to many divisions of the highway department.   Under such a plan, traffic engineers might be found in the design department, the maintenance department, the construction department, and in general administrative divisions, in addition to the operations or traffic division.   This is desirable provided it does not remove the traffic engineering work as a separate agency or division of the department.

Consolidated with Highway-planning Surveys.   In about 25 per cent of the state highway departments, traffic engineering has been consolidated with highway-planning survey activities to form a division of traffic and planning. In general this has been quite successful. While such units (Fig. 33-11) must devote efforts to collection of considerable data and to statistical analysis, they also have major operating functions.

A desirable plan of organization in most state highway departments entails creation of a division of planning at top staff level. This division would coordinate and direct planning in all areas, of which the department of traffic is only one. Each organizational unit would carry on its own planning functions and would provide basic information and data required by the top-staff-level planning unit. If this is done and if some of the other functions of the highway-planning survey organizations which do not deal with traffic were assigned to other existing agencies, such as the construction division and the maintenance division, then the remaining functions, which are largely those having to do with continuing collection of traffic facts and some traffic research, would very logically become a part of the traffic engineering unit.

In summary, it is desirable to have the traffic engineering work of the state placed in a separate division of the highway department in most states. For most effective results through traffic engineering principles and techniques and for the greatest efficiency and economy, this has proven to be the preferred type of state traffic engineering organization.

33-5. Traffic Engineering in Other Branches of State Government. Traffic engineering has not been limited to the state highway departments at the state level of government. In some cases a traffic engineer is found in the motor vehicle division and is usually responsible for motor vehicle inspections and for the examination of automotive equipment and accessories. He is not likely to have broad duties or responsibilities, but may serve as representative of the motor vehicle department in work with the state highway department, or a state traffic commission, or other divisions of state government.

A few state police organizations have employed traffic engineers, chiefly to represent the police with the state highway department in work of mutual interest and to make technical studies and investigations of accident records and reports submitted by personnel of the police.

If the state police are principally responsible for the collection and basic analysis of accident records, and where the records are housed with the police authority, it may be beneficial to assign full-time traffic engineers to work with the police in procuring accident data needed by the state highway department. The state of Virginia, for example, has found this to be a very effective means of using accident records and of maintaining working relationships with the police.

In a few states, traffic commissions or committees have been legally constituted and have been assigned broad powers to regulate traffic activities throughout the state. A traffic engineer is a logical person to

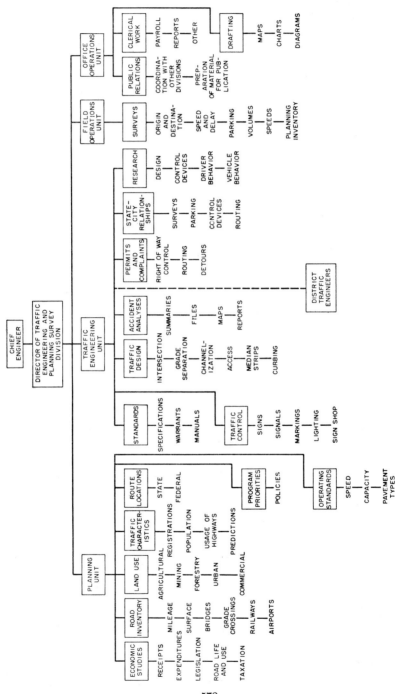

Fig. 33-11. Functional diagram of traffic engineering and planning survey as a joint division.

direct the technical activities of such a committee, which is usually constituted of the state highway department head, the state police head, and the motor vehicle commissioner. Where the commission has the right to approve traffic control devices erected by the various city and county governmental units, it is important to have a traffic engineer on the commission staff. The work of the commissions might also include the examination of requests for special regulations and for the applications of technical treatments, in which case a traffic engineer would be a logical member of the commission staff. Again, however, the assignment of a traffic engineer to a state traffic commission does not relieve the necessity for making traffic engineering a principal activity of the highway department.

Headquarters and Field Organizations. The discussion of state traffic engineering organizations thus far has been concerned primarily with headquarters activities. In most states, the work of the highway department is performed in highway divisions or districts with general control and direction by the headquarters unit. In some states, the highway organization is highly centralized, while in others a decentralized plan of organization prevails, with the district offices having important powers and responsibilities. If the headquarters organization is largely duplicated at the district or division level, a district traffic engineer should be provided in each of the field offices to assume responsibility for the same duties and activities as those assigned to the state traffic engineer.

In most states, adequate and effective traffic engineering jobs can be achieved by limiting the powers of the state traffic engineers largely to staff functions. Under such a concept, the importance of district traffic engineers would be minimized. If the major traffic engineering activities cannot be achieved solely through staff responsibilities, and subordinated line responsibilities are given the traffic engineer, then it becomes necessary to have adequate district organizations to carry through the required line activities.

The traffic engineer of many states can perform some line activities very effectively with only a headquarters organization. Traffic surveys and studies can be conducted with field parties reporting directly to headquarters, since work on planning and design largely involves activities at headquarters. Line work is required principally on signs, signals, marking, and the application of traffic regulations.

County Organizations. Some large urban counties throughout the country have traffic and highway problems which closely parallel those of large cities. Some have conditions which are similar to both city and state problems. In these counties, where many of the principal traffic arteries are under the jurisdiction of the county road agency, it is desirable to establish traffic engineering as a principal function of county government. Since the organization structure of county highway departments usually parallels closely the plan of the organization in state highway departments, there is little need to differentiate between the place of the traffic engineer in the counties and the traffic engineer at the state level. A county traffic engineer should in most instances be more thoroughly grounded and experienced in urban traffic affairs than is necessary with some state traffic engineers.

Other Traffic Engineering Assignments. The work of the traffic engineer is not limited to governmental agencies. Public authorities and commissions frequently employ traffic engineers if they have as their primary responsibilities traffic and parking matters. It is common to find a traffic engineer the director of a city parking authority. Other public authorities, such as the Port of New York Authority, and bridge authorities and commissions seek the services of traffic engineers. With the widespread application of the public authority in traffic and highway areas, opportunities for traffic engineers have greatly increased.

Consulting engineers employ many traffic engineers, so that they can provide traffic engineering services and make available traffic engineering skills and techniques in other areas of highway engineering work.

Business groups employing traffic engineers include insurance companies, banking interests, transit companies, and retail stores.

The Federal government, particularly the U.S. Bureau of Public Roads, is concerned with traffic engineering matters and employs many persons trained and experienced in traffic engineering.

Automobile manufacturers, traffic and safety associations, foundations, automobile clubs, and research organizations are among the other groups employing traffic engineers. The increasing applications of traffic engineering and the widespread employment of traffic engineers typifies the breadth of interest and the scope of traffic engineering work.

33-6. Financing Traffic Engineering Activities. The cost of providing traffic engineering services varies with the size of the department and the delegated functions and responsibilities. It is fortunate for the traffic engineer that because of the nature of the work in which he is engaged, expenditures can be supported and justified in terms of savings which are effected. Although it is difficult to assign accurate monetary values, it is known that a reduction in accidents, a decrease in delays due to traffic congestion, and other improvements which might come directly from traffic engineering approaches have values that are easily recognized by the public and by public officials and that these values bear a relationship to the cost of providing the correctives. Increasingly it is being recognized and acknowledged by public officials that traffic conditions are of such magnitude that they justify substantial outlays of public funds. Public-opinion polls invariably indicate that traffic is one of the topics which is foremost in the minds of the public and the voters of most communities. Thus it is natural for public officials to feel that it is no longer a case of being "able to afford traffic engineering" but rather one of not being able "not to afford it."

The downtown invested interest can rightly point out that accessibility, parking, and other traffic matters are primary concerns of their welfare and future stability. It then follows that the large proportion of the total tax load which they carry should at least in part be used for improving the traffic situation.

The highway users who pay large sums of taxes, who contribute to parking meters, who pay for off-street parking, and who pay tolls for operations on certain types of facilities have a right to ask whether or not adequate provision is being made for safe and efficient movements.

Commercial vehicle operators know that the efficiency of their business is related in large degree to the freedom of movement on streets and highways, and they can demand proper public actions to create the greatest efficiency in the movement of goods and purchases.

Traffic engineering heads should be constantly on the alert for examples to show that the work which they are accomplishing is entirely justifiable from the standpoint of quality and services rendered. They should use before-and-after accident studies, before-and-after traffic surveys, objectives indices, and all the other factual measurements which are available to them.

Traffic engineering budgets, of course, vary greatly. In cities with only staff activities to perform, work is largely carried on by a single traffic engineer at a cost of perhaps only a few thousand dollars. In large cities where major items such as street lighting and some physical roadway improvements are a part of the traffic engineering budget, expenses may be many thousands of dollars. There is also a great variation in the size of state budgets, because of the division of staff and line responsibilities in establishing traffic engineering functions. Typical traffic engineering budgets for cities are shown in Tables 33-1 and 33-2. Traffic engineering expenditures for states are illustrated in Tables 33-3 and 33-4.

The traffic engineer must acquaint himself with standard budgetary procedures, and he must exercise the necessary controls over the expenditure of funds. The pressure of certain types of traffic improvements sometimes makes it difficult to adhere to budget allocation, as do unexpected requirements for the replacement of equipment and devices and for major improvements. However, the traffic engineer must anticipate need within reason and must follow the priority lists, time schedules, and general budgetary controls available to him. The whole traffic engineering program depends upon a sound plan of financing and the balancing of revenues and expenditures. Whether or not the budget is of the lump-sum type, segregated type, or the allotment type, the traffic engineer should acquaint himself with all the requirements and should endeavor to administer carefully his over-all program of work so as to perform

TABLE 33-1. TRAFFIC ENGINEERING BUDGET FOR CITY OF 600,000 POPULATION

Personnel Services
A-1	Salaries of regular employees	$ 69,840.00	
A-2	Wages of regular employees	31,500.00	
A-9	Provision for salary increases	3,488.00	
	Total Personnel Services		$104,828.00

Contractual Services
B-2.4	Drayage, freight, express, and demurrage	$ 20.00	
B-2.6	Traveling expenses	500.00	
B-3.3	Photographing and blueprinting	650.00	
B-3.4	Printing, duplicating, and multilithing	200.00	
B-4.1	Fire and extended coverage insurance	42.00	
B-4.7	Surety, forgery, and burglary bonds	18.00	
B-5.2	Electric light and power (traffic lights)	67,560.00	
B-5.4	Telegraph	10.00	
B-5.5	Telephones	175.00	
B-6.5	Repair of operating, maintenance, and construction equipment	250.00	
B-6.6	Office-equipment repair and maintenance service	25.00	
B-7.3	Rent of automotive equipment	4,373.00	
B-9.3	Dues and memberships	85.00	
B-9.10	Contract work not otherwise specified	1,500.00	
	Total	$ 75,408.00	

Contractual Services (Street Lights)
B-5.1	Gas	30,000.00	
B-5.2	Electric light and power	470,000.00	
	Total	$500,000.00	
	Total Contractual Services		$575,408.00

Commodities
C-1.1	General office supplies	$ 950.00	
C-2.4	Cleaning, toilet, and sanitation supplies	50.00	
C-2.5	Concrete and aggregate materials	75.00	
C-2.8	Fuel (other than motor vehicle)	75.00	
C-2.9	Household and institutional supplies	75.00	
C-2.10	License plates, badges, tags, etc	1,200.00	
C-2.12	Lumber, wood products, etc	750.00	
C-2.14	Minor equipment, hand tools, etc	25.00	
C-2.17	Paints and painting supplies	12,000.00	
C-2.18	Plumbing, sewage, and drainage supplies	200.00	
C-2.20	Structural steel, iron, and related metals	2,400.00	
	Total Commodities		$ 17,800.00

Capital Outlays
E-3.2	Engineering and scientific equipment	$ 1,000.00	
E-3.8	Office equipment	573.00	
	Total Capital Outlays		$ 1,573.00
Total			$699,609.00

TABLE 33-2. EXPENDITURES FOR SMALL CITY TRAFFIC ENGINEERING DIVISION

Salaries...		$6,400.00
Traffic engineer...........................	$4,500.00	
Secretary.................................	1,900.00	
Office supplies and equipment...........................		200.00
Drafting and photographic supplies and expense..........		250.00
Transportation expense................................		300.00
Depreciation on equipment (reserve)....................		375.00
Expense account (public relations).....................		200.00
Total...		$7,725.00

TABLE 33-3. BUDGET FOR SMALL STATE TRAFFIC ENGINEERING UNIT (FIRST YEAR)
(Centralized operation is indicated)

Expenditures

Personnel
  Headquarters

State traffic engineer...	$ 6,600
2 Assistant traffic engineers......................................	8,400
8 Engineering aides...	26,880
2 Draftsmen..	6,960
1 Stenographer...	2,400

  District

1 Assistant traffic engineer.......................................	4,200
Total Personnel...	$55,440

Equipment

8 Executive-type desks, chairs.....................................	$ 1,600
10 Filing cabinets...	500
Drafting equipment and supplies.................................	600
5 Drafting tables...	250
Typewriters..	150
Office supplies...	1,000
Traffic control devices..	20,000

Special Surveys

Miscellaneous Expenses—library, traveling expenses, publications, telephones...	5,000
Total equipment special surveys, miscellaneous....................	$41,100
Total expenditures...	$96,540

TABLE 33-4. SUMMARY OF EXPENDITURES OF STATE TRAFFIC ENGINEERING UNIT

Salaries...........................	$140,000	
Field		
Vehicle operation....................	3,600	
Personnel expense accounts..........	13,000	
Office supplies......................	2,000	
Public relations.....................	2,000	
Total...........................	$160,600	

within budget restrictions. Of course, there must be a distinction between the operating budget and capital-improvement budget.

Many state traffic engineering functions have been approved for financing under the earmarked *planning fund* which has been available to states for a number of years as a part of the Federal grant and aids. Recently, this amount has been 1½ per cent of the total Federal-aid allotment to the state. While these planning funds were basically intended to support the work of the highway-planning survey, the activities of traffic engineering are so closely related to the highway-survey objectives that they fall clearly within the scope of activities approved for the use of such funds.

In some cities where ordinance or state law requires that parking-meter revenues be used for traffic improvement purposes, it has been held that traffic engineering is a proper function to charge against meter receipts.

In most instances it is unnecessary to find a list of earmarked revenues to support traffic engineering activities. Traffic improvements are considered such an important part of the over-all responsibility of government that they can be readily justified from the general funds of the city, county, or state.

33-7. Extending the Program of Traffic Engineering. If traffic engineering activities grow only in proportion to the increase in traffic volumes, a substantial growth is indicated. In most cases, the past development of the traffic engineering bureau has lagged behind the need to such an extent that a growth much out of proportion to the traffic growth is possible in the future. It is the responsibility of the traffic engineer to determine and support the proper rate of advancement and growth to meet the needs and, at the same time, to maintain sound and stable development of the department. There are many administrative tools which can be used in this connection.

Departmental Records. Numerous factual and statistical records can be prepared, varying from simple-trend survey to complex estimates of highway and traffic needs. It is generally a sound policy to prepare complete reports for departmental distribution and for public dissemination, on major achievements of the traffic engineering department. Such reports have definite sales values, and they have lasting values as reference. They also have great educational value for other members of the government, executive heads, and elected officials. In certain instances, these reports can have a very positive influence on the public in general.

Staff Meetings. The traffic engineer should endeavor to present at staff meetings of highway-department heads or of city officials interesting and objective reports on the activities of his division. He should use such meetings as a medium for exchanging ideas and plans between his department and other departments, and he should endeavor to achieve the greatest possible cooperation and sponsorship of his objectives by other departments.

In large traffic engineering operations, staff meetings are quite desirable within the traffic engineering units. The exchange of ideas and the opportunities which such meetings provide for discussing problems and programs can be invaluable to the over-all traffic engineering program.

In-service Training. The growth of the traffic engineering department, if it is to be sound, cannot exceed the abilities of the staff available to carry on the functions at all levels. Because there is a lack of qualified and experienced traffic engineering in most sections, owing to the newness of the profession, it is essential to establish sound in-service training programs.

The traffic engineer should be alert to procure all available training aids, including a library of traffic literature, the preparation and dissemination of memorandums, the procurement of factual data, and visual aids that can be used in augmenting the abilities of personnel in his department.

Research. Research is important in accelerating the over-all traffic engineering program. It can become costly both in time and money, but there are so many areas of traffic work in which conclusive answers are not available that it is important that every effort be exerted by traffic engineers to carry on research programs of both the pure and applied types.

Research endeavors are valuable in relations with other divisions of state or city government. The traffic engineer has many opportunities to carry on research projects jointly with other groups. In some cities the traffic engineer collaborates and pools resources with the planning division in certain research areas.

The traffic engineer in collecting materials from other agencies should be alert to research results, and in turn he should make available through technical publications, speeches, or otherwise significant means results of his own research endeavor.

Traffic Engineering and the Police. If the traffic engineer does not maintain close relations with the enforcement group, a sound program of over-all traffic control and improvement cannot be achieved.

It is of utmost importance that the traffic engineer maintain a day-to-day working relationship with the police, whether his functions are those of a city, county, or state government. Practically all activities of the traffic engineer that have to do with control devices or with regulations involve the police. In most cases the police need technical assistance and advice from the traffic engineer on the installation and operation of the traffic control devices. They need the help of the traffic engineer in channelizing and controlling traffic in complex and irregular intersections. The traffic engineer should measure the need for parking regulations and for assisting parking treatments, such as the installations of meters. However, the success of regulations and parking devices is dependent upon the cooperation of the police.

In such matters as speed control, the traffic engineer can be of great assistance to the police, particularly in the establishment of reasonable speed limits through zoning and in measuring the actual speed conditions. The traffic engineer can be especially helpful in overcoming the strong "restrictive" views held by the police with regard to such actions as speeding. For example, slow-speed drivers are in some cases more hazardous than fast drivers. The traffic engineer has available to him survey techniques and traffic data which can result in the proper evaluation of such matters.

In many instances the police actually furnish the traffic engineers manpower and other assistance in the conduct of surveys and studies which have mutual interests or activities.

In accident investigations the traffic engineer must work very closely with the police to be sure that the type of information which is most useful to him is available from police records and reports.

There are numerous areas in which the police have the right to expect financial aid and assistance from the traffic engineer, including (1) plans for special events, (2) selective enforcements, (3) issuance of permits, and (4) applications of technical or scientific approaches to traffic problems.

Managerial Essentials. The essentials for sound management and executive actions are in general:

1. A definite objective or goal should be established with complete plans for proper stage development. As this development takes place, a proper balance must be maintained throughout the department and a natural sequence of growth should be the aim. Functions should be added as the department grows. The plan should be modified and changed in the light of changing conditions and requirements.

2. Proper distinction should be made between headquarters and field activities. Centralized policy decision must be combined with decentralized administrative execution of the decision. Also, in traffic engineering it is difficult to distinguish between staff and line functions from the standpoint of what various people do. It is not important to make these distinctions, but it is important to have a balanced line and staff organization.

3. Public relations must be constantly maintained. This responsibility cannot be easily delegated and always requires substantial effort from the traffic engineer himself. It is an activity which is essential to every organization, and it should be constantly emphasized that propaganda and advertising are not sound public relations. Properly informing the public and seeking their support is the basic objective.

4. Over-all planning should include all activities of the traffic engineer-

ing department. Everyone in the organization is involved, and a constant review of plans and activities is necessary.

5. Performance should be measured, both for sound planning and for over-all support of the traffic engineering program.

6. Authority and responsibilities should be delegated as the traffic engineering function grows, so that the traffic engineer need not concern himself with all details and facets of the operation.

7. Close supervision and control must be maintained over all personnel and all activities of the department.

8. Basic factors in executive leadership must be recognized. Qualities and abilities of the individuals, actions and strategy involving all units of the department, political influences, and judgment are important keys in a leadership and direction.

9. Optimum size is a factor which all traffic engineers should consider in developing and expanding their activities. A traffic engineering unit may become so large that it is ineffective or inefficient.

10. Managerial relationships in traffic engineering involve so many different groups and activities that the traffic engineer must work closely with public officials, including the city council and the Board of Aldermen on the municipal levels.

CHAPTER 34

TRAFFIC FACTORS
IN ADMINISTRATIVE DECISIONS

Traffic engineering techniques and data are becoming increasingly important to street and highway officials in objective decisions and in policy formation. As highway needs grow rapidly and greatly exceed the resources available to meet them, factual evaluations become especially significant. Traffic factors provide the best basis for proper distribution and application of available funds and manpower, and perhaps equally important, they support requests for additional resources.

In the early years of automotive transportation, road needs were apparent and were largely measured by the mileage of unpaved roads. Today, since construction of a "basic" system of all-weather roads, road needs are not so apparent to the public, and support is urgently needed for additional highway aids. Farm blocs or rural-road groups are primarily interested in secondary feeder and land-service roads. Urban groups want street widening, bypasses, and expressways. Trucking groups want supports aimed principally at the primary rural-road system. National defense needs, military and evacuation, favor concentrations of highway efforts on certain key routes. These factors, plus the conflicting views of road users, land owners, business competitors, civic interests, and others, produce a complex picture of highway needs. General tax increases, diversion of road-user revenues, large increases in road costs (both capital and maintenance), and strong opposition to additional motor vehicle taxes further add to the problem of highway needs. Administrative decisions become more difficult, and legislative actions may become uncertain and inadequate.

Many of the data collected by traffic engineers and by state highway-planning survey agencies have direct application in administrative consideration of highway needs. The highway-planning surveys were developed in the thirties with this as a primary purpose. They provide traffic inventories, including blanket traffic counts; traffic classifications; loadometer records; vehicle-performance studies; physical roadway inventories; detailed base maps; highway and transportation maps; route-improvement maps; special road-use maps; traffic density diagrams and road-life studies; fiscal data, including such things as cumulative costs, tax rates and trends, and highway income. Facts compiled by these organizations are especially helpful in determining roadway needs and in projecting such needs. The many special studies and researches, including such things as speed and placement, road-use practices, truck performance, and accidents, undertaken by the planning survey units, usually with Federal-aid funds, are also useful to administrators. The comprehensive studies made of traffic origins and destinations, parking, and mode of travel are well known in highway and traffic fields. The O-D data are proving useful in almost every type of route planning. They are perhaps the most valuable of the traffic surveys in planning and administration.

The information accumulated through these day-to-day activities is providing the objective basis for key administrative decisions in state highway departments.   Their values are also recognized in urban traffic and street work.   Through the efforts of traffic engineers, and in some cases with the aid of the state, continuing traffic volume data and other planning data are being collected in cities.   Both immediate and long-range highway- and street-improvement programs are being developed.

34-1. Needs and Evaluations. Traffic facts are put to much use in studies of highway needs and in the evaluation of highway adequacies. Since World War II many highway-need studies have been made, often at the instigation of state legislative bodies, seeking to establish programs of short- and long-range needs.   A fiscal study is usually made concurrently to establish proper and equitable means of meeting the needs.

Most needs studies involve:

1. A general analysis of the historical development of highways in the state, emphasizing the plan of organization, operations, and road-system classifications, and reviewing intergovernmental relations.

2. An analysis of financial conditions and plans in the highway department, examining sources of all revenues and determining distribution of revenues by types of roadways for state and local areas (state, county, city, and others).

3. A discussion of the relationships of highway transportation to the state's economy.

4. Review of traffic and travel trends on the different road systems, as a basis for projecting travel growths.

5. An evaluation of services rendered by existing streets and highways, using accidents and congestion to show standards of service.

6. Classification of routes according to services performed, indicating relative responsibilities of governmental agencies involved.

7. Definitions and relative values of various classes of roads.

8. A program of road needs and priorities, relating desirable standards to existing routes and services to determine inadequate facilities and using life expectancies to show estimates of replacement needs. In addition to capital-expenditure needs, maintenance and administrative needs are anticipated, and cost estimates are prepared for different classes of road needs.

9. A suggested program of management and administration.

In addition, highway-sufficiency studies might include a plan for highway-department reorganization, covering such things as type of organization, assignments of principal functions, job classifications, personnel policies, and salary schedules. Economic justification of proposed improvements might be shown by relative costs to road-user benefits and to other benefits. Some surveys have established sufficiency ratings to obtain a numerical comparison of needed improvements.

Economic Comparisons of Routes. A balance must be maintained between revenues and costs in considering benefits, yet theoretical economic values should often be tempered by other factors of highway economics so complex that it is impossible to be validly conclusive in broad areas. Facts must be acted upon with reason. Factors are numerous because highway transportation is related to so many non-highway economic values. No phase of modern life is unaffected by highway transportation. For example, a roadway in western New York might have significant effects upon the port of New York. It follows that all highways cannot be solvent when solvency is measured by *direct* benefits. Many indirect and intangible benefits must be reckoned, and collecting this information is involved and uncertain—yet challenging. Traffic and highway officials reduce all possible measurements to monetary values for a common unit of comparisons.

A primary intent is to show economic dividends from funds invested. Another aim is to show whether or not highway incomes are distributed so as to produce greatest returns in needed services. But, again, the services are extensive and both direct and indirect. All services may not be supplied, public desires may not be met, but the highway official cannot provide beyond the willingness, or ability, of people to pay. Early policies of pursuing road-building programs on hit-or-miss personal opinions are giving way to sound economic procedures. Traffic and accident data are a principal plank in economic analyses and comparisons. Most economic studies deal with highways as they serve traffic.

As indicated previously, revenues must balance costs, but costs must be supported by benefits to maintain public sanction and support. In considering the economic values, it must be realized that highway transport differs from other general transportation systems in some aspects. Principally, the vehicles are owned and often operated by the beneficiaries. Some highway revenues come from other sources than the users. In highway transport, private and public benefits go far beyond pure transportation benefits; they apply to government, to road users, to the public, and to property owners.

Some of the economic features of highway administration are (1) selection of highway routes and priorities; the highway system must be expanded along sound economic lines; (2) equitable assessment of charges to persons and agencies benefited, including different classes of vehicles:

(3) economical design and construction of roadways; (4) proper allocation of funds to all the different classes of streets and highways.

Benefit-cost Analysis. The benefit-cost analysis is one of the most used economic measurements in establishing priorities, solvency, and general comparisons. User benefits include time savings, mileage savings, reductions in operating costs, reductions in accidents, and even intangible benefits (ease of driving, uniform speeds, fewer stops, etc.). They are reduced to monetary values, and even though the unit values have not been well established, consistent reasonable values permit good comparisons, if not good absolute findings. On the other hand, costs are more accurately determined and include the actual estimate of providing various levels of traffic services on a given project or costs of a particular project.

The benefits, or savings, have no relationship to the revenues to be derived, and they do not add to the public treasury for road building. However, when related to costs or required expenditures, they provide a sound basis for rating projects and for projected improvements.

Common values used in benefit-cost comparisons are given in Chapter 2, where the ratio is discussed as a means of comparing construction projects. Because of wide variances in cost and income per vehicle-mile, a ton-mile basis is often used to reduce variations. Detailed breakdown of vehicles in a given traffic stream by type is obviously necessary in calculating the benefits.

Cost-income Ratios. The annual income derived from a highway facility can be related to the annual costs to determine whether or not it is a self-liquidating project. Such calculations and comparisons deal with actual income values, i.e., fuel taxes, tolls, registrations (apportioned). The costs represent average annual values covering capital charges and interest, maintenance, and operating charges.

With increased emphasis on toll roads, more use is being made of cost-income comparisons in rating projects and in fixing highway needs. Obviously, lightly traveled roads are not found to be self-liquidating with this approach. Variables are numerous, because the useful life of a facility is debatable, apportionments of some incomes are difficult, and nonuser contributions to the incomes are ignored.

Indices and Ratings. In addition to the sufficiency index and ratios discussed, other numerical values are used to show needs. Most of these are derived from arbitrary assignments of numerical values to important highway and traffic components. The figures have little value except to position projects relatively and thereby provide the chief administrator a somewhat objective approach to major highway decisions.

Special Highway Problems. Defense highways, evacuation routes, military roads, and other special-type highways involve the use of traffic data to ensure that they are properly located and designed for adequate capacity. The U.S. Bureau of Public Roads, working with state high-way agencies, has put traffic volume, origin and destination, and com-mercial-vehicle-loadings data to good use in planning and designing such special routes. An outstanding example of the application of traffic facts to a special highway problem was the planning of the Interregional High-way System. It was based largely upon the travel-desires and road-needs data of the state highway-planning surveys. The programming of construction on this system takes into account the type and magnitude of traffic assigned to each section.

Innumerable problems arise in the development of a highway program that challenges highway officials who are required to make decisions. What effect will a bypass have on businesses along existing routes and on the central business district of the city? Are controlled-access highways harmful to land values? How can funds be properly distributed between the various systems and classes of roadways? These are typical of the questions that can only be measured by traffic and economic studies, and many interesting findings are available.

34-2. Development of Standards. Traffic engineering techniques and the data they produce are invaluable in developing standards for roadway improvements. Most standards are for the purpose of providing safe, efficient travel for the present and for anticipated traffic loads, as eco-nomically as possible. There is naturally an optimum value for the standards: when too high, the investments may exceed traffic needs, resulting in excessive costs. When too low, hazards, delays, and exces-sive maintenance costs develop. The objective is to derive minimum standards which will accommodate essential service requirements well over a reasonable roadway life span.

The traffic characteristics which the traffic engineer can ascertain are essential in establishing standards for sight distances, curvature, align-ment, grades, roadway widths, and even some structural aspects. The value of controls over entering and intersecting traffic, relations of volume fluctuations to road capacity, accident expectancies, and over-all effects of road conditions on economic and social conditions of the state or city are other factors of direct concern to the traffic engineer. Careful analyses of traffic growths and of potential growths are a major responsi-bility in forestalling a repetition of inadequacies in road standards which have been so conspicuous in the past. Long-range forecasts of traffic growths are difficult because they involve changes in over-all economic conditions, shifts in populations and businesses, and competitive modes of transportation. The traffic engineer now relies on population trends and vehicle-ownership ratios more than on a projection of past travel trends to forecast future traffic conditions and to provide thereby figures on which to base highway standards.

Benefits to road users and to others play a part in the development of standards; for example, it has been shown that average speeds on urban streets in the interregional highway system are about 20 mph and that by increasing their speeds to an average of 35 mph, benefits to motorists in time savings alone would amount to substantially the cost of the necessary improvements.   However, while such benefits should be determined and considered in preparing standards, they do not represent real fiscal values in the financing of roadway improvements.

Before-and-after studies of accidents and the approximation of accident expectancies are essential to development of sound standards.   Measurements of speeds and the determination of speed desires are necessary to avoid overrestrictive conditions in adopting standards.   Effects of various loads and volumes of commercial vehicles on pavements and structures are difficult to determine and involve many controversial approaches, but such knowledge is needed to develop proper roadway and vehicle standards.

The American Association of State Highway Officials and the U.S. Bureau of Public Roads take the leadership in developing and promulgating standards.   These agencies give full consideration to traffic characteristics and to service requirements in establishing the standards.   To a great degree, basic data must come from the state and local traffic agencies.   Some standards for roadway design are given in Chapters 24 to 28.

34-3. Intergovernmental Relations.   The demand for highway transportation does not generally relate to political boundaries, but in the areas of finance and regulation there are many interrelationships between states, counties, and cities.

With approximately one-half of the total automobile travel (about 600 billion vehicle-miles per year) on city streets, it is apparent that large expenditures are required to achieve necessary roadway developments and modernization.   Out of highway development, several distinct road classifications have evolved.   Town and city streets may be a separate road class in some states, and they are almost always considered apart from other roads in allocations of state funds and, in some instances, in traffic controls and regulations.

The expeditious and safe movement of vehicles in cities has become a major problem, recognized by state, city, and Federal officials.   While some relief can be provided by control measures, lasting improvements can come only from corrective physical actions.

The shift in character and priority of highway-improvement needs from rural to urban locations is generally recognized.   Emphasis was given to this fact by the 1944 Federal Aid Highway Act, which specifically earmarked some Federal-aid funds for expenditure in urban areas.   This has had far-reaching effects in the affairs of state and local governments. In many instances, the relationships had been so slight, prior to passage of the Federal Aid Act, that no effective program of joint activity was apparent. It is obviously necessary to have cooperative planning and coordination to develop the specific area and responsibility of the state in urban work, to prepare uniform policies and procedures, to develop design standards acceptable to both agencies, and to effect the proper degree of state control over traffic regulations. An adequate network of roads and streets throughout the country cannot be achieved except by close state and city relations, since city streets serve as continuations of state highways and are thus essential links in the over-all road system.

Basis for State Control and Action in Cities. Under the basic plan for government, cities are agents of the state. Even though there have been many waves of insistence for local control over government—"self-government"—and the clamor for "home rule," it has always been found desirable and necessary for the state to maintain a high degree of control over basic municipal functions. Traffic may first appear to have only municipal interest, but soon becomes a matter of state-wide concern and must be controlled, regulated, and aided by the state. It is, therefore, necessary for the state to extend controls in cities from the basic items of taxation and borrowing powers to many other day-to-day affairs.

Such controls can be exercised either as *legislative controls* or as *administrative controls*. Under the first plan, the state assigns a sphere of action to the city, within which the state government may not legally enter and in which the city is free to act without state controls. However, by assignment in the state constitution (home rule) the legislature may be prevented from interfering with certain matters specifically entrusted to cities. The extent of such control and the manner whereby it is exercised vary greatly among the states. In general, the courts and other actions have had a tendency to keep the controls narrow. This has not been the case with regard to traffic and highway matters. In recent years, there has been a strong trend from legislative to administrative controls. Under this plan the legislature grants broad authorities to administrative agencies of the state to exercise certain controls over affairs of the municipalities. It has the advantages of minimizing political interferences, and it places the controls in the hands of agencies that are capable of developing professional and expert assistance for cities.

The control of traffic and the development of a more complete roadway system involve both legislative and administrative techniques. The state highway department is the logical party to exercise most state controls over highway and traffic affairs in cities. Authorizations for work by the state highway organizations come principally from constitutional amendments and, most commonly, from acts of the legislature.

In a few states, the constitution describes in detail the activities of the highway department and fixes its responsibilities in the cities. This method is not very flexible, and it is difficult to make changes in line with changing conditions and unusual needs. In almost every state, highway departments are given less right to exercise authority in cities than they have for road development and traffic control in rural areas.

The extent of benefits and aids to cities and of the control and regulation of traffic varies markedly. Sometimes there is considerable variance in cities of different size within the same state. In traffic control, the rights of the state may be entirely or principally limited to streets used as state highways. In the matter of financial aids and physical work on the roadways, aids may be limited to primary state routes, or to any state route, or in some cases they may cover all city streets. Whether the funds or aids are given to the cities to apply and administer or are applied directly by the state, the eligibility of the streets on which they can be used is usually prescribed by administrative actions if not by specific legislation. Occasionally there is a specific contract for a single activity. Often it is found that the state will enter into continuing agreements with the cities relative to action therein. With its power to give or deny certain financial and other aids to cities, it is found that the state frequently in the agreements requires the cities to grant it certain powers over traffic control and operations which may not be specifically granted by legislation. These are usually in the best interest of the public, and no basic fault can be found with them.

Extent of Controls. States can provide substantial aids to cities in highway matters. Through these aids they are able to exercise strong controls over certain municipal activities. In some instances, states merely give to the cities funds which are to be used for highway purposes. Sometimes the exercise of the power over the use of these funds is substantial; in other cases, the state is simply the banking agent. In addition to the granting of funds, the state may provide certain construction in cities or may maintain routes which are used as state highways or even other routes. In some instances, they provide traffic control devices and traffic equipment, such as pavement markings and signs. It is apparent that the state can exercise controls over the standards of construction, the numbers and location of right of ways, the levels of maintenance, and the type of traffic control equipment both as to operating characteristics and needs.

Allocations for construction activities in cities may be fixed by legislative formulas, or they may be based on decisions of the highway commission or some other executive officer. In the latter case the traffic engineer and others engaged in traffic work are usually called upon to fit the needs into over-all plans and to relate the needs to traffic requirements. The basic tools of traffic measurement and the projection of traffic needs into the future by the traffic engineer are highly useful in the determination of proper allotments of activities, materials, or equipment.

In general administrative fields with basic legislative controls, the state may only provide substitute administration in construction and maintenance activities in which the state participates or for which the state provides funds.   On the other hand, the state may actually inspect the work and may supervise the filing and checking of routine reports, give general advice and technical services, approve construction plans and specifications, aid in development of minimum technical standards, local audits over the use of funds, and the proper allocation of funds, and many other special requirements having to do with the caliber and accuracy of the activities which are within the scope of control by the state.

In traffic operations, the state administrative supervision and controls exercised might include traffic ordinances, traffic control devices, and general surveys and services.

Traffic Controls by State.   It has been generally held that the responsibility for traffic control on public ways rests primarily with the state. Through legislative and administrative processes, broad controls are maintained, leaving the regulations of details to the localities.   State control improves uniformity in regulation, enforcement, punishments, and general standards of road use.   Public benefits can be measured in terms of reasonable regulations, simplicity of control devices and regulations, and a proper balance between movement and safety objectives. In some states, it has been found that centralized purchasing results in economic savings in the procurement of control devices and traffic control equipment.   Some equalization of rural and urban road interests can be achieved through traffic services and traffic control.   In some instances, the traffic services provided by the states are incidental to more basic objectives, such as the planning, location, design, construction, and maintenance of roadways.

The most common type of traffic service and traffic control exercised by state highway departments in cities is the erection and maintenance of route markets on streets used as state highways.   This can be extended to include all types of signs and markings except perhaps those having to do with parking.   In some instances, the pavement markings are placed and maintained by the state.   Some states only maintain the markings which have to do with moving traffic, such as center lines and lane lines, leaving the markings of parking stalls to the municipality.   In some states, the highway department provides traffic signals in cities on streets used as state highways.   In a few instances, the state maintains the signals but usually presents the signals to the cities for maintenance and operation.   The provision of street lighting on urban extensions has been a very controversial phase of state participation.

Organization for State-City Relations.   Most state highway departments are creating special units or positions charged with the sole responsibility for developing and supervising state-city relations.   Many of the state and city highway problems and activities are so new that management procedures and organization plans frequently appear haphazard and inadequate. In some cases, it is not deemed necessary to provide a specialized unit to handle work in cities. In these cases, it is usually found that the over-all organization of the state highway department is capable and adequately staffed to take on all of the responsibilities which are required for an effective relationship. The job, however, is not simple, and it must be recognized by both state and city officials that the complex relationships which they should have in highway and traffic matters require the establishment of a sound plan of organization to deal with them. Of principal importance is the appointment of personnel experienced in city traffic and urban street standards to the highway organizational unit to perform functions in the cities. On the other hand, the cities should delegate authority to existing engineering units, or should establish new units and authorize them to coordinate the activities of all city agencies involved in work with the state on highway and traffic matters. Increasingly, the traffic engineer is assuming principal responsibility for negotiations, both for the state and for the city in the many matters involving both jurisdictions. This is logical because so many of the decisions and so many of the needs have a direct relationship to traffic matters.

Consideration should be given to the development of clearly defined and uniformly applied working procedures. Wherever possible, it is wise to leave as many administrative decisions and final actions at the local level as can be justified.

For sound working relations, the state should take the lead in holding state-wide meetings with city officials to discuss future problems. The state should keep close contact with cities in formulating policies and procedures. The state also should make use of manuals and other guides for local officials which outline procedures for carrying on the cooperative work.

In conclusion, it is apparent that for sound working procedures (1) the state, through the state highway department, must assume a large share of responsibilities for major highway improvements and for traffic betterments in cities; (2) the problems must be looked upon as mutual problems with the desires and limitations of each agency acknowledged; (3) a complete and continuous interchange of information between the state and city groups involved must be maintained; (4) comprehensive state policies must be effected and uniformly administered; and (5) both the state and city agencies responsible must equip themselves administratively and technically to perform the duties with which they are charged.

Relations in Other Areas. Intergovernmental relationships in highway and traffic matters are centered principally around the affairs of cities involving the state. There are, however, other governmental and political subdivisions with which the state and, in some instances, the cities must effect working relationships. Relationships between state and county government and between county and city government are of paramount importance in some cases. Most county governments are diminishing their activities in highway and traffic matters, and in some states the county highway systems have been absorbed by urban areas or by the state highway department. In large urbanized counties, however, traffic problems are of much concern and are a principal responsibility of county officials. In such instances it can be expected that the counties are concerned with relationships with the state in about the same degree as are the larger cities. The extent to which the states participate in the affairs of the counties is usually limited to the allocation and distribution of highway funds.

In some large metropolitan areas, peculiar traffic conditions arise because of the differences in laws and regulations between different cities and political subdivisions making up the metropolitan community. Traffic control and highway development can become so complicated in the metropolitan district that it is desirable to establish boards or metropolitan governments empowered to carry on over-all actions in certain fields such as highways and traffic. A notable example is in the area of Toronto, Canada, where planning and highway-development matters are under the control of a metropolitan governmental unit. To properly integrate highway activities and to ensure the proper allocation of highway funds, as well as to ensure uniformity and traffic control, it is apparent that increasing attention must be given to the creation of special governmental units to supervise activities in metropolitan areas.

34-4. Mass Transportation. The part played by mass transit in urban transportation is one of the chief interests and concerns of urban leaders and city traffic engineers. It has become increasingly necessary for state highway officials and engineers to attain knowledge of mass-transportation needs and characteristics because of the work of the state departments in cities. In almost every community, there is a trend away from mass transportation and toward the private automobile. The proportion of the population using mass transportation increases as the size of the city or metropolitan community increases, because of high-density residential areas, traffic congestion, and more extensive off-street parking facilities.

Mass transportation must have an important place in the activities of a traffic engineering program. To make mass transportation services more attractive and more economical, the traffic engineer must have a thorough working knowledge of mass transportation and its objectives. Express and limited bus services, skip stops, requirements in establishing bus stops, transit loading zones and islands, bus terminals, routings, and staggered hours are only a few of the matters having direct bearing on the transit operations with which the traffic engineer must concern himself.

Traffic engineering offers one of the major possibilities for improving transit services. A constant effort must be exerted to maintain proper balance between transit operations and other users of the public ways. In the application of traffic regulations, particularly curb-parking controls, signal timing, and one-way streets, much can be done to improve transit speeds, reduce operating costs, and provide a better service for transit patrons. Substantial savings have often been shown when peak-hour parking bans were applied to principal transit streets and rigidly enforced. In some cases, it has been found advisable to effect turning regulations and parking controls so that some key transit streets are given over principally to transit operations. The timing of traffic signals to meet transit demands and other regulatory steps which remove the delays and irritants to transit patrons are usually desirable.

Many other questions arise in which the traffic engineer is involved in decisions relative to mass transportation. For example, it has been advocated in some cities that private vehicles should be banned from the central business district during most hours of the day. A few cities are now applying extensive parking bans as an aid to mass transportation and to eliminate congestion for all classes of vehicular movements. In Philadelphia, the elimination of parking in the downtown area is reported to have increased transit speeds by from 15 to 20 per cent. Such efficiencies in transit movement are highly desirable and should be effected, but care must be exercised not to effect them to the detriment of other forms of travel.

It is also pointed out that the efficiencies of mass transportation in the movement of persons greatly exceed those of the private vehicle. This is an area for considerable research and development to determine the extent to which persons can be moved more efficiently in mass carriers and private vehicles, taking into account time and vehicular-headway factors, rather than simple area comparisons, as has been the case principally in the past.

It should be recognized by traffic engineers that some of the improvements which they can make in transit operations are as real contributions to the satisfaction and economic well-being of the transit company as are benefits from fare increases.

Whenever petitions are made for financial benefits to transit companies in the form of lower taxes, removal of various types of fees, or increases in riding rates, the public administrator needs to measure the caliber of transit services provided. In most cities, the traffic engineer is the logical person to undertake these measurements. He can determine the adequacy of the area coverage, the adequacy of time coverage (schedules), the standard of operating practices, the character and condition of equipment, and other things which are vital to an objective measurement of over-all transit services. In many cities, the traffic engineering bureau is being given the responsibility for approving transit stops, transit routes, and other operating practices of transit companies. It is desirable and proper that the traffic engineer should maintain a close supervision and audit over the level of services rendered the community by the transit company. This will require continuing measurements and observations of transit by the traffic engineering bureau.

In the development of highway facilities, there are many policy points to be decided concerning mass transportation. One of the most common has to do with the provision of transit facilities, particularly bus runouts and stops, on major expressways and limited-access roadways. The importance of such facilities in the over-all transportation picture of the community, the proper distribution of costs for the facilities, and rules and regulations for their use should be based upon measurements of the needs for express and limited bus services. Relations to the general riding practices of the community and the effects on distribution of funds for the roadway improvements must be studied.

Sometimes the traffic engineer is called upon to examine fully the manner in which transit spends its time in traffic. In such examinations the traffic engineer is apt to observe the bad practices and conditions which he can either correct directly or can call to the attention of proper city officials for correction. For example, laxity on the part of the police in keeping private vehicles and trucks out of bus stops obviously decreases the efficiency of the operation and makes loading and unloading of transit vehicles hazardous. It is just as hazardous and just as necessary for correction when it is found that the drivers of the transit vehicles by their own choice elect not to pull into the curb at designated loading zones.

The objectives and at most times the decisions by public authorities are the proper integration of transit and automobile usage on the public ways. This must be the case because the freedom of choice in transportation must be retained. Obviously, many decisions of individuals with regard to the choice of transportation types have no basis in sound economics. Most people who elect to drive their private automobiles could travel much more economically by mass transportation. Some results come from an evaluation of the demands for use of particular streets. In some cases, there is a much greater need for transit services than in others. Decisions relative to operations and other matters must be made to conform with these relative demands. It is difficult to achieve the proper integration of transit and automobile usage unless the transit officials, the key public officials, and the traffic authorities coordinate their objectives and studies and assume a progressive viewpoint regarding the over-all needs of the community and employ technical engineers for full-time work on their staffs. A mere repetition of what appears to be equitable in the choice of transportation is no way to solve the problems. They must be carefully measured and intelligently acted upon before public cooperation and support at all levels of government can be expected.

Conversion from streetcars to buses, or to trackless trolleys, is another major policy decision in which the traffic engineer can be helpful. He can measure the time and magnitude of peak transit demands and can also observe the interferences which are caused by fixed-wheel vehicles traveling on heavily used downtown streets. In evaluating the need for transit services on through ways and expressways, thought must be given not only to existing transit routes, but to routes which might be established on an improved roadway. With proper coordination, transit and highway officials can plan jointly to provide maximum over-all service to the community. Particular attention should be given to mass-transportation service between chief concentrations of population and between the central business district and other business districts of the community, as well as to special traffic generators such as airports, rail terminals, and other places where large numbers of persons concentrate.

Frequently it is found that downtown business interests and other groups have a distorted view of transportation of persons by means of mass carriers. It is easy to measure the relative importance of mass transportation to different downtown groups, and such measurement should be made regularly by the traffic engineering department to support recommended changes and improvements in over-all traffic control and in top transportation policies.

34-5. Research and Development. Research in traffic operations is not included in much of the work of fact finding and planning now conducted by city and state highway departments. Consequently, many factors controlling the operations and behavior of individual drivers and of vehicles remain unmeasured. Once determined, many of the factors relating to individual traffic units can be fitted to a composite traffic stream to obtain practical situations. When a more definite knowledge of these fundamentals becomes available, broader and sounder applications of other traffic facts will be possible in planning and operations.

Practically every highway organization needs to expand its research activities to cover traffic operation subjects more adequately. Physical researches on materials and equipment have resulted in vastly improved pavements and structures. Studies *from the surface down* have done much to make possible the construction of economical, structurally sound, and durable roadways. Researches *from the surface up* will yield facts which are just as essential to the determination of proper types and locations of routes, as well as to the regulation and control of vehicles and pedestrians on both old and new routes. Field investigations are essential to the determination of practical values. In some cases, traffic research may be advantageously correlated with other highway researches. For example, the problem of a *pavement performance study* could include a consideration of design, construction, maintenance, soils, and economics, along with traffic.

Traffic operations research should become as integral a part of highway departments as the work of their testing laboratories.

A few projects are listed below to illustrate the breadth and fundamental character of the information desired:

A. Traffic characteristics
  1. Induced traffic on improved roadway facilities
  2. Effect of streetcar and bus operation on street capacities
  3. Acceptance and discharge rates of off-street parking facilities
B. Traffic control devices
  1. Practical criteria for stop-sign installations
  2. Before-and-after experience with regulatory devices
  3. Delineation methods and their effects on traffic speed and placement
C. Traffic regulations
  1. Effect of left turns on traffic movements on main thoroughfares
  2. Patterns and characteristics of one-way streets
  3. Influence of street parking on street capacity and accidents
  4. Experience with and methods of control of entrances and exits to roadside establishments
  5. Effect of unregulated pedestrian movements on vehicle capacity at intersections in concentrated areas
D. Traffic design
  1. Variation in rates of volume flow of merging vehicular movements
  2. Effect of shoulders and shoulder types on traffic accidents
  3. Transition characteristics of vehicles where roadways narrow or widen

It is essential that research information derived from special studies or from day-to-day activity be published and disseminated. All state highway departments in the country, and many local road agencies, undoubtedly have information and facts that would be useful to others. In many instances, important pieces of an unfinished and complicated puzzle might be furnished, yet many valuable studies and findings of individual traffic agencies are never made available outside the department; often they do not go beyond the traffic engineering or planning division.

The exchange of findings and experience between highway departments always results in the development of improved methods and uniform patterns for studies. Such exchanges also make possible the coordination of findings and developments so that a minimum of duplication will occur.

The elements of successful traffic research do not differ substantially from those applicable to other highway researches, but they are frequently overlooked, nevertheless. As state and city highways and traffic agencies approach a problem involving research, they may make any one of several common errors. First, the initial project may be defined in such general terms that the scope becomes too comprehensive. Unless the project is broken down during its early stages into workable size components for concentrated study, the venture may grow to cumbersome proportions and eventually fail because of internal complications. Since funds, manpower, and facilities are usually limited, it is only a matter of good organization and practicality to set down in the most definite terms possible the nature of the fundamental problem and the particular segments of that problem to which research effort is to be devoted.

There are misunderstandings in some quarters as to the very nature of research itself. Certainly the mere presentation of one's thoughts concerning a particular traffic problem cannot be regarded as research unless there is involved some fundamental, factual contribution. The restatement of theories and general dissertations on traffic matters, even though they be of considerable interest, does not constitute research. Research is characterized by the presentation or development of facts and the synthesis of those facts in a manner logical and pertinent to the problem under consideration.

Not all engineers have the investigative yen that distinguishes the accomplished researcher, and it is important that personnel be carefully selected for traffic research tasks. Their activities should be sufficiently prescribed within the highway department to avoid the pitfalls of excessive transfer back and forth between research and a wholly different class of activity. This constancy of study will show rewards not only in the greater rate of progress on individual projects, but also in the more successful applications of the entire research program to the ever-changing traffic situations and environment.

Negative findings from research are often as valuable as positive ones. It may be a case of simply separating a falsehood from the truth. Certainly "pay dirt" in the form of revolutionary facts or developments cannot be expected with every research attempt. Like most researches, traffic investigations yield conclusive new facts or findings in only a small percentage of the total onslaughts made against the problems presented.

Whatever the results, but particularly in those instances where the findings are apt to be of general interest, a carefully drawn and brief summary of the outstanding facts should be prepared. Few highway administrators have time, opportunity, or much inclination to study lengthy reports on subjects of research, and there is always a better chance for acceptance if the pertinent material is condensed. Those interested in the fuller report will read it anyway—which leads to a final and important point.

Most traffic research is effective only to the extent that it is applied in the more efficient solution of practical Monday-to-Friday problems of highway engineering. The research specialist must temper a natural desire for a certain amount of pure research with the genuine needs for information that exist in his particular department. His results must be in tune with requirements and in such form that they will be of direct and practical value. The rate of intensification of traffic research in the 48 highway departments will be likely to have a direct relation to the degree of adherence to these simple principles. With sympathetic and intelligent highway administration, traffic research has the opportunity to push back frontiers to the end that motorists will have the right to pass—the right to go and come, expeditiously, conveniently, and safely.

CHAPTER 35

TRAFFIC AND PLANNING

While good city and regional planning makes full use of traffic engineering data and disciplines, good traffic engineering recognizes planning objectives and policies. Planning and traffic engineering functions can be distinctly separated organizationally, and clearly defined areas of work can be delegated to each; yet for efficiency and effectiveness there should be close collaboration. The traffic specialist and the planning specialist will inevitably arrive at points in their work when divergent views occur. These should be reconciled by close working relationships. The planner should be familiar with traffic engineering functions and objectives. The traffic engineer should have a broad appreciation of planning.

35-1. Major Streets and Expressways. Transportation facilities form the skeleton of every city plan. A major street plan could be principally developed by the traffic engineer, but this is not desirable. The basic plan should be prepared by the planning agency, with the cooperation and assistance of the traffic engineer.

Major street plans should take into account all types of traffic. Through-traffic needs should be measured and accommodated. Commercial traffic desires might be markedly different from other traffic movements. Peak-hour requirements might vary from street to street, and so forth. Traffic needs that are measurable through continual traffic engineering studies should always be made before altering and extending the major street plans. The desires and needs of an existing land-use pattern can be measured by volume counts and by origin-and-destination surveys. The amount of traffic in the central business district that has no basic desire to be there and that could be better accommodated by new routes around the district could be readily ascertained. The amount of traffic of each type that could be bypassed around the entire city could be measured. The relative importance of each form of transportation will be indicated. When these and other things are known about travel in a community, roadway improvements and new roadway developments can be evaluated in terms of aids to existing traffic. Then an important planning function arises. Roads cannot be built for only immediate needs, so the controls of land use, the projections of land development, population shifts and trends, and many other planning data apply to future needs.

The traffic engineer relates his knowledge of present traffic needs to the growth and general urban-development problems of the planner to prepare a major-street system for the community—a system capable of meeting immediate and long-range needs.

There are other things the traffic engineer can contribute to the major street plan. He can calculate road-capacity values for a given route design; he can assist in preparing functional or geometric design standards; he can help locate routes for traffic needs and for economical construction.

The street plan might include expressways, major through ways, and local-service streets. As the city grows and traffic volumes increase, through ways or thoroughfares become a necessity, and finally a system of expressways must be superimposed on the whole. As each new-type facility comes into being, provision should be made for modernization and reconstruction of portions of existing networks when traffic requires better accommodations. The thoroughfares provide a system of feeder streets to the expressways; they also provide means for major traffic flows of the area. By proper location, the thoroughfares can be made to serve large volumes of traffic at reasonable speeds, even though they are not of the controlled-access-type facility. Adequate right of way to provide good alignments and sufficient lanes is another essential of good thoroughfares. Major urban-improvement programs, such as public housing, slum clearance, urban-redevelopment projects, construction of civic centers, and park acquisitions afford opportunities to plan and locate thoroughfares to great advantage.

A principle in thoroughfare planning is to establish direct connections between residential areas and between rural routes and the heart of the city. The character and development of the city often make the preferred locations self-evident. Origin-and-destination surveys confirm the correctness of a given location trafficwise. The basic pattern of thoroughfares usually assumes a radial or spider-web characteristic. The radials connect the center of the city with outlying areas; the circumferentials allow complete interchange between the radials and between industrial and commercial centers outside the central business district.

Expressway patterns are likely to be different in each city. To best serve the central area of the city, many expressways are now planned to carry traffic into, or immediately adjacent to, the central business district. Modern expressways are essential to good traffic flow in large cities. The expressway can accommodate both local and through traffic. It can benefit the neighborhoods through which it passes. Differences of opinion on expressway location between the planners and traffic engineers might have to be resolved. For example, the traffic engineer wants the expressway near the heavy desire lines of travel. The planner is likely to be concerned with "Chinese wall" effects and would locate the route more in the terms of land uses. The two views should be compromised.

No phase of highway development requires more careful analysis and study than the urban expressway. Only through comprehensive knowledge of the origin and destination of traffic in the metropolitan area can the route be properly located and designed. These traffic desires must be considered in light of development of new subdivisions, industrial areas, business areas, and changes in land uses (zoning). Urban expressways block and cross existing streets, and these affect the street services of different areas. Location of interchanges has a bearing on street closures and on the land-use characteristics. Without further discussion of expressway requirements, it should be apparent that this type of facility calls for special collaboration between planners and traffic engineers.

35-2. Metropolitan-system Planning. The increasing need for considering metropolitan roadway needs has been emphasized in several previous chapters. An integrated road plan for a metropolitan community is necessary if highway funds are to be most effectively approved. The earliest concepts of metropolitan government started with the need for metropolitan planning. Zoning and planning laws have long given cities jurisdiction for considerable distances beyond the corporate limits, thereby acknowledging that a broad planning function cannot be confined to a single city's limits. In developing the metropolitan plan, as is being done in many areas by special agreements and legislation, roadways again are a principal consideration. Basically, the objectives and approaches of the traffic engineer are no different from those used in large cities. However, the administrative aspects may be complicated by the divided activities and complex intergovernmental relations.

Arrangements for proper traffic facilities for an area or region, such as a metropolitan district, challenge the traffic engineers and planners to work together, even though it might be more difficult than working under a common governmental jurisdiction.

35-3. Land-use Controls. Through zoning, the planner is able to control land uses in the best interest of the community. These controls obviously have direct bearing on the amount and type of traffic to be developed in a given area of the city. It should not be a function of the traffic engineer to establish zoning regulations, but he needs full information on them to effect sound traffic plans.

Zoning plans and changes are major factors in forecasting the traffic potentials of any given route. Land development is largely controlled by zoning. Traffic loads are directly related to land development.

35-4. Special Land Uses. Special land uses, principally a planning decision and function, have many effects on traffic planning. Locations of parks and parkways, urban-redevelopment projects, schools, civic centers, and other public areas, have traffic implications. In new land-use plans, the traffic engineer should indicate the types of transportation facilities that will best serve the uses and which will fit best into the basic street and highway patterns.

The extensive urban-redevelopment projects now under way in many cities include extensive area and thereby provide opportunities to extend key streets, to locate expressways and major thoroughfares, and to construct terminal facilities.

35-5. Mass Transportation. Locations of transit facilities, the extent to which transit services should be provided on expressways and thoroughfares, what type of transit services are best, how subways should augment surface transit, and relationships between transit and private-automobile usage are involved in most urban planning.

It is often found that there is a divergence of views between planners and traffic engineers covering the use of mass transportation. Planners are likely to be more inclined to advocate rigid controls over vehicle uses in congested areas of cities. The traffic engineer probably feels that, in spite of the recognized essentials of mass transportation, provisions must be made for both in central city plans. It would be foolish for the planner to plan on the basis that private autos would be barred from the central business district, and for the traffic authorities to proceed with their work on the basis that transit and autos would be served by the streets. No indications are available to suggest that a ban of private autos from any area of the city would succeed under the present desires and concepts of individual motorists. It would very likely do much harm to the central area of the city. However, from the standpoint of street capacity and economy of transportation, there is strong argument for bans on private autos. The matter must be objectively studied, but it obviously cannot be arbitrarily decided without risk of damages that cannot be recovered.

The balance of expressways and rapid transit is another area of mutual concern to traffic and planning groups. Problems differ greatly from city to city, so no single answer will apply broadly. Sometimes the location and design of the expressway might allow a rapid-transit development that would not be economically feasible.

Adequate provisions for bus stops or turnouts and pedestrian facilities on some expressways are other mutual-concern areas. Demands for transfers, changes from auto to bus, and terminal desires are important points that should be resolved.

In some sections of the country studies are now being made to ascertain the desirability of integrating auto and railroad transportation. Truck trailers are being hauled on railroad flatcars between big population centers. New railroad stations are being developed on the fringes of large metropolitan areas to give spacious free parking to motorists who travel by rail from the area to work or for other purposes. All such developments involve careful study and planning and intricate economic investigations.

35-6. Control of Access. Controlling access to principal roadways from a traffic operations point of view was discussed in Chapters 29 to 31. Such controls obviously have a bearing on land uses and therefore concern the planner. Planning such roads offers an excellent opportunity to integrate the work of traffic engineers and planners. A comparison between the land-use services and the traffic services might be required.

For certain types of developments, limited access is essential. Rigid controls may greatly restore land values. Proper relationship of land uses to access controls and to roadway designs can cause unusual appreciation of property adjacent to major roadways.

The traffic engineer cannot close his eyes to the needs and desires of property owners and businesses. Neither should the planner minimize the safety and convenience factors afforded by good design (which frequently limits access) to the road users.

35-7. Project Priorities. When a plan has been developed, the task of determining the proper schedules and priorities for effecting the plan remains. The highway parts of the plan can be scheduled in terms of priorities objectively established from traffic studies. Traffic trends and user benefits point the way to establishment of priority listing. These needs must then be weighed in terms of other public needs and capital-expenditure programs, which are, again, the chief interest of the planning agencies. By close communication, the traffic engineer can be very helpful to the planner in setting up priorities for road improvements which will best serve traffic as each section materializes.

35-8. Terminals. The importance of planning in the development of off-street parking terminals has been discussed in Chapter 32. Most of the reasons for close cooperation between traffic and planning groups in roadway matters apply equally to terminal developments. The terminal is an essential part of the transportation system, and its development cannot be neglected.

Because some parking facilities have a strong influence on land uses in their immediate vicinity, they have broader interests for planners than just as to their location.

Application of zoning powers to require new traffic generators like shopping centers to provide off-street parking is one of the most valuable planning adjuncts to terminal matters and to traffic problems.

35-9. Nontraffic Factors Essential in Traffic Engineering. The traffic engineer needs much information of a "nontraffic" character to evaluate traffic data properly. These facts have to do principally with traffic-generating factors and physical characteristics. They are facts which are usually available from the planning agencies and in which the planners have equal, or greater, interest.

Topographic data have major effects upon highway transportation. Rivers, valleys, hills, lakes, etc., might impose hardships in planning highways and traffic improvements or they might afford advantages. A lake front can be an economical site for a major traffic way. A bluff or a hill might pose serious impediments to major flows within a city or region. While topographic conditions often increase transportation costs and efficiencies, there are many opportunities to take advantage of these conditions. They warrant careful study and full consideration by the highway authorities and the planners.

Population distributions are valuable to the traffic engineers, since people make traffic. Traffic agencies need data on the distribution of populations, income groupings, ages, races, modes of travel, ages of residences, and trends. From simple evaluations of the adequacy of traffic services to the design and location of major routes, these data are valuable.

Vehicle ownership and use data are often used by traffic engineers. These data are related to populations and to traffic studies to determine travel-generation characteristics and potentials. They also serve the traffic engineer in other ways in dealing with general planning matters.

Economic factors have the most direct bearing on the capacity of the community to undertake highway and traffic improvements. Over-all information on the income, tax rates, debt condition, and general economy should be thoroughly understood in applying traffic facts.

Studies of other transportation media and terminals, commonly made by the planner, have applications in traffic engineering. The city planning agency will usually develop the information as part of the over-all city plan. The traffic and highway agencies often desire to subject it to different analyses and to put it to different uses—uses which integrate all forms of transportation.

Standards for roadways should be reviewed by traffic engineers to assure safe conditions and long-range capacity. This applies particularly to the following: curb cuts, corner setbacks, subdivision controls, roadside plantings and median plantings, sidewalks, and driveways.

The planner can well use the knowledge of the traffic engineer in developing and administering these and other standards which affect vehicle operations and pedestrian safety. Location of schools and other public buildings involves standards which relate directly to traffic.

CHAPTER 36

OFF-STREET PARKING

The parking needs of all cities require many off-street spaces. These needs must be handled *locally*, yet the state and the Federal government also have direct interests. The state cannot overlook parking on key highways in cities—especially those on which state funds are expended—in relation to factors of safety and efficiency of traffic movement. The distribution of Federal-aid funds to cities should involve a careful study of parking and movement relationships. The lack of adequate planning of street systems of communities is the principal reason for many of today's terminal difficulties. Communities are faced with the continued rapid growth in traffic, competition for street space between moving and parked vehicles, and economic difficulties in providing effective, and therefore adequate, parking and loading spaces. The economic health of the city may slip unless parking and loading spaces are provided, especially in the central business districts.

While the import of the parking problem is generally recognized, actions vary widely in both scope and depth. In some cases the problems are faced with indifference, and little is done. In others, superficial and very short-range activity is observed. In some cities, changing circumstances have been recognized, and bold, progressive actions are being taken to provide parking commensurate with needs and the ability to pay. In these cases, terminal needs are being carefully integrated with roadway developments, with mass-transportation plans, and with over-all urban-planning redevelopment. Public policies on parking are essential to objective approaches in long-range planning and community development.

Parking programs are complicated by the peculiar and selfish desires of individual road users. Durations of stay, time of demands, walking distances, rates, attractiveness of facilities and services, and other factors are essential elements in plans and policies. Many questions arise: What is the responsibility of the parking "generator" in relation to the over-all or general demands? To what extent should government "compete" with private parking developments? Is it proper to make tax and other concessions for parking? Should off-street and curb parking be considered as a "system"? What method of finance is best if public development is pursued?

36-1. Measuring Parking Needs. Parking surveys have become commonplace. As the problems become more chronic and as differences in opinions relative to approaches arise, the need for facts about parking is increasingly evident. Through the use of Federal planning funds, numerous surveys have been made of parking needs by state highway organizations with the assistance of the U.S. Bureau of Public Roads. About one hundred cities have been so surveyed. Many other surveys have been made by city traffic engineers, local planning groups, civic bodies, and by consulting engineers. Where revenue-bond financing is contemplated, reports by recognized consultants are usually required.

Most comprehensive surveys involve (1) a physical inventory of available spaces, (2) measurements of size of both curb and off-street facilities, (3) ascertainment of demand from determination of places where motorists are destined (and where, therefore, they desire to park), (4) observance of basic characteristics such as fees paid, time parked, and distances walked, and (5) projection of demands.

Space Inventories. Observations are made of spaces in use at the curb and in off-street facilities. Time restrictions are recorded. Areas of parking prohibitions and special curb uses (bus stops, loading zones, crosswalks, etc.) are recorded. Meters, meter rates, and off-street rates are indicated, so that a completed record of the location and character of available spaces is known. Information might also be recorded on the areas and streets on which additional spaces can be developed.

Space Usage. The turnover of curb and off-street space usage is observed. If parkers are interviewed, an accurate record of space occupancy is available for all areas. With other survey methods, it might be necessary to record usage characteristics by license-plate checks or by sampling processes.

Demands. Parking demands take into account not only the existing parking practices but also the destinations of motorists. The destinations and duration of stay provide the best indications of where people would like to park and how long they might park, if attractive spaces were available. Demands can be measured from interviews of parkers or from cordon origin-and-destination surveys. Detailed land-use studies also have direct relations to demands.

Characteristics and Habits. From the basic parking survey, the characteristics of parkers can be determined for a given city or for a specific area of a city. How far motorists walk from point of parking to principal destination is easily determined. Relationships of fees to parking time, to distance walked, and the time of day are important in calculating space needs.

Projections. Projections of demands to future years require collection of data on traffic trends, vehicle-registration trends, population growths, and land-use changes.

Summary and Analysis. In studies of parking for a central business district, the data on practices and demands are usually summarized by blocks. Unless the survey is quite small, the data are usually put on mechanical-tabulation cards. This gives quick tabulations and many cross references. In small areas, hand tallies of the field information may be adequate. By relating demands to existing spaces, total deficiencies must be *assigned* to specific sites to calculate the use and income that can be expected from development of a parking facility at a particular location. In estimating parking demands, consideration is given to violations and to enforcement difficulties in use of existing spaces. The separate requirements of different types of vehicles must be ascertained.

In interpreting survey findings and in preparation of parking programs, it is essential to take account of legal and administrative aspects of parking. Proper laws and regulations must be provided if any approach to solving the problem is to succeed. It seems necessary, therefore, as a part of the over-all survey, to review and interpret existing parking laws and administrative machinery.

In analyzing parking data, it is customary to deal with various categories of parkers, based on the length of time parked. Invariably parkers are classified according to long-time and short-time parkers. Generally it is agreed that *short time* should apply to parkers desiring to park in a given space for not more than 3 hr. Parking of 3 hr or longer is usually classed as *long time.* Another class of parkers is the *overnight* user.

Planing the Surveys. Care must be exercised in planning the surveys to ensure accuracy and breadth of information needed to effect a parking program. Special attention must be given to the requirements for financing, and data needed in developing a bond prospectus must be procured. Defining survey areas must be carefully done. In this connection both present and future needs must be considered. Sometimes the areas are made so large as to cause the needless expenditure of monies. In other instances the scope of the survey is so limited as to make it valueless in comprehensive planning and development.

In planning the scope of surveys, it is essential to know the interests of principal groups in the city and adjoining areas. On any substantial plan or survey it is essential that the public be brought face to face with the plans, desires, and working procedures. It is a simple matter for people to get disinterested and uncooperative, unless the public is kept informed of objectives and basic desires.

Assignments. In assigning parking needs to a specific site, it is usually assumed that existing competing facilities will be used to capacity. This is conservative, but desirable in supporting income calculations. No mathematical formulas or graphical means are known for the objective assignment of demands. In general, the assignments are made on the basis of distance-cost relationships which can be measured for existing facilities.

36-2. Responsibility and Authority. Responsibility for a comprehensive parking program involves many agencies and individuals: Federal and state governments, because parking is so intimately related to the movement and efficiency of the over-all automotive transportation system; local government, because it has a responsibility related to land values, economic vitality of the central area of the city, services to citizens, and an integrated transportation plan; many businesses and private groups, because generators of parking demands might be expected to provide off-street parking, while, on the other hand, most businesses and generators might expect assistance from the city in developing parking as a public service. It is natural that those who have entered the parking field as a private business should feel that they are assuming a responsibility for which they deserve consideration and some protection.

In most cities, the responsibilities for developing parking should be divided, with the city assuming direction. Experiences indicate that for proper coordination and planning, it is usually essential that the city assume the leadership, although this does not mean that the city must actively enter the field of developing and operating off-street facilities. Neither does it mean that any *one* agency of city government has a preferred position for the city's responsibilities.

When the city has decided whether there are parking needs, the decision must then be made as to the type of program that is best for the community. The proper direction of the program is usually apparent. Advantages and disadvantages of the different areas of responsibility must be carefully examined in light of local conditions.

The question of responsibility is closely related to matters of legal authority. Many legal questions can be raised.

It has been generally established that off-street parking serves a public purpose. Most courts have held that cities have a right, with proper legislation, to develop and operate parking facilities as a public purpose. They are thereby allowed to use powers of eminent domain to acquire property for parking. This right of cities usually prevails whether the parking facility is to be operated by the city or is to be leased for private operations, assuming that the lease arrangement includes some controls over operating procedures and rates.

It appears that no significant legal difficulty has arisen over the right of municipalities to contract with private developers both for the construction and operation of garages on public lands. By designing the structures, as in San Francisco, Los Angeles, and Kansas City, so that public parks as a basic purpose are retained atop the underground garage, no questions have arisen as to whether the public purpose of the parking facility was being violated. Again, this reflects the attitudes of the courts that with modern traffic conditions, off-street parking is in the public interest.

Some cities with strong "home-rule" government have all the needed powers to engage in off-street parking development. Other cities are dependent on enabling state legislation. It is essential that such legislation allow cities to exercise powers of eminent domain in acquisition of parking sites. Without such powers, the cities can do little more than develop park areas and other publicly owned properties for parking. Under pressure of varied interests, some state enabling laws contain notable limitations. For example, several specifically forbid the sale of services and other nonparking functions in municipally developed facilities, on the basis that these are not necessary to fulfill the public purpose. Contrasted with this are the laws that allow large-scale retail and other realty development as an adjunct to the garage structure. In the granting of rights of eminent domain, the question is often raised as to whether the rights should apply to private property already devoted to parking. From a technical point of view, the powers should not be restricted in this regard.

36-3. Approaches and Programs. Numerous types of parking programs have been developed. All have their advantages and disadvantages. All have been made to work satisfactorily. It becomes necessary, therefore, to select the approach that is best suited to the local community, taking into account the legal authority, political pressures, financial possibilities, and the character and magnitude of needs. Most cities approach the matter on a basic assumption that every effort should be made to have the job assumed by private enterprise before active participation by the city. In this connection, the interests and needs of *two* private-enterprise groups must be considered: (1) there is the group of owners and operators of parking lots and garages; but (2) there is also the much larger private-enterprise group made up of merchants, property owners, and others who are dependent upon attractive automotive transportation for the success of their business interests. One group should not be considered to the exclusion of the other.

Developments by Parking Enterprise. Parking facilities developed by private interests are of three types: (1) lots and garages operated as pure business ventures; (2) those developed by businesses as an adjunct or ancillary part of a commercial establishment; and (3) those provided by special combines and groups representing several business interests.

Almost every city and town has privately operated parking facilities.

Lots and garages developed and operated for a profit are still the principal contributors of off-street spaces. Some are attractive and reasonable in price. Others are poorly constructed and operated. Many have rates that are considered too high by the average parker. The success of these facilities is attested to by the large number of parking corporations found in cities.

Numerous businesses, like department stores, food markets, and banks, have developed lots and garages as a service to their customers and subsidize the parking activity. Often the parking is free. Sometimes it is free when tickets are validated after a visit to the store or a purchase. In some cases, small fees are charged all parkers whether they patronize the store or not.

Office buildings, businesses, and industries are increasingly developing parking spaces for employees. In some instances, it has been found necessary to provide parking to attract employees in the competitive labor markets. New office buildings usually include parking for tenants and often for visitors. The Cafritz Building in Washington is often cited as an example of an office building which incorporates parking in its basic design.

The following might be considered as advantages in the private-ownership approach:

1. It maintains and encourages the free enterprise system, with its initiative and competitive qualities.
2. It is easier to combine parking activity with other business activities.
3. Taxes are paid, as with other businesses (in a few instances tax relief has been granted parking developments to create incentive).
4. Use of private capital removes fiscal burdens from the city.
5. The level of service usually bears direct relation to parking demands.
6. Complete automotive services can be provided.

Disadvantages of private-enterprise parking include:

1. Parking lots might be temporary in character and would, therefore, not be considered a part of the permanent program of parking.
2. Rates might be excessive.
3. Unattractive locations might be necessary because of the inability of a private developer to secure key property.
4. Some operators are irresponsible and careless, not even assuming financial responsibility for damages to vehicles.
5. Some lots and garages are poorly constructed, maintained, and lighted.

Public Assistance to Private Development. Public assistance can be given to private-enterprise developments. In Buffalo, N.Y., tax exemptions are allowed on garage structures for a 15-year period. In Washington, D. C., the public parking body provides private interests technical assistance and other services. In Baltimore, Md., public monies are loaned to individuals at very reasonable terms for garage development.

Plans which involve the purchase of land by cities and the leasing of these lands for private parking developments are especially significant. It is becoming a common practice for cities to make publicly owned lands, such as parks, which are strategically located, suitable for private development for parking. Public squares in Los Angeles and San Francisco are outstanding examples. Legislation has been passed to make this possible under the Commons of Boston and under a park in Memphis. Long-term leases allow private construction and operation under strict controls by public agencies. While the details vary, the joint achievements are similar in most cases:

Advantages of this approach include:

1. Ideally located sites are made accessible.
2. The city's powers of eminent domain can be used to acquire sites.
3. Low-rate public funds can be used partially to ensure the payments.
4. Major responsibility and risk are assumed by private capital.
5. Efficiencies associated with private operation are encouraged.
6. Tax aids can readily be provided.
7. Low parking rates are often provided.
8. Parking is given permanence and can be positioned in an over-all plan.

Disadvantages can be cited:

1. Many cities do not have ideally located public land parcels for parking.
2. Criticism might arise from "competing" private operators, and the claims might be that the city is unfair in aiding the particular development involved.
3. Extensive private financing of the venture might result in high rates.
4. Necessary enabling legislation might be difficult to obtain.
5. Opposition might arise where powers of eminent domain are used to procure lands for private development and operation.

Municipal Developments. Numerous cities have provided off-street parking facilities when it became apparent that private interests would not adequately meet the problem. Perhaps the most common approach to development is through the *public authority.*

Parking authorities are like other public authorities in that they are set apart from the normal governmental structure and are given broad business powers. The degree of authority is varied by the state and local legislation creating the authority. The most successful authorities have autonomy of operation, powers to acquire land and to construct and operate parking facilities, and the right to issue revenue bonds. These qualities usually permit the parking authority to go about its duties in a direct and effective businesslike manner, while at the same time enjoying the powers normally held only by governmental agencies. Authorities are immune to taxation.

In opposition to the authority, it can be pointed out that the powers and extent of activity might exceed the basic concepts of the legislative bodies creating it. It is also possible for the agency to overemphasize parking in relation to other basic components of city planning and over-all transportation requirements.

Some public officials claim that no greater efficiencies can be attained through authorities than through regular governmental channels. They might also feel that authorities constitute an undemocratic approach to a *public* problem.

About 50 cities are now using the authority device to cope with their parking problems, and the popularity of the approach is growing rapidly.

Legal problems often arise in the creation of parking authorities. They must be sanctioned by the state legislature. This is sometimes difficult to obtain. Sometimes the legislation imposes restrictions which make it difficult to obtain, because of opposition from private parking owners and operators. Sometimes the legislation imposes restrictions which make it difficult or impossible to operate, such as withholding the right of eminent domain, banning automotive services and retail outlets, deny-ing the right to acquire any property operated for parking purposes, etc. In spite of some narrow enabling laws and some adverse court decisions, the majority of the questions raised have been decided in favor of allowing authorities to develop and direct programs of parking. The decisions invariably involve the point that parking is looked upon as a *public purpose.*

Parking programs directed by *regular governmental agencies* are often very effective. Broad powers and responsibilities are delegated by the legislative bodies to an existing city department, or department head, to undertake the provision of needed parking spaces. These powers can be as great as those given an authority, or they can do little more than make the department the operating agency for the legislative body. The assignment is usually given to the traffic engineer or to the director of public works.

The principal disadvantage to this approach is that the department head might be so occupied and burdened with other functions that he does not give the parking problems the attention they require. It might be difficult to concentrate on parking when other and older activities are pressing.

Parking boards or commissions have been established in some cities. They are common only in cities where other governmental functions are assigned to similar boards. Sometimes it is felt that boards can be more effective in gaining public support than can individual agencies, through civic leaders who might serve on them. The disadvantages of parking boards are similar to those of traffic commissions. They might be domi-nated by one or two members; they might take narrow and selfish views; they are subject to political influences; and they might not properly coordinate efforts with interested governmental agencies and offices.

Buffalo, N. Y., provides one of the best examples of an effective parking program developed by a board of parking.

Advantages and Disadvantages of Municipal Programs. Some *advantages* of municipal parking programs and operation, in addition to those already given, are:

1. Parking can be effectively integrated with all highway and traffic activities. Planning is simplified.
2. Off-street parking can be tied to curb parking as a *system of parking.* Strong facilities can be made to help support the weak ones.
3. Full use of all governmental benefits enable rapid and economical development of space. Low rates are possible.
4. Many arrangements for financing are possible.

Some *disadvantages* in municipal parking activities, in addition to those discussed under the various plans, are:

1. Politics might hamper and interfere, as in all municipal activity. Confusion and delay might result.
2. Direct competition can develop with other approaches to the parking problem.
3. Some consider them an overextension of municipal activity.

The service of municipal parking activities is attested by the popularity of this approach to modern parking and terminal problems. It is apparent that about one-half of all cities of over 10,000 population now own off-street parking facilities. Both large and small cities are included, and the activity flourishes throughout the country. A key to the programs is adequate legislative activity and public support.

36-4. Financing Parking Facilities. Discussion of finance of parking facilities is limited to the public developments, since financing of private facilities does not differ from the financing of other private business. Numerous methods have been employed by cities to procure funds for off-street parking. Some include direct appropriations from the general funds; others entail on-street meter revenues which are accumulated into a substantial working capital; and some assign surpluses from other municipal endeavors to parking programs. In the great majority of cases, however, funds are borrowed by the city for the activity.

Regardless of the method, it is important to determine the proper distribution of costs for parking, so that as equitable an arrangement as possible can be effected. It is obviously also necessary to establish the proper legislative and legal base for the particular type of financing employed.

Practically all cities with substantial programs charge patrons for the privilege of parking. Even though some subsidization might be required, most efforts have as an objective the application of fees adequate to support the parking venture.

General-obligation Bonds. Some cities consider parking so essential that they resort to financing by the issuance of general-obligation bonds—bonds supported by the credit and general taxing power of the municipality. This method would very likely be more popular except for the fact that the debt limits of cities are usually fixed by the state. In many cities it is necessary to obtain the approval of citizens through a referendum. Laws sometimes provide that general-obligation bonds issued for self-liquidating projects are not applicable to the statutory debt limits, so that some parking projects might be undertaken without affecting the city's borrowing power.

Financing with general-obligation bonds offers several advantages. They usually bear the lowest interest rates available to the city, making the money the "cheapest" that can be acquired for the parking activity. The bonds can be issued without surveys and prospectuses, thereby reducing preliminary costs and saving time.

Revenue Bonds. Currently, the most popular method of financing parking programs is the issuance of revenue bonds—bonds payable from the revenues produced by the parking project. Many state laws allowing cities to engage in off-street parking allow them to issue such bonds. In general, the issuance of revenue bonds does not involve special approval by local voters. They do not affect the city's debt limit or borrowing powers. One of the greatest appeals of this type of financing is that funds can be quickly procured to develop much needed parking spaces without incurring municipal debt which adds to the tax burden of the city. The public likes this.

While much is heard about revenue-bond financing in the parking field, there are actually only a few cities that have sold such bonds without pledging some potential revenue other than the revenue from the parking facility. A common method is to pledge all, or a fixed part, of the curb-meter receipts to the repayment of the bonds. This plan has met with much favor, and some authorities advocate a system of parking which integrates the curb and off-street parking activities, with regard to supervision, financing, and regulation.

To produce added revenue, many facilities constructed with revenue-bond money contain retail areas, offices, and other building uses that produce substantial revenues. This has value in supporting the financial structure and in maintaining retail values in a particular area. On the other hand, it means constructing parking generators which might induce a substantial demand for the new parking spaces developed and thereby allow little net gain in parking for the general area. Questions have also been raised as to the validity of condemning private property for non-parking purposes or whether tax exemptions allowed for garages should apply to other uses of the total structure. However, the courts have generally held that such benefits are allowable if needed to develop the parking as a needed public purpose.

Automotive services and the sale of gasoline in public garages as an added source of income are also debatable. It has already been pointed out that some legislation prohibits such "nonstorage" activities.

Long-term lease plans are used to support revenue bonds for parking. In a few cases, leases providing an annual income for the entire period of the bond issue have been allowed.

In a few cases, mortgaging of city property has been employed to support the revenue bonds.

It has even been suggested that taxes collected for parking violations might be properly set aside for application to debt requirements in revenue-bond financing.

General requirements in the issuance of revenue bonds differ from those for general-obligation bonds. Because of the greater risks involved, the banking interests set forth specific requirements before revenue bonds are bought. It is desired to have an average earnings ratio, or coverage, in excess of 1.5—the average annual net income available for debt service exceeds the average annual debt service by 1.5 times. Covenants may also be required. These cover rate schedules, sustained employment of technical consultants, insurance, budget and fiscal procedures, and establishment of special funds (maintenance, extraordinary charges, etc.). If curb-parking meter revenues are pledged, covenants might control the rights of the city to remove curb meters or to reduce the number of meters until the bonds are retired. However, this control of the basic police power on which meters are installed at the curb raises many interesting legal questions, and it should, therefore, be carefully examined when it is advocated in a particular parking program.

Benefit Assessments. Some parking facilities are financed by issuance of special-assessment bonds. The full faith in credit of the governmental agency supports the bonds, but properties within certain bounds or certain areas of benefit are assessed especially to retire the debt obligations.

The plans can be modified in many ways. For example, it can be required by a governmental agency that businesses providing their own off-street parking are exempt from the taxes which are assessed against other properties which must depend upon the publicly developed facilities. As with other special-assessment programs, the taxpayers usually have a right to elect, to accept, or reject proposals to develop parking for which they are to be taxed. Also, the degree to which they can be assessed is usually controlled by state legislation. Usually the general receipts of the government are used to defray a part of the costs, with the benefit assessments covering only a fixed percentage. This quantity does not normally exceed 80 or 90 per cent of the total. This plan of financing has had only limited application. It usually breaks down primarily because of the strong opposition raised by downtown interests. Its primary application is in those cases where non-revenue-producing parking facilities are desired. Another difficulty in using special assessments also arises in connection with the proper distribution of the assessments in relation to benefits. Experiences in cost distributions vary widely.

Construction of Private Facilities on Publicly Owned Property. Under this plan of garage development, the city enters into a long-term contract with the private group or corporation whereby the costs of constructing the facilities may be borne by the private interests. Whether the structure is developed with private or public funds, the contract arrangements usually ensure the city adequate income for debt charges to offset tax losses and to receive a reasonable return upon property. It is usually further provided that upon the termination of the basic agreement the facilities become the property of the city.

Principal benefits of the plan are that it allows the development of parking at ideal locations and at the same time permits a substantial contribution in the form of land by the city. For underground garages the costs of construction are invariably so great that it is not usually possible to finance them as self-liquidating facilities except with the contribution of the land or with other substantial assistance from the city.

The application of the plan is limited because of the few locations where such facilities can be developed and because there is still strong resistance in many quarters to the disruption of long-established parks, monuments, and other urban landmarks.

The Union Square Garage in San Francisco set a precedent in the development of this type of off-street parking. An underground garage was built on a public square by a private corporation. The park atmosphere was restored on top of the four-level underground garage building. The success of this garage, due in part to its ideal location in the heart of a section of the city which generates a very high parking demand, has led to other similar developments.

36-5. Zoning for Parking. There are ways in which the city can provide off-street parking other than through the acquisition of land and the development of facilities. One of these is the application of zoning powers to require the generators of traffic to provide off-street spaces commensurate with the needs developed. This approach does not develop appreciable immediate parking capacity, but it ensures a satisfactory long-range solution in expanding and growing areas and a marked aid even in the old section, such as the central business district. It has been well established legally, and no serious economic hardships are imposed. It can even be shown that intelligent and reasonable zoning requirements are definitely in the best interests of the property developers.

More than 300 municipalities have zoning ordinances which require off-street parking and/or truck-loading facilities. The requirements vary greatly as to standards for spaces and areas and circumstances of application. While the standards might, in some instances, appear crude, the encouraging factor is that the idea is well established and is being enthusiastically pursued throughout the country.

Requirements for Spaces. By amending the zoning ordinances, cities can require provision of off-street parking for new buildings and for buildings that are being substantially reconstructed. These requirements should be directly related to the normal space needs of the particular type of generator. Obviously, the spaces necessary vary widely for different types of generators and to some extent for similar generators in different locations. In some instances, such as office buildings, the number of spaces required probably has a more direct relationship to the square feet of floor area in the building than to other criteria. An auditorium or theater, on the other hand, generates space need in proportion to the number of seats. For a hospital, the number of beds might be the best basis for fixing parking requirements. In a hotel, it might be the number of rooms, and in an industrial plant, the anticipated number of employees is probably the best gauge available for fixing space requirements. Table 36-1 gives some space requirements contained in typical zoning ordinances.

When space may be readily obtainable, such as at buildings in new areas, more liberal off-street parking is apt to be provided than the zoning ordinances require. For example, the developers of large regional shopping centers realize that an abundance of attractive parking is an absolute essential to their success. Accordingly, they follow high standards, aiming for a parking index of 10 (10 spaces per 1,000 sq ft of retail floor area). They rarely provide less than an index of 6, and an average is probably 8.

Loading Facilities. Zoning laws cover the requirements for loading and unloading of both passenger and commercial vehicles, as well as conventional parking spaces. Since the use of public ways for loading operations by trucks causes serious delays and traffic-capacity restriction, it is in some instances even more important to require off-street truck berths for parking. Standards for off-street terminal facilities are also in their formative stages.

Developers and business leaders are increasingly providing their own off-street parking and loading areas for customers, tenants, employees, and visitors. They are developing such facilities in many cases on a voluntary basis. When this is not the case, zoning regulations can require reasonable minimum amounts of both loading and parking spaces.

Again, it is important to emphasize that the schedule of spaces required should be based on thorough studies of present parking conditions and practices, or characteristics, in the particular community.

Retroactive Zoning. Zoning usually applies to future conditions—to conditions that do not exist at the time of first passage of the regulations. As already acknowledged, zoning regulations for parking and loading are principally effective in developing long-range solutions. However, the question is often raised as to whether or not the powers can be applied to the short-range program of off-street parking.

TABLE 36-1. SUMMARY OF PARKING REQUIREMENTS FOR VARIOUS BUILDING
TYPES AND RECOMMENDED BASIC UNITS
(One parking space for each)

Building types	Recommended basic unit	Modal values	Range of middle two-thirds values	Number of cities represented
Single-family dwelling.....	Family unit	1 unit	....	94
Multifamily dwelling......	Family unit	1 unit	1 unit	176
Theater.................	Seat	10 seats	5–10 seats	104
Hotel...................	Guest room	3 guest rooms	2–6 guest rooms	88
Place of public assembly...	Seat	10 seats	5–10 seats	93
Retail store..............	Sq ft gross floor area	200 sq ft	200–1,000 sq ft	56
Office building...........	Sq ft gross floor area	200 sq ft	200–1,000 sq ft	49
Hospital.................	Bed	4 beds	2–5 beds	38
Industrial plant..........	Employees	5 employees	2–5 employees	37
Wholesale establishment...	Employees	3 and 5 employees	2–5 employees	16
Restaurant..............	Seat	4 and 5 seats	3–10 seats	13

SOURCE: E. Mogren and W. Smith, *Zoning and Traffic*, The Eno Foundation for Highway Traffic Control, Inc., Saugatuck, Conn., 1952, p. 70.

The courts have held that retroactive zoning can be applied in some cases. These cases have always had a direct relation to public health and safety and do not strongly indicate that retroactive zoning can be applied to parking. Many legal advisers seem to feel that the retroactive application of zoning for parking and loading will not be favorably considered in most courts. Accordingly, it might be detrimental to the whole program of zoning for parking if the point is overstressed. Omission of retroactive features would probably be desirable in first instances of the zoning approach.

If parking is developed by new buildings in a uniform manner stipulated in zoning ordinances, it will not be long before economic laws and competitive forces act upon businesses and buildings which do not have similar, or better, parking facilities. The buildings with the desirable parking and loading facilities will be so much more attractive, other things being equal, that many of the older buildings will be forced to conform, so that the zoning regulations will have been effective in providing more short-range relief than might be realized. The laws of survival can be stronger than an attempt to include in the zoning ordinance retroactive features that are of doubtful validity.

Other Features of Zoning. There has not been very strong opposition to parking and loading requirements in most cities, *except* in the central business districts. Many hold that in the old and densely developed districts the zoning requirements cannot be realistically applied. It is claimed that the needs for parking are so intrinsically related to many generators that rules applicable to individual generators cannot be realistic. It is further shown that the costs of developing and operating parking facilities by individual generators in central areas are so great as to impose impossible financial burdens. Because of the high incomes from taxes that are derived from the central business districts, it is argued that the city has a more important responsibility for parking and loading than in other areas. The city must see to the health of the central district in order to protect the over-all health and future of the entire city. Loading facilities, it can be claimed, are in effect nothing more than an extension of the street services and, therefore, fall within the sphere of municipal responsibility; this reasoning would apply equally to other sections of the city.

In favor of extending zoning requirements to cover the central area is the point that these areas are the principal generators of parking and loading demands. If there is a parking problem, it certainly will be found in a central business district. Therefore, to eliminate the district from the zoning regulations is to eliminate a major part of the total problem. Parking is more needed to place the central districts in a favorable competitive position with outlying districts than anything else. It is essential to check rapid decentralization and decay of these old districts. To exempt the central district is to overlook the most serious part of the problem. To overcome some of the arguments, thought should be given to writing the ordinances so that cooperative facilities can be allowed.

When strong opposition occurs in a city, it is wise to adopt the sections of proposed ordinances not applicable to the central area. Exempting the area might not be desirable, but getting the regulations partially applied to the city is usually a step forward.

Another partial approach to the parking and loading problems is through subdivision controls. In approving plans for new subdivisions, the public authorities can require block sizes and arrangements and roadway layouts that take into account parking and loading needs.

The legal propositions concerning zoning for parking and loading are no different from those applicable to zoning in general. They involve the questions of safety, comfort, health, and welfare of the community. Whether the ordinance involves in effect a public "acquisition" of private property or whether it is arbitrary and discriminatory must be answered. Most courts now presume the reasonableness of zoning ordinances, and their unreasonableness must, therefore, be clearly demonstrated if the ordinances are to be set aside. There seems to be little doubt that zoning ordinances requiring off-street parking will be looked upon by the courts as valid. Relief of traffic congestion has been held to be in the public interest. Parking and loading are intricately associated with traffic movement. Over-all applications of zoning have developed that the public powers of governmental agencies include the right to control activities that are major exigencies of government; the modern traffic problems, including parking, must certainly be included. It can be readily shown that poor traffic conditions affect adversely the economic conditions of cities and reduce the attractiveness of the city as a place to live and enjoy basic ways of life. The extensions of police powers to require off-street parking and loading as a means of freeing the streets for traffic flow can be expected to be legally acceptable. However, the newness of the concept and the necessity of showing "reasonableness" make it desirable to support proposed zoning ordinances with extensive facts. These facts should be aimed at demonstrating the public necessity and at showing the relationships of the parking and loading proposals to relief of traffic congestion. Adverse conditions which can be expected to develop if the parking and traffic problems are further neglected should be a substantial part of the supporting facts.

36-6. Municipal Regulation of Private Facilities. It is commonplace for municipalities to regulate certain aspects of private parking enterprises. Such regulations can provide protection for the public and can increase the over-all attractiveness of the city's parking program. At the same time, the regulations should not be so restrictive as to discourage private development of needed facilities.

Where controls are applied, the cities usually license the lots. The right to license lots and garages is well founded; legal questions, if any, are likely to arise only with regard to standards prescribed for licensing. The standards must be carefully determined by both the legislative and administrative branches of government.

Annual fees collected for the licensing of parking facilities range from a few dollars per year to several hundred dollars annually. The ordinances fixing the licensing procedures and prescribing the fees are universally upheld as a proper utilization of the city's police powers.

Regulations usually cover:

Signs. Locations, letter sizes, illumination, over-all size, and mounting standards are involved.

Surface. Dustproof, all-weather surfaces are required in most cities.

Marginal barriers. Fencing, wheel stops, and bumper guards are required to reduce the chance of encroachment on other properties, on alleys, and on sidewalks. Physical nature of the barriers varies widely.

Shelters. Shelters may be required for the convenience of customers and employees. Size and basic physical character are covered by the regulations.

Lighting. When lots are operated at night, it is proper to prescribe minimum-lighting standards.

Ingress and egress. The maximum number of contact points can be prescribed. The general plan is to give broad powers to some city department, such as traffic engineering, to review and approve the locations and numbers of entrances and exits.

Other physical requirements. Numerous other aspects of the physical features of lots and garages might be regulated. Encroachments on sidewalks, snow removal, and attire of attendants are included in some regulations.

Damage to Vehicles and Theft. Protection is provided patrons of lots and garages by regulations which require operators to assume reasonable responsibility for losses from damages or theft. In some instances, the operators must post bonds or insurance coverage.

Some operators attempt to relieve themselves of responsibility for loss or damage by printing *conditions* on the parking tickets. If allowed at all, the courts are apt to give only small relief to operators from such notices; they are not often construed to constitute legal contracts.

Rates. There are no known cases in which cities have attempted to prescribe the rates to be charged for parking. The control might mean placing parking in the category of public utilities; although some officials think this would not be necessary to enforce price controls for parking, it can be demonstrated that such controls are necessary for the public interest. It is still questionable whether price fixing for parking is necessary for the public interest.

Rate controls thus far do not extend beyond the requirements of cities that rate schedules be filed with a public agency and that the rates can be changed only after a minimum notice. This type of control is to prevent operations for gouging the public when unusual heavy demands arise for a short period of time.

Other regulations can be applied. These illustrate the approaches that are considered in the public interest. They do not impose unreasonable hardships on the owners and operators, and because they develop uniformity and attractiveness, they are likely to improve the over-all business of the private operators: they serve to encourage advantageous use of off-street parking and to make local motorists more off-street-conscious. To this end, they serve as a distinct aid to the total program of parking and relief of traffic congestion.

36-7. An Over-all Parking Program. Regardless of the local policies adopted, a long-range integrated program is essential to achievement of effective and lasting cures for the parking problems. The city must render the aid required of it in locating and developing facilities where they cannot otherwise be provided. The program of the city should be well meshed with the activities of private operators and other private-interest developments.

Strategically located lots operated by private interests should be carefully watched and acquired by the city for parking if change in use is contemplated. Key sites, especially at the periphery of the central business district, should be perpetuated for parking, even if governmental action is needed.

This is being done to some extent in a few cities, such as Milwaukee, through zoning. Special parking zones are established, and the use of the property for nonparking purposes requires a change in the zoning ordinances.

The total parking program might involve the consideration of parking operations in relation to other public and safety factors. Through permits or zoning, parking facilities might be excluded from the areas around schools, from places where special fire hazards might develop, and from locations where bad traffic conditions might be developed because of the parking operations.

Extensive parking programs extend beyond the corporate limits. The metropolitan area is of concern. It is difficult to prescribe the approach to metropolitan parking problems. They differ in each community and are largely controlled by legal authority. Obviously, however, planning of an entire program is a first essential. Planning on a metropolitan basis involves many factors other than these for a single area. The *needs*, on a metropolitan basis, may differ from those for any component of the metropolitan district.

Metropolitan planning and action in parking is hampered by the lack of experience, precedents, and basic governmental interrelationships. The conditions are receiving wide attention, and it is likely that closer cooperation and formal combined actions will occur in the near future because so many complex situations are arising which point to the need for metropolitan planning and action. Expressways, urban redevelopment, and many other activities accentuate the need. A major roadway or a major generator can alter the street and parking needs in several communities other than those in which they are situated. Extensive new traffic regulations in one city of the metropolitan community can alter parking demands in other cities of the community. Improvements of all kinds in cities can have an impact upon parking requirements for the metropolitan area.

An over-all plan of parking should consider the integration of curb and off-street spaces. It was pointed out that in revenue-bond financing, the pledge of receipts from curb meters is often required or is desirable for more favorable interest rates. This and other situations make it desirable to consider the curb and off-street spaces in the same program. If the city has a parking authority or otherwise creates a parking body, some thought should be given to delegating to this body the direction of the curb as well as off-street facilities.

Another reason for bringing the curb and off-street parking under one agency is the possible use of the pricing mechanism. If curb rates are increased, it is likely that the fees charged can be adjusted to achieve a desired usage. Through pricing, the type of parking and parking time at the curb can be affected, and so the relative use of curb and off-street spaces can be achieved.

The integration of the two types of parking develops the *system* idea. However, care must be exercised so that revenues from the curb do not unduly influence the prohibition of curb parking when the street capacity is more urgently needed for traffic movement.

Some extensive parking plans have attempted to relate parking development to civil defense facilities—bomb shelters. This idea has not become very popular because of the changes in concept of values of street shelters in modern warfare. While Federal funds for planning civil defense facilities made available money for planning underground parking, little of value has been reported from these efforts. Developers of some garage structures claim bomb-protective features. Again, however, the true values of these facilities, in relation to costs of conventional structures, are still to be ascertained.

While little can be gained by presenting parking as a part of the civil defense program, it is very important to consider parking as a part of wider redevelopment programs. The Federal funds now available to cities for wider redevelopment are sometimes available when a substantial part of the project is for parking. Certainly, all redevelopment projects should adequately provide for an abundance of parking for the new buildings and land uses that are developed on the urban land.

For completeness, the urban redevelopment program, the public housing program, and even the civil defense programs should be considered in the over-all plans for parking and loading facilities for both the community and the metropolitan area.

No mention has been made of mechanical garages, yet every parking program must consider them. Policies must be established as to the place of mechanical garages in relation to lots and open-deck ramp-type garages. Many types of patented garages have appeared in recent years. Strong claims are made for them, but thus far experiences have not been extensive. Many parking experts think that some type of mechanical garage will prove to be very efficient and attractive to parkers and will, thereby, play a major role in the long-range parking solutions. It is not apparent whether one or more of the present types will take such a place in the parking program.

CHAPTER 37

TRAFFIC ENGINEERING INTERESTS
IN HIGHWAY FINANCE

Most cities, counties, and states face great deficiencies in revenues for reasonable, immediate needs for highway construction and modernization. This is a direct outgrowth of the tremendous increase in automobile usage and the large increases in costs of constructing, maintaining, and operating highway systems. Because of his familiarity with traffic and related data, the traffic engineer finds himself in a prominent position in many questions having to do with highway economics.

37-1. Highway Funds. In the early days of highway development, construction costs were covered by nominal road taxes. As construction requirements increased, benefits assessed against abutting properties and general revenues were used to support construction programs in most cities, and large bond issues were frequently needed. These issues were backed by the general faith and credit of the municipalities or counties, in almost every instance. However, funds generally fell far short of requirements for highway and street developments, even in local communities, and other sources had to be sought.

The gas or fuel tax, enacted first in Oregon in 1919, is currently the greatest source of highway revenues. All states now have taxes on motor vehicle fuels, varying from 2 cents per gallon to 9 cents per gallon. In addition, the Federal government imposes a 2 cents per gallon tax on all gasoline sold for motor vehicle purposes, and in a few states the counties and cities have the right to impose additional taxes.

In some states motor vehicle registration fees produce a substantial amount of highway revenue. Usually fees for commercial vehicles are substantial, but in some states the fees for private vehicles cover little more than the expense of issuing licenses and maintaining the necessary records on registrations, transfers, and title certification.

Mileage taxes on commercial vehicles, especially trucks, are becoming commonly used to impose a tax in terms of ton-miles of use of the vehicle. In most states where these taxes have been tested, courts have held that the right to impose such taxes rests with the state legislature, but much opposition is usually registered by commercial vehicle operators on the grounds of the difficulties of uniform applications and administration and the high costs of collecting. Many legislators and public officials consider, however, that mileage or ton-mile taxes are justified by the damages done by trucks to roadways. The question as to what is an equitable tax base for commercial vehicles has not been settled. Views vary widely. Until highway officials can more objectively establish the degree to which various types of commercial vehicles increase road construction and maintenance costs in relation to the requirements imposed by all traffic, weather, and other factors, it will not be possible to prove the many arguments or to settle those which arise concerning equitable levies on commercial vehicles in their use of the public ways.

The importance of equity of taxation of commercial vehicles is reflected by the extensive tests which are now being made in several sections of the country. Road tests made in Maryland in 1952 indicated considerable damage to concrete pavement and increased maintenance cost from heavily loaded trucks. These tests were largely considered inconclusive, and similar tests are being conducted in other sections of the country on several types of pavements under different conditions of vehicular loadings. The tests are sponsored principally by the highway officials and the manufacturers of commercial vehicles and accessories. They are coordinated and directed by the Highway Research Board. It is hoped that when the results of the current tests are available, a sound basis can be agreed upon for a more equitable tax formula for different types of vehicles.

Gross receipts taxes have been levied in some instances against commercial vehicles, with the receipts used for highway improvement and maintenance purposes. As with haulage and ton-mile taxes, however, collection and administration may become so cumbersome and costly that benefits are questionable.

Special license taxes, such as those issued by public utility commissions against trucks and public carriers, produce some revenues, but the amounts are small in relation to those produced from fuel taxes and other sources. The principal purpose of such taxes is effective regulation of for-hire and public carrier vehicles. In most states the net financial gains from the exercise of such taxes is not an important consideration.

Property taxes as a source of financing highway needs are rapidly disappearing at the state level, but they are still a principal source of revenue for highway purposes at city and county levels. They usually have a direct relationship to special and benefit assessments for road and street purposes in cities.

Personal-property taxes levied on automobiles by cities and counties are rarely related to highway revenues or made a part of special highway demands. Almost without exception, they go into the general funds and are never considered as a source of income to the political jurisdictions available for highway activities.

Driver-license taxes are collected in all except one state. These revenues are often earmarked for highway-improvement purposes and are frequently used to defray a portion of the expenses of state highway patrols, state police, or general-safety education. In other instances, they are pooled with other highway-user revenues into the general highway fund. In most cases, however, fees are so nominal that the net available for expenditures on highway and traffic improvements is not appreciable.

Motor vehicle departments also collect fees for certification of motor vehicle titles, transfers, the issuance of duplicate licenses, and other charges for special services. Again, however, the net income from these sources is small.

In some cases, special local taxes are assessed for highway purposes, including local motor fuel taxes, franchise fees, and other charges against mass carriers and special types of road users. Curb parking meters are a source of considerable revenue in cities, and in some states revenues from parking meters are earmarked by legislation for traffic improvement purposes.

In a few states, local vehicle-registration licenses are still issued. In such cases these fees are an appreciable part of the local revenues taken from road users. New York City, for example, collects several million dollars each year by this method. These monies go into the general funds of the city.

37-2. Federal Aid for Highways. A major source of revenue for highway purposes is the Federal government, which appropriates funds to the states through the U.S. Bureau of Public Roads. In addition to the Federal gasoline tax, there are special excise taxes on such automotive necessities as oil, motor vehicle accessories, and on new automobiles. These revenues are not placed in a special highway fund, but become a part of general funds to be dispersed for highway purposes in relation to needs and demands or for other Federal expenditures. The 1954 Federal Aid Act for highways greatly increased the amounts of Federal monies available to the states and to local jurisdictions within the various states. The following are some of the more significant features of the bill.

The Federal Aid Highway Act of 1954 provides the greatest amount of Federal money ever authorized for highway purposes. A total amount of almost 1 billion dollars is authorized for the fiscal year ending June 30, 1956, and a like sum is authorized for the fiscal year ending June 30, 1957. In the appropriation an amount of 175 million dollars is earmarked for the national system of interstate highways, and 81 million dollars for forest highways, forest roads and trails, park roads, parkways, Indian roads, and roads through public lands.

The emphasis on the interstate system of highways is especially significant. It changes the standard allocations formula so that the Federal share of construction costs, including the purchase of rights of way, can be up to 60 per cent. In addition to the earmarked funds for interstate highways, the states are authorized to use other regular Federal-aid funds for development of the interstate highway system.

TABLE 37-1. COMPARISON OF AUTHORIZATIONS IN FEDERAL AID HIGHWAY ACTS OF 1952 AND 1954

	Federal Aid Act of 1952	Federal Aid Act of 1954
	For fiscal years ending June 30, 1954, and June 30, 1955	For fiscal years ending June 30, 1956, and June 30, 1957
"Regular" Federal aid authorizations		
Primary Federal aid system.........	$247,500,000	$315,000,000
Secondary Federal aid system........	165,000,000	210,000,000
Urban Federal aid system...........	137,500,000	175,000,000
Subtotal......................	$550,000,000	$700,000,000
Interstate system..................	25,000,000	175,000,000
Total "regular" Federal aid systems	$575,000,000	$875,000,000
Special authorizations		
Forest highways...................	$ 22,500,000	$ 22,500,000
Forest roads and trails.............	22,500,000	24,000,000
Park roads and trails..............	10,000,000	12,500,000
Parkways........................	10,000,000	11,000,000
Indian roads.....................	10,000,000	10,000,000
Roads through public lands.........	2,500,000	1,000,000
Total......................	$ 77,500,000	$ 81,000,000
Total of regular and special authorizations..............	$652,500,000	$956,000,000

SOURCE: American Automobile Association, Washington.

Broad provisions have been made in the Federal Aid Highway Act for highway researches.   Studies are to be made of size and weight standards and the feasibility of uniformity among the states, problems posed by necessary relocation and reconstruction of public utility services resulting from highway improvements, all phases of highway financing including toll financing, and the effects of toll roads upon Federal-aid highway programs and the coordination thereof.

Comparisons of authorizations in the Federal Aid Highway Acts of 1952 and 1954 are given in Table 37-1.

It appears that the 1954 capital-outlay expenditures on all roads and streets amounted to almost 4 billion dollars. Of this amount approximately 700 million dollars was for toll facilities. In addition, much was spent on highway-debt retirements, so that the total disbursements for road purposes in 1954 reached approximately $6\frac{1}{2}$ billion dollars. The trend toward credit financing of highway improvements is continuing at a rapid pace with approximately one-half of the anticipated capital-outlay program being financed by long-term bond issues. There is every indication that the expenditures in 1955 will substantially exceed those of 1954.

Since the end of World War II, there has been a rapid development of toll roads. Such roads are generally constructed under a special enabling act of the legislature and are financed with revenue bonds. The toll road has been developed as a means of providing high-type highways in heavy traffic corridors without encroachment on general funds normally available for free highway development at the state and local levels. It is difficult to forecast the future of the development of toll roads, but currently it is a rapidly growing activity and one in which intense interest is being shown.

37-3. Methods of Highway Finance. A few states still operate their highway programs on a pay-as-you-go basis. This system avoids conditions where present services are paid for by future governments, and it provides an even rate of highway expenditures. On the other hand, it may require the construction of less permanent types of roads which result in early maintenance and excessive operating cost, so that net efficiencies are difficult to achieve.

The more common current method of state highway finance is through the issuance of bonds. The urgency of needs and road emergencies make this "immediate" type of financing desirable. It is rarely possible to meet large-scale immediate needs through the pay-as-you-go plan of highway financing. Government borrowing has become as common in highway circles as in many other areas of public finance. When it is necessary to attain the approval of the public through a referendum, favorable votes are usually received at the state level, but quite frequently special bond issues for highways and other purposes fail at the local level. It is important to recognize state controls over debt limits in municipalities when the high cost of highway improvements and bond financing for highways and traffic is considered.

It is apparent that traffic values and traffic investigations which can be made by the traffic engineer are a primary aid to public administrators and to others in determining fiscal policies and particularly the equity of various types of highway charges or assessments.

Highway in Relation to Other Public Needs. Special types of roads and road needs enter into the consideration of highway funds. There must be improvements to the interregional highway system so that there will be an integrated plan of regional and nationwide highways. Other highway needs have a direct bearing upon civilian defense; still others must be checked against military plans for national defense. At the local level, some funds for highway development are based primarily upon property improvements.

A complete economic evaluation of highway needs will properly position highway transportation with other major public requirements. It is not wise, for example, to improve highways to the serious detriment of the public school system. On the other hand, the apportionment of funds for highways should not be neglected. Roadways open up new recreational areas, new sources of raw materials, new land, and sales of special products, thus occupying a somewhat specialized place in the over-all economy. This is particularly true of highway improvements, for they are primarily supported by highway users. This peculiar characteristic of highway finance should not unduly influence decisions. To consider the development of highways and the appropriation of funds for highway operations independently of other public responsibilities, even though some of them might also be independently supported, would not be sound.

Toll-highway Facilities.   Toll roads have made tremendous advancements since the first section of the Pennsylvania Turnpike was opened in 1940 between Harrisburg and Pittsburgh. This road, built under favorable conditions of Federal-government finance, has been remarkably successful. Its easy grades in rough terrain, its limited-access features, and other characteristics which make high uniform speeds possible have attracted many passenger vehicles and a high percentage of through truck traffic.

The New Jersey Turnpike, completed in record time in 1951, has also been most successful. The traffic volumes on this route have exceeded many times the estimated traffic, and at many sections its capacity was taxed when it was only a few years old. The success of this facility and of the Pennsylvania Turnpike stimulated nationwide interest. Public officials looked to the toll road as a means of providing high-type facilities in heavy-traffic corridors without interfering with other major highway developments within the state. Banking and investment groups found the revenue bonds, supported by tolls, attractive investments.

The Maine Turnpike, the New Hampshire Turnpike, the Denver-Boulder Turnpike, Oklahoma Turner Turnpike, and the parkways in Westchester County, N. Y., and in Connecticut are all well-known toll facilities. Roads of similar types are under construction or are in final states of planning in many other states: Ohio, Indiana, Illinois, Nebraska, Michigan, New York, Florida, West Virginia, Virginia, Arkansas, and Texas. Many routes studied will undoubtedly be found unfeasible as self-liquidating facilities.

The growing popularity of toll facilities has raised questions as to whether the nationwide system of free roads has been upset, but it is more a case of the inadequacy of financing to cope with the heavy demands for more roadway capacity and better roadway operations. Toll roads represent another effort to provide first-class facilities which are needed and desired by the motoring public. Some look upon the roads as providing a choice of first-class transportation, since they are almost always paralleled by free roads. Controversies arise because some feel that all of the principal routes developed as toll roads should more properly be developed as integral parts of the free highway system. Many arguments can be advanced for and against toll roads. In plans and studies, outside legislative and policy questions, the traffic engineer is playing an important part in the traffic surveys and in the traffic and earnings studies which are prerequisite to financing.

37-4. Traffic Assignment. The assignment of traffic to roadway facilities has become an important task of the traffic engineer. Conducting traffic studies and interpreting the data obtained in light of other factors are necessary steps in estimating the amount and type of traffic which can be expected to use any particular section of roadway or a bridge or tunnel facility. These studies are the basis for *traffic and earnings reports* for toll-highway facilities.

Origin-and-destination surveys are made on competitive routes to determine existing travel characteristics. Speed-and-delay studies give trip times over the different routes. Assuming operating speeds on proposed facilities (usually based on design speeds, but lower), it is possible to calculate the time savings or losses which can be expected from use of the new facility. This is usually done for all trips that have alternate routes or a selection of several routes.

When time differentials are determined for the different trips (found from the origin-and-destination studies), mileage differentials are recorded. The existing trips in the traffic corridor that would save time or distance, or both, by using the proposed facility, are apparent.

Traffic assignments to toll facilities can now be done more accurately than by time savings alone. To develop a common base, time and mileage are reduced to monetary values, or costs. Wide ranges will be found in use. Mileage values for passenger cars vary from 2 cents per mile to 6 cents per mile in traffic assignments. Time values vary from 1 cent to 3 cents per minute for passenger cars and are higher for trucks and other vehicles.

With the monetary values, the time and distance relationships can be combined to determine whether there will be a gain or a loss by travel over a given route in relation to travel over alternate routes. Where toll facilities are being considered, the toll charges must be deducted from the savings that might otherwise be derived from use of the facility. A diversion curve can be developed for assignment purposes which takes into account relative costs and permits ready conversion of any speed ratio and equivalent-cost ratio, for a given origin-and-destination grouping, to per cent diversion, i.e., a *diversion factor*.

In a few reports, traffic assignment has been made on the basis of free roads,[1] and arbitrary percentages of the traffic found to be potential to a new free facility are assumed to be assignable to a toll facility. The percentage of traffic for toll facility is usually taken as about 40 per cent of the free-road traffic. Obviously this method is not as accurate as methods which take into account other cost and savings factors.

Induced or Generated Traffic. New and improved facilities create traffic in addition to that diverted from existing facilities. Studies made on both old and new facilities show a substantial total increase in traffic using both roads when the new facility is completed. These increases continue for several years, but at diminishing rates.

Since generated traffic comes from the provision of a greater freedom of movement and may result from greater use of vehicles, its volume will be determined by the magnitude of existing inadequacies, the relative attractiveness of the new facility, and the amount of toll.

The increased traffic factors vary widely. On some bridges, for example, the induced traffic has been as great as 100 per cent of the diverted traffic flows. On most toll roads, the generated or induced traffic is about 25 to 50 per cent of the diverted traffic. First-year values of 20 to 25 per cent are considered conservative for most projected toll roads, but, as indicated, the values are higher when the facility is superior to existing routes. The induced-traffic ratio usually drops below 10 per cent for the second year's operation of a toll road.

Land-development traffic is another type of generated traffic. By providing a new and convenient route of access for land not previously served by highways, building and development are likely to occur. Intensified commercial, industrial, and residential use can be expected. Experiences with toll roads and major free roads indicate that traffic increases attributable to land development will average as much as 10 per cent per year for about 5 years and then gradually diminish to zero in the next five-year period.

Future Traffic. The extent to which traffic can be expected to increase during the life of the bond issue is a major consideration in planning toll roads. The revenue will bear a direct relation to traffic volume, and proper design must take into account the future traffic requirements. Roadway capacities should be adequate to accommodate traffic increases likely to occur within a reasonable period of time, or provision should be

[1] Studies published by the Highway Research Board and the American Association of State Highway Officials Subcommittee on Factual Surveys are valuable references in such assignments.

made in the grading and structures so that additional lanes can be economically added as traffic requires them.

To estimate future traffic, studies are usually made of the trends in (1) population, (2) vehicle ownership, (3) vehicle-miles, (4) commercial–passenger-vehicle ratios, and (5) general development of the area and region served.   Increasingly, it appears that the vehicle-population ratio is one of the best guides to traffic trends in a given area.   This means that where more people own automobiles, the traffic growth potentials are greater than when the ratios of vehicle ownership are low.

Obviously, the data from permanent traffic-count stations are useful in calculating future traffic because they permit the plotting of travel desires on given routes (see Chapter 6).

The use of automobiles has a direct relationship to the over-all economy.   This is reflected by studies of the gross national product.   The product agrees closely with travel trends, except in war years or in other periods of unusual economy.   Because of this condition, estimates of traffic growths for toll roads must be kept on the conservative side. Growths of from 3 to 5 per cent per year are most commonly used, even though growths consistently in excess of these values might have occurred for many years on primary highways in the same area.

Some traffic engineers feel that the great fluctuations in travel caused by economic conditions make it unwise to attempt to forecast traffic growths for more than 20 years.   Many revenue estimates therefore remain constant for each year after 20 years of operation, even though the bonds are outstanding for 20, 30, or even 40 years.

37-5. Costs and Revenues.   Since the success of the toll facility will be indicated by its economic feasibility, very careful consideration must be given to costs of construction, maintenance, operations, and financing.

Capital costs include the following items:
1. Right-of-way
2. Clearing
3. Earthwork
4. Pavement (toll road and feeder roads)
5. Shoulders
6. Drainage
7. Structures
8. Landscaping
9. Appurtenances (right-of-way markers, guardrail, etc.)
10. Lighting
11. Utilities
12. Traffic control (signs and markings)
13. Buildings and equipment (administration, maintenance, and concessions)
14. Interchanges (ramps and extra structures)
15. Toll collection (booths, equipment)
16. Engineering services
17. Contingencies

18. Administrative and legal
19. Interest during construction

Maintenance costs include the usual costs for roadway and structure maintenance and replacement: mowing, roadside clearing, painting, roadway-surface repairs, centerline and lane striping, sign replacements, snow removal, maintenance of drainage facilities, supervision, and equipment replacement.

Operating costs cover toll-collection costs, including salaries of supervisors and collectors, utilities, and accounting. Police services are usually furnished by the state police organization and paid by the toll-road authority. Administrative expenses, covering salaries of time-operating staff, liability insurance, consultants' fees, office supplies, and legal services, can also be included in operating costs.

Revenues from toll roads come principally from the fees collected from motorists for travel over the route. Incomes from concessions (service stations and restaurants) are also a part of the total income and in some cases are substantial.

Financial Feasibility. The costs for maintenance and operation are deducted from the gross income to determine the net amount available for debt service. The ratio of the amount available for debt service to the debt service gives a *coverage factor* and indicates how much the anticipated net income exceeds the amount required to meet interest and amortization charges. Methods of computing coverage vary. Sometimes it is taken as the average over-all net income divided by the average annual debt charges. Sometimes it is calculated by dividing the total income for the period of indebtedness by the income anticipated during the bond life.

Thus far in toll-road financing, where revenue bonds are issued and the indenture is not backed by the faith and credit of the state or by other funds, coverages of at least 1.5 are usually required. In most cases the earnings anticipated (gross less maintenance and operations) exceed the total debt service charges (interest plus bond retirement) by 1.7.

The cost of financing roads and structures with revenue bonds is substantially higher than with general-obligation bonds or bonds backed by general highway revenues. Interest rates on the revenue bonds range from 3 per cent to over 4 per cent, depending on the coverage and the earnings experiences of the area.

Comparisons of Routes. Sometimes the exact location of a route is not fixed when toll-road studies are initiated. This might make it necessary to assign traffic to several routes and to estimate the relative revenues and costs for alternate routes. While the economic comparisons must be the principal factors influencing the choice of location, there are other factors which should also be considered. If two or more locations prove economically desirable, then the best route should be determined by comparisons of volumes of traffic served, relief provided existing traffic facilities, user benefits, and possibilities for integration into present and future highway systems.

37-6. Other Factors in Toll-road Development. There are several additional factors to be considered in planning, financing, and operating toll facilities.

The type of collection system to be used is important both to toll authorities and to motorists. The common practice is to collect at interchanges. If cards are used, motorists can be checked at points of entry and departure and charged accordingly. On bridges, at terminals, and in metropolitan areas, tolls are usually collected at *barrier* stations. All vehicles passing the point of the barrier are charged a fixed fee. Under the barrier plan, it is often possible for some motorists to use portions of the facility without payment of tolls. Analyses frequently reveal, however, that the toll losses are more than offset by the savings in capital and collection costs. Traffic engineering studies of toll facilities must obviously consider the two methods and determine which is best for the particular facility.

More attention is now being given to the collection of tolls on roads financed from the general funds or from the state highway funds. For example, one extensive highway development in Connecticut is to be a toll road, but the legislature provides that any monies needed for meeting debt and operating costs, in addition to the revenues produced by the road, will be taken from the general highway funds of the state. Such plans produce better interest rates than those for which pure revenue bonds are sold. It seems likely that the "partial" financing by tolls will become increasingly popular, with the possible application of Federal-aid highway funds to such endeavors.

Other Interests in Highway Finance. The traffic engineer can become involved in many phases of highway finance. The traffic engineering department in state highway organizations might develop and direct the budget function, because of the direct relationships of traffic facts to expenditures and receipts.

In cities, the traffic engineer is intimately connected with public off-street parking programs. These are discussed in Chapter 36. The studies and basic calculations for revenue-bond financing of off-street parking bear close resemblance to studies for toll roads. In both instances, major burdens are assumed in forecasting uses and, therefore, revenues.

The proper distribution of the highway-transportation dollar depends upon road-sufficiency studies and analyses of services rendered to income. Again, traffic movements and investigations are essential.

INDEX